Applications of Tensor Analysis in Continuum Mechanics

Applications of Tensor Analysis in Continuum Mechanics

Victor A Eremeyev
Gdańsk University of Technology, Poland

Michael J Cloud
Lawrence Technological University, USA

Leonid P Lebedev
National University of Colombia, Colombia

World Scientific

NEW JERSEY • LONDON • SINGAPORE • BEIJING • SHANGHAI • HONG KONG • TAIPEI • CHENNAI

Published by

World Scientific Publishing Co. Pte. Ltd.

5 Toh Tuck Link, Singapore 596224

USA office: 27 Warren Street, Suite 401-402, Hackensack, NJ 07601

UK office: 57 Shelton Street, Covent Garden, London WC2H 9HE

Library of Congress Cataloging-in-Publication Data

Names: Eremeyev, Victor A., author. | Cloud, Michael J., author. | Lebedev, L. P. author.
Title: Applications of tensor analysis in continuum mechanics /
Victor A Eremeyev, Gdańsk University of Technology, Poland,
Michael J Cloud, Lawrence Technological University, USA,
Leonid P Lebedev, National University of Colombia, Colombia.
Description: New Jersey : World Scientific, [2018] | Includes bibliographical references and index.
Identifiers: LCCN 2018028400 | ISBN 9789813238961 (hardback : alk. paper)
Subjects: LCSH: Continuum mechanics. | Calculus of tensors. | Engineering mathematics.
Classification: LCC QA808.2 .E699 2018 | DDC 531.01515/63--dc23
LC record available at https://lccn.loc.gov/2018028400

British Library Cataloguing-in-Publication Data

A catalogue record for this book is available from the British Library.

For any available supplementary material, please visit
https://www.worldscientific.com/worldscibooks/10.1142/10959#t=suppl

Desk Editor: Amanda Yun

Preface

Each science has its own tools and language through which it is developed and expressed. In its initial stages of formulation, classical mechanics depended on force and displacement vectors as well as on certain tensorial quantities such as the inertia tensor. Later, linear elasticity began to employ strain and tension tensors written mainly in Cartesian coordinate systems. But the requirements of nonlinear continuum mechanics, where we must describe the medium in two states — an initial state and a deformed state — and relate them through the constitutive equations, prompted mechanicists to elaborate a general language by which all the old results could be presented via elegant and easily remembered formulas. This was the language of tensor analysis. It has proved useful in all areas of physics that deal with continuum media, not just in mechanics.

The notion of vector is centuries old. Students are accustomed to it, working with vectors written in Cartesian coordinates and using the dot and cross products. They are aware that a vector is an object independent of the basis of the space. Few consider the issue that in three dimensions the position vectors $\mathbf{a}, \mathbf{b}$ and their cross product $\mathbf{a} \times \mathbf{b}$ have different units (m vs. m^2). Indeed, $\mathbf{a} \times \mathbf{b}$ represents the area of the parallelogram spanned by $\mathbf{a}$ and $\mathbf{b}$. So $\mathbf{a} \times \mathbf{b}$ does not belong to the space of position vectors. Aside from the issue of units, the vector $\mathbf{a} \times \mathbf{b}$ has the problem of orientation: the direction of $\mathbf{a} \times \mathbf{b}$ depends on the orientation of the basis of $\mathbb{R}^3$; it reverses direction when the basis changes orientation. So $\mathbf{a} \times \mathbf{b}$ is not an object independent of the basis of the space. This does not complicate the problems of classical mechanics or linear elasticity, as in these problems changes in the handedness of the basis normally do not arise. However, the difference between ordinary (polar) vectors and pseudovectors such as the cross product is relevant to modern models of materials — linear and nonlinear

— which are considered in the last chapter of this book. This difference is also quite important in various physical problems of electrodynamics.

This book is a considerably revised and extended version of our earlier title *Tensor Analysis with Applications in Mechanics* (World Scientific, 2010). The first four chapters have been transformed into the new Chapters 1 and 2 with some significant additions; in particular, the new Chapter 2 features sections covering tensor groups, the theory of invariants, and the distinction between polar vectors and axial vectors. The fifth chapter has been rewritten slightly to become the new Chapter 3, with the former sections on geodesics and contact of curves omitted to streamline the presentation. The sixth and seventh chapters, now appearing as Chapters 4 and 5, remain substantially unchanged. Finally, a new Chapter 6 on generalized media has been added. The exercises, problems, and hints have also been thoroughly revised. These changes, aimed at an increased emphasis on applications to continuum mechanics, seemed to warrant a new title and a change in the author sequence. But the goal remains the same: a clear, concise, self-contained introduction to tensors, tensor fields, and their applications to mechanics. We would like to thank World Scientific editors Amanda Yun and Jason Lim for prompting the new version, and are pleased to once again acknowledge our prior editors Tjan Kwang Wei and Sook-Cheng Lim.

Victor A. Eremeyev
Gdańsk University of Technology

Michael J. Cloud
Lawrence Technological University

Leonid P. Lebedev
National University of Colombia

Contents

Chapter 1

Vectors and Transformations

1.1 Change of Basis and the Reciprocal Basis

The crucial characteristic of a vector is its nature as an *objective entity*. That is, a given vector $\mathbf{x}$ is independent of any coordinate system used to describe it or any basis used to expand it. The components x^1, x^2, x^3 of a three-dimensional vector $\mathbf{x}$ relative to a basis $\{\mathbf{e}_1, \mathbf{e}_2, \mathbf{e}_3\}$ will certainly differ from its components $\tilde{x}^1, \tilde{x}^2, \tilde{x}^3$ relative to some other basis $\{\tilde{\mathbf{e}}_1, \tilde{\mathbf{e}}_2, \tilde{\mathbf{e}}_3\}$. However, the relations between (x^1, x^2, x^3) and $(\tilde{x}^1, \tilde{x}^2, \tilde{x}^3)$ must be such that the expressions

$$x^1\mathbf{e}_1 + x^2\mathbf{e}_2 + x^3\mathbf{e}_3 \quad \text{and} \quad \tilde{x}^1\tilde{\mathbf{e}}_1 + \tilde{x}^2\tilde{\mathbf{e}}_2 + \tilde{x}^3\tilde{\mathbf{e}}_3$$

both describe the same vector $\mathbf{x}$.

How do we find these relations? This question is important in applications and it leads to a consideration of reciprocal bases. First, however, let's agree to write down vector expansions in shorter and clearer ways.

Index notation; Einstein's convention

The two labor-saving rules are as follows.

1. Range convention. A nonrepeated index takes on all values $1, 2, \ldots, n$ where n is the dimension of the space in which we are working. In $\mathbb{R}^3$, for instance, the notation x_i will stand for the three quantities x_1, x_2, x_3. (In large part we are interested in applications of the vector theory in $\mathbb{R}^3$ and $\mathbb{R}^2$. Hence we will concentrate on these cases even though the extension of formulas to any n is often easy.) Here the index i was a subscript, but indices in tensor analysis can be superscripts as well (and the fact that they are raised will not indicate raising of variable to powers — more on this below). We call a nonrepeated index a *free index*. Note that x_k or x_n

would refer to the same three quantities x_1, x_2, x_3, as long as k and n are free indices.

2. Summation convention. When an index occurs exactly twice in a term — once as a subscript and once as a superscript — then summation over that index is understood. This convention is often called *Einstein's summation convention.* In $\mathbb{R}^n$, for instance, the notation $x_j y^j$ stands for $\sum_{j=1}^n x_j y^j$. We call a repeated index a *bound index.* Note that $x_k y^k$ would refer to exactly the same sum, as k is just another bound index playing the same role as j above. The summation rule is modified for a Cartesian basis: the repeated index may occur twice in the same position as a subscript. This case can require caution, however, and we will elucidate further when the time arrives.

For instance, in $\mathbb{R}^3$ the term $x_i y_j z^j$ stands for the three quantities

$$x_1(y_1 z^1 + y_2 z^2 + y_3 z^3), \quad x_2(y_1 z^1 + y_2 z^2 + y_3 z^3), \quad x_3(y_1 z^1 + y_2 z^2 + y_3 z^3).$$

We could also denote them by $x_k y_p z^p$. The term $x_i y_j z_k$, on the other hand, stands for the 27 quantities $x_1 y_1 z_1$, $x_1 y_1 z_2$, ..., $x_3 y_3 z_3$.

Back to bases

In elementary physics they typically use the Cartesian (i.e., orthogonal unit) basis $\{\mathbf{i}, \mathbf{j}, \mathbf{k}\}$. We will denote this same basis as $\{\mathbf{i}_1, \mathbf{i}_2, \mathbf{i}_3\}$ instead, so that we can use the range convention to refer to the whole set of three vectors as $\{\mathbf{i}_k\}$. The reader is familiar with the properties of this basis,[1]

$$\begin{aligned} \mathbf{i}_1 \cdot \mathbf{i}_1 = \mathbf{i}_2 \cdot \mathbf{i}_2 = \mathbf{i}_3 \cdot \mathbf{i}_3 = 1, \\ \mathbf{i}_1 \cdot \mathbf{i}_2 = \mathbf{i}_2 \cdot \mathbf{i}_3 = \mathbf{i}_3 \cdot \mathbf{i}_1 = 0, \end{aligned} \qquad \begin{aligned} \mathbf{i}_1 \times \mathbf{i}_2 = \mathbf{i}_3, \\ \mathbf{i}_2 \times \mathbf{i}_3 = \mathbf{i}_1, \\ \mathbf{i}_3 \times \mathbf{i}_1 = \mathbf{i}_2, \end{aligned}$$

as well as the expressions for dot and cross products in this basis:

$$\mathbf{a} \cdot \mathbf{b} = a_1 b_1 + a_2 b_2 + a_3 b_3, \qquad \mathbf{a} \times \mathbf{b} = \begin{vmatrix} \mathbf{i}_1 & \mathbf{i}_2 & \mathbf{i}_3 \\ a_1 & a_2 & a_3 \\ b_1 & b_2 & b_3 \end{vmatrix}.$$

However, we will be very interested in bases consisting of nonorthogonal vectors not necessarily of unit magnitude (Fig. 1.1).

Let us denote such a basis by $\{\mathbf{e}_j\}$. Any linearly independent set of n such vectors can serve as a basis in $\mathbb{R}^n$. Of course, this means it is possible to express an arbitrary vector $\mathbf{x} \in \mathbb{R}^n$ in the form

$$\mathbf{x} = x^j \mathbf{e}_j, \tag{1.1}$$

[1] But see § 2.10 for some peculiarities of the cross product.

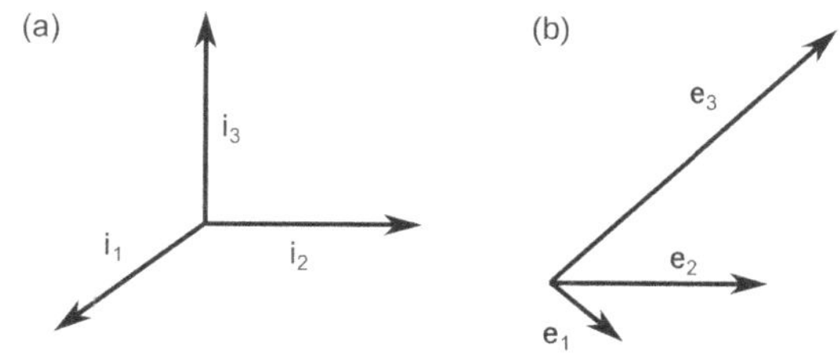

Fig. 1.1 Some basis sets in $\mathbb{R}^3$: (a) an orthogonal unit basis; (b) a more general basis.

where the summation convention is in effect and x^j is the jth component of $\mathbf{x}$ relative to the basis $\{\mathbf{e}_j\}$. It is clear that superscripts are not meant to denote exponents in this situation; instead they are simply tags that number the components of $\mathbf{x}$.

Given $\mathbf{x}$ and a basis $\{\mathbf{e}_j\}$, how can we find the components x^j in (1.1)? This is where the *reciprocal basis* $\{\mathbf{e}^j\}$ comes in (note the superscript tag on these vectors). The vectors $\mathbf{e}^j$ are defined by the requirement that

$$\mathbf{e}_j \cdot \mathbf{e}^k = \begin{cases} 1, & j = k, \\ 0, & j \neq k, \end{cases} \tag{1.2}$$

where $\cdot$ is the ordinary vector dot product. For example, in $\mathbb{R}^3$ we have

$$\mathbf{e}_1 \cdot \mathbf{e}^1 = \mathbf{e}_2 \cdot \mathbf{e}^2 = \mathbf{e}_3 \cdot \mathbf{e}^3 = 1, \tag{1.3}$$

$$\mathbf{e}_1 \cdot \mathbf{e}^2 = \mathbf{e}_1 \cdot \mathbf{e}^3 = \mathbf{e}_2 \cdot \mathbf{e}^1 = \mathbf{e}_2 \cdot \mathbf{e}^3 = \mathbf{e}_3 \cdot \mathbf{e}^1 = \mathbf{e}_3 \cdot \mathbf{e}^2 = 0. \tag{1.4}$$

When the basis $\mathbf{e}_j$ is orthogonal, $\mathbf{e}^j$ is codirected with $\mathbf{e}_j$ and $|\mathbf{e}^j| = |\mathbf{e}_j|^{-1}$. So the orthogonal unit basis $\{\mathbf{i}_k\}$ is self-reciprocal.

Exercise 1.1. Show that (a) $\mathbf{e}^k$ is determined uniquely by the requirement that $x^k = \mathbf{x} \cdot \mathbf{e}^k$ for every $\mathbf{x}$, and (b) the reciprocal basis to $\{\mathbf{e}^k\}$ is $\{\mathbf{e}_k\}$.

How is the reciprocal basis employed? Before addressing this, let us introduce another helpful space saver: the *Kronecker delta* symbol. This is just shorthand for the right side of (1.2):

$$\delta_j^k = \begin{cases} 1, & j = k, \\ 0, & j \neq k. \end{cases}$$

But it is a powerful tool when combined with the index conventions.[2]

[2]Although the Kronecker symbol looks like the components of a second-order tensor, it is not a tensor, just as an n-tuple is not a vector without a basis understood. *Kronecker's*

Dotting $\mathbf{e}^k$ into both sides of $\mathbf{x} = x^j \mathbf{e}_j$, we get

$$\mathbf{x} \cdot \mathbf{e}^k = x^j \mathbf{e}_j \cdot \mathbf{e}^k = x^j \delta_j^k = x^k$$

and hence $x^k = \mathbf{x} \cdot \mathbf{e}^k$. (Note the action of δ_j^k as a *substitution multiplier* here.) So we can find the components x^k if we know the reciprocal basis vectors $\mathbf{e}^k$. The latter are easily constructed in $\mathbb{R}^3$. Indeed, the requirements (1.4) are met by writing

$$\mathbf{e}^1 = \alpha_1 \mathbf{e}_2 \times \mathbf{e}_3, \quad \mathbf{e}^2 = \alpha_2 \mathbf{e}_3 \times \mathbf{e}_1, \quad \mathbf{e}^3 = \alpha_3 \mathbf{e}_1 \times \mathbf{e}_2$$

where the α_i are nonzero constants. We then satisfy (1.3) by setting

$$\mathbf{e}_1 \cdot \mathbf{e}^1 = \alpha_1 \mathbf{e}_1 \cdot (\mathbf{e}_2 \times \mathbf{e}_3) = 1,$$
$$\mathbf{e}_2 \cdot \mathbf{e}^2 = \alpha_2 \mathbf{e}_2 \cdot (\mathbf{e}_3 \times \mathbf{e}_1) = 1,$$
$$\mathbf{e}_3 \cdot \mathbf{e}^3 = \alpha_3 \mathbf{e}_3 \cdot (\mathbf{e}_1 \times \mathbf{e}_2) = 1,$$

and finding that $\alpha_1, \alpha_2, \alpha_3$ all equal $1/V$ where

$$V = \mathbf{e}_1 \cdot (\mathbf{e}_2 \times \mathbf{e}_3). \tag{1.5}$$

Geometrically, $|V|$ is the volume of the frame parallelepiped spanned by the basis $\{\mathbf{e}_i\}$. The sign of V also shows the orientation of the triplet $\mathbf{e}_1, \mathbf{e}_2, \mathbf{e}_3$: it is right-handed[3] if $V > 0$. Similarly,

$$\mathbf{e}_1 = \frac{1}{V'} \mathbf{e}^2 \times \mathbf{e}^3, \quad \mathbf{e}_2 = \frac{1}{V'} \mathbf{e}^3 \times \mathbf{e}^1, \quad \mathbf{e}_3 = \frac{1}{V'} \mathbf{e}^1 \times \mathbf{e}^2$$

where

$$V' = \mathbf{e}^1 \cdot (\mathbf{e}^2 \times \mathbf{e}^3). \tag{1.6}$$

In Problem 1.3 the reader will be invited to show that

$$V' = 1/V. \tag{1.7}$$

Verification that $\{\mathbf{e}^k\}$ is a linearly independent set (and hence a basis) is straightforward. We must show that the equation $\alpha_1 \mathbf{e}^1 + \alpha_2 \mathbf{e}^2 + \alpha_3 \mathbf{e}^3 = \mathbf{0}$ implies $\alpha_1 = \alpha_2 = \alpha_3 = 0$. To get $\alpha_1 = 0$, for example, simply dot-multiply the equation by $\mathbf{e}_1$ and use $\mathbf{e}_i \cdot \mathbf{e}^j = \delta_i^j$.

As the dot product is commutative, the basis reciprocal to $\{\mathbf{e}^j\}$ is $\{\mathbf{e}_j\}$. So we immediately have $\mathbf{x} = x_k \mathbf{e}^k$ where $x_k = \mathbf{x} \cdot \mathbf{e}_k$. It is now clear why

symbols do not depend on any basis, they are only constants defined by the indices. However, for the metric (or unit) tensor, as we will see later, δ_j^k are the components in the so-called mixed basis. Also note that in the literature one can encounter Kronecker symbols with the indices in any positions: δ^{ij}, δ_{ij}, and δ_i^j.

[3] As we use the right-hand definition of the cross product, we call a frame $(\mathbf{e}_1, \mathbf{e}_2, \mathbf{e}_3)$ right-handed if $\mathbf{e}_1 \cdot (\mathbf{e}_2 \times \mathbf{e}_3) > 0$.

the summation convention requires both a superscript and a subscript. The only exception is when we use the canonical basis $\{\mathbf{i}_k\}$, in which case the convention holds even when we write $x_k\mathbf{i}_k$.

Contravariant and covariant components; change to the reciprocal basis

We have seen that, given a basis $\{\mathbf{e}_k\}$, a vector $\mathbf{x}$ has two representations:

(1) $\mathbf{x} = x^k\mathbf{e}_k$ where $x^k = \mathbf{x}\cdot\mathbf{e}^k$,
(2) $\mathbf{x} = x_k\mathbf{e}^k$ where $x_k = \mathbf{x}\cdot\mathbf{e}_k$.

We call x^k the *contravariant components* of $\mathbf{x}$, and x_k the *covariant components* of $\mathbf{x}$ with respect to the basis $\{\mathbf{e}_k\}$. It is easy to switch between these two representations of $\mathbf{x}$. Writing $x^i\mathbf{e}_i = x_j\mathbf{e}^j$, we have $x^i\mathbf{e}_i\cdot\mathbf{e}^k = x_j\mathbf{e}^j\cdot\mathbf{e}^k$ which implies that $x^i\delta_i^k = x_j\mathbf{e}^j\cdot\mathbf{e}^k$ and hence

$$x^k = x_j g^{jk} \quad \text{where} \quad g^{ji} = \mathbf{e}^j\cdot\mathbf{e}^k. \tag{1.8}$$

Similarly, $x^i\mathbf{e}_i\cdot\mathbf{e}_k = x_j\mathbf{e}^j\cdot\mathbf{e}_k$ gives us

$$x^i g_{ik} = x_k \quad \text{where} \quad g_{ik} = \mathbf{e}_i\cdot\mathbf{e}_k. \tag{1.9}$$

The quantities g_{ij} and g^{ij} are known as *metric coefficients*. Note the symmetry in the indices: $g_{ij} = g_{ji}$ and $g^{ij} = g^{ji}$. By substituting $x^i = x_j g^{ij}$ into $x_j = x^i g_{ij}$, we find that $x_j = x_q g^{iq} g_{ij}$ which, by arbitrariness of $\mathbf{x}$, implies

$$g_{ij}g^{iq} = \delta_j^q. \tag{1.10}$$

This equality shows that the matrices (g_{ij}) and (g^{ij}) are mutually inverse.

Here we should make a simple but important observation. Notation is one of the beauties of tensor analysis, and we can begin to appreciate this through equations (1.8)–(1.10). The equation $x^k = x_j g^{jk}$ shows that multiplication of x_j by g^{jk} (with the implied sum over j) has the effect of *raising the index* j into the upper position (as the k in x^k). Similarly, $x^i g_{ik} = x_k$ shows that multiplication of x^i by g_{ik} (with the implied sum over i) has the effect of *lowering an index*. These raising and lowering processes also work with the basis vectors. Indeed, setting $\mathbf{x} = \mathbf{e}^j$ in the relation $\mathbf{x} = x^i\mathbf{e}_i = (\mathbf{x}\cdot\mathbf{e}^i)\mathbf{e}_i$, we get

$$\mathbf{e}^j = (\mathbf{e}^j\cdot\mathbf{e}^i)\mathbf{e}_i = g^{ji}\mathbf{e}_i. \tag{1.11}$$

Analogously, setting $\mathbf{x} = \mathbf{e}_j$ in the relation $\mathbf{x} = x_i\mathbf{e}^i = (\mathbf{x}\cdot\mathbf{e}_i)\mathbf{e}^i$, we get

$$\mathbf{e}_j = (\mathbf{e}_j\cdot\mathbf{e}_i)\mathbf{e}^i = g_{ji}\mathbf{e}^i. \tag{1.12}$$

So we can use the metric coefficients to raise and lower indices (and we will see that this can be done with any tensor). Also observe the *index balance* present in equations (1.8)–(1.12). Either an index is repeated in a term, or that index appears on both sides of an equation in the same position (upper or lower).

Finally, let us examine the dot product $\mathbf{x} \cdot \mathbf{y}$ under various representations of the two vectors $\mathbf{x}$ and $\mathbf{y}$. Using contravariant components for both vectors, we get

$$\mathbf{x} \cdot \mathbf{y} = x^i \mathbf{e}_i \cdot y^j \mathbf{e}_j = x^i y^j g_{ij}. \tag{1.13}$$

Using covariant components, we get

$$\mathbf{x} \cdot \mathbf{y} = x_i \mathbf{e}^i \cdot y_j \mathbf{e}^j = x_i y_j g^{ij}. \tag{1.14}$$

Observe the appearance of the metric coefficients each time. We can avoid this through the use of *mixed* representations:

$$\mathbf{x} \cdot \mathbf{y} = x^i \mathbf{e}_i \cdot y_j \mathbf{e}^j = x^i y_j \delta_i^j = x^i y_i, \tag{1.15}$$

$$\mathbf{x} \cdot \mathbf{y} = x_i \mathbf{e}^i \cdot y^j \mathbf{e}_j = x_i y^j \delta_j^i = x_i y^i. \tag{1.16}$$

Note that (1.15)–(1.16) could be obtained from (1.13)–(1.14) directly by using the metric coefficients to raise and lower indices as mentioned above.

More general change of basis

If $\{\mathbf{e}_i\}$ and $\{\mathbf{e}_{i'}\}$ are bases[4] for the same vector space, then the vectors $\mathbf{e}_{i'}$ are expressible as linear combinations of the vectors $\mathbf{e}_i$:

$$\mathbf{e}_{i'} = A_{i'}^j \mathbf{e}_j, \tag{1.17}$$

where the $A_{i'}^j$ are coefficients describing the transformation between the two bases. Dotting both sides of (1.17) with $\mathbf{e}^k$, we have

$$\mathbf{e}_{i'} \cdot \mathbf{e}^k = A_{i'}^j \mathbf{e}_j \cdot \mathbf{e}^k = A_{i'}^j \delta_j^k = A_{i'}^k$$

and therefore[5]

$$A_{i'}^j = \mathbf{e}_{i'} \cdot \mathbf{e}^j.$$

[4] The prime on the index is necessary here in order to distinguish between the two bases. So, just as i is a dummy index in $\mathbf{e}_i$, so also i' is a dummy index in $\mathbf{e}_{i'}$. When replacing it with another dummy index, we must retain the prime (e.g., $\mathbf{e}_{j'}$) to maintain proper association of the vector with the second basis.

[5] Another comment about notation may be in order here. Since the dot product is commutative ($\mathbf{e}_{i'} \cdot \mathbf{e}^j = \mathbf{e}^j \cdot \mathbf{e}_{i'}$), the order of indices in $A_{i'}^j$ is unimportant. In cases where the order *is* important, we will use horizontal positioning to specify it (e.g., $a_i^{\ j}$ vs. $a^i_{\ j}$ on p. 16).

The reverse basis transformation is described by the matrix inverse to $A^j_{i'}$:

$$\mathbf{e}_i = A^{j'}_i \mathbf{e}_{j'}. \tag{1.18}$$

That is, we have

$$\mathbf{e}_i = A^{j'}_i \mathbf{e}_{j'} = A^{j'}_i A^k_{j'} \mathbf{e}_k \quad \text{so that} \quad A^{j'}_i A^k_{j'} = \delta^k_i, \tag{1.19}$$

$$\mathbf{e}_{i'} = A^j_{i'} \mathbf{e}_j = A^j_{i'} A^{k'}_j \mathbf{e}_{k'} \quad \text{so that} \quad A^j_{i'} A^{k'}_j = \delta^{k'}_{i'}. \tag{1.20}$$

Similarly, we can write

$$\mathbf{e}^{i'} = B^{i'}_j \mathbf{e}^j \quad \text{and} \quad \mathbf{e}^i = B^i_{j'} \mathbf{e}^{j'}$$

for a suitable set of transformation coefficients between the two reciprocal bases. But

$$\delta^k_i = \mathbf{e}_i \cdot \mathbf{e}^k = A^{m'}_i \mathbf{e}_{m'} \cdot B^k_{j'} \mathbf{e}^{j'} = A^{m'}_i B^k_{j'} \delta^{j'}_{m'} = A^{j'}_i B^k_{j'},$$

$$\delta^{k'}_{i'} = \mathbf{e}_{i'} \cdot \mathbf{e}^{k'} = A^m_{i'} \mathbf{e}_m \cdot B^{k'}_j \mathbf{e}^j = A^m_{i'} B^{k'}_j \delta^j_m = A^j_{i'} B^{k'}_j,$$

and comparison with (1.19)–(1.20) shows that $B^k_{j'} = A^k_{j'}$ and $B^{k'}_j = A^{k'}_j$. Therefore

$$\mathbf{e}_{i'} = A^j_{i'} \mathbf{e}_j, \qquad \mathbf{e}_i = A^{j'}_i \mathbf{e}_{j'}, \qquad \mathbf{e}^{i'} = A^{i'}_j \mathbf{e}^j, \qquad \mathbf{e}^i = A^i_{j'} \mathbf{e}^{j'}.$$

Now we examine how the components of a vector

$$\mathbf{x} = x^i \mathbf{e}_i = x_i \mathbf{e}^i = x^{i'} \mathbf{e}_{i'} = x_{i'} \mathbf{e}^{i'}$$

must transform when we change basis. We have

$$x^j \mathbf{e}_j = x^{i'} \mathbf{e}_{i'} = x^{i'} A^j_{i'} \mathbf{e}_j \implies x^j = A^j_{i'} x^{i'},$$

$$x^{j'} \mathbf{e}_{j'} = x^i \mathbf{e}_i = x^i A^{j'}_i \mathbf{e}_{j'} \implies x^{j'} = A^{j'}_i x^i,$$

$$x_j \mathbf{e}^j = x_{i'} \mathbf{e}^{i'} = x_{i'} A^{i'}_j \mathbf{e}^j \implies x_j = A^{i'}_j x_{i'},$$

$$x_{j'} \mathbf{e}^{j'} = x_i \mathbf{e}^i = x_i A^i_{j'} \mathbf{e}^{j'} \implies x_{j'} = A^i_{j'} x_i.$$

Comparing these formulas with those for transforming the basis vectors, we see why the components are called covariant and contravariant: the covariant components change like the principal basis vectors, while the contravariant components change like the reciprocal basis vectors.

1.2 The Cross Product in Index Notation

The cross product, like any other vector, can be expanded in a basis. Let us write

$$\mathbf{c} = \mathbf{a} \times \mathbf{b} = c_i \mathbf{e}^i$$

and seek c_i in the form

$$c_i = \epsilon_{ijk} a^j b^k. \tag{1.21}$$

To determine the coefficients ϵ_{ijk} we write $\mathbf{a} = a^j \mathbf{e}_j$ and $\mathbf{b} = b^k \mathbf{e}_k$ and use the fact that the cross product is distributive over addition to obtain

$$\mathbf{c} = a^j \mathbf{e}_j \times b^k \mathbf{e}_k = (\mathbf{e}_j \times \mathbf{e}_k) a^j b^k.$$

Then

$$\mathbf{e}_i \cdot \mathbf{c} = \mathbf{e}_i \cdot c_m \mathbf{e}^m = c_i = [\mathbf{e}_i \cdot (\mathbf{e}_j \times \mathbf{e}_k)] a^j b^k$$

and comparison with (1.21) shows that

$$\epsilon_{ijk} = \mathbf{e}_i \cdot (\mathbf{e}_j \times \mathbf{e}_k).$$

The possibilities regarding the three indices i, j, k are conveniently handled by considering permutations of the ordered triple $(1, 2, 3)$. A permutation of $(1, 2, 3)$ is called *even* if it can be effected through an even number of interchanges of pairs of these numbers; it is *odd* if it can be effected through an odd number of interchanges. For instance, the permutations $(2, 3, 1)$ and $(3, 1, 2)$ are even, while $(1, 3, 2)$, $(3, 2, 1)$, and $(2, 1, 3)$ are odd. With this terminology we can write

$$\epsilon_{ijk} = \begin{cases} +V, & \text{if } (i, j, k) \text{ is an even permutation of } (1, 2, 3), \\ -V, & \text{if } (i, j, k) \text{ is an odd permutation of } (1, 2, 3), \\ 0, & \text{if two or more indices are equal,} \end{cases}$$

where V gives the volume of the frame parallelepiped defined by (1.5). We will call ϵ_{ijk} a *permutation* or *Levi–Civita symbol.* Another such quantity ϵ^{ijk} will be introduced below.

Gram determinant

We recall the two properties of determinants

$$\det A = \det A^T, \qquad \det AB = \det A \det B,$$

where A and B are square matrices and A^T is the transpose of A. Hence, using a Cartesian frame we can write

$$[\mathbf{a}\cdot(\mathbf{b}\times\mathbf{c})][\mathbf{u}\cdot(\mathbf{v}\times\mathbf{w})] = \begin{vmatrix} a_1 & a_2 & a_3 \\ b_1 & b_2 & b_3 \\ c_1 & c_2 & c_3 \end{vmatrix} \begin{vmatrix} u_1 & u_2 & u_3 \\ v_1 & v_2 & v_3 \\ w_1 & w_2 & w_3 \end{vmatrix} = \begin{vmatrix} a_1 & a_2 & a_3 \\ b_1 & b_2 & b_3 \\ c_1 & c_2 & c_3 \end{vmatrix} \begin{vmatrix} u_1 & v_1 & w_1 \\ u_2 & v_2 & w_2 \\ u_3 & v_3 & w_3 \end{vmatrix}$$

$$= \begin{vmatrix} a_1u_1 + a_2u_2 + a_3u_3 & a_1v_1 + a_2v_2 + a_3v_3 & a_1w_1 + a_2w_2 + a_3w_3 \\ b_1u_1 + b_2u_2 + b_3u_3 & b_1v_1 + b_2v_2 + b_3v_3 & b_1w_1 + b_2w_2 + b_3w_3 \\ c_1u_1 + c_2u_2 + c_3u_3 & c_1v_1 + c_2v_2 + c_3v_3 & c_1w_1 + c_2w_2 + c_3w_3 \end{vmatrix}$$

so that

$$[\mathbf{a}\cdot(\mathbf{b}\times\mathbf{c})][\mathbf{u}\cdot(\mathbf{v}\times\mathbf{w})] = \begin{vmatrix} \mathbf{a}\cdot\mathbf{u} & \mathbf{a}\cdot\mathbf{v} & \mathbf{a}\cdot\mathbf{w} \\ \mathbf{b}\cdot\mathbf{u} & \mathbf{b}\cdot\mathbf{v} & \mathbf{b}\cdot\mathbf{w} \\ \mathbf{c}\cdot\mathbf{u} & \mathbf{c}\cdot\mathbf{v} & \mathbf{c}\cdot\mathbf{w} \end{vmatrix}. \tag{1.22}$$

Although our intermediate steps occurred in a Cartesian frame, the result is frame independent.

Using (1.22) we can write

$$V^2 = [\mathbf{e}_1\cdot(\mathbf{e}_2\times\mathbf{e}_3)]^2 = \begin{vmatrix} \mathbf{e}_1\cdot\mathbf{e}_1 & \mathbf{e}_1\cdot\mathbf{e}_2 & \mathbf{e}_1\cdot\mathbf{e}_3 \\ \mathbf{e}_2\cdot\mathbf{e}_1 & \mathbf{e}_2\cdot\mathbf{e}_2 & \mathbf{e}_2\cdot\mathbf{e}_3 \\ \mathbf{e}_3\cdot\mathbf{e}_1 & \mathbf{e}_3\cdot\mathbf{e}_2 & \mathbf{e}_3\cdot\mathbf{e}_3 \end{vmatrix} = \begin{vmatrix} g_{11} & g_{12} & g_{13} \\ g_{21} & g_{22} & g_{23} \\ g_{31} & g_{32} & g_{33} \end{vmatrix},$$

hence

$$V^2 = g \tag{1.23}$$

where $g = |g_{ij}|$ is called the *Gram determinant.* If the Gram determinant vanishes, the $\mathbf{e}_i$ are linearly dependent. For Cartesian frames $|V| = 1$.

Important Remark. It is a standard trick to deduce general formulas for vectors and tensors using some special basis — usually a Cartesian basis. If both the initial and final versions of the formula are presented via vectors and tensors in basis-free form with use of dot products, then the basis in which the intermediate calculations occurred becomes irrelevant.

Back to permutation symbols and the cross product

The quantity ϵ_{ijk} is useful in writing formulas. For example, the determinant of a matrix $A = (a_{ij})$ can be expressed succinctly as

$$\det A = \epsilon_{ijk} a_{1i} a_{2j} a_{3k}.$$

Much more than a notational device however, ϵ_{ijk} represents a pseudotensor (the so-called *Levi–Civita tensor*). We discuss this further in Chapter 2.

Exercise 1.2. Show that when ϵ_{ijk} multiplies an expression symmetric in the indices j and k, the product is zero.

We can also express $\mathbf{a} \times \mathbf{b}$ in contravariant components:

$$c^i = \epsilon^{ijk} a_j b_k.$$

To determine the coefficients ϵ^{ijk}, write $\mathbf{a} = a_j \mathbf{e}^j$, $\mathbf{b} = b_k \mathbf{e}^k$ so that

$$\mathbf{c} = a_j \mathbf{e}^j \times b_k \mathbf{e}^k = (\mathbf{e}^j \times \mathbf{e}^k) a_j b_k.$$

Then

$$\mathbf{e}^i \cdot \mathbf{c} = \mathbf{e}^i \cdot c^m \mathbf{e}_m = c^i = [\mathbf{e}^i \cdot (\mathbf{e}^j \times \mathbf{e}^k)] a_j b_k$$

and we have

$$\epsilon^{ijk} = \mathbf{e}^i \cdot (\mathbf{e}^j \times \mathbf{e}^k) = \begin{cases} +V', & \text{if } (i,j,k) \text{ is an even permutation of } (1,2,3), \\ -V', & \text{if } (i,j,k) \text{ is an odd permutation of } (1,2,3), \\ 0, & \text{if two or more indices are equal.} \end{cases}$$

We now have

$$\mathbf{a} \times \mathbf{b} = \mathbf{e}^i \, \epsilon_{ijk} \, a^j \, b^k = \mathbf{e}_i \, \epsilon^{ijk} \, a_j \, b_k.$$

The special cases

$$\mathbf{e}_i \times \mathbf{e}_j = \epsilon_{kij} \mathbf{e}^k \qquad \text{and} \qquad \mathbf{e}^i \times \mathbf{e}^j = \epsilon^{kij} \mathbf{e}_k$$

are often useful.

Setting $\mathbf{a} = \mathbf{e}_i$, $\mathbf{b} = \mathbf{e}_j$, $\mathbf{c} = \mathbf{e}_k$, $\mathbf{u} = \mathbf{e}^p$, $\mathbf{v} = \mathbf{e}^q$, and $\mathbf{w} = \mathbf{e}^r$ in (1.22), we get

$$\epsilon_{ijk} \epsilon^{pqr} = \begin{vmatrix} \delta_i^p & \delta_i^q & \delta_i^r \\ \delta_j^p & \delta_j^q & \delta_j^r \\ \delta_k^p & \delta_k^q & \delta_k^r \end{vmatrix}. \tag{1.24}$$

A very important special case is

$$\epsilon_{ijk} \epsilon^{pqk} = \delta_i^p \delta_j^q - \delta_i^q \delta_j^p \tag{1.25}$$

from which we can obtain

$$\epsilon_{ijk} \epsilon^{pjk} = 2\delta_i^p \qquad \text{and} \qquad \epsilon_{ijk} \epsilon^{ijk} = 6. \tag{1.26}$$

As another application of (1.25) we have

$$\begin{aligned}
\mathbf{a} \times (\mathbf{b} \times \mathbf{c}) &= \mathbf{e}^i \epsilon_{ijk} a^j \epsilon^{kpq} b_p c_q \\
&= \mathbf{e}^i \epsilon_{ijk} \epsilon^{pqk} a^j b_p c_q \\
&= \mathbf{e}^i (\delta_i^p \delta_j^q - \delta_i^q \delta_j^p) a^j b_p c_q \\
&= (\mathbf{e}^i b_i)(a^q c_q) - (\mathbf{e}^i c_i)(a^p b_p) \\
&= \mathbf{b}(\mathbf{a} \cdot \mathbf{c}) - \mathbf{c}(\mathbf{a} \cdot \mathbf{b})
\end{aligned}$$

for any vectors **a**, **b**, and **c**. This is the familiar *vector triple product identity*.

1.3 Problems

1.1 A Cartesian frame is rotated about its third axis to give a new Cartesian frame. Find the matrix of transformation.

1.2 Show that $|\mathbf{a} \times \mathbf{b}|^2 = (\mathbf{a} \cdot \mathbf{a})(\mathbf{b} \cdot \mathbf{b}) - (\mathbf{a} \cdot \mathbf{b})^2$.

1.3 Find $\epsilon^{ijk} a_i b_j c_k$.

1.4 Prove (1.7).

1.5 Derive the formulas $\mathbf{e}^k = \frac{1}{2}\epsilon^{kij}(\mathbf{e}_i \times \mathbf{e}_j)$ and $\mathbf{e}_k = \frac{1}{2}\epsilon_{kij}(\mathbf{e}^i \times \mathbf{e}^j)$.

1.6 Show that if **x** is any vector and **n** is a unit vector, then **x** has the representation $\mathbf{x} = (\mathbf{n} \cdot \mathbf{x})\mathbf{n} + (\mathbf{n} \times \mathbf{x}) \times \mathbf{n}$.

1.7 Prove that $\mathbf{a} \times (\mathbf{b} \times \mathbf{c}) + \mathbf{b} \times (\mathbf{c} \times \mathbf{a}) + \mathbf{c} \times (\mathbf{a} \times \mathbf{b}) = \mathbf{0}$.

1.8 Prove that

$$(\mathbf{a} \times \mathbf{b}) \cdot (\mathbf{c} \times \mathbf{d}) = \begin{vmatrix} \mathbf{a} \cdot \mathbf{c} & \mathbf{a} \cdot \mathbf{d} \\ \mathbf{b} \cdot \mathbf{c} & \mathbf{b} \cdot \mathbf{d} \end{vmatrix}.$$

1.9 Show that $\mathbf{e}_k \times \mathbf{e}^k = \mathbf{0}$.

1.10 Prove: $(\mathbf{a} \times \mathbf{b}) \cdot (\mathbf{c} \times \mathbf{d}) + (\mathbf{b} \times \mathbf{c}) \cdot (\mathbf{a} \times \mathbf{d}) + (\mathbf{c} \times \mathbf{a}) \cdot (\mathbf{b} \times \mathbf{d}) = 0$.

Chapter 2

Tensors and Tensor Fields

2.1 Dyadic Quantities

Let $\mathbf{a}$ and $\mathbf{b}$ be two vectors in $\mathbb{R}^3$ (although, as noted previously, most of the following statements hold more generally for $\mathbf{a}, \mathbf{b} \in \mathbb{R}^n$). The *tensor product* of $\mathbf{a}$ and $\mathbf{b}$, denoted $\mathbf{a} \otimes \mathbf{b}$, is defined to satisfy many expected properties of a product. For all real α we have

$$\begin{aligned}(\alpha\mathbf{a}) \otimes \mathbf{b} &= \mathbf{a} \otimes (\alpha\mathbf{b}) = \alpha(\mathbf{a} \otimes \mathbf{b}),\\ (\mathbf{a} + \mathbf{b}) \otimes \mathbf{c} &= \mathbf{a} \otimes \mathbf{c} + \mathbf{b} \otimes \mathbf{c},\\ \mathbf{a} \otimes (\mathbf{b} + \mathbf{c}) &= \mathbf{a} \otimes \mathbf{b} + \mathbf{a} \otimes \mathbf{c}.\end{aligned}$$

But the tensor product is not commutative: if $\mathbf{a}$ and $\mathbf{b}$ are not proportional, then

$$\mathbf{a} \otimes \mathbf{b} \neq \mathbf{b} \otimes \mathbf{a}.$$

Although we use the term "product", it is (say, unlike the cross product of vectors) not a vector itself; rather it is a completely new and independent object. Henceforth, for brevity in long formulas, we will omit the $\otimes$ symbol, write $\mathbf{a} \otimes \mathbf{b}$ as simply $\mathbf{ab}$, and refer to the latter quantity as a *dyad* of the vectors $\mathbf{a}$ and $\mathbf{b}$. We therefore have

$$\begin{aligned}(\alpha\mathbf{a})\mathbf{b} &= \mathbf{a}(\alpha\mathbf{b}) = \alpha(\mathbf{ab}),\\ (\mathbf{a} + \mathbf{b})\mathbf{c} &= \mathbf{ac} + \mathbf{bc}, \qquad (2.1)\\ \mathbf{a}(\mathbf{b} + \mathbf{c}) &= \mathbf{ab} + \mathbf{ac},\end{aligned}$$

where α is a real scalar. A linear combination (weighted sum) of dyads is known as a *dyadic* or a *second-order tensor*.

Some operations with dyads

As in any finite dimensional vector space, in the space of dyads we can introduce a *scalar product* between dyads $\mathbf{ab}$ and $\mathbf{cd}$:

$$\mathbf{ab} \bullet \mathbf{cd} = (\mathbf{a} \cdot \mathbf{c})(\mathbf{b} \cdot \mathbf{d}), \tag{2.2}$$

where the smaller dot symbol occurring twice on the right is the ordinary dot product in $\mathbb{R}^3$. It is a commutative operation:

$$\mathbf{ab} \bullet \mathbf{cd} = (\mathbf{a} \cdot \mathbf{c})(\mathbf{b} \cdot \mathbf{d}) = (\mathbf{c} \cdot \mathbf{a})(\mathbf{d} \cdot \mathbf{b}) = \mathbf{cd} \bullet \mathbf{ab}.$$

Frequently $\mathbf{ab} \bullet \mathbf{cd}$ is denoted by $\mathbf{ab} : \mathbf{cd}$.

We will see that the scalar product serves as a suitable *inner product* on the space of tensors.

The scalar product of dyads is to be carefully distinguished from the *dot product* of $\mathbf{ab}$ and $\mathbf{cd}$, defined by

$$\mathbf{ab} \cdot \mathbf{cd} = (\mathbf{b} \cdot \mathbf{c})\mathbf{ad}. \tag{2.3}$$

Note that the scalar product of dyads yields a scalar, whereas the dot product of dyads yields a dyad. The dot product of dyads is clearly *not* commutative.

Another operation between dyads is *double dot multiplication.* If $\mathbf{ab}$ and $\mathbf{cd}$ are dyads, we define

$$\mathbf{ab} \cdot\cdot\, \mathbf{cd} = (\mathbf{b} \cdot \mathbf{c})(\mathbf{a} \cdot \mathbf{d}).$$

This time we first dot the near-standing vectors, then the remaining vectors. The resulting scalar proves useful in applications.

The *transpose* of a dyad $\mathbf{ab}$ is the dyad $\mathbf{ba}$. We write

$$(\mathbf{ab})^T = \mathbf{ba}. \tag{2.4}$$

We will see below that when extended to more general second-order tensors, this operation bears a relation to the familiar transpose operation for a matrix. Clearly the scalar product and the double dot product are related:

$$\mathbf{ab} \bullet \mathbf{cd} = \mathbf{ab} \cdot\cdot\, (\mathbf{cd})^T.$$

Exercise 2.1. Show that $\mathbf{ab} \cdot\cdot\, (\mathbf{cd})^T = (\mathbf{ab})^T \cdot\cdot\, \mathbf{cd} = (\mathbf{b} \cdot \mathbf{d})(\mathbf{a} \cdot \mathbf{c})$.

We can also introduce dot products between a dyad $\mathbf{ab}$ and a vector $\mathbf{c}$. The rules are

$$\mathbf{ab} \cdot \mathbf{c} = (\mathbf{b} \cdot \mathbf{c})\mathbf{a}, \tag{2.5}$$

$$\mathbf{c} \cdot \mathbf{ab} = (\mathbf{c} \cdot \mathbf{a})\mathbf{b}. \tag{2.6}$$

In the first case the result is a vector along $\mathbf{a}$; in the second it is along $\mathbf{b}$. For example, if $\mathbf{e}$ is a fixed unit vector then a dyad of the form $\mathbf{ee}$ is called a *projection dyad* because

$$\mathbf{a} \cdot \mathbf{ee} = \mathbf{ee} \cdot \mathbf{a} = (\mathbf{a} \cdot \mathbf{e})\mathbf{e}.$$

Here the dot product (from either side) produces the vector component of a given vector $\mathbf{a}$ along $\mathbf{e}$, i.e., the orthogonal projection of $\mathbf{a}$ on $\mathbf{e}$. The vector-dyadic dot product introduced here is a linear operation with

$$\begin{aligned}
\mathbf{wx} \cdot (\lambda \mathbf{a} + \mu \mathbf{b}) &= \lambda \mathbf{wx} \cdot \mathbf{a} + \mu \mathbf{wx} \cdot \mathbf{b}, \\
(\lambda \mathbf{a} + \mu \mathbf{b}) \cdot \mathbf{wx} &= \lambda \mathbf{a} \cdot \mathbf{wx} + \mu \mathbf{b} \cdot \mathbf{wx}, \\
\mathbf{a} \cdot (\lambda \mathbf{wx} + \mu \mathbf{yz}) &= \lambda \mathbf{a} \cdot \mathbf{wx} + \mu \mathbf{a} \cdot \mathbf{yz}, \\
(\lambda \mathbf{wx} + \mu \mathbf{yz}) \cdot \mathbf{a} &= \lambda \mathbf{wx} \cdot \mathbf{a} + \mu \mathbf{yz} \cdot \mathbf{a},
\end{aligned}$$

for all $\mathbf{a}, \mathbf{b}, \mathbf{w}, \mathbf{x}, \mathbf{y}, \mathbf{z} \in \mathbb{R}^3$ and $\lambda, \mu \in \mathbb{R}$.

When $\mathbf{a}, \mathbf{b}, \mathbf{c} \in \mathbb{R}^3$, we can define the cross product between two dyads:

$$\mathbf{ab} \times \mathbf{cd} = \mathbf{a}(\mathbf{b} \times \mathbf{c})\mathbf{d}. \tag{2.7}$$

However, this operation yields a *triad* — a type of quantity not considered further until § 2.3. Cross products between a dyad and a vector are accomplished as follows:

$$(\mathbf{ab}) \times \mathbf{c} = \mathbf{a}(\mathbf{b} \times \mathbf{c}), \qquad \mathbf{c} \times (\mathbf{ab}) = (\mathbf{c} \times \mathbf{a})\mathbf{b}. \tag{2.8}$$

Both results are dyads.

Exercise 2.2. Show that $(\mathbf{y} \cdot \mathbf{ab}) \times \mathbf{x} = \mathbf{y} \cdot (\mathbf{ab} \times \mathbf{x})$ for any $\mathbf{a}, \mathbf{b}, \mathbf{x}, \mathbf{y} \in \mathbb{R}^3$.

Exercise 2.3. Show that

$$\begin{aligned}
&(\mathbf{a} \times \mathbf{b}) \times (\mathbf{c} \times \mathbf{d}) = \mathbf{ba} \cdot (\mathbf{c} \times \mathbf{d}) - \mathbf{ab} \cdot (\mathbf{c} \times \mathbf{d}), \\
&(\mathbf{a} \times \mathbf{b}) \times (\mathbf{c} \times \mathbf{d}) = \mathbf{cd} \cdot (\mathbf{a} \times \mathbf{b}) - \mathbf{dc} \cdot (\mathbf{a} \times \mathbf{b}), \\
&\mathbf{ab} \cdot (\mathbf{c} \times \mathbf{d}) - \mathbf{ba} \cdot (\mathbf{c} \times \mathbf{d}) + \mathbf{cd} \cdot (\mathbf{a} \times \mathbf{b}) - \mathbf{dc} \cdot (\mathbf{a} \times \mathbf{b}) = \mathbf{0},
\end{aligned}$$

for any vectors $\mathbf{a}, \mathbf{b}, \mathbf{c}, \mathbf{d} \in \mathbb{R}^3$.

The operations introduced in this section are summarized for convenience in Table 2.1.

Table 2.1 Some basic operations involving dyads.

scalar product of dyads	$\mathbf{ab} \bullet \mathbf{cd} = (\mathbf{a} \cdot \mathbf{c})(\mathbf{b} \cdot \mathbf{d})$
dot product of dyads	$\mathbf{ab} \cdot \mathbf{cd} = (\mathbf{b} \cdot \mathbf{c})\mathbf{ad}$
double dot product of dyads	$\mathbf{ab} \cdot\cdot\, \mathbf{cd} = (\mathbf{b} \cdot \mathbf{c})(\mathbf{a} \cdot \mathbf{d})$
transpose of dyad	$(\mathbf{ab})^T = \mathbf{ba}$
dot product of dyad and vector	$\mathbf{ab} \cdot \mathbf{c} = (\mathbf{b} \cdot \mathbf{c})\mathbf{a}$ or $\mathbf{c} \cdot \mathbf{ab} = (\mathbf{c} \cdot \mathbf{a})\mathbf{b}$
cross product of dyads	$\mathbf{ab} \times \mathbf{cd} = \mathbf{a}(\mathbf{b} \times \mathbf{c})\mathbf{d}$
cross product of dyad and vector	$(\mathbf{ab}) \times \mathbf{c} = \mathbf{a}(\mathbf{b} \times \mathbf{c})$ or $\mathbf{c} \times (\mathbf{ab}) = (\mathbf{c} \times \mathbf{a})\mathbf{b}$

2.2 Second-Order Tensors

By expanding both $\mathbf{a}$ and $\mathbf{b}$ in a basis $\mathbf{e}_i$ of $\mathbb{R}^3$, and using the properties (2.1), we get

$$\mathbf{ab} = a^i \mathbf{e}_i \, b^j \mathbf{e}_j = a^i b^j \, \mathbf{e}_i \mathbf{e}_j.$$

This is a representation of the dyad $\mathbf{ab}$ in terms of nine *basis dyads* $\mathbf{e}_i \mathbf{e}_j$ with components $a^i b^j$. The dyadic basis $\{\mathbf{e}_i \mathbf{e}_j\}$ spans a nine-dimensional linear space known as the *space of second-order Euclidean tensors*. (The space is n^2-dimensional if $\mathbf{a}, \mathbf{b} \in \mathbb{R}^n$.) An element of this space, a second-order tensor $\mathbf{A}$, has the representation

$$\mathbf{A} = a^{ij} \mathbf{e}_i \mathbf{e}_j$$

where the a^{ij} are the *contravariant components* of $\mathbf{A}$ relative to the given basis. Besides the simple basis $\mathbf{e}_i \mathbf{e}_j$, three other types of simple basis related to $\mathbf{e}_k$ and its reciprocal basis $\mathbf{e}^j$ are possible: $\mathbf{e}^i \mathbf{e}^j$, $\mathbf{e}_i \mathbf{e}^j$, and $\mathbf{e}^i \mathbf{e}_j$. Hence a second-order tensor also has representations

$$\mathbf{A} = a^{ij} \mathbf{e}_i \mathbf{e}_j = a_{ij} \mathbf{e}^i \mathbf{e}^j = a_i^{\ j} \mathbf{e}^i \mathbf{e}_j = a^i_{\ j} \, \mathbf{e}_i \mathbf{e}^j. \tag{2.9}$$

We term a_{ij} the *covariant components* of $\mathbf{A}$, and $a_i^{\ j}$ and $a^i_{\ j}$ the two types of *mixed components* of $\mathbf{A}$. Note the unambiguous positioning of the indices in the notation for the mixed components. In some textbooks, to show the order, the mixed components are denoted with dots: $a_{i}^{\cdot j}$ and $a^{j}_{\cdot i}$.

In each basis, the components of a tensor $\mathbf{A}$ can be presented in the form of a 3×3 matrix analogous to the column matrix for a vector. We have introduced the operations with dyads (and thus with tensors) and vectors in such a way that, as we will see, they will allow us to mimic actions familiar from matrix algebra. However, the use of multiple bases with tensors will require us to incorporate the components of the metric tensor.

Transformation of second-order tensor components

Let us revisit the frame transformation of Chapter 1,

$$\mathbf{e}_i = A_i^{j'}\mathbf{e}_{j'}, \quad \mathbf{e}_{i'} = A_{i'}^{j}\mathbf{e}_j, \quad \mathbf{e}^i = A_{j'}^{i}\mathbf{e}^{j'}, \quad \mathbf{e}^{i'} = A_j^{i'}\mathbf{e}^j,$$

and investigate how the components of a second-order tensor $\mathbf{A}$ must transform to ensure that $\mathbf{A}$ is an objective entity. Of course, there are many ways in which we can represent $\mathbf{A}$:

$$\begin{aligned}\mathbf{A} &= a^{ij}\mathbf{e}_i\mathbf{e}_j = a_{kl}\mathbf{e}^k\mathbf{e}^l = a_i{}^j\mathbf{e}^i\mathbf{e}_j = a^k{}_l\,\mathbf{e}_k\mathbf{e}^l \\ &= a^{i'j'}\mathbf{e}_{i'}\mathbf{e}_{j'} = a_{k'l'}\mathbf{e}^{k'}\mathbf{e}^{l'} = a_{i'}{}^{j'}\mathbf{e}^{i'}\mathbf{e}_{j'} = a^{k'}{}_{l'}\,\mathbf{e}_{k'}\mathbf{e}^{l'}.\end{aligned}$$

Let us begin with the first of these representation types. Since $\mathbf{A}$ is a frame-independent mathematical object, we must have

$$a^{ij}\mathbf{e}_i\mathbf{e}_j = a^{i'j'}\mathbf{e}_{i'}\mathbf{e}_{j'}.$$

Predotting with $\mathbf{e}^k$ and postdotting with $\mathbf{e}^m$, we get

$$\begin{aligned}\mathbf{e}^k \cdot a^{ij}\mathbf{e}_i\mathbf{e}_j \cdot \mathbf{e}^m &= \mathbf{e}^k \cdot a^{i'j'}\mathbf{e}_{i'}\mathbf{e}_{j'} \cdot \mathbf{e}^m, \\ \delta_i^k\, a^{ij}\,\delta_j^m &= A_{i'}^k\, a^{i'j'}\, A_{j'}^m, \\ \Rightarrow\ a^{km} &= A_{i'}^k A_{j'}^m a^{i'j'}.\end{aligned}$$

Alternatively, predotting with $\mathbf{e}^{k'}$ and postdotting with $\mathbf{e}^{m'}$, we get

$$a^{k'm'} = A_i^{k'} A_j^{m'} a^{ij}.$$

In this way we obtain the first line of Table 2.2. The remaining entries are left to the reader.

Table 2.2 Second-order tensors under change of basis.

components a^{ij}	contravariant	$a^{i'j'} = A_k^{i'} A_l^{j'} a^{kl}$	$a^{ij} = A_{k'}^{i} A_{l'}^{j} a^{k'l'}$
components a_{ij}	covariant	$a_{i'j'} = A_{i'}^{k} A_{j'}^{l} a_{kl}$	$a_{ij} = A_i^{k'} A_j^{l'} a_{k'l'}$
components $a_i{}^j$	mixed	$a_{i'}{}^{j'} = A_{i'}^{k} A_l^{j'} a_k{}^l$	$a_i{}^j = A_i^{k'} A_{l'}^{j} a_{k'}{}^{l'}$
components $a^i{}_j$	mixed	$a^{i'}{}_{j'} = A_k^{i'} A_{j'}^{l} a^k{}_l$	$a^i{}_j = A_{k'}^{i} A_j^{l'} a^{k'}{}_{l'}$

Transformation to the reciprocal frame can be considered as a special case; alternatively, it can be investigated from scratch by imposing the requirement

$$a_{ij}\mathbf{e}^i\mathbf{e}^j = a^{ij}\mathbf{e}_i\mathbf{e}_j$$

and taking dot products to obtain

$$a^{km} = g^{ki}g^{jm}a_{ij} \quad \text{and} \quad a_{km} = a^{ij}g_{ki}g_{jm}.$$

We observe once again the index raising and lowering properties of the metric coefficients.

We stress that we can find the components of a tensor $\mathbf{A}$ in any basis through simple dot product operations:

$$a^{ij} = \mathbf{e}^i \cdot \mathbf{A} \cdot \mathbf{e}^j, \qquad a^i{}_j = \mathbf{e}^i \cdot \mathbf{A} \cdot \mathbf{e}_j,$$

$$a_{ij} = \mathbf{e}_i \cdot \mathbf{A} \cdot \mathbf{e}_j, \qquad a_i{}^j = \mathbf{e}_i \cdot \mathbf{A} \cdot \mathbf{e}^j.$$

Operations with second-order tensors

Two second-order tensors $\mathbf{A}$ and $\mathbf{B}$ can be added through their representations relative to a basis. If the chosen basis is $\mathbf{e}_i\mathbf{e}_j$, for example, we have

$$\mathbf{A} + \mathbf{B} = a^{ij}\mathbf{e}_{ij} + b^{ij}\mathbf{e}_{ij} = (a^{ij} + b^{ij})\mathbf{e}_i\mathbf{e}_j.$$

Tensor addition is obviously commutative:

$$\mathbf{A} + \mathbf{B} = \mathbf{B} + \mathbf{A}.$$

More generally, the linear combination $\alpha\mathbf{A} + \beta\mathbf{B}$ can be done as

$$\alpha\mathbf{A} + \beta\mathbf{B} = (\alpha a^{ij} + \beta b^{ij})\mathbf{e}_i\mathbf{e}_j. \tag{2.10}$$

Subtraction can be done by taking $\alpha = 1$ and $\beta = -1$:

$$\mathbf{A} - \mathbf{B} = (a^{ij} - b^{ij})\mathbf{e}_i\mathbf{e}_j. \tag{2.11}$$

Any product of the form $0\mathbf{A}$ gives a *zero tensor* denoted $\mathbf{0}$.

The scalar product of two dyads, defined by (2.2), can be extended to two general second-order tensors $\mathbf{A}$ and $\mathbf{B}$ via linearity:

$$\mathbf{A} \bullet \mathbf{B} = a^{ij}b^{kl}\,\mathbf{e}_i\mathbf{e}_j \bullet \mathbf{e}_k\mathbf{e}_l = a^{ij}b^{kl}(\mathbf{e}_i \cdot \mathbf{e}_k)(\mathbf{e}_j \cdot \mathbf{e}_l) = a^{ij}b^{kl}g_{ik}g_{jl} = a^{ij}b_{ij}.$$

The same approach applies to the dot product,

$$\mathbf{A} \cdot \mathbf{B} = a^{ij}b^{kl}\,\mathbf{e}_i\mathbf{e}_j \cdot \mathbf{e}_k\mathbf{e}_l = a^{ij}g_{jk}b^{kl}\,\mathbf{e}_i\mathbf{e}_l,$$

the double dot product,

$$\mathbf{A} \cdot\cdot\, \mathbf{B} = a^{ij}b^{kl}\,\mathbf{e}_i\mathbf{e}_j \cdot\cdot\, \mathbf{e}_k\mathbf{e}_l = a^{ij}b^{kl}g_{jk}g_{il} = a_l{}^j b_j{}^l,$$

and the various other product operations of Table 2.1. Note that the dot product is not commutative (in general $\mathbf{A}\cdot\mathbf{B}\neq\mathbf{B}\cdot\mathbf{A}$), but it is associative and distributive over addition:

$$(\mathbf{A}\cdot\mathbf{B})\cdot\mathbf{C}=\mathbf{A}\cdot(\mathbf{B}\cdot\mathbf{C})=\mathbf{A}\cdot\mathbf{B}\cdot\mathbf{C},$$
$$(\mathbf{A}+\mathbf{B})\cdot\mathbf{C}=\mathbf{A}\cdot\mathbf{C}+\mathbf{B}\cdot\mathbf{C}.$$

Exercise 2.4. (a) We have seen that the tensor $\mathbf{C}=\mathbf{A}\cdot\mathbf{B}$ has contravariant components $c^{ij}=a^{in}g_{nm}b^{mj}$. Obtain the other component representations of $\mathbf{C}$. (b) Obtain some representations of $c=\mathbf{A}\cdot\cdot\,\mathbf{B}$. (c) Write down some component expressions for $\mathbf{A}\times\mathbf{x}$ and $\mathbf{x}\times\mathbf{A}$. (d) Show that if $\mathbf{a}\cdot\mathbf{X}\cdot\mathbf{b}=0$ for all $\mathbf{a},\mathbf{b}\in\mathbb{R}^3$, then $\mathbf{X}=\mathbf{0}$.

Let us consider the transpose operation for a second-order tensor $\mathbf{A}$. The various representations

$$\mathbf{A}=a^{ij}\mathbf{e}_i\mathbf{e}_j=a_{ij}\mathbf{e}^i\mathbf{e}^j=a_i^{\ j}\mathbf{e}^i\mathbf{e}_j=a^i_{\ j}\mathbf{e}_i\mathbf{e}^j$$

lead to transpose expressions

$$\begin{aligned}\mathbf{A}^T&=a^{ij}\mathbf{e}_j\mathbf{e}_i=a_{ij}\mathbf{e}^j\mathbf{e}^i=a_i^{\ j}\mathbf{e}_j\mathbf{e}^i=a^i_{\ j}\mathbf{e}^j\mathbf{e}_i\\&=a^{ji}\mathbf{e}_i\mathbf{e}_j=a_{ji}\mathbf{e}^i\mathbf{e}^j=a_j^{\ i}\mathbf{e}_i\mathbf{e}^j=a^j_{\ i}\mathbf{e}^i\mathbf{e}_j\end{aligned}$$

and reveal that $\mathbf{A}^T$ may be taken in any of the following ways:

$$\begin{aligned}\mathbf{A}&=a^{ij}\mathbf{e}_i\mathbf{e}_j&&\Longrightarrow&\mathbf{A}^T&=a^{ji}\mathbf{e}_i\mathbf{e}_j,\\\mathbf{A}&=a_{ij}\mathbf{e}^i\mathbf{e}^j&&\Longrightarrow&\mathbf{A}^T&=a_{ji}\mathbf{e}^i\mathbf{e}^j,\\\mathbf{A}&=a_i^{\ j}\mathbf{e}^i\mathbf{e}_j&&\Longrightarrow&\mathbf{A}^T&=a^j_{\ i}\mathbf{e}^i\mathbf{e}_j,\\\mathbf{A}&=a^i_{\ j}\mathbf{e}_i\mathbf{e}^j&&\Longrightarrow&\mathbf{A}^T&=a_j^{\ i}\mathbf{e}_i\mathbf{e}^j.\end{aligned}$$

Note that the indices are merely swapped in the contravariant and covariant components. This is not the case with the mixed components; rather, the transpose operation introduces the other type of mixed component. Clearly

$$(\mathbf{A}^T)^T=\mathbf{A}.$$

Exercise 2.5. Show that (a) $\mathbf{A}\cdot\mathbf{x}=\mathbf{x}\cdot\mathbf{A}^T$ for any vector $\mathbf{x}$; (b) $(\mathbf{A}\cdot\mathbf{B})^T=\mathbf{B}^T\cdot\mathbf{A}^T$; (c) if $\mathbf{a}$ and $\mathbf{b}$ are vectors and $\mathbf{C}$ is a second-order tensor, then $\mathbf{a}\cdot\mathbf{C}^T\cdot\mathbf{b}=\mathbf{b}\cdot\mathbf{C}\cdot\mathbf{a}$; (d) $\mathbf{A}\bullet\mathbf{B}=\mathbf{A}\cdot\cdot\,\mathbf{B}^T=\mathbf{A}^T\cdot\cdot\,\mathbf{B}$; (e) $(\mathbf{A}\times\mathbf{B})\cdot\cdot\,\mathbf{B}^T=\mathbf{0}$.

Exercise 2.6. Show that for second-order tensors $\mathbf{A}$, $\mathbf{U}$, and $\mathbf{X}$, we have $(\mathbf{A}\cdot\mathbf{U})\cdot\cdot\,\mathbf{X}=(\mathbf{U}\cdot\mathbf{X})\cdot\cdot\,\mathbf{A}$ and $\mathbf{X}\cdot\cdot\,(\mathbf{A}\cdot\mathbf{U})=\mathbf{U}\cdot\cdot\,(\mathbf{X}\cdot\mathbf{A})$.

Identity, metric, or unit tensor

We define a special tensor $\mathbf{E}$ by

$$\mathbf{E} = \mathbf{e}_i\mathbf{e}^i. \tag{2.12}$$

This is called the *identity*, *metric*, or *unit tensor*. Expressed variously as

$$\mathbf{E} = \mathbf{e}^i\mathbf{e}_i = g_{ij}\mathbf{e}^i\mathbf{e}^j = g^{ij}\mathbf{e}_i\mathbf{e}_j = \sum_{i=1}^{3} \mathbf{i}_i\mathbf{i}_i, \tag{2.13}$$

it has the properties that

$$\mathbf{x} \cdot \mathbf{E} = \mathbf{E} \cdot \mathbf{x} = \mathbf{x}$$

for any vector $\mathbf{x}$ and

$$\mathbf{A} \cdot \mathbf{E} = \mathbf{E} \cdot \mathbf{A} = \mathbf{A}$$

for any second-order tensor $\mathbf{A}$. As a simple application of $\mathbf{E}$ we can formulate the *reflection tensor* with respect to a plane having unit normal $\mathbf{e}$:

$$\mathbf{T} = \mathbf{E} - 2\mathbf{e}\mathbf{e}.$$

This tensor maps a given vector $\mathbf{x}$ into the vector

$$\mathbf{T} \cdot \mathbf{x} = \mathbf{x} - 2(\mathbf{e} \cdot \mathbf{x})\mathbf{e}$$

as shown in Fig. 2.1.

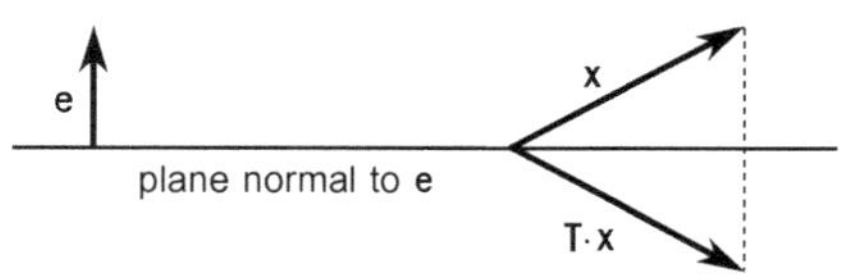

Fig. 2.1 Action of the reflection tensor $\mathbf{T}$ on a given vector $\mathbf{x}$.

From (2.13), we see that the matrix representation of $\mathbf{E}$ in a Cartesian basis is the identity matrix:

$$I = \begin{pmatrix} 1 & 0 & 0 \\ 0 & 1 & 0 \\ 0 & 0 & 1 \end{pmatrix}.$$

The same is true of $\mathbf{E}$ in mixed components for a general basis induced by a basis of $\mathbb{R}^3$, since the mixed components of $\mathbf{E}$ are $e^i{}_j = \delta^i_j$ and $e_j{}^i = \delta^i_j$. It is not the case for the non-mixed components, however, as they are metric coefficients.

Exercise 2.7. (a) Write down the matrix of the reflection tensor $\mathbf{T}$ in a Cartesian basis. (b) Find $\mathbf{A}\cdot\cdot\mathbf{E}$. (c) Show that $\mathbf{E}\cdot\cdot\mathbf{E} = 3$ and $\mathbf{E}\cdot\cdot\mathbf{vv} = \mathbf{v}\cdot\mathbf{v}$. (d) Show that $\mathbf{E}\times\mathbf{v} = \mathbf{v}\times\mathbf{E}$ for any vector $\mathbf{v}$.

Inverse of a nonsingular tensor

Let $\mathbf{A}$ be a second-order tensor. If there is a tensor $\mathbf{B}$ such that

$$\mathbf{A}\cdot\mathbf{B} = \mathbf{B}\cdot\mathbf{A} = \mathbf{E},$$

then we call $\mathbf{B}$ the *inverse* of $\mathbf{A}$ and denote it by $\mathbf{A}^{-1}$. In this case $\mathbf{A}$ is said to be *invertible* or *nonsingular*. Hence for a nonsingular tensor

$$\mathbf{A}\cdot\mathbf{A}^{-1} = \mathbf{A}^{-1}\cdot\mathbf{A} = \mathbf{E}.$$

If $\mathbf{A}$ and $\mathbf{B}$ are both nonsingular, then we have both

$$(\mathbf{A}\cdot\mathbf{B})\cdot(\mathbf{B}^{-1}\cdot\mathbf{A}^{-1}) = \mathbf{A}\cdot(\mathbf{B}\cdot\mathbf{B}^{-1})\cdot\mathbf{A}^{-1} = \mathbf{A}\cdot\mathbf{E}\cdot\mathbf{A}^{-1} = \mathbf{E},$$
$$(\mathbf{B}^{-1}\cdot\mathbf{A}^{-1})\cdot(\mathbf{A}\cdot\mathbf{B}) = \mathbf{B}^{-1}\cdot(\mathbf{A}^{-1}\cdot\mathbf{A})\cdot\mathbf{B} = \mathbf{B}^{-1}\cdot\mathbf{E}\cdot\mathbf{B} = \mathbf{E},$$

and therefore

$$(\mathbf{A}\cdot\mathbf{B})^{-1} = \mathbf{B}^{-1}\cdot\mathbf{A}^{-1}. \tag{2.14}$$

Of course, $(\mathbf{A}^{-1})^{-1} = \mathbf{A}$.

We can also characterize a nonsingular second-order tensor $\mathbf{A}$ by stating that $\mathbf{A}\cdot\mathbf{x} = \mathbf{0}$ only if $\mathbf{x} = \mathbf{0}$. Hence $\mathbf{A}$ is singular if $\mathbf{A}\cdot\mathbf{x} = \mathbf{0}$ for some nonzero $\mathbf{x}$. An example of a singular tensor is the projection dyad $\mathbf{ee}$, as any unit vector $\mathbf{n}$ perpendicular to $\mathbf{e}$ yields $\mathbf{ee}\cdot\mathbf{n} = \mathbf{0}$.

Exercise 2.8. Show that $(\mathbf{A}^T)^{-1} = (\mathbf{A}^{-1})^T$.

Determinant of a second-order tensor

The determinant of a second-order tensor $\mathbf{A}$ is the determinant of the matrix of its mixed components:

$$\det\mathbf{A} = |a_i^{\cdot j}| = |a^i_{\cdot j}|.$$

In terms of non-mixed components we can write

$$\det\mathbf{A} = |a_i^{\cdot j}| = |g_{ik}a^{kj}| = |g_{ik}||a^{kj}| = g|a^{kj}|,$$
$$\det\mathbf{A} = |a_i^{\cdot j}| = |a_{ik}g^{kj}| = |a_{ik}||g^{kj}| = g^{-1}|a_{ik}|.$$

It can also be shown that

$$\det \mathbf{A} = \tfrac{1}{6}\epsilon_{ijk}\epsilon^{mnp}a_m{}^i a_n{}^j a_p{}^k.$$

Exercise 2.9. (a) Verify that $|a_i{}^j| = |a^k{}_m|$, and that $\det \mathbf{A}$ is independent of the choice of basis in which the mixed components are taken. (b) Show that $\det \mathbf{A}^{-1} = (\det \mathbf{A})^{-1}$.

As for a matrix, a tensor is nonsingular if and only if its determinant is nonzero.

Trace of a second-order tensor

The *trace* of a second-order tensor $\mathbf{A}$ is denoted by $\operatorname{tr} \mathbf{A}$ and defined by

$$\operatorname{tr} \mathbf{A} = a_i{}^i. \tag{2.15}$$

This quantity is independent of the choice of basis in which the mixed components are taken:

$$a_i{}^i = A_i^{k'} A_{l'}^i a_{k'}{}^{l'} = \delta_{l'}^{k'} a_{k'}{}^{l'} = a_{k'}{}^{k'}.$$

In coordinate-free form

$$\operatorname{tr} \mathbf{A} = \mathbf{E} \cdot\cdot\, \mathbf{A} = \mathbf{E} \bullet \mathbf{A}. \tag{2.16}$$

Some other identities involving the trace are listed on p. 328.

Exercise 2.10. (a) Verify (2.16). (b) Calculate the trace of the dyad $\mathbf{ab}$.

Tensors raised to powers

Positive integer powers of a second-order tensor $\mathbf{A}$ are defined using the dot product. That is,

$$\mathbf{A}^0 = \mathbf{E}, \qquad \mathbf{A}^n = \mathbf{A} \cdot \mathbf{A}^{n-1} \text{ for } n = 1, 2, 3, \ldots.$$

Obviously

$$\mathbf{A}^m \cdot \mathbf{A}^n = \mathbf{A}^{m+n}, \qquad (\mathbf{A}^m)^n = \mathbf{A}^{mn}.$$

Negative integer powers of $\mathbf{A}$ are obtained by raising $\mathbf{A}^{-1}$ to positive integer powers:

$$\mathbf{A}^{-n} = \mathbf{A}^{-n+1} \cdot \mathbf{A}^{-1} \text{ for } n = 2, 3, 4, \ldots.$$

The usefulness of raising tensors to powers becomes clear when the idea of a Taylor series is introduced. We may write, for instance,

$$e^{\mathbf{A}} = \sum_{k=0}^{\infty} \frac{\mathbf{A}^k}{k!}. \tag{2.17}$$

The question of convergence will be delayed until we have covered the notion of a tensor norm.

Symmetric and antisymmetric tensors

A second-order tensor $\mathbf{A}$ is said to be *symmetric* if

$$\mathbf{A} = \mathbf{A}^T.$$

In this case

$$\mathbf{A} \cdot \mathbf{x} = \mathbf{x} \cdot \mathbf{A} \qquad \text{and} \qquad (\mathbf{A} \cdot \mathbf{x}) \cdot \mathbf{y} = \mathbf{x} \cdot (\mathbf{A} \cdot \mathbf{y}).$$

Important examples of symmetric tensors are the identity tensor $\mathbf{E}$ and a projection dyad, but many examples arise in the physical sciences as well. In linear elasticity, for instance, the stress and strain tensors are both symmetric. In $\mathbb{R}^3$ a second-order tensor $\mathbf{A}$ will have nine components. If $\mathbf{A}$ is symmetric then only six of these components can be specified independently, with $a^{mn} = a^{nm}$ and $a_{nm} = a_{mn}$. From

$$a_k{}^s \mathbf{e}^k \mathbf{e}_s = (a_k{}^s \mathbf{e}^k \mathbf{e}_s)^T = a^s{}_k \mathbf{e}^k \mathbf{e}_s$$

we see that the mixed components satisfy $a_k{}^s = a^s{}_k$.

A second-order tensor $\mathbf{A}$ is said to be *antisymmetric* or *skew* if

$$\mathbf{A} = -\mathbf{A}^T.$$

In this case only three components of $\mathbf{A}$ can be specified independently. As $a^{mn} = -a^{nm}$, $a_{nm} = -a_{mn}$, and $a_m{}^n = -a^n{}_m$, we can take as independent, say, a^{12}, a^{13}, and a^{23}. The remaining components are $a^{11} = 0$, $a^{22} = 0$, $a^{33} = 0$, $a^{21} = -a^{12}$, $a^{31} = -a^{13}$, and $a^{32} = -a^{23}$.

Exercise 2.11. Write out all the other relations for the components of $\mathbf{A}$ in terms of a^{12}, a^{13}, and a^{23}.

Exercise 2.12. Prove: (a) if $\mathbf{A}$ is symmetric and $\mathbf{B}$ is antisymmetric, then $\mathbf{A} \cdot\cdot \mathbf{B} = 0$ and $\mathbf{A} \bullet \mathbf{B} = 0$; (b) if $\mathbf{A}$ is symmetric, then $\mathbf{a} \cdot \mathbf{A} \cdot \mathbf{b} = \mathbf{b} \cdot \mathbf{A} \cdot \mathbf{a}$ for any vectors $\mathbf{a}$ and $\mathbf{b}$; (c) if $\mathbf{A}$ is antisymmetric, then $\mathbf{a} \cdot \mathbf{A} \cdot \mathbf{b} = -\mathbf{b} \cdot \mathbf{A} \cdot \mathbf{a}$; (d) $\mathbf{E} \times \mathbf{v}$ is antisymmetric for any vector $\mathbf{v}$.

For any second-order tensor $\mathbf{A}$, the two tensors

$$\tfrac{1}{2}\left(\mathbf{A}+\mathbf{A}^T\right) \qquad \text{and} \qquad \tfrac{1}{2}\left(\mathbf{A}-\mathbf{A}^T\right)$$

are symmetric and antisymmetric, respectively. Since the sum of these two tensors is $\mathbf{A}$, we can state (leaving the uniqueness assertion as an exercise)

Theorem 2.1. *Every second-order tensor can be uniquely written as a sum of a symmetric tensor and an antisymmetric tensor.*

A simple example of such a decomposition is

$$\mathbf{ab} = \tfrac{1}{2}(\mathbf{ab}+\mathbf{ba}) + \tfrac{1}{2}(\mathbf{ab}-\mathbf{ba}).$$

Exercise 2.13. Show that the quadratic form $\mathbf{x}\cdot\mathbf{A}\cdot\mathbf{x}$ does not change if $\mathbf{A}$ is replaced by its symmetric part. What happens to $\mathbf{x}\cdot\mathbf{A}\cdot\mathbf{x}$ if $\mathbf{A}$ is antisymmetric?

Conjugate vector

We have seen that the tensors $\mathbf{E}\times\mathbf{v}$ and $\mathbf{v}\times\mathbf{E}$ are antisymmetric and equal for any vector $\mathbf{v}$. It turns out that any antisymmetric tensor $\mathbf{C}$ in three-dimensional space has a representation of the form

$$\mathbf{C} = \mathbf{E}\times\boldsymbol{\omega} = \boldsymbol{\omega}\times\mathbf{E} \tag{2.18}$$

where $\boldsymbol{\omega}$ is known as the *conjugate vector* for $\mathbf{C}$. Indeed, if we set

$$\boldsymbol{\omega} = -\tfrac{1}{2}\epsilon^{ksp}C_{ks}\mathbf{e}_p \tag{2.19}$$

then

$$\begin{aligned}
\mathbf{E}\times\boldsymbol{\omega} &= -\tfrac{1}{2}\epsilon^{ksp}C_{ks}\mathbf{e}^n(\mathbf{e}_n\times\mathbf{e}_p) \\
&= -\tfrac{1}{2}\epsilon^{ksp}\epsilon_{snp}C_{ks}\mathbf{e}^n\mathbf{e}^s \\
&= \tfrac{1}{2}\epsilon^{spk}\epsilon_{spn}C_{ks}\mathbf{e}^n\mathbf{e}^s \\
&= \delta^k_n C_{ks}\mathbf{e}^n\mathbf{e}^s \\
&= \mathbf{C}.
\end{aligned}$$

It is easy to verify that

$$\mathbf{C}\cdot\mathbf{x} = \boldsymbol{\omega}\times\mathbf{x}, \qquad \mathbf{x}\cdot\mathbf{C} = \mathbf{x}\times\boldsymbol{\omega}, \tag{2.20}$$

where $\mathbf{x}$ is an arbitrary vector.

Exercise 2.14. Do this.

Eigenvalues and eigenvectors of a symmetric tensor

Let $\mathbf{A}$ be a symmetric, second-order tensor. If the equation

$$\mathbf{A} \cdot \mathbf{x} = \lambda \mathbf{x} \tag{2.21}$$

holds for some nonzero vector $\mathbf{x}$, then $\mathbf{x}$ is said to be an *eigenvector*[1] corresponding to an *eigenvalue* λ. Intuitively, $\mathbf{A}$ operates on $\mathbf{x}$ to yield a vector along $\mathbf{x}$ (i.e., an eigenvector $\mathbf{x}$ is codirected with the image $\mathbf{A} \cdot \mathbf{x}$ of $\mathbf{x}$ under $\mathbf{A}$). We may refer to the pair $(\lambda, \mathbf{x})$ as an *eigenpair* of $\mathbf{A}$. Note that if $\mathbf{x}$ is an eigenvector of $\mathbf{A}$ then so is $\alpha\mathbf{x}$ for any nonzero α.

Since $\mathbf{x} = \mathbf{E} \cdot \mathbf{x}$, we may write (2.21) as

$$(\mathbf{A} - \lambda \mathbf{E}) \cdot \mathbf{x} = \mathbf{0}$$

or, in terms of mixed components of $\mathbf{A}$ and contravariant components of $\mathbf{x}$,

$$({a^i}_k - \lambda \delta^i_k) x^k = 0.$$

This constitutes[2] a system of simultaneous linear equations in the variables x^k, and a nontrivial solution exists if and only if

$$\begin{vmatrix} {a^1}_1 - \lambda & {a^1}_2 & {a^1}_3 \\ {a^2}_1 & {a^2}_2 - \lambda & {a^2}_3 \\ {a^3}_1 & {a^3}_2 & {a^3}_3 - \lambda \end{vmatrix} = 0. \tag{2.22}$$

This is called the *characteristic equation* for $\mathbf{A}$. Although it has been expressed in terms of the mixed components of $\mathbf{A}$, it characterizes $\mathbf{A}$ itself in the sense that the eigenvalues of $\mathbf{A}$ are frame independent. It can also be written in the form

$$-\lambda^3 + I_1(\mathbf{A})\lambda^2 - I_2(\mathbf{A})\lambda + I_3(\mathbf{A}) = 0, \tag{2.23}$$

from which we see that it is cubic in λ and there can be at most three distinct eigenvalues λ_1, λ_2, and λ_3. These eigenvalues are invariants of $\mathbf{A}$: they do not depend on the basis in which $\mathbf{A}$ is represented. The coefficients $I_1(\mathbf{A})$, $I_2(\mathbf{A})$, and $I_3(\mathbf{A})$ are called the *principal invariants* (specifically, the

[1] More precisely, $\mathbf{x}$ is a *right eigenvector* of $\mathbf{A}$. The notion of *left eigenvector* can also be defined, but the distinction disappears for symmetric tensors and only the latter will be needed for our chosen applications in mechanics.

[2] Note that this is a traditional eigenvalue problem for the matrix constituted of the mixed components of $\mathbf{A}$. This and similar relations demonstrate the relation between second-order tensors and linear operators in $\mathbb{R}^3$ about which we will speak later in more detail.

first, second, and third invariants) of $\mathbf{A}$; they can be expressed in terms of the eigenvalues by the *Viète formulas*

$$\begin{aligned} I_1(\mathbf{A}) &= \lambda_1 + \lambda_2 + \lambda_3, \\ I_2(\mathbf{A}) &= \lambda_1\lambda_2 + \lambda_2\lambda_3 + \lambda_3\lambda_1, \\ I_3(\mathbf{A}) &= \lambda_1\lambda_2\lambda_3. \end{aligned} \tag{2.24}$$

Exercise 2.15. Derive (2.24) from the fact that $\lambda_1, \lambda_2, \lambda_3$ are the three roots of (2.23).

We also have

$$\begin{aligned} I_1(\mathbf{A}) &= \operatorname{tr}\mathbf{A}, \\ I_2(\mathbf{A}) &= \tfrac{1}{2}[\operatorname{tr}^2\mathbf{A} - \operatorname{tr}\mathbf{A}^2], \\ I_3(\mathbf{A}) &= \det\mathbf{A}. \end{aligned} \tag{2.25}$$

It can be shown that if $\mathbf{A}$ is symmetric and invertible, then

$$I_1(\mathbf{A}^{-1}) = \frac{I_2(\mathbf{A})}{I_3(\mathbf{A})}, \qquad I_2(\mathbf{A}^{-1}) = \frac{I_1(\mathbf{A})}{I_3(\mathbf{A})}, \qquad I_3(\mathbf{A}^{-1}) = I_3^{-1}(\mathbf{A}).$$

Until now we have considered vectors and tensors over the real space $\mathbb{R}^3$. But the characteristic equation, as a cubic equation, can have complex roots λ_k. So we momentarily consider the situation where $\mathbf{A}$ can have complex λ_k and complex components of an eigenvector $\mathbf{x}$. If $\mathbf{A}$ is a *real* symmetric tensor considered as an operator in the complex space $\mathbb{C}^3$, then for any vectors $\mathbf{x}$ and $\mathbf{y}$

$$(\mathbf{A}\cdot\mathbf{x})\cdot\overline{\mathbf{y}} = \mathbf{x}\cdot\overline{(\mathbf{A}\cdot\mathbf{y})} \tag{2.26}$$

where the overbar denotes the complex conjugate operation. We have

Theorem 2.2. *The eigenvalues and eigenvector components of a real symmetric tensor are real, and eigenvectors corresponding to distinct eigenvalues are orthogonal.*

Proof. Let $(\lambda, \mathbf{x})$ be an eigenpair of a real symmetric tensor $\mathbf{A}$. Dotting both sides of the equation $\mathbf{A}\cdot\mathbf{x} = \lambda\mathbf{x}$ with $\overline{\mathbf{x}}$, we obtain

$$\lambda = \frac{(\mathbf{A}\cdot\mathbf{x})\cdot\overline{\mathbf{x}}}{\mathbf{x}\cdot\overline{\mathbf{x}}}.$$

The denominator $|\mathbf{x}|^2$ is real and positive. The numerator is real because

$$(\mathbf{A}\cdot\mathbf{x})\cdot\overline{\mathbf{x}} = \overline{\mathbf{x}\cdot\overline{(\mathbf{A}\cdot\mathbf{x})}} = \overline{(\mathbf{A}\cdot\mathbf{x})\cdot\overline{\mathbf{x}}}$$

by (2.26). Hence λ is real. Because the components of the eigenvectors satisfy a linear system of simultaneous equations having real coefficients, they are also real. Now suppose $\mathbf{A} \cdot \mathbf{x}_1 = \lambda_1 \mathbf{x}_1$ and $\mathbf{A} \cdot \mathbf{x}_2 = \lambda_2 \mathbf{x}_2$ where $\lambda_2 \neq \lambda_1$. We get

$$\lambda_1 \mathbf{x}_1 \cdot \mathbf{x}_2 = (\mathbf{A} \cdot \mathbf{x}_1) \cdot \mathbf{x}_2, \qquad \lambda_2 \mathbf{x}_2 \cdot \mathbf{x}_1 = (\mathbf{A} \cdot \mathbf{x}_2) \cdot \mathbf{x}_1,$$

and then by subtraction

$$(\lambda_1 - \lambda_2)\mathbf{x}_1 \cdot \mathbf{x}_2 = \mathbf{x}_2 \cdot \mathbf{A} \cdot \mathbf{x}_1 - \mathbf{x}_1 \cdot \mathbf{A} \cdot \mathbf{x}_2 = 0.$$

The orthogonality property

$$\mathbf{x}_1 \cdot \mathbf{x}_2 = 0 \tag{2.27}$$

follows. From (2.27) we may obtain another property

$$\mathbf{x}_1 \cdot \mathbf{A} \cdot \mathbf{x}_2 = 0.$$

known as *generalized orthogonality* of the eigenvectors $\mathbf{x}_1$ and $\mathbf{x}_2$. □

Below we will show that for any symmetric tensor $\mathbf{A}$, there always exist three eigenvectors constituting an orthonormal basis of $\mathbb{R}^3$.

Exercise 2.16. Show that for any second-order tensor $\mathbf{A}$ (not necessarily symmetric), eigenvectors corresponding to distinct eigenvalues are linearly independent.

Orthogonal representation of a tensor

The characteristic equation (2.22) may yield three or fewer distinct solutions for λ. When they are all distinct, we can find three orthonormal eigenvectors $\mathbf{x}_k$, use these to compose a Cartesian frame, and then express $\mathbf{A}$ in terms of its components a_{ij} relative to this frame:

$$\mathbf{A} = \sum a_{ij} \mathbf{x}_i \mathbf{x}_j \quad \text{where} \quad a_{ij} = \mathbf{x}_i \cdot \mathbf{A} \cdot \mathbf{x}_j.$$

Dotting the relation $\mathbf{A} \cdot \mathbf{x}_j = \lambda_j \mathbf{x}_j$ with $\mathbf{x}_i$ from the left, we obtain

$$\mathbf{x}_i \cdot \mathbf{A} \cdot \mathbf{x}_j = \mathbf{x}_i \cdot \lambda_j \mathbf{x}_j = \lambda_j \delta_i^j.$$

Hence the coefficients of the dyads $\mathbf{x}_i \mathbf{x}_j$ in $\mathbf{A}$ are nonzero only on the main diagonal of the matrix representation of $\mathbf{A}$. Since these diagonal entries are the eigenvalues of $\mathbf{A}$, we have

$$\mathbf{A} = \sum_{i=1}^{3} \lambda_i \mathbf{x}_i \mathbf{x}_i. \tag{2.28}$$

This dyadic expansion is called the *orthogonal* or *spectral representation* of $\mathbf{A}$. The eigenvectors composing the coordinate frame determine the *principal axes* of $\mathbf{A}$, and the process of referring the tensor to its principal axes is called *diagonalization*. The diagonal form is regarded as the simplest form a given tensor can take.

If (2.22) has fewer than three distinct solutions, the repeated eigenvalue is said to be *degenerate*. With $\mathbf{A}$ symmetric, we may represent $\mathbf{A}$ in a Cartesian basis and call on some facts from the theory of symmetric matrices. To each multiple root of (2.22), there corresponds a multi-dimensional subspace spanned by the respective eigenvectors. Any vector of the subspace is orthogonal to the eigenvectors corresponding to other eigenvalues. Supposing $\lambda_1 = \lambda_2 \neq \lambda_3$ so λ_1 is a double root, the subspace is two-dimensional and we can select an orthonormal pair of normalized eigenvectors $\mathbf{x}_1, \mathbf{x}_2$ which are also orthogonal to the third normalized eigenvector $\mathbf{x}_3$ as $\lambda_1 \neq \lambda_3$. In this case we call λ_1 a *multiple eigenvalue* of multiplicity two. Note that in this case the representation (2.28) is also valid, but unlike the case of non-repeated eigenvalues it is not unique as we can select infinitely many different orthonormal pairs $\mathbf{x}_1, \mathbf{x}_2$ from the subspace of eigenvectors corresponding to $\lambda_1 = \lambda_2$. However, the tensor $\mathbf{x}_1\mathbf{x}_1 + \mathbf{x}_2\mathbf{x}_2 = \mathbf{E}_2$ is uniquely defined; as $\mathbf{E}_2$ is a unit tensor on the subspace that is unique, we can represent it using any orthonormal pair of eigenvectors of the subspace. In this case the unique spectral representation of $\mathbf{A}$ is

$$\mathbf{A} = \lambda_1 \mathbf{E}_2 + \lambda_3 \mathbf{x}_3. \tag{2.29}$$

If $\lambda_1 = \lambda_2 = \lambda_3$, then any vector is an eigenvector; choosing a Cartesian frame we would have an eigenvalue of multiplicity three. But this case arises only if $\mathbf{A}$ is a *ball tensor*, i.e., it is proportional to $\mathbf{E}$ with $\mathbf{A} = \lambda_1 \mathbf{E}$. Clearly we can take any Cartesian trihedron and get the representation (2.28), which is $\mathbf{A} = \lambda_1 \mathbf{i}_k \mathbf{i}_k$.

Exercise 2.17. Show that a second-order tensor (not necessarily symmetric) having three distinct eigenvalues cannot have more than one linearly independent eigenvector corresponding to each eigenvalue.

Cayley–Hamilton theorem

We know that every square matrix satisfies its own characteristic equation. Similarly, a second-order tensor $\mathbf{A}$ satisfies

$$-\mathbf{A}^3 + I_1(\mathbf{A})\mathbf{A}^2 - I_2(\mathbf{A})\mathbf{A} + I_3(\mathbf{A})\mathbf{E} = \mathbf{0}. \tag{2.30}$$

The proof is easy if $\mathbf{A}$ is symmetric. By (2.28) we have

$$\mathbf{A}^2 = \sum_{i=1}^{3} \lambda_i \mathbf{i}_i \mathbf{i}_i \cdot \sum_{j=1}^{3} \lambda_j \mathbf{i}_j \mathbf{i}_j = \sum_{i,j=1}^{3} \lambda_i \mathbf{i}_i \lambda_j \mathbf{i}_j \delta_j^i = \sum_{i=1}^{3} \lambda_i^2 \mathbf{i}_i \mathbf{i}_i.$$

Similarly

$$\mathbf{A}^3 = \sum_{i=1}^{3} \lambda_i^3 \mathbf{i}_i \mathbf{i}_i.$$

This and (2.23) yield

$$\begin{aligned}
\mathbf{A}^3 &= \sum_{i=1}^{3} [I_1(\mathbf{A})\lambda_i^2 - I_2(\mathbf{A})\lambda_i + I_3(\mathbf{A})]\mathbf{i}_i \mathbf{i}_i \\
&= I_1(\mathbf{A}) \sum_{i=1}^{3} \lambda_i^2 \mathbf{i}_i \mathbf{i}_i - I_2(\mathbf{A}) \sum_{i=1}^{3} \lambda_i \mathbf{i}_i \mathbf{i}_i + I_3(\mathbf{A}) \sum_{i=1}^{3} \mathbf{i}_i \mathbf{i}_i \\
&= I_1(\mathbf{A})\mathbf{A}^2 - I_2(\mathbf{A})\mathbf{A} + I_3(\mathbf{A})\mathbf{E},
\end{aligned}$$

as desired. The Cayley–Hamilton theorem lets us write $\mathbf{A}^3$ in terms of $\mathbf{A}^2$, $\mathbf{A}$, and $\mathbf{E}$:

$$\mathbf{A}^3 = I_1(\mathbf{A})\mathbf{A}^2 - I_2(\mathbf{A})\mathbf{A} + I_3(\mathbf{A})\mathbf{E}. \tag{2.31}$$

Dot multiplying (2.30) with $\mathbf{A}$, we can write $\mathbf{A}^4$ in terms of $\mathbf{A}^3$, $\mathbf{A}^2$, and $\mathbf{A}$:

$$\begin{aligned}
\mathbf{A}^4 &= I_1(\mathbf{A})\mathbf{A}^3 - I_2(\mathbf{A})\mathbf{A}^2 + I_3(\mathbf{A})\mathbf{A} \\
&= I_1(\mathbf{A})[I_1(\mathbf{A})\mathbf{A}^2 - I_2(\mathbf{A})\mathbf{A} + I_3(\mathbf{A})\mathbf{E}] - I_2(\mathbf{A})\mathbf{A}^2 + I_3(\mathbf{A})\mathbf{A} \\
&= [I_1^2(\mathbf{A}) - I_2(\mathbf{A})]\mathbf{A}^2 + [I_3(\mathbf{A}) - I_1(\mathbf{A})I_2(\mathbf{A})]\mathbf{A} + I_1(\mathbf{A})I_3(\mathbf{A})\mathbf{E}.
\end{aligned}$$

Continuing in this way we can write any desired power of $\mathbf{A}$ in terms of $\mathbf{E}$, $\mathbf{A}$, and $\mathbf{A}^2$. Through the use of Taylor expansions for analytic functions we are thereby able to introduce functions of tensorial arguments. A useful example is the exponential function involving a constant second-order tensor $\mathbf{A}$,

$$e^{t\mathbf{A}} = \sum_{k=0}^{\infty} \frac{(t\mathbf{A})^k}{k!},$$

which yields a solution $e^{t\mathbf{A}} \cdot \mathbf{x}_0$ to the initial-value problem

$$\frac{d\mathbf{x}(t)}{dt} = \mathbf{A} \cdot \mathbf{x}(t), \qquad \mathbf{x}(0) = \mathbf{x}_0.$$

If $\mathbf{A}$ is nonsingular so that $\mathbf{A}^{-1}$ exists and $I_3(\mathbf{A}) \neq 0$, we can dot both sides of (2.31) with $\mathbf{A}^{-1}$ to get

$$\mathbf{A}^{-1} \cdot \mathbf{A}^3 = \mathbf{A}^2 = I_1(\mathbf{A})\mathbf{A} - I_2(\mathbf{A})\mathbf{E} + I_3(\mathbf{A})\mathbf{A}^{-1}$$

and then solve for $\mathbf{A}^{-1}$:

$$\mathbf{A}^{-1} = I_3^{-1}(\mathbf{A})[\mathbf{A}^2 - I_1(\mathbf{A})\mathbf{A} + I_2(\mathbf{A})\mathbf{E}].$$

Rotation tensors

A second-order tensor $\mathbf{Q}$ is said to be *orthogonal* if

$$\mathbf{Q} \cdot \mathbf{Q}^T = \mathbf{Q}^T \cdot \mathbf{Q} = \mathbf{E} \qquad \text{or equivalently} \qquad \mathbf{Q}^T = \mathbf{Q}^{-1}.$$

Such a tensor has $\det \mathbf{Q} = \pm 1$. Indeed, $\det \mathbf{Q}$ is determined by the determinant of the matrix of mixed components of $\mathbf{Q}$. By the properties of determinants, we have for second-order tensors $\mathbf{A}$ and $\mathbf{B}$

$$\det(\mathbf{A} \cdot \mathbf{B}) = \det \mathbf{A} \det \mathbf{B}, \qquad \det \mathbf{A} = \det \mathbf{A}^T.$$

Hence $1 = \det \mathbf{E} = \det \mathbf{Q} \det \mathbf{Q}^T = (\det \mathbf{Q})^2$ as desired. If $\det \mathbf{Q} = +1$, we say that $\mathbf{Q}$ is a *proper* orthogonal tensor. Otherwise we say that it is *improper*.

Exercise 2.18. Show that if $\mathbf{Q}$ is orthogonal then so is $\mathbf{Q}^n$ for each positive integer n.

As an operator, an orthogonal tensor has the special property of preserving both the magnitudes of vectors and the angles between them. This follows from the relation

$$(\mathbf{Q} \cdot \mathbf{x}) \cdot (\mathbf{Q} \cdot \mathbf{y}) = (\mathbf{x} \cdot \mathbf{Q}^T) \cdot (\mathbf{Q} \cdot \mathbf{y}) = \mathbf{x} \cdot \mathbf{Q}^T \cdot \mathbf{Q} \cdot \mathbf{y} = \mathbf{x} \cdot \mathbf{E} \cdot \mathbf{y} = \mathbf{x} \cdot \mathbf{y}$$

and its immediate consequence

$$|\mathbf{Q} \cdot \mathbf{x}|^2 = |\mathbf{x}|^2.$$

A proper orthogonal tensor $\mathbf{Q}$ rotates all the vectors of the space over which it is defined. An improper orthogonal tensor also causes an axis reflection that changes the "handedness" of the frame; an example is $-\mathbf{E}$, the tensor of inversion. In this way we can state that each orthogonal tensor is either a rotation tensor or the product of a rotation tensor and a reflection tensor.

It is important to note that certain quantities remain invariant under rotation. One is the first invariant $I_1(\mathbf{A})$ which, as we saw, happens to equal $\operatorname{tr}\mathbf{A}$. Indeed, consider the tensor

$$\begin{aligned}\mathbf{Q}\cdot\mathbf{A}\cdot\mathbf{Q}^T &= \mathbf{Q}\cdot(a^{ij}\mathbf{e}_i\mathbf{e}_j)\cdot\mathbf{Q}^T\\ &= a^{ij}(\mathbf{Q}\cdot\mathbf{e}_i)(\mathbf{e}_j\cdot\mathbf{Q}^T)\\ &= a^{ij}(\mathbf{Q}\cdot\mathbf{e}_i)(\mathbf{Q}\cdot\mathbf{e}_j).\end{aligned}$$

This is a representation of the "rotated" tensor, derived by applying $\mathbf{Q}$ to each vector of the dyadic components of $\mathbf{A}$. Its trace is

$$\mathbf{E}\cdot\cdot(\mathbf{Q}\cdot\mathbf{A}\cdot\mathbf{Q}^T) = \mathbf{E}\cdot\cdot(\mathbf{Q}\cdot a_i^{\ j}\mathbf{e}^i\mathbf{e}_j\cdot\mathbf{Q}^T) = \mathbf{E}\cdot\cdot\, a_i^{\ j}(\mathbf{Q}\cdot\mathbf{e}^i)(\mathbf{Q}\cdot\mathbf{e}_j).$$

Under the action of $\mathbf{Q}$ the frame $\mathbf{e}_i$ transforms to the frame $\mathbf{Q}\cdot\mathbf{e}_i$, which has reciprocal basis $\mathbf{Q}\cdot\mathbf{e}^i$. Hence $\mathbf{E}$ can be represented as

$$\mathbf{E} = (\mathbf{Q}\cdot\mathbf{e}_i)(\mathbf{Q}\cdot\mathbf{e}^i) = (\mathbf{Q}\cdot\mathbf{e}^j)(\mathbf{Q}\cdot\mathbf{e}_j).$$

It follows that

$$\begin{aligned}\mathbf{E}\cdot\cdot(\mathbf{Q}\cdot\mathbf{A}\cdot\mathbf{Q}^T) &= \big[(\mathbf{Q}\cdot\mathbf{e}^k)(\mathbf{Q}\cdot\mathbf{e}_k)\big]\cdot\cdot\, a_i^{\ j}(\mathbf{Q}\cdot\mathbf{e}^i)(\mathbf{Q}\cdot\mathbf{e}_j)\\ &= a_i^{\ j}\big[(\mathbf{Q}\cdot\mathbf{e}_k)\cdot(\mathbf{Q}\cdot\mathbf{e}^i)\big]\big[(\mathbf{Q}\cdot\mathbf{e}^k)\cdot(\mathbf{Q}\cdot\mathbf{e}_j)\big]\\ &= a_i^{\ j}\big[(\mathbf{e}_k\cdot\mathbf{Q}^T)\cdot(\mathbf{Q}\cdot\mathbf{e}^i)\big]\big[(\mathbf{e}^k\cdot\mathbf{Q}^T)\cdot(\mathbf{Q}\cdot\mathbf{e}_j)\big]\\ &= a_i^{\ j}(\mathbf{e}_k\cdot\mathbf{Q}^T\cdot\mathbf{Q}\cdot\mathbf{e}^i)\left(\mathbf{e}^k\cdot\mathbf{Q}^T\cdot\mathbf{Q}\cdot\mathbf{e}_j\right)\\ &= a_i^{\ j}(\mathbf{e}_k\cdot\mathbf{E}\cdot\mathbf{e}^i)(\mathbf{e}^k\cdot\mathbf{E}\cdot\mathbf{e}_j)\\ &= a_i^{\ j}(\mathbf{e}_k\cdot\mathbf{e}^i)(\mathbf{e}^k\cdot\mathbf{e}_j) = a_i^{\ j}\delta_k^i\delta_j^k = a_k^{\ k} = \operatorname{tr}\mathbf{A}\end{aligned}$$

and invariance of the trace is established.

Let us also show that $\mathbf{Q}$ preserves (in a certain sense that we will discuss for the problem of physical invariance of functions under rotations) the eigenvalues of a tensor $\mathbf{A}$ but rotates the eigenvectors. That is, if $\mathbf{A}$ has eigenpair $(\lambda_i, \mathbf{x}_i)$ then $\mathbf{Q}\cdot\mathbf{A}\cdot\mathbf{Q}^T$ has corresponding eigenpair $(\lambda_i, \mathbf{Q}\cdot\mathbf{x}_i)$:

$$\begin{aligned}(\mathbf{Q}\cdot\mathbf{A}\cdot\mathbf{Q}^T)\cdot(\mathbf{Q}\cdot\mathbf{x}_i) &= \mathbf{Q}\cdot\mathbf{A}\cdot\mathbf{Q}^T\cdot\mathbf{Q}\cdot\mathbf{x}_i = \mathbf{Q}\cdot\mathbf{A}\cdot\mathbf{E}\cdot\mathbf{x}_i\\ &= \mathbf{Q}\cdot(\mathbf{A}\cdot\mathbf{x}_i) = \mathbf{Q}\cdot(\lambda_i\mathbf{x}_i) = \lambda_i(\mathbf{Q}\cdot\mathbf{x}_i).\end{aligned}$$

If we have two orthonormal bases $\mathbf{i}_k$ and $\mathbf{i}'_k$, then $\mathbf{Q} = \mathbf{i}_k\mathbf{i}'_k$ is orthogonal. Moreover, an orthogonal tensor $\mathbf{Q}$ transforms an orthonormal basis $\mathbf{i}_k$ into another orthonormal basis $\mathbf{i}'_k = \mathbf{i}_k\cdot\mathbf{Q}$.

It can be shown that every proper orthogonal tensor $\mathbf{Q}$ has the representation via the *Gibbs' formula*

$$\mathbf{Q} = \mathbf{E}\cos\omega + (1-\cos\omega)\mathbf{e}\mathbf{e} - \mathbf{e}\times\mathbf{E}\sin\omega, \tag{2.32}$$

where $\mathbf{e}$ is the axis of rotation determined by $\mathbf{Q}$ and ω is the angle of the rotation about $\mathbf{e}$.

From Gibbs' formula it follows that $\mathbf{e}$ is the eigenvector corresponding to the real eigenvalue 1 of $\mathbf{Q}$ such that $\mathbf{Q}\cdot\mathbf{e} = \mathbf{e}$ and

$$\cos\phi = \tfrac{1}{2}(\operatorname{tr}\mathbf{Q} - 1), \quad \sin\phi\,\mathbf{e} = \tfrac{1}{2}\left(\mathbf{Q} - \mathbf{Q}^T\right)_\times . \tag{2.33}$$

Gibbs' formula offers the possibility of calculating rotations, as seen in the next exercise.

Exercise 2.19. For a given vector $\mathbf{e}$ and angle ω, find the matrix of the corresponding orthogonal tensor. Take $\mathbf{e} = e_1\mathbf{i}_1 + e_2\mathbf{i}_2 + e_3\mathbf{i}_2$ with

$$e_1 = 0.61571932895865500144,$$
$$e_2 = 0.48041947641511894334,$$
$$e_3 = -0.62456931931349826910$$

and a rotation angle of 129.70 degrees (about 2.2637 radians).

For convenience we present the solution here rather than near the back of the book. It is easily verified that $\mathbf{e}$ is a unit vector by computing $(e_1^2+e_2^2+e_3^2)^{1/2}$. For the Cartesian basis vectors $\mathbf{i}_1, \mathbf{i}_2, \mathbf{i}_3$, the tensor $\mathbf{Q}$ is given by

$$\mathbf{Q} = Q_{mn}\mathbf{i}_m\mathbf{i}_n.$$

Here $Q_{mn} = \mathbf{i}_m\cdot\mathbf{Q}\cdot\mathbf{i}_n$.

The first term in (2.32), $\mathbf{E}\cos\omega$, gives us the diagonal matrix $\delta_{mn}\cos\omega$. The term $(1-\cos\omega)\mathbf{e}\mathbf{e}$ is associated with the matrix $(1-\cos\omega)e_me_n$. To calculate $\mathbf{e}\times\mathbf{E}$, we need the formula $\mathbf{i}_p\times\mathbf{i}_n = \epsilon_{pnm}\mathbf{i}_m$. We remind the reader that since $\mathbf{i}_k$ is an orthonormal basis, here and below we use Einstein's summation rule over repeated subscripts. So

$$\mathbf{e}\times\mathbf{E} = e_p\mathbf{i}_p\times\mathbf{i}_n\mathbf{i}_n = e_p\epsilon_{pnm}\mathbf{i}_m\mathbf{i}_n.$$

We also see that $\mathbf{e}\times\mathbf{E}\sin\omega$ corresponds to the skew matrix $e_p\epsilon_{pnm}\sin\omega$.

Thus the matrix of components of $\mathbf{Q}$ takes the form

$$Q_{mn} = \delta_{mn}\cos\omega + (1-\cos\omega)e_me_n - e_p\epsilon_{pnm}\sin\omega.$$

These are

$$Q_{11} = \cos\omega + (1 - \cos\omega)e_1^2,$$
$$Q_{22} = \cos\omega + (1 - \cos\omega)e_2^2,$$
$$Q_{33} = \cos\omega + (1 - \cos\omega)e_3^2,$$

$$Q_{12} = (1 - \cos\omega)e_1e_2 - e_p\epsilon_{p21}\sin\omega = (1 - \cos\omega)e_1e_2 + e_3\sin\omega,$$
$$Q_{21} = (1 - \cos\omega)e_1e_2 - e_p\epsilon_{p12}\sin\omega = (1 - \cos\omega)e_1e_2 - e_3\sin\omega,$$

and

$$Q_{13} = (1 - \cos\omega)e_1e_3 - e_2\sin\omega,$$
$$Q_{31} = (1 - \cos\omega)e_1e_3 + e_2\sin\omega,$$
$$Q_{23} = (1 - \cos\omega)e_2e_3 + e_1\sin\omega,$$
$$Q_{32} = (1 - \cos\omega)e_2e_3 - e_1\sin\omega.$$

Calculating, we get

$$(Q_{mn}) = \begin{pmatrix} -0.01749787426481 & 0.00421498160336 & -0.99983801604374 \\ 0.96529534283395 & -0.26054020749805 & -0.01799170308610 \\ -0.26057383886200 & -0.96545379703378 & 0.00049019767986 \end{pmatrix}.$$

Vectorial decomposition of a rotation tensor

In shell theory and in rigid-body dynamics we encounter other representations of a rotation tensor. We begin with the most popular of these. Introducing the *finite rotation vector* $\boldsymbol{\theta} = 2\mathbf{e}\tan\omega/2$ and using

$$\cos\omega = \frac{1 - \tan^2\omega/2}{1 + \tan^2\omega/2}, \qquad \sin\omega = \frac{2\tan\omega/2}{1 + \tan^2\omega/2},$$

we can get a representation of $\mathbf{Q}$ devoid of trigonometric functions:

$$\mathbf{Q} = \frac{1}{(4 + \theta^2)}\left[(4 - \theta^2)\mathbf{E} + 2\boldsymbol{\theta}\boldsymbol{\theta} - 4\mathbf{E}\times\boldsymbol{\theta}\right], \quad \theta^2 = \boldsymbol{\theta}\cdot\boldsymbol{\theta}. \tag{2.34}$$

Other known vectorial parameterizations of an orthogonal tensor are summarized in [Pietraszkiewicz and Eremeyev (2009)]. For the vectorial parameterization of $\mathbf{Q}$ we introduce a scalar function $p(\omega)$ and the generalized finite rotation vector $\mathbf{p}$:

$$\mathbf{p} = p(\omega)\,\mathbf{e}. \tag{2.35}$$

The generating function $p(\omega)$ in (2.35) should be an odd function of ω with the limit behavior $\lim\limits_{\omega\to 0}\frac{p(\omega)}{\omega} = \kappa$, where κ is a positive real normalization factor (usually 1 or $\frac{1}{2}$), and $p(0) = 0$. In terms of (2.35), tensor $\mathbf{Q}$ takes the form

$$\mathbf{Q} = \cos\omega\,\mathbf{E} + \frac{1-\cos\omega}{p^2}\,\mathbf{pp} - \frac{\sin\omega}{p}\,\mathbf{p}\times\mathbf{E}. \tag{2.36}$$

The literature contains various definitions of $\mathbf{p}$. Common forms for $\mathbf{p}$ are the Cayley–Gibbs vector $\boldsymbol{\theta}$, the linear vector $\boldsymbol{\varphi}$ (also called the exponential map), the Rodrigues rotation vector $\boldsymbol{\rho}$, the vector $\boldsymbol{\vartheta}$, the Euler–Rodrigues vector $\boldsymbol{\sigma}$, the Wiener–Milenkovic vector $\boldsymbol{\mu}$, and the Bauchau–Trainelli vector $\boldsymbol{\beta}$. These are defined by the formulas

$$\boldsymbol{\theta} = 2\tan\frac{\omega}{2}\,\mathbf{e}, \quad \boldsymbol{\varphi} = \omega\,\mathbf{e}, \quad \boldsymbol{\vartheta} = \sin\omega\,\mathbf{e}, \quad \boldsymbol{\rho} = \tan\frac{\omega}{2}\,\mathbf{e}, \tag{2.37}$$

and

$$\boldsymbol{\sigma} = 2\sin\frac{\omega}{2}\,\mathbf{e}, \quad \boldsymbol{\mu} = 4\tan\frac{\omega}{4}\,\mathbf{e}, \quad \boldsymbol{\beta} = 4\sin\frac{\omega}{4}\,\mathbf{e}. \tag{2.38}$$

For small ω, we can replace $\sin\omega$ by ω, $\tan\omega$ by ω, etc., and the corresponding finite rotation vectors reduce to

$$\boldsymbol{\theta} = \boldsymbol{\varphi} = \boldsymbol{\vartheta} = \boldsymbol{\sigma} = \boldsymbol{\mu} = \boldsymbol{\beta} = \omega\,\mathbf{e}, \quad \boldsymbol{\rho} = \frac{1}{2}\omega\,\mathbf{e}.$$

So for small rotation angles it suffices to know only the *linear rotation vector* $\boldsymbol{\omega} = \omega\,\mathbf{e}$, and from (2.34) or (2.36) we get $\mathbf{Q} = \mathbf{E} - \mathbf{E}\times\boldsymbol{\omega}$.

For non-small values of ω, the finite rotation vectors differ. For example, $|\boldsymbol{\theta}| \to \infty$ if $\omega \to \pm\pi$ whereas $|\boldsymbol{\mu}| \to \infty$ if $\omega \to \pm 2\pi$. So singularities may occur for some values of ω following from the singularities of the corresponding generating functions $p(\omega)$, when $p \to \infty$. Another type of singularity in a vectorial parametrization arises from the singularities of the inverse relations $\mathbf{p} = \mathbf{p}(\mathbf{Q})$. This inverse relation follows from (2.33)

$$\cos\omega = \frac{1}{2}\left(\operatorname{tr}\mathbf{Q} - 1\right), \quad \sin\omega\,\mathbf{e} = \frac{1}{2}\mathbf{Q}_\times, \quad \tan\omega\mathbf{e} = \frac{1}{\operatorname{tr}\mathbf{Q} - 1}\mathbf{Q}_\times$$

and the trigonometric dependencies. The range of validity for ω, which is the domain of the rotation angle for which the vectorial representation of the rotation tensor is not singular, is summarized in Table 2.3 for various forms of the vector.

Obviously, when considering rigid body rotation one can select an arbitrary value of the rotation angle without taking into account the range. On the other hand, due to the periodicity of sine and cosine functions we

Table 2.3 Rotation vectors and ranges of ω.

$\mathbf{p}$	range of validity for ω
$\boldsymbol{\theta} \equiv 2\tan\frac{\omega}{2}\,\mathbf{e}$	$(-\pi, \pi)$
$\boldsymbol{\varphi} \equiv \omega\,\mathbf{e}$	$(-2\pi, 2\pi)$
$\boldsymbol{\vartheta} \equiv \sin\omega\,\mathbf{e}$	$(-\pi, \pi)$
$\boldsymbol{\rho} \equiv \tan\frac{\omega}{2}\,\mathbf{e}$	$(-\pi, \pi)$
$\boldsymbol{\sigma} \equiv 2\sin\frac{\omega}{2}\,\mathbf{e}$	$(-\pi, \pi)$
$\boldsymbol{\mu} \equiv 4\tan\frac{\omega}{4}\,\mathbf{e}$	$(-2\pi, 2\pi)$
$\boldsymbol{\beta} \equiv 4\sin\frac{\omega}{4}\,\mathbf{e}$	$(-2\pi, 2\pi)$

have $\mathbf{Q}(\omega) = \mathbf{Q}(\omega \pm 2\pi)$. So the rotation tensors for angles ω and $\omega \pm 2\pi$ coincide but the real motions differ. To calculate the rotation angle in real motion using the rotation tensor or finite rotation vectors, one must pay special attention to transitions over singular points. For applications of the rotation vector we refer, e.g., to [Bauchau (2010)] and [Wiśniewski (2010)].

Polar decomposition theorem

Now we will show how to express a nonsingular second-order tensor as a product of a positive definite symmetric tensor and an orthogonal tensor. We say that a symmetric tensor is *positive definite* if all its eigenvalues are positive. We know that such a tensor $\mathbf{S}$ may be expanded as

$$\mathbf{S} = \lambda_1\,\mathbf{i}_1\mathbf{i}_1 + \lambda_2\,\mathbf{i}_2\mathbf{i}_2 + \lambda_3\,\mathbf{i}_3\mathbf{i}_3,$$

where the $\lambda_i > 0$ and the eigenvectors $\mathbf{i}_k$ form an orthonormal basis of $\mathbb{R}^3$. The representation is unique if none of the eigenvalues of $\mathbf{S}$ are repeated.

If $\mathbf{A}$ is nonsingular, the tensor $\mathbf{A}\cdot\mathbf{A}^T$ is symmetric and positive definite. Symmetry follows from the equation

$$(\mathbf{A}\cdot\mathbf{A}^T)^T = (\mathbf{A}^T)^T\cdot\mathbf{A}^T = \mathbf{A}\cdot\mathbf{A}^T.$$

To see positive definiteness of $\mathbf{A}\cdot\mathbf{A}^T$, we start with the definition of an eigenvalue λ:

$$(\mathbf{A}\cdot\mathbf{A}^T)\cdot\mathbf{x} = \lambda\mathbf{x}.$$

Dotting this from the left $\mathbf{x}$ and dividing through by $|\mathbf{x}|^2$, we get

$$\lambda = \frac{\mathbf{x}\cdot(\mathbf{A}\cdot\mathbf{A}^T)\cdot\mathbf{x}}{|\mathbf{x}|^2}.$$

Because

$$\mathbf{x} \cdot (\mathbf{A} \cdot \mathbf{A}^T) \cdot \mathbf{x} = (\mathbf{x} \cdot \mathbf{A}) \cdot (\mathbf{A}^T \cdot \mathbf{x}) = (\mathbf{x} \cdot \mathbf{A}) \cdot (\mathbf{x} \cdot \mathbf{A}) = |\mathbf{x} \cdot \mathbf{A}|^2,$$

the numerator is positive. Hence all the eigenvalues λ of $\mathbf{A} \cdot \mathbf{A}^T$ are positive and $\mathbf{A} \cdot \mathbf{A}^T$ is positive definite. With these facts in hand, we may turn to the *polar decomposition theorem.*

Theorem 2.3. *Any nonsingular second-order tensor* $\mathbf{A}$ *may be written as a product of an orthogonal tensor and a positive definite symmetric tensor. There is the* ***left polar decomposition***

$$\mathbf{A} = \mathbf{S} \cdot \mathbf{Q} \tag{2.39}$$

and the ***right polar decomposition***

$$\mathbf{A} = \mathbf{Q} \cdot \mathbf{S}', \tag{2.40}$$

where $\mathbf{Q}$ *is an orthogonal second-order tensor, and* $\mathbf{S}$ *and* $\mathbf{S}'$ *are positive definite and symmetric. Each form of decomposition is unique.*

Proof. Since $\mathbf{A} \cdot \mathbf{A}^T$ is positive definite and symmetric, we have the expansion

$$\mathbf{A} \cdot \mathbf{A}^T = \lambda_1 \, \mathbf{i}_1\mathbf{i}_1 + \lambda_2 \, \mathbf{i}_2\mathbf{i}_2 + \lambda_3 \, \mathbf{i}_3\mathbf{i}_3,$$

where the $\mathbf{i}_k$ are orthonormal and all $\lambda_k > 0$. We define

$$\mathbf{S} \equiv \left(\mathbf{A} \cdot \mathbf{A}^T\right)^{1/2} = \sqrt{\lambda_1} \, \mathbf{i}_1\mathbf{i}_1 + \sqrt{\lambda_2} \, \mathbf{i}_2\mathbf{i}_2 + \sqrt{\lambda_3} \, \mathbf{i}_3\mathbf{i}_3.$$

This square root of $\mathbf{A} \cdot \mathbf{A}^T$ is clearly symmetric and positive definite as well. We also see that $\mathbf{S}^{-1}$ exists and can verify directly that

$$\mathbf{S}^{-1} = \frac{1}{\sqrt{\lambda_1}} \mathbf{i}_1\mathbf{i}_1 + \frac{1}{\sqrt{\lambda_2}} \mathbf{i}_2\mathbf{i}_2 + \frac{1}{\sqrt{\lambda_3}} \mathbf{i}_3\mathbf{i}_3.$$

Now we set $\mathbf{Q} \equiv \mathbf{S}^{-1} \cdot \mathbf{A}$. The tensor $\mathbf{Q}$ is orthogonal because

$$\begin{aligned}
\mathbf{Q} \cdot \mathbf{Q}^T &= (\mathbf{S}^{-1} \cdot \mathbf{A}) \cdot (\mathbf{S}^{-1} \cdot \mathbf{A})^T = (\mathbf{S}^{-1} \cdot \mathbf{A}) \cdot (\mathbf{A}^T \cdot (\mathbf{S}^{-1})^T) \\
&= \mathbf{S}^{-1} \cdot (\mathbf{A} \cdot \mathbf{A}^T) \cdot (\mathbf{S}^T)^{-1} = \mathbf{S}^{-1} \cdot \mathbf{S}^2 \cdot \mathbf{S}^{-1} = \mathbf{E}.
\end{aligned}$$

Thus we have deduced the needed representation (2.39). We obtain (2.40) by writing

$$\mathbf{A} = \mathbf{S} \cdot \mathbf{Q} = \mathbf{E} \cdot \mathbf{S} \cdot \mathbf{Q} = \mathbf{Q} \cdot \mathbf{Q}^T \cdot \mathbf{S} \cdot \mathbf{Q}$$

and taking $\mathbf{S}' \equiv \mathbf{Q}^T \cdot \mathbf{S} \cdot \mathbf{Q}$, which also is definite positive and symmetric.

Now suppose to the contrary that the left polar decomposition is not unique; i.e., there exists another left polar decomposition $\mathbf{A} = \mathbf{U} \cdot \mathbf{H}$, where $\mathbf{H}$ is an orthogonal second-order tensor, $\mathbf{U}$ is positive definite and symmetric, and $\mathbf{H} \neq \mathbf{S}$ and $\mathbf{U} \neq \mathbf{S}$. We see that

$$\mathbf{A} \cdot \mathbf{A}^T = \mathbf{U} \cdot \mathbf{H} \cdot \mathbf{H}^T \cdot \mathbf{U}^T = \mathbf{U}^2$$

and so $\mathbf{U}^2 = \mathbf{S}^2$. But $\mathbf{U} \neq \mathbf{S}$. Let us change $\mathbf{U}$ to the diagonal form

$$\mathbf{U} = u_1\mathbf{e}_1\mathbf{e}_1 + u_2\mathbf{e}_2\mathbf{e}_2 + u_3\mathbf{e}_3\mathbf{e}_3 \qquad (u_k > 0).$$

Squaring both sides and recalling that $\mathbf{U}^2 = \mathbf{S}^2$, we get

$$\mathbf{U}^2 = u_1^2\mathbf{e}_1\mathbf{e}_1 + u_2^2\mathbf{e}_2\mathbf{e}_2 + u_3^2\mathbf{e}_3\mathbf{e}_3 = \lambda_1\,\mathbf{i}_1\mathbf{i}_1 + \lambda_2\,\mathbf{i}_2\mathbf{i}_2 + \lambda_3\,\mathbf{i}_3\mathbf{i}_3 = \mathbf{S}^2.$$

If no eigenvalue is repeated, then the eigenvalues u_k^2 and λ_k of $\mathbf{U}$ and $\mathbf{S}$ coincide and the corresponding normalized eigenvectors coincide up to their directions. So $u_k = \sqrt{\lambda_k}$ and $\mathbf{e}_k\mathbf{e}_k = \mathbf{i}_k\mathbf{i}_k$, and hence the left polar decomposition is unique:

$$\mathbf{U} = u_1\mathbf{e}_1\mathbf{e}_1 + u_2\mathbf{e}_2\mathbf{e}_2 + u_3\mathbf{e}_3\mathbf{e}_3 = \sqrt{\lambda_1}\,\mathbf{i}_1\mathbf{i}_1 + \sqrt{\lambda_2}\,\mathbf{i}_2\mathbf{i}_2 + \sqrt{\lambda_3}\,\mathbf{i}_3\mathbf{i}_3 = \mathbf{S}.$$

In case of multiplicity two of the eigenvalues, say $\lambda_1 = \lambda_2 \neq \lambda_3$, we get the unique representation (2.29) for $\mathbf{U}^2$ and $\mathbf{S}^2$

$$\mathbf{U}^2 = u_1^2\mathbf{E}_2' + u_3^2\mathbf{e}_3\mathbf{e}_3, \qquad \mathbf{S}^2 = \lambda_1\mathbf{E}_2 + \lambda_3\mathbf{i}_3,$$

and from the equality $\mathbf{U}^2 = \mathbf{S}^2$ we find that the eigenpairs should be equal (up to the signs of the eigenvectors) and so $\mathbf{U} = \mathbf{S}$ and the left polar representation is unique. Uniqueness for the case of multiplicity three of an eigenvalue is proved similarly. We recall that in the case of multiple eigenvalues, the form of the spectral of representation for $\mathbf{S}$ is not unique but $\mathbf{S}$ is defined uniquely.

Thus, we proved that the tensor $\mathbf{S}$ is uniquely defined. The orthogonal tensor $\mathbf{H}$ is also uniquely defined as

$$\mathbf{H} = \mathbf{U}^{-1} \cdot \mathbf{A} = \mathbf{S}^{-1} \cdot \mathbf{A} = \mathbf{Q}.$$

So uniqueness is proved completely.

Existence and uniqueness of the right polar decomposition can be proved in a similar manner and we leave this to the reader. □

From the proof of the theorem it follows that $\mathbf{Q}$ is determined uniquely, whereas $\mathbf{S}$ may have different but equivalent representations in the case of repeated eigenvalues. For example, if $\lambda_1 = \lambda_2 = \lambda \neq \lambda_3$ then for $\mathbf{S}$ we have representations

$$\mathbf{S} = \sqrt{\lambda}\,(\mathbf{i}_1\mathbf{i}_1 + \mathbf{i}_2\mathbf{i}_2) + \sqrt{\lambda_3}\,\mathbf{i}_3\mathbf{i}_3 = \sqrt{\lambda}\,(\mathbf{i}_1'\mathbf{i}_1' + \mathbf{i}_2'\mathbf{i}_2') + \sqrt{\lambda_3}\,\mathbf{i}_3\mathbf{i}_3,$$

where $\mathbf{i}'_1, \mathbf{i}'_2$ are new unit vectors orthogonal to $\mathbf{i}_3$.

Note that in the theorem on polar decomposition, the representation contains two positive definite tensors $\mathbf{S}$ and $\mathbf{S}'$ but a single orthogonal tensor $\mathbf{Q}$ for which there is an interesting interpretation arising from the polar decomposition of $\mathbf{A}$. The tensors $\mathbf{A}\cdot\mathbf{A}^T$ and $\mathbf{A}^T\cdot\mathbf{A}$ are both positive definite and symmetric with the same eigenvalues, so in addition we have

$$\mathbf{A}^T \cdot \mathbf{A} = \lambda_1\, \mathbf{i}'_1\mathbf{i}'_1 + \lambda_2\, \mathbf{i}'_2\mathbf{i}'_2 + \lambda_3\, \mathbf{i}'_3\mathbf{i}'_3$$

where the $\mathbf{i}'_k$ are new orthonormal basis vectors. Therefore we introduce

$$\mathbf{S}' \equiv \left(\mathbf{A}^T \cdot \mathbf{A}\right)^{1/2} = \sqrt{\lambda_1}\, \mathbf{i}'_1\mathbf{i}'_1 + \sqrt{\lambda_2}\, \mathbf{i}'_2\mathbf{i}'_2 + \sqrt{\lambda_3}\, \mathbf{i}'_3\mathbf{i}'_3.$$

From $\mathbf{S}' \equiv \mathbf{Q}^T \cdot \mathbf{S} \cdot \mathbf{Q}$ it follows that

$$\mathbf{Q} = \mathbf{i}_1\mathbf{i}'_1 + \mathbf{i}_2\mathbf{i}'_2 + \mathbf{i}_3\mathbf{i}'_3.$$

This means that $\mathbf{Q}$ transforms basis vectors $\mathbf{i}_k$ into $\mathbf{i}'_k$, whereas $\mathbf{S}$ and $\mathbf{S}'$ have the same set of eigenvalues but different eigenvectors. We may say that $\mathbf{Q}$ "rotates" $\mathbf{S}$ into $\mathbf{S}'$.

Considering a second-order tensor $\mathbf{A}$ as a linear operator, we may describe its action in $\mathbb{R}^3$ as follows. By (2.39) we have

$$\begin{aligned}\mathbf{A} \cdot \mathbf{x} &= \mathbf{S} \cdot \mathbf{Q} \cdot \mathbf{x} = \mathbf{S} \cdot \mathbf{Q} \cdot x_k \mathbf{i}'_k = \mathbf{S} \cdot x_k \mathbf{i}_k \\ &= \sqrt{\lambda_1}\, x_1 \mathbf{i}_1 + \sqrt{\lambda_2}\, x_2 \mathbf{i}_2 + \sqrt{\lambda_3}\, x_3 \mathbf{i}_3,\end{aligned}$$

where we used the representation $\mathbf{x} = x_k \mathbf{i}'_k$. So we first rotate the basis vectors $\mathbf{i}'_k$ into $\mathbf{i}_k$ and then stretch the vector space along the $\mathbf{i}_k$. The linear transformation related with the right polar decomposition consists of stretching along the $\mathbf{i}'_k$ and then rotating $\mathbf{i}'_k$ into $\mathbf{i}_k$.

If we do not require the polar decomposition to be positive definite, the decomposition also exists for singular tensors but is not unique. For example, for the zero tensor $\mathbf{0}$ we have the polar decomposition in the form

$$\mathbf{0} = \mathbf{0} \cdot \mathbf{Q}$$

with an arbitrary orthogonal tensor $\mathbf{Q}$, where $\mathbf{0}$ plays the role of a symmetric tensor in the decomposition.

Let us now consider the case of $\mathbf{A} \cdot \mathbf{A}^T$ having a zero eigenvalue. Let $\lambda_3 = 0$, so the spectral representation of $\mathbf{A} \cdot \mathbf{A}^T$ is

$$\mathbf{A} \cdot \mathbf{A}^T = \lambda_1\, \mathbf{i}_1\mathbf{i}_1 + \lambda_2\, \mathbf{i}_2\mathbf{i}_2.$$

We see that $\mathbf{i}_3 \cdot \mathbf{A} = 0\mathbf{i}_3 = \mathbf{0}$, so $\mathbf{i}_3$ is a left eigenvector of $\mathbf{A}$ with corresponding zero eigenvalue. Now

$$\mathbf{S} \equiv (\mathbf{A} \cdot \mathbf{A}^T)^{1/2} = \sqrt{\lambda_1}\,\mathbf{i}_1\mathbf{i}_1 + \sqrt{\lambda_2}\,\mathbf{i}_2\mathbf{i}_2$$

and obviously $\mathbf{S}^{-1}$ does not exist. But we can define an "inverse" of $\mathbf{S}$ as follows. Considering $\mathbf{S}$ as two-dimensional tensor in the vector space spanned by $\mathbf{i}_1$ and $\mathbf{i}_2$, we define

$$\mathbf{S}^{-1} = \frac{1}{\sqrt{\lambda_1}}\,\mathbf{i}_1\mathbf{i}_1 + \frac{1}{\sqrt{\lambda_2}}\,\mathbf{i}_2\mathbf{i}_2.$$

Let $\mathbf{Q} \equiv \mathbf{S}^{-1} \cdot \mathbf{A} \pm \mathbf{i}_3\mathbf{i}_3$. Then $\mathbf{Q}$ is orthogonal because

$$\begin{aligned}
\mathbf{Q} \cdot \mathbf{Q}^T &= (\mathbf{S}^{-1} \cdot \mathbf{A} \pm \mathbf{i}_3\mathbf{i}_3) \cdot (\mathbf{S}^{-1} \cdot \mathbf{A} \pm \mathbf{i}_3\mathbf{i}_3)^T \\
&= (\mathbf{S}^{-1} \cdot \mathbf{A} \pm \mathbf{i}_3\mathbf{i}_3) \cdot [\mathbf{A}^T \cdot (\mathbf{S}^{-1})^T \pm \mathbf{i}_3\mathbf{i}_3] \\
&= \mathbf{S}^{-1} \cdot (\mathbf{A} \cdot \mathbf{A}^T) \cdot (\mathbf{S}^T)^{-1} + \mathbf{i}_3\mathbf{i}_3 = \mathbf{i}_1\mathbf{i}_1 + \mathbf{i}_2\mathbf{i}_2 + \mathbf{i}_3\mathbf{i}_3 = \mathbf{E}.
\end{aligned}$$

Dot-multiplying $\mathbf{Q} = \mathbf{S}^{-1} \cdot \mathbf{A} \pm \mathbf{i}_3\mathbf{i}_3$ from the left by $\mathbf{S}$, we get the needed left polar decomposition $\mathbf{A} = \mathbf{S} \cdot \mathbf{Q}$.

Similarly, we introduce the polar decomposition in the case when $\mathbf{A} \cdot \mathbf{A}^T$ has double zero eigenvalue. Assuming $\lambda_2 = \lambda_3 = 0$ and $\lambda_1 > 0$, we have

$$\mathbf{A} \cdot \mathbf{A}^T = \lambda_1\,\mathbf{i}_1\mathbf{i}_1, \quad \mathbf{S} = \sqrt{\lambda_1}\,\mathbf{i}_1\mathbf{i}_1, \quad \mathbf{S}^{-1} = \frac{1}{\sqrt{\lambda_1}}\,\mathbf{i}_1\mathbf{i}_1.$$

Here we again used an "inverse" to the singular tensor $\mathbf{S}$, considering its restriction to the one-dimensional subspace spanned by $\mathbf{i}_1$. Now any unit vector orthogonal to $\mathbf{i}_1$ is a left eigenvector of $\mathbf{A}$ with zero eigenvalue. So $\mathbf{A}$ has a dyad form: $\mathbf{A} = \lambda_1\mathbf{i}_1\mathbf{i}_1'$. For the corresponding orthogonal tensor, we get

$$\mathbf{Q} \equiv \pm\mathbf{Q}(\mathbf{i}_1) \cdot \mathbf{S}^{-1} \cdot \mathbf{A}$$

where $\mathbf{Q}(\mathbf{i}_1)$ is the rotation tensor about the axis $\mathbf{i}_1$ defined by

$$\mathbf{Q}(\mathbf{i}_1) \equiv \mathbf{E}\cos\omega + (1 - \cos\omega)\mathbf{i}_1\mathbf{i}_1 - \mathbf{i}_1 \times \mathbf{E}\sin\omega$$

with an arbitrary angle ω. Using the identity $\mathbf{Q}(\mathbf{i}_1) \cdot \mathbf{i}_1 = \mathbf{i}_1 \cdot \mathbf{Q}(\mathbf{i}_1) = \mathbf{i}_1$, we prove its orthogonality as follows:

$$\begin{aligned}
\mathbf{Q} \cdot \mathbf{Q}^T &= (\pm\mathbf{Q}(\mathbf{i}_1) \cdot \mathbf{S}^{-1} \cdot \mathbf{A}) \cdot (\pm\mathbf{Q}(\mathbf{i}_1) \cdot \mathbf{S}^{-1} \cdot \mathbf{A})^T \\
&= \mathbf{Q}(\mathbf{i}_1) \cdot (\mathbf{S}^{-1} \cdot \mathbf{A}) \cdot (\mathbf{A}^T \cdot (\mathbf{S}^{-1})^T) \cdot \mathbf{Q}^T(\mathbf{i}_1) \\
&= \mathbf{Q}(\mathbf{i}_1) \cdot \mathbf{S}^{-1} \cdot (\mathbf{A} \cdot \mathbf{A}^T) \cdot (\mathbf{S}^T)^{-1} \cdot \mathbf{Q}^T(\mathbf{i}_1) \\
&= \mathbf{Q}(\mathbf{i}_1) \cdot \mathbf{i}_1\mathbf{i}_1 \cdot \mathbf{Q}^T(\mathbf{i}_1) = \mathbf{E}.
\end{aligned}$$

Thus, for a singular second-order tensor $\mathbf{A}$ there is a polar decomposition with uniquely defined symmetric tensors $\mathbf{S}$ and $\mathbf{S}'$ and a non-uniquely defined orthogonal tensor $\mathbf{Q}$.

Let us also note a similarity between the polar decomposition of a second-order tensor and the polar representation of a complex number z: $z = re^{i\varphi}$, where $r = |z|$, $i^2 = -1$, and φ is a polar angle. Here r plays the role of a symmetric positive definite tensor $\mathbf{S}$, while $e^{i\varphi}$ is analogous to $\mathbf{Q}$.

Exercise 2.20. Show that an equivalent definition of positive definiteness for a symmetric second-order tensor $\mathbf{A}$ is that $\mathbf{x} \cdot \mathbf{A} \cdot \mathbf{x} > 0$ for all $\mathbf{x} \neq 0$.

Exercise 2.21. Use Theorem 2.3 to show that a nonsingular tensor $\mathbf{A}$ transforms the unit sphere into an ellipsoid.

Square roots and other roots of a symmetric positive tensor

In the polar decomposition theorem we implicitly introduced a square root of a positive definite tensor and proved

Theorem 2.4. *Let* $\mathbf{S}$ *be a symmetric, positive definite tensor. Then there exists a unique tensor* $\mathbf{U}$ *such that*

$$\mathbf{U}^2 = \mathbf{S},$$

where $\mathbf{U}$ *is also symmetric and positive definite.*

Proof. Following [Stephenson (1980)] we present an independent and elegant proof of the theorem. It is clear that for $\mathbf{S} = \lambda_1 \mathbf{i}_1\mathbf{i}_1 + \lambda_2 \mathbf{i}_2\mathbf{i}_2 + \lambda_3 \mathbf{i}_3\mathbf{i}_3$ the tensor $\mathbf{U} = \sqrt{\lambda_1}\, \mathbf{i}_1\mathbf{i}_1 + \sqrt{\lambda_2}\, \mathbf{i}_2\mathbf{i}_2 + \sqrt{\lambda_3}\, \mathbf{i}_3\mathbf{i}_3$ is a square root of $\mathbf{S}$ having the required properties.

To prove uniqueness of such a $\mathbf{U}$, let us use the following one-to-one correspondence between the eigenvalues and eigenvectors of $\mathbf{S}$ and $\mathbf{U}$. If $(\lambda, \mathbf{e})$ is an eigenpair of $\mathbf{S}$, then $(\sqrt{\lambda}, \mathbf{e})$ is an eigenpair of $\mathbf{U}$. Indeed, by hypothesis we have

$$\mathbf{S} \cdot \mathbf{e} = \mathbf{U}^2 \cdot \mathbf{e} = \lambda \mathbf{e}, \quad \mathbf{e} \neq \mathbf{0}, \quad \lambda > 0.$$

Hence

$$(\mathbf{U}^2 - \lambda \mathbf{E}) \cdot \mathbf{e} = (\mathbf{U} + \sqrt{\lambda}\mathbf{E}) \cdot (\mathbf{U} - \sqrt{\lambda}\mathbf{E}) \cdot \mathbf{e} = \mathbf{0}.$$

So

$$(\mathbf{U} + \sqrt{\lambda}\mathbf{E}) \cdot \mathbf{p} = \mathbf{0} \tag{2.41}$$

where $\mathbf{p} = (\mathbf{U} - \sqrt{\lambda}\mathbf{E}) \cdot \mathbf{e}$. Suppose $\mathbf{p} \neq \mathbf{0}$. Dot-multiplication of (2.41) by $\mathbf{p}$ results in

$$\mathbf{p} \cdot \mathbf{U} \cdot \mathbf{p} = -\sqrt{\lambda}\mathbf{p} \cdot \mathbf{p} < 0,$$

contradicting the positive definiteness of $\mathbf{U}$. Hence $\mathbf{p} = \mathbf{0}$ and consequently

$$\mathbf{U} \cdot \mathbf{e} = \sqrt{\lambda}\mathbf{e}. \tag{2.42}$$

Conversely, if $(\sqrt{\lambda}, \mathbf{e})$ is an eigenpair of $\mathbf{U}$, then $(\lambda, \mathbf{e})$ is an eigenpair of $\mathbf{S}$. From (2.42) we have

$$(\mathbf{U} - \sqrt{\lambda}\mathbf{E}) \cdot \mathbf{e} = \mathbf{0},$$

and by multiplying the latter equation by $\mathbf{U} + \sqrt{\lambda}\mathbf{E}$ we obtain

$$\mathbf{0} = (\mathbf{U} + \sqrt{\lambda}\mathbf{E}) \cdot (\mathbf{U} - \sqrt{\lambda}\mathbf{E}) \cdot \mathbf{e} = (\mathbf{U}^2 - \lambda\mathbf{E}) \cdot \mathbf{e} = (\mathbf{S} - \lambda\mathbf{E}) \cdot \mathbf{e}$$

so that $\mathbf{S} \cdot \mathbf{e} = \lambda\mathbf{e}$ as required. □

Here we denote a square root of a symmetric positive definite tensor $\mathbf{S}$ as follows:

$$\mathbf{U} = \mathbf{S}^{1/2} \equiv \sqrt{\mathbf{S}} \equiv \sqrt{\lambda_1}\,\mathbf{i}_1\mathbf{i}_1 + \sqrt{\lambda_2}\,\mathbf{i}_2\mathbf{i}_2 + \sqrt{\lambda_3}\,\mathbf{i}_3\mathbf{i}_3,$$

where the $\mathbf{i}_k$ are orthonormal eigenvectors of $\mathbf{S}$ and the $\lambda_k > 0$ are its eigenvalues.

In a similar way we introduce other roots of a symmetric positive definite tensor $\mathbf{S}$,

$$\mathbf{S}^{1/n} \equiv \sqrt[n]{\mathbf{S}} \equiv \sqrt[n]{\lambda_1}\,\mathbf{i}_1\mathbf{i}_1 + \sqrt[n]{\lambda_2}\,\mathbf{i}_2\mathbf{i}_2 + \sqrt[n]{\lambda_3}\,\mathbf{i}_3\mathbf{i}_3,$$

and even a general tensor-function of a symmetric positive definite tensor:

$$\mathbf{f}(\mathbf{S}) \equiv f(\lambda_1)\,\mathbf{i}_1\mathbf{i}_1 + f(\lambda_2)\,\mathbf{i}_2\mathbf{i}_2 + f(\lambda_3)\,\mathbf{i}_3\mathbf{i}_3,$$

where $f(x)$ is a function of one variable. Let us note that with this definition one can define non-analytic functions of a symmetric positive definite tensor.

If we consider all the solutions of the equation $\mathbf{U}^2 = \mathbf{S}$, then we arrived at all square roots of a symmetric positive definite tensor. These solutions clearly take the form

$$\mathbf{U} = \pm\sqrt{\lambda_1}\,\mathbf{i}_1\mathbf{i}_1 \pm \sqrt{\lambda_2}\,\mathbf{i}_2\mathbf{i}_2 \pm \sqrt{\lambda_3}\,\mathbf{i}_3\mathbf{i}_3,$$

so we have six square roots of $\mathbf{S}$, of which only one is positive definite.

The operations of polar decomposition and calculation of the square root are often used in nonlinear elasticity. Unfortunately, the formulas presented above require calculation of eigenvalues and eigenvectors of a symmetric tensor. Theoretically the problem is simple from a numerical standpoint, but sometimes we need exact formulas for these operations.

We start in $\mathbb{R}^2$. To find the solution of $\mathbf{U}^2 = \mathbf{S}$, we use the two-dimensional variant of the Cayley–Hamilton theorem

$$\mathbf{S}^2 - I_1(\mathbf{S})\mathbf{S} + I_2(\mathbf{S})\mathbf{E}_2 = \mathbf{0}, \tag{2.43}$$

where $I_1(\mathbf{S}) = \operatorname{tr}\mathbf{S}$, $I_2(\mathbf{S}) = \frac{1}{2}(\operatorname{tr}^2\mathbf{S} - \operatorname{tr}\mathbf{S}^2) = \det\mathbf{S}$, and $\mathbf{E}_2$ is the two-dimensional unit tensor. Note that here det is calculated as the determinant of a two-dimensional tensor. Obviously $\mathbf{U}$ satisfies

$$\mathbf{U}^2 - I_1(\mathbf{U})\mathbf{U} + I_2(\mathbf{U})\mathbf{E}_2 = \mathbf{0}. \tag{2.44}$$

Substituting the formulas $\mathbf{U}^2 = \mathbf{S}$ and $I_2(\mathbf{U}) = \sqrt{I_2(\mathbf{S})}$ into (2.44), we transform the latter into

$$\mathbf{S} - I_1(\mathbf{U})\mathbf{U} + \sqrt{I_2(\mathbf{S})}\mathbf{E}_2 = \mathbf{0}. \tag{2.45}$$

Let us take the trace of both sides of (2.45):

$$\operatorname{tr}\mathbf{S} - \operatorname{tr}^2\mathbf{U} + 2\sqrt{I_2(\mathbf{S})} = 0.$$

It follows that

$$\operatorname{tr}\mathbf{U} = \sqrt{\operatorname{tr}\mathbf{S} + 2\sqrt{I_2(\mathbf{S})}} = \sqrt{I_1(\mathbf{S}) + 2\sqrt{I_2(\mathbf{S})}}. \tag{2.46}$$

For positive definite $\mathbf{U}$ we have $\operatorname{tr}\mathbf{U} > 0$, so substituting (2.46) into (2.45) we find the required expression for $\mathbf{U}$ as a function of $\mathbf{S}$:

$$\mathbf{U} = \frac{1}{\sqrt{I_1(\mathbf{S}) + 2\sqrt{I_2(\mathbf{S})}}}(\mathbf{S} + \sqrt{I_2(\mathbf{S})}\mathbf{E}_2). \tag{2.47}$$

Multiplying (2.44) by $\mathbf{U}^{-1}$, we obtain

$$\mathbf{U} - I_1(\mathbf{U})\mathbf{E}_2 + I_2(\mathbf{U})\mathbf{U}^{-1} = \mathbf{0},$$

which leads to the equation for the inverse of $\mathbf{U}$:

$$I_2(\mathbf{U})\mathbf{U}^{-1} = I_1(\mathbf{U})\mathbf{E}_2 - \mathbf{U}.$$

Finally,

$$\mathbf{U}^{-1} = \frac{1}{\sqrt{I_2(\mathbf{S})}}\left[\sqrt{I_1(\mathbf{S}) + 2\sqrt{I_2(\mathbf{S})}}\mathbf{E}_2 - \mathbf{U}\right]. \tag{2.48}$$

Formulas (2.47) and (2.48) may also be useful for three-dimensional tensors having a special structure. Let us consider a symmetric positive definite tensor $\mathbf{S}$ of the form

$$\mathbf{S} = \mathbf{S}_0 + \mathbf{nn},$$

where $\mathbf{n}$ is a unit vector and $\mathbf{S}_0 \cdot \mathbf{n} = \mathbf{0}$ with $\mathbf{S}_0 = \mathbf{S}_0^T$. Obviously, for such $\mathbf{S}$ we have

$$\mathbf{S}^2 = \mathbf{S}_0^2 + \mathbf{nn}, \quad \operatorname{tr}\mathbf{S} = \operatorname{tr}\mathbf{S}_0 + 1, \quad I_2(\mathbf{S}) = I_2(\mathbf{S}_0) + \operatorname{tr}\mathbf{S}_0, \quad \det\mathbf{S} = \det\mathbf{S}_0,$$

so all invariants of $\mathbf{S}$ are expressed through those of $\mathbf{S}_0$ and vice versa. Looking for the solution of $\mathbf{U}^2 = \mathbf{S}$ in the form $\mathbf{U} = \mathbf{U}_0 + \mathbf{nn}$, where $\mathbf{U}_0$ is a symmetric tensor such that $\mathbf{U}_0 \cdot \mathbf{n} = \mathbf{0}$, we find that $\mathbf{U}_0$ must satisfy

$$\mathbf{U}_0^2 = \mathbf{S}_0,$$

which was solved above. Its solution follows from (2.47):

$$\mathbf{U}_0 = \frac{1}{\sqrt{I_1(\mathbf{S}_0) + 2\sqrt{I_2(\mathbf{S}_0)}}}\left(\mathbf{S}_0 + \sqrt{I_2(\mathbf{S}_0)}\mathbf{E}_2\right). \tag{2.49}$$

To find the inverse to $\mathbf{U}$, we use a similar representation $\mathbf{U}^{-1} = \mathbf{U}_0^{-1} + \mathbf{nn}$ and formula (2.48).

Exercise 2.22. Find the positive square root and its inverse for the tensor $\mathbf{S} = \mathbf{S}_0 + \lambda\mathbf{nn}$, where $\mathbf{n}$ is a unit vector, $\lambda > 0$, and $\mathbf{S}_0 \cdot \mathbf{n} = \mathbf{0}$, $\mathbf{S}_0 = \mathbf{S}_0^T$.

Formulas (2.47) and (2.48), together with Exercise 2.22, give the square root and polar decomposition of a second-order tensor in case we know one eigenvalue and corresponding eigenvector in advance.

Let us now consider the general case. For brevity we denote the principal invariants of a positive definite second-order tensor $\mathbf{S}$ by I_1, I_2, I_3 and the invariants of its square root $\mathbf{U}$ by J_1, J_2, J_3. So

$$\begin{aligned} I_1 &= \lambda_1 + \lambda_2 + \lambda_3, \\ I_2 &= \lambda_1\lambda_2 + \lambda_1\lambda_3 + \lambda_2\lambda_3, \\ I_3 &= \lambda_1\lambda_2\lambda_3, \end{aligned}$$

and

$$\begin{aligned} J_1 &= \sqrt{\lambda_1} + \sqrt{\lambda_2} + \sqrt{\lambda_3}, \\ J_2 &= \sqrt{\lambda_1\lambda_2} + \sqrt{\lambda_1\lambda_3} + \sqrt{\lambda_2\lambda_3}, \\ J_3 &= \sqrt{\lambda_1\lambda_2\lambda_3}. \end{aligned}$$

Using the elementary identity

$$(a+b+c)^2 = a^2 + b^2 + c^2 + 2ab + 2ac + 2bc,$$

we find the relations between I_k and J_k:

$$I_1 = J_1^2 - 2J_2, \quad I_2 = J_2^2 - 2J_1J_3, \quad I_3 = J_3^2. \tag{2.50}$$

Zubov and Rudev [Zubov and Rudev (1996)] deduced that

$$\begin{aligned} J_1 &= \sqrt{x} + \sqrt{I_1 - x + 2\sqrt{I_3/x}}, \\ J_2 &= \sqrt{I_3/x} + \sqrt{x}\sqrt{I_1 - x + 2\sqrt{I_3/x}}, \\ J_3 &= \sqrt{I_3}, \end{aligned} \tag{2.51}$$

where

$$\begin{aligned} x &= \tfrac{1}{3}\left[I_1 + 2\sqrt{y}\cos\tfrac{\varphi}{3}\right], & \varphi &= \arccos\left(yz^{-3/2}\right), \\ y &= I_1^2 - 3I_2, & z &= \tfrac{1}{2}(2I_1^3 - 9I_1I_2 + 27I_3). \end{aligned}$$

From these we can deduce that $\mathbf{U}$ takes the form

$$\mathbf{U} = (J_1J_2 - J_3)^{-1}\left[J_1J_3\mathbf{E} + (J_1^2 - J_2)\mathbf{S} - \mathbf{S}^2\right]. \tag{2.52}$$

Exercise 2.23. Prove (2.51) and (2.52) by straightforward calculation.

Deviator and ball tensor representation

A representation of the form

$$\mathbf{A} = \tfrac{1}{3}I_1(\mathbf{A})\mathbf{E} + \operatorname{dev}\mathbf{A} \tag{2.53}$$

also proves useful in applications. This equation defines the tensor $\operatorname{dev}\mathbf{A}$, called the *deviator* of $\mathbf{A}$. It has the same eigenvectors as $\mathbf{A}$, but eigenvalues that differ from those of $\mathbf{A}$ by $\frac{1}{3}I_1(\mathbf{A})$:

$$\tilde{\lambda}_i = \lambda_i - \tfrac{1}{3}\operatorname{tr}\mathbf{A}.$$

The first term on the right-hand side of (2.53) is, of course, a ball tensor.

An example of a tensor coinciding with its deviator is a dyad $\mathbf{ab}$ with orthogonal vectors $\mathbf{a}$ and $\mathbf{b}$: $\mathbf{a}\cdot\mathbf{b} = 0$.

A symmetric second-order tensor $\mathbf{S}$ with $\operatorname{tr}\mathbf{S} = 0$ has the representation [Zubov and Rudev (1998)]

$$\mathbf{S} = \operatorname{dev}\mathbf{S} = \tau_1(\mathbf{i}_1\mathbf{i}_2 + \mathbf{i}_2\mathbf{i}_1) + \tau_2(\mathbf{i}_1\mathbf{i}_3 + \mathbf{i}_3\mathbf{i}_1) + \tau_3(\mathbf{i}_3\mathbf{i}_2 + \mathbf{i}_2\mathbf{i}_3),$$

where $\mathbf{i}_k$ are orthonormal vectors and τ_k are some numbers.

Exercise 2.24. Show that $\mathbf{A}\cdot\cdot\,\mathbf{B} = \frac{1}{3}I_1(\mathbf{A})I_1(\mathbf{B}) + \operatorname{dev}\mathbf{A}\cdot\cdot\operatorname{dev}\mathbf{B}$.

Unimodular tensors

A second-order tensor $\mathbf{U}$ is said to be *unimodular* if

$$|\det \mathbf{U}| = 1.$$

Obviously, any orthogonal tensor is unimodular. Another example of a unimodular tensor is

$$\mathbf{U} = \lambda_1 \mathbf{i}_1 \mathbf{i}_1 + \lambda_2 \mathbf{i}_2 \mathbf{i}_2 + \frac{1}{\lambda_1 \lambda_2} \mathbf{i}_3 \mathbf{i}_3.$$

The dot product of two or more unimodular tensors is unimodular as well.

Exercise 2.25. Verify the preceding statements.

It can be shown that any nonsingular second-order tensor $\mathbf{A}$ has a unique representation as the dot product of a positive definite ball tensor $\mathbf{B}$ and a unimodular tensor $\mathbf{U}$:

$$\mathbf{A} = \mathbf{B} \cdot \mathbf{U} = \mathbf{U} \cdot \mathbf{B}, \quad \mathbf{B} = |\det \mathbf{A}|\, \mathbf{E}, \quad \mathbf{U} = \frac{1}{|\det \mathbf{A}|} \mathbf{A}. \tag{2.54}$$

The additive representation (2.53) is widely used in mechanics for small strains; the multiplicative representation (2.54) is used for finite deformations, for example in nonlinear elasticity.

Groups of second-order tensors

For the moment we have introduced a few types of second-order tensors. Among them are rotation tensors, deviators, unimodular tensors, and certain others. These tensors can be characterized using the mathematical definition of a *group*.

We call a set of objects a group G (the objects are called elements of G) if there exists an operation $*$ between the elements of G (called, for simplicity, multiplication) for which the following four conditions hold:

(1) If a and b are any two elements of G, then the product $a * b$ is also an element of G. So G is closed under multiplication of its elements.

(2) Multiplication is associative: $(a * b) * c = a * (b * c)$.

(3) In G there is an unique identity element I such that $I * a = a * I = a$ for each $a \in G$.

(4) Each element a of G has an inverse, denoted by a^{-1}, such that $a * a^{-1} = a^{-1} * a = I$.

A trivial example of a group is the group consisting only of the identity element I.

Exercise 2.26. Show that the following four tensors in $\mathbb{R}^2$ constitute a group with respect to the dot product:

$$\mathbf{I} = \mathbf{E}_2 \equiv \mathbf{i}_1\mathbf{i}_1 + \mathbf{i}_2\mathbf{i}_2, \quad \mathbf{A} = -\mathbf{E}_2, \quad \mathbf{B} = \mathbf{i}_1\mathbf{i}_2 + \mathbf{i}_2\mathbf{i}_1, \quad \mathbf{C} = -\mathbf{B}.$$

This is a representation of the Vierergruppe (from German — the group of four elements). Here $\mathbf{I}$ is the identity element, $\mathbf{A}^{-1} = \mathbf{A}$, $\mathbf{B}^{-1} = \mathbf{B}$, and $\mathbf{C}^{-1} = \mathbf{C}$, so each element of the group coincides with its inverse.

Exercise 2.26 presents an example of a *discrete group*: one containing finitely many elements. A group with infinitely many elements indexed by a continuously variable parameter is a *continuous group.*

The set of all orthogonal tensors constitutes a group. The dot product plays the role of group multiplication; the product of two orthogonal tensors is also an orthogonal tensor, and $\mathbf{E}$ is the identity element. Several tensor groups are summarized in Table 2.4. Many other tensor groups are used, for example, in crystallography.

Table 2.4 Some tensor groups.

Notation	Description	Operation	Identity
Orth or O(3)	Full orthogonal group	$\cdot$	$\mathbf{E}$
Orth_+ or SO(3)	Rotation group	$\cdot$	$\mathbf{E}$
Unim	Group of all unimodular tensors	$\cdot$	$\mathbf{E}$
Unim_+	Subgroup of unimodular tensors with positive determinants	$\cdot$	$\mathbf{E}$
Lin	Set of all second-order tensors	$+$	$\mathbf{0}$
InvLin	Set of all invertible second-order tensors	$\cdot$	$\mathbf{E}$
Skw	Set of all skew tensors	$+$	$\mathbf{0}$
Sym	Set of all symmetric tensors	$+$	$\mathbf{0}$
DevSym	Set of all symmetric deviators	$+$	$\mathbf{0}$

Exercise 2.27. Verify the group properties for each tensor group listed in Table 2.4.

We can describe a group in terms of its *generators*, which constitute a minimal set of the group elements that could produce any element of the group. For example, DevSym can be produced using the generators $\mathbf{i}_1\mathbf{i}_2 + \mathbf{i}_2\mathbf{i}_1$, $\mathbf{i}_1\mathbf{i}_3 + \mathbf{i}_3\mathbf{i}_1$, and $\mathbf{i}_3\mathbf{i}_2 + \mathbf{i}_2\mathbf{i}_3$. Another example is the *triclinic group* containing only the two elements $\{\mathbf{E}, -\mathbf{E}\}$.

Exercise 2.28. Check the group properties for the triclinic group.

Let $\mathbf{R}(\omega\mathbf{e})$ be a rotation tensor given by Gibbs' formula (2.32), where ω is the rotation angle, $\mathbf{e}$ is the rotation axis, and $\mathbf{R}_\mathbf{i} = \mathbf{E} - 2\mathbf{i}\mathbf{i}$ is a reflection tensor. For a fixed axis we have a property of rotational tensors that

$$\mathbf{R}(\omega_1\mathbf{e}) \cdot \mathbf{R}(\omega_2\mathbf{e}) = \mathbf{R}((\omega_1 + \omega_2)\mathbf{e}).$$

One can easily check that the following sets, produced by the multiplications of the listed generators, constitute groups:

(1) $\{\mathbf{R}(\omega\mathbf{e})\}$ for arbitrary rotation angle,

(2) $\{\mathbf{R}(\omega\mathbf{e}), \mathbf{R}_\mathbf{e}\}$,

(3) $\{\mathbf{E}, \mathbf{R}(\frac{\pi}{3}\mathbf{e})\}$,

(4) $\{\mathbf{E}, -\mathbf{E}, \mathbf{R}(\frac{\pi}{2}\mathbf{e})\}$.

Exercise 2.29. Find the groups whose elements keep invariant (a) a square, (b) a circular cone, and (c) a circular cylinder. What is the difference between the groups for the cone and the cylinder?

2.3 Third- and Higher-Order Tensors

Third-order tensors

What about a tensor of order three? Here we encounter basic elements called *triads*, which are quantities of the type $\mathbf{R} = \mathbf{a} \otimes \mathbf{b} \otimes \mathbf{c} \equiv \mathbf{abc}$ where $\mathbf{a}$, $\mathbf{b}$, and $\mathbf{c}$ are vectors. Expanding these vectors in terms of a basis $\mathbf{e}_i$ we obtain

$$\mathbf{R} = a^i b^j c^k \mathbf{e}_i \mathbf{e}_j \mathbf{e}_k.$$

We then consider a linear space whose basis is the set of 27 quantities $\mathbf{e}_i\mathbf{e}_j\mathbf{e}_k$ and call it the space of third-order tensors. We continue to refer to the numerical values $a^i b^j c^k$ (and, more generally, to r^{ijk} in the expansion $r^{ijk}\mathbf{e}_i\mathbf{e}_j\mathbf{e}_k$) as the tensor's components. A general element $\mathbf{R}$ of this space has the representation

$$\mathbf{R} = r^{ijk}\, \mathbf{e}_i \mathbf{e}_j \mathbf{e}_k.$$

The property of objectivity of $\mathbf{R}$ remains paramount, leading to the requirement that the components transform appropriately when we change the frame. The usual procedure of setting

$$r^{i'j'k'}\,\mathbf{e}_{i'}\mathbf{e}_{j'}\mathbf{e}_{k'} = r^{mnp}\,\mathbf{e}_m\mathbf{e}_n\mathbf{e}_p$$

under the frame transformation $\mathbf{e}_i = A_i^{j'}\,\mathbf{e}_{j'}$ gives

$$r^{i'j'k'} = r^{mnp}A_m^{i'}A_n^{j'}A_p^{k'}\,.$$

Exercise 2.30. Write down some other representations and frame transformation rules for third-order tensors.

Dot products involving triads behave in expected ways, as summarized in Table 2.5.

Table 2.5 Operations involving triads.

dot product of triad with vector	$\mathbf{abc}\cdot\mathbf{x} = \mathbf{ab}(\mathbf{c}\cdot\mathbf{x})$
double dot product of triad with dyad	$\mathbf{abc}\cdot\cdot\,\mathbf{xy} = \mathbf{a}(\mathbf{c}\cdot\mathbf{x})(\mathbf{b}\cdot\mathbf{y})$
triple dot product between triads	$\mathbf{abc}\cdot\cdot\cdot\,\mathbf{xyz} = (\mathbf{c}\cdot\mathbf{x})(\mathbf{b}\cdot\mathbf{y})(\mathbf{a}\cdot\mathbf{z})$
scalar product of triads	$\mathbf{abc}\bullet\mathbf{xyz} = (\mathbf{a}\cdot\mathbf{x})(\mathbf{b}\cdot\mathbf{y})(\mathbf{c}\cdot\mathbf{z})$

It has become clear that the order of a tensor is equal to the number of free indices required to specify its components. We should mention the tensor whose components ϵ_{ijk} were used to treat the cross product. Known as the *Levi–Civita tensor*, it is a third-order tensor in three-dimensional Euclidean space given by

$$\boldsymbol{\mathcal{E}} = \epsilon_{ijk}\,\mathbf{e}^i\mathbf{e}^j\mathbf{e}^k.$$

Exercise 2.31. Show that

(a) $\boldsymbol{\mathcal{E}} = -\mathbf{E}\times\mathbf{E}$,
(b) $\boldsymbol{\mathcal{E}}\cdot\cdot\cdot\,\mathbf{zyx} = \mathbf{x}\cdot(\mathbf{y}\times\mathbf{z})$,
(c) $\boldsymbol{\mathcal{E}}\cdot\cdot\,\mathbf{xy} = \mathbf{y}\times\mathbf{x}$,
(d) $\boldsymbol{\mathcal{E}}\cdot\mathbf{x} = -\mathbf{x}\times$.
(e) $\boldsymbol{\mathcal{E}} = \mathbf{i}_1\mathbf{i}_2\mathbf{i}_3 + \mathbf{i}_2\mathbf{i}_3\mathbf{i}_1 + \mathbf{i}_3\mathbf{i}_1\mathbf{i}_2 - \mathbf{i}_1\mathbf{i}_3\mathbf{i}_2 - \mathbf{i}_3\mathbf{i}_2\mathbf{i}_1 - \mathbf{i}_2\mathbf{i}_1\mathbf{i}_3$.

The notation in part (d) calls for clarification. The result of applying $\boldsymbol{\mathcal{E}}$ to a vector $\mathbf{x}$ is a second-order tensor $\boldsymbol{\mathcal{E}}\cdot\mathbf{x}$. When $\boldsymbol{\mathcal{E}}\cdot\mathbf{x}$ is applied to another vector $\mathbf{y}$ (i.e., dot multiplied by $\mathbf{y}$), it becomes equivalent to the cross product $\mathbf{y}\times\mathbf{x}$. This is the meaning of (d).

Fourth-order tensors and beyond

We could continue to extend our treatment to tensors of any desired order. The essential points can be illustrated using fourth-order tensors.

A fourth-order tensor $\mathbf{C}$ can be represented by several types of components, e.g.,

$$c^{ijkl}, \qquad c_i{}^{jkl}, \qquad c_{ij}{}^{kl}, \qquad c_{ijk}{}^{l}, \qquad c_{ijkl}.$$

These components represent $\mathbf{C}$ with respect to various bases; for instance,

$$\mathbf{C} = c^{ijkl}\, \mathbf{e}_i\mathbf{e}_j\mathbf{e}_k\mathbf{e}_l.$$

Dot products with vectors can be taken as before: the rule is that we simply dot multiply the basis vectors positioned nearest to the dot. Such operations yield results of various kinds. A dot product between a fourth-order tensor and a vector yields a third-order tensor:

$$\mathbf{R} = \mathbf{C}\cdot\mathbf{x} = c^{ijkl}\, \mathbf{e}_i\mathbf{e}_j\mathbf{e}_k\mathbf{e}_l \cdot x_m\mathbf{e}^m = c^{ijkl}\, \mathbf{e}_i\mathbf{e}_j\mathbf{e}_k\, \delta_l^m x_m = c^{ijkl} x_l\, \mathbf{e}_i\mathbf{e}_j\mathbf{e}_k.$$

A dot product between a third-order tensor and a vector gives a second-order tensor. Wherever the dot product is used, it continues to enjoy its expected linearity properties.

An example of a fourth-order tensor is the tensor of elastic moduli in linear elasticity.

Note that if $\mathbf{A}$ is of order m and $\mathbf{B}$ is of order n, then $\mathbf{A}\cdot\mathbf{B}$ is of order $m+n-2$ and $\mathbf{A}\times\mathbf{B}$ is of order $m+n-1$.

An operation known as the *full product* of tensors [Zubov (1982); Zubov and Karyakin (2006)] will appear in the definition of tensor differentiation in § 2.8. Let us introduce the operation by means of some examples. If

$$\mathbf{X} = x^{ijk}\mathbf{e}_i\mathbf{e}_j\mathbf{e}_k \quad \text{and} \quad \mathbf{Y} = y_{lm}\mathbf{e}^l\mathbf{e}^m,$$

then the full product $\mathbf{X}\odot\mathbf{Y}$ is computed as

$$\begin{aligned}\mathbf{X}\odot\mathbf{Y} &= x^{ijk}\mathbf{e}_i\mathbf{e}_j\mathbf{e}_k \odot y_{lm}\mathbf{e}^l\mathbf{e}^m \\ &= x^{ijk}y_{lm}\mathbf{e}_i(\mathbf{e}_j\cdot\mathbf{e}^l)(\mathbf{e}_k\cdot\mathbf{e}^m) \\ &= x^{ijk}y_{lm}\mathbf{e}_i\delta_j^l\delta_k^m \\ &= x^{ijk}y_{jk}\mathbf{e}_i.\end{aligned}$$

Observe the way in which the constituent vectors of the triad $\mathbf{e}_i\mathbf{e}_j\mathbf{e}_k$ and the dyad $\mathbf{e}^l\mathbf{e}^m$ were paired for taking dot products ("last with last, then next-to-last with next-to-last, and so on, until the vectors in the second tensor were exhausted"). Similarly

$$\mathbf{A} \odot \mathbf{B} = a^{ijkl}\mathbf{e}_i\mathbf{e}_j\mathbf{e}_k\mathbf{e}_l \odot b_{mn}\mathbf{e}^m\mathbf{e}^n = a^{ijkl}b_{kl}\mathbf{e}_i\mathbf{e}_j.$$

In general, the full product $\mathbf{X} \odot \mathbf{Y}$ is defined for tensors $\mathbf{X}, \mathbf{Y}$ of respective orders p, q provided that $p \geq q$. The result is a tensor of order $p - q$. Note that if $\mathbf{X}, \mathbf{Y}$ are second-order tensors, then $\mathbf{X} \odot \mathbf{Y} = \mathbf{X} \bullet \mathbf{Y}$.

Some authors employ other notations for tensor operations. For example, instead of $\bullet$ they use : for second-order tensors and :: for fourth-order tensors. They may write

$$\begin{aligned}
&\mathbf{ab} : \mathbf{xy} = (\mathbf{a} \cdot \mathbf{x})(\mathbf{b} \cdot \mathbf{y}),\\
&\mathbf{abc} : \mathbf{xy} = \mathbf{a}(\mathbf{b} \cdot \mathbf{x})(\mathbf{c} \cdot \mathbf{y}),\\
&\mathbf{abcd} : \mathbf{xy} = \mathbf{ab}\,(\mathbf{c} \cdot \mathbf{x})(\mathbf{d} \cdot \mathbf{y}),\\
&\mathbf{ab} : \mathbf{xyzu} = (\mathbf{a} \cdot \mathbf{x})(\mathbf{b} \cdot \mathbf{y})\,\mathbf{zu},\\
&\mathbf{abcd} :: \mathbf{xyzu} = (\mathbf{a} \cdot \mathbf{x})(\mathbf{b} \cdot \mathbf{y})(\mathbf{c} \cdot \mathbf{z})(\mathbf{d} \cdot \mathbf{u}),
\end{aligned}$$

etc. Such notation may be useful to emphasize the orders of the tensors under consideration, but they are less than convenient in the general case.

Exercise 2.32. For $\mathbf{a}, \mathbf{b}, \mathbf{c}, \mathbf{d}, \mathbf{e}, \mathbf{f} \in \mathbb{R}^3$, compute (a) $\mathbf{ab} \odot \mathbf{c}$, (b) $\mathbf{ab} \odot \mathbf{cd}$, (c) $\mathbf{abcd} \odot \mathbf{df}$, and (d) $\mathbf{a} \odot \mathbf{b}$.

2.4 Isotropic Tensors

A tensor is said to be *isotropic* if its individual components are invariant under all possible rotations and mirror reflections in $\mathbb{R}^3$.

Any scalar quantity is isotropic. Clearly, the only isotropic vector is $\mathbf{0}$. Let $\mathbf{A}$ be a second-order tensor so that in a basis $\mathbf{e}_i$ we have

$$\mathbf{A} = a^{ij}\,\mathbf{e}_i\mathbf{e}_j.$$

Recall that any rotation or mirror reflection of $\mathbb{R}^3$ is uniquely defined by an orthogonal tensor $\mathbf{Q}$. Let us apply $\mathbf{Q}$ to each basis vector:

$$\tilde{\mathbf{e}}_k = \mathbf{Q} \cdot \mathbf{e}_k.$$

In the new basis we have $\mathbf{A} = \tilde{a}^{ij}\,\tilde{\mathbf{e}}_i\tilde{\mathbf{e}}_j$. If $\mathbf{A}$ is isotropic, then $a^{ij} = \tilde{a}^{ij}$ and

$$\mathbf{A} = \tilde{a}^{ij}\,\tilde{\mathbf{e}}_i\tilde{\mathbf{e}}_j = a^{ij}\mathbf{Q} \cdot \mathbf{e}_i\mathbf{Q} \cdot \mathbf{e}_j = \mathbf{Q} \cdot (a^{ij}\,\mathbf{e}_i\mathbf{e}_j) \cdot \mathbf{Q}^T.$$

Hence $\mathbf{A}$ *is isotropic if and only if*

$$\mathbf{A} = \mathbf{Q} \cdot \mathbf{A} \cdot \mathbf{Q}^T$$

for any orthogonal tensor $\mathbf{Q}$.

The identity tensor $\mathbf{E}$ is, of course, isotropic. Indeed, for any orthogonal $\mathbf{Q}$ we have $\mathbf{Q} \cdot \mathbf{Q}^T = \mathbf{E}$ and hence $\mathbf{E} = \mathbf{Q} \cdot \mathbf{E} \cdot \mathbf{Q}^T$. Clearly any ball tensor of the form $\lambda \mathbf{E}$ is isotropic. Although nontrivial, the converse holds as well: a second-order tensor $\mathbf{A}$ is isotropic if and only if it is a ball tensor.

Exercise 2.33. Verify directly that the mixed components of $\mathbf{E}$ are invariant under coordinate transformation.

Under an orthogonal transformation $\mathbf{Q}$ of $\mathbb{R}^3$, a fourth-order tensor

$$\mathbf{C} = c^{ijmn} \, \mathbf{e}_i \mathbf{e}_j \mathbf{e}_m \mathbf{e}_n$$

takes the form

$$\mathbf{C} = \tilde{c}^{ijmn} (\mathbf{Q} \cdot \mathbf{e}_i)(\mathbf{Q} \cdot \mathbf{e}_j)(\mathbf{Q} \cdot \mathbf{e}_m)(\mathbf{Q} \cdot \mathbf{e}_n).$$

By definition, it is isotropic if $\tilde{c}^{ijmn} = c^{ijmn}$ for all $\mathbf{Q}$.

2.5 Tensor-Valued Functions of Tensor Arguments

We now begin to discuss mappings between spaces of tensors. We could, for instance, consider scalar-valued mappings of the forms $f = f(\mathbf{x})$ or $g = g(\mathbf{X})$ where $\mathbf{x}$ is a vector and $\mathbf{X}$ is a second-order tensor. In any such case we shall require that the function be independent of how the argument (here $\mathbf{x}$ or $\mathbf{X}$) is represented in a basis. We can also consider vector- or tensor-valued functions of scalar, vector, or tensor arguments. Examples include a time-dependent force vector $\mathbf{f} = \mathbf{f}(t)$, and static vector and tensor fields as functions of position in space: $\mathbf{g} = \mathbf{g}(\mathbf{r})$ and $\mathbf{A} = \mathbf{A}(\mathbf{r})$.

Linear functions

The familiar requirement of linearity will appear in such forms as

$$f(\lambda \mathbf{a} + \mu \mathbf{b}) = \lambda f(\mathbf{a}) + \mu f(\mathbf{b}),$$

$$f(\lambda \mathbf{A} + \mu \mathbf{B}) = \lambda f(\mathbf{A}) + \mu f(\mathbf{B}),$$

$$\mathbf{f}(\lambda \mathbf{a} + \mu \mathbf{b}) = \lambda \mathbf{f}(\mathbf{a}) + \mu \mathbf{f}(\mathbf{b}),$$

where λ and μ are arbitrary scalars. For a linear, scalar-valued function of a vector argument, we have the following representation theorem.

Theorem 2.5. *If f is a scalar-valued linear function of a vector argument* $\mathbf{x}$*, there is a unique vector* $\mathbf{c}$ *such that* $f(\mathbf{x}) = \mathbf{c} \cdot \mathbf{x}$ *for all* $\mathbf{x}$.

Proof. The relation $f(\mathbf{x}) = f(x^k \mathbf{e}_k) = x^k f(\mathbf{e}_k) = f(\mathbf{e}_k)\mathbf{e}^k \cdot \mathbf{x}$ shows that we can take $\mathbf{c} = f(\mathbf{e}_k)\mathbf{e}^k$. Supposing the existence of two vectors $\mathbf{c}_1$ and $\mathbf{c}_2$ such that $\mathbf{c}_1 \cdot \mathbf{x} = \mathbf{c}_2 \cdot \mathbf{x}$, and putting $\mathbf{x} = \mathbf{c}_2 - \mathbf{c}_1$, we obtain $\mathbf{c}_1 = \mathbf{c}_2$. □

Next we consider a scalar-valued function whose argument is a second-order tensor $\mathbf{A}$. Clearly, for a second-order tensor $\mathbf{B}$ the function

$$f(\mathbf{A}) = \operatorname{tr}(\mathbf{B} \cdot \mathbf{A}^T) = \mathbf{B} \cdot\cdot\, \mathbf{A}^T$$

is linear.

Exercise 2.34. Show that $\mathbf{B} \cdot\cdot\, \mathbf{A}^T = \mathbf{A} \cdot\cdot\, \mathbf{B}^T$.

Proof of the following theorem is left to the reader.

Theorem 2.6. *For any scalar-valued linear function f depending on a second-order tensor* $\mathbf{A}$*, there is a unique second-order tensor* $\mathbf{B}$ *such that for all* $\mathbf{A}$*,*

$$f(\mathbf{A}) = \operatorname{tr}(\mathbf{B} \cdot \mathbf{A}^T). \tag{2.55}$$

Exercise 2.35. Write out the proof.

The representation of any tensor-valued function of a tensorial argument is similar.

Theorem 2.7. *Let* $\mathbf{F} = \mathbf{F}(\mathbf{A})$ *be a linear function acting in the set of second-order tensors. There is a unique fourth-order tensor* $\mathbf{C}$ *such that for all* $\mathbf{A}$*,*

$$\mathbf{F}(\mathbf{A}) = \mathbf{C} \cdot\cdot\, \mathbf{A}^T. \tag{2.56}$$

Proof. Write $\mathbf{A} = a^{mn}\mathbf{e}_m\mathbf{e}_n$ and introduce $\mathbf{C} = \mathbf{F}(\mathbf{e}_m\mathbf{e}_n)\mathbf{e}^m\mathbf{e}^n$. Then

$$\mathbf{C} \cdot\cdot\, \mathbf{A}^T = \mathbf{F}(\mathbf{e}_m\mathbf{e}_n)\mathbf{e}^m\mathbf{e}^n \cdot\cdot\, a^{ij}\,\mathbf{e}_j\mathbf{e}_i = \mathbf{F}(\mathbf{e}_m\mathbf{e}_n)a^{mn} = \mathbf{F}(\mathbf{A}).$$

Supposing the existence of two tensors $\mathbf{C}_1$ and $\mathbf{C}_2$ such that $\mathbf{C}_1 \cdot\cdot\, \mathbf{A}^T = \mathbf{C}_2 \cdot\cdot\, \mathbf{A}^T$, and putting $\mathbf{D} = \mathbf{C}_1 - \mathbf{C}_2$, we get

$$0 = \mathbf{D} \cdot\cdot\, \mathbf{A}^T = d^{ij}\mathbf{e}_i\mathbf{e}_j \cdot\cdot\, a^{mk}\mathbf{e}_k\mathbf{e}_m = d^{ij}a^{mk}g_{jk}g_{im} = d^{ij}a_{ij}$$

which holds for all $\mathbf{A}$ if and only if $\mathbf{D} = \mathbf{0}$. □

Exercise 2.36. Let $\mathbf{B}$ be a fixed second-order tensor and consider the function $f(\mathbf{A}) = \mathbf{B} \cdot \mathbf{A}^T$. Show that $f(\mathbf{A}) = \mathbf{C} \cdot\cdot\, \mathbf{A}^T$ where $\mathbf{C} = \mathbf{B} \cdot \mathbf{e}^n\mathbf{E}\mathbf{e}_n$.

It was first proved in a Cartesian frame and then extended to the general case that the general form of a fourth-order isotropic tensor is

$$\alpha\,\mathbf{E}\mathbf{E} + \beta\,\mathbf{e}_k\mathbf{E}\mathbf{e}^k + \gamma\mathbf{I} \tag{2.57}$$

where α, β, γ are scalars and $\mathbf{I} = \mathbf{e}_k\mathbf{e}_m\mathbf{e}^k\mathbf{e}^m$ [Jeffreys (1931)]. This is an important result in linear elasticity, as it provides the general form of the elastic moduli tensor for an isotropic material. The addends, three independent isotropic tensors, have properties exhibited in the following exercise.

Exercise 2.37. Verify the following representations per Theorem 2.7:

(a) $\mathbf{F}(\mathbf{A}) = \mathbf{A}$ and $\mathbf{C} = \mathbf{I}$,
(b) $\mathbf{F}(\mathbf{A}) = \mathbf{A}^T$ and $\mathbf{C} = \mathbf{e}_k\mathbf{E}\mathbf{e}^k$,
(c) $\mathbf{F}(\mathbf{A}) = (\operatorname{tr}\mathbf{A})\mathbf{E}$ and $\mathbf{C} = \mathbf{E}\mathbf{E}$.

Isotropic scalar-valued functions

A scalar function of a tensor is said to be *isotropic* if it retains its form under any orthogonal transformation of the space (or equivalently, of its basis).

A scalar-valued function $f(\mathbf{A})$ of a second-order argument $\mathbf{A}$ is isotropic if and only if for any orthogonal tensor $\mathbf{Q}$ we have

$$f(\mathbf{A}) = f\left(\mathbf{Q}\cdot\mathbf{A}\cdot\mathbf{Q}^T\right).$$

Any eigenvalue λ of a second-order tensor $\mathbf{A}$, when considered as a scalar-valued function of $\mathbf{A}$ (i.e., $\lambda = f(\mathbf{A})$), is isotropic. Indeed, an eigenvalue satisfies the characteristic equation $\det(\mathbf{A} - \lambda\mathbf{E}) = 0$. The eigenvalues of $\mathbf{Q}\cdot\mathbf{A}\cdot\mathbf{Q}^T$ satisfy

$$\det(\mathbf{Q}\cdot\mathbf{A}\cdot\mathbf{Q}^T - \lambda\mathbf{E}) = 0.$$

The equality

$$\begin{aligned}\det(\mathbf{Q}\cdot\mathbf{A}\cdot\mathbf{Q}^T - \lambda\mathbf{E}) &= \det(\mathbf{Q}\cdot\mathbf{A}\cdot\mathbf{Q}^T - \lambda\mathbf{Q}\cdot\mathbf{E}\cdot\mathbf{Q}^T)\\ &= \det\left[\mathbf{Q}\cdot(\mathbf{A} - \lambda\mathbf{E})\cdot\mathbf{Q}^T\right]\\ &= (\det\mathbf{Q})^2\det(\mathbf{A} - \lambda\mathbf{E}) = \det(\mathbf{A} - \lambda\mathbf{E})\end{aligned}$$

shows that the eigenvalues of $\mathbf{A}$ and $\mathbf{Q}\cdot\mathbf{A}\cdot\mathbf{Q}^T$ satisfy the same equation and therefore coincide.

Exercise 2.38. Show that $I_1(\mathbf{A})$, $I_2(\mathbf{A})$, and $I_3(\mathbf{A})$ are isotropic functions of $\mathbf{A}$.

It can be shown that any scalar-valued isotropic function of a second-order tensor is a function of its invariants [Lurie (1990); Ogden (1997); Truesdell and Noll (2004)].

By Theorem 2.6, a scalar-valued linear function has the representation

$$f(\mathbf{A}) = \mathbf{B} \cdot\cdot\, \mathbf{A}^T \tag{2.58}$$

for some second-order tensor $\mathbf{B}$. The next theorem states that such a function is isotropic if and only if $\mathbf{B}$ is a ball tensor.

Theorem 2.8. *The function* (2.58) *is isotropic if and only if* $\mathbf{B}$ *is isotropic.*

Proof. Assuming f is isotropic, we have for any orthogonal $\mathbf{Q}$

$$\begin{aligned}
\mathbf{B} \cdot\cdot\, \mathbf{A}^T &= \mathbf{B} \cdot\cdot\, (\mathbf{Q} \cdot \mathbf{A} \cdot \mathbf{Q}^T)^T = \operatorname{tr}(\mathbf{B} \cdot (\mathbf{Q} \cdot \mathbf{A} \cdot \mathbf{Q}^T)^T) \\
&= \operatorname{tr}(\mathbf{B} \cdot (\mathbf{Q}^T)^T \cdot (\mathbf{Q} \cdot \mathbf{A})^T) = \operatorname{tr}(\mathbf{B} \cdot \mathbf{Q} \cdot \mathbf{A}^T \cdot \mathbf{Q}^T) \\
&= \operatorname{tr}((\mathbf{B} \cdot \mathbf{Q} \cdot \mathbf{A}^T) \cdot \mathbf{Q}^T) = \operatorname{tr}(\mathbf{Q}^T \cdot (\mathbf{B} \cdot \mathbf{Q} \cdot \mathbf{A}^T)) \\
&= \operatorname{tr}((\mathbf{Q}^T \cdot \mathbf{B} \cdot \mathbf{Q}) \cdot \mathbf{A}^T) = (\mathbf{Q}^T \cdot \mathbf{B} \cdot \mathbf{Q}) \cdot\cdot\, \mathbf{A}^T.
\end{aligned}$$

This relation holds for all $\mathbf{A}$ if and only if $\mathbf{B} = \mathbf{Q}^T \cdot \mathbf{B} \cdot \mathbf{Q}$ for all $\mathbf{Q}$, i.e., if and only if $\mathbf{B}$ is isotropic. □

Exercise 2.39. Show that a scalar-valued, linear, isotropic function of a second-order tensor $\mathbf{A}$ is a linear function of $\operatorname{tr}\mathbf{A}$. In other words, demonstrate that $f(\mathbf{X}) = \lambda \operatorname{tr}\mathbf{X}$.

Isotropic tensor-valued functions

Now we consider a function $\mathbf{F}$ acting in the set of second-order tensors. We call $\mathbf{F}(\mathbf{A})$ *isotropic* if the components of its image value do not change from one basis to another. So $\mathbf{F}$ is isotropic if

$$\mathbf{F}\left(\mathbf{Q} \cdot \mathbf{A} \cdot \mathbf{Q}^T\right) = \mathbf{Q} \cdot \mathbf{F}(\mathbf{A}) \cdot \mathbf{Q}^T$$

for any orthogonal tensor $\mathbf{Q}$. An example is the function $\mathbf{F}(\mathbf{A}) = \lambda\mathbf{A}$.

As with a scalar-valued function, it can be shown that a tensor-valued linear function $\mathbf{F}$ represented in terms of a fourth-order tensor $\mathbf{C}$,

$$\mathbf{F}(\mathbf{A}) = \mathbf{C} \cdot\cdot\, \mathbf{A}^T,$$

is isotropic if and only if $\mathbf{C}$ is isotropic and therefore given by (2.57). So $\mathbf{F}$ takes the form

$$\mathbf{F}(\mathbf{A}) = \alpha\mathbf{E} \operatorname{tr}\mathbf{A} + \beta\mathbf{A}^T + \gamma\mathbf{A}.$$

Mechanicists often use functions that act in sets of symmetric tensors. This places additional restrictions on the form of $\mathbf{C}$. Indeed, let $\mathbf{A} = \mathbf{A}^T$ and $\mathbf{F}(\mathbf{A}) = \mathbf{F}(\mathbf{A})^T$. Using the representation

$$\mathbf{F}(\mathbf{A}) = c^{ijmn} a_{mn} \mathbf{i}_i \mathbf{i}_j$$

with a Cartesian basis $\mathbf{i}_i$, we see that

$$c^{ijmn} = c^{jimn} = c^{ijnm}.$$

In the general case, $\mathbf{C}$ has 81 independent components. But, in view of symmetry, $\mathbf{C}$ has only 36 independent components. An isotropic fourth-order tensor $\mathbf{C}$ satisfying the symmetry conditions takes the form

$$\alpha \mathbf{E}\mathbf{E} + \beta(\mathbf{e}_k \mathbf{E} \mathbf{e}^k + \mathbf{I}), \qquad \mathbf{I} = \mathbf{e}_k \mathbf{e}_m \mathbf{e}^k \mathbf{e}^m.$$

So the general form of an isotropic linear function satisfying the symmetry condition is

$$\mathbf{F}(\mathbf{A}) = \alpha \mathbf{E} \operatorname{tr} \mathbf{A} + 2\beta \mathbf{A}.$$

In terms of linear elasticity, this means that all components of the elastic moduli tensor $\mathbf{C}$ are expressed through two constants. Young's modulus E and Poisson's ratio ν are commonly employed.

2.6 Norms for Vectors and Tensors

The dot product has all the properties of an inner product on a linear space. That is, for $\mathbf{x}, \mathbf{y}, \mathbf{z} \in \mathbb{R}^3$ and $\alpha \in \mathbb{R}$ we have[3]

(i) $\mathbf{x} \cdot \mathbf{x} \geq 0$, with $\mathbf{x} \cdot \mathbf{x} = 0$ if and only if $\mathbf{x} = \mathbf{0}$,
(ii) $\mathbf{x} \cdot \mathbf{y} = \mathbf{y} \cdot \mathbf{x}$,
(iii) $(\alpha \mathbf{x} + \mathbf{y}) \cdot \mathbf{z} = \alpha \mathbf{x} \cdot \mathbf{z} + \mathbf{y} \cdot \mathbf{z}$.

The corresponding induced norm, given by

$$\|\mathbf{x}\| = (\mathbf{x} \cdot \mathbf{x})^{1/2}, \tag{2.59}$$

is the usual magnitude of the vector $\mathbf{x}$. Computable in the various forms

$$\|\mathbf{x}\| = \left(x^i x^j g_{ij}\right)^{1/2} = \left(x_i x_j g^{ij}\right)^{1/2} = \left(x^i x_i\right)^{1/2},$$

it is typically used as a measure of the size or intensity of $\mathbf{x}$. The properties

(i) $\|\mathbf{x}\| \geq 0$ and $\|\mathbf{x}\| = 0$ if and only if $\mathbf{x} = 0$,

[3] These are the inner product axioms for a *real* linear space. Eigenproblems are often discussed in complex spaces, as we saw earlier.

(ii) $\|\alpha\mathbf{x}\| = |\alpha|\,\|\mathbf{x}\|$,
(iii) $\|\mathbf{x}+\mathbf{y}\| \leq \|\mathbf{x}\| + \|\mathbf{y}\|$,

hold by definition of the induced norm. We also have the *Schwarz inequality*

$$|\mathbf{x}\cdot\mathbf{y}| \leq \|\mathbf{x}\|\,\|\mathbf{y}\| \tag{2.60}$$

for any $\mathbf{x}, \mathbf{y} \in \mathbb{R}^3$, with equality if and only if $\mathbf{x} = \lambda\mathbf{y}$ for some $\lambda \in \mathbb{R}$.

Similarly, the scalar product $\mathbf{A} \bullet \mathbf{B}$ of two second-order tensors satisfies the inner product axioms. Hence for the induced norm

$$\|\mathbf{A}\| = (\mathbf{A} \bullet \mathbf{A})^{1/2}$$

we have

(i) $\|\mathbf{A}\| \geq 0$, with $\|\mathbf{A}\| = 0$ if and only if $\mathbf{A} = \mathbf{0}$;
(ii) $\|\alpha\mathbf{A}\| = |\alpha|\,\|\mathbf{A}\|$,
(iii) $\|\mathbf{A}+\mathbf{B}\| \leq \|\mathbf{A}\| + \|\mathbf{B}\|$,

along with the Schwarz inequality

$$|\mathbf{A} \bullet \mathbf{B}| \leq \|\mathbf{A}\|\,\|\mathbf{B}\|\,.$$

Exercise 2.40. (a) Verify that the scalar product of two second-order tensors satisfies the first inner product axiom. (b) Find $\|\mathbf{E}\|$.

Additionally, if $\mathbf{A}$ and $\mathbf{B}$ are second-order tensors and $\mathbf{x}$ is a vector, then

$$\|\mathbf{A}\cdot\mathbf{x}\| \leq \|\mathbf{A}\|\,\|\mathbf{x}\| \qquad \text{and} \qquad \|\mathbf{A}\cdot\mathbf{B}\| \leq \|\mathbf{A}\|\,\|\mathbf{B}\|\,.$$

By the second property we have

$$\|\mathbf{A}^k\| \leq \|\mathbf{A}\|^k \qquad (k = 1, 2, \ldots) \tag{2.61}$$

which allows us to prove convergence of series such as (2.17).

Exercise 2.41. Prove (2.61). (b) Let $\mathbf{A}$ be a second-order tensor with $\|\mathbf{A}\| = q < 1$. Show that $(\mathbf{E}-\mathbf{A})^{-1}$ exists and is equal to $\mathbf{E}+\mathbf{A}+\mathbf{A}^2+\mathbf{A}^3+\cdots+\mathbf{A}^n+\cdots$. Also show that $(\mathbf{E}+\mathbf{A})^{-1} = \mathbf{E}-\mathbf{A}+\mathbf{A}^2-\mathbf{A}^3+\cdots+(-1)^n\mathbf{A}^n+\cdots$.

Matrix norms

In the theory of matrices there exist many other so-called matrix norms [Horn and Johnson (1990)]. Let us present the most popular expressions.

(1) $\|(a_{ij})\|_1 = \max\limits_{j=1,\ldots,n} \sum\limits_{i=1}^{n} |a_{ij}|$ is the maximum column sum norm;

(2) $\|(a_{ij})\|_\infty = \max\limits_{i=1,\ldots,n} \sum\limits_{j=1}^{n} |a_{ij}|$ is the maximum row sum norm;

(3) $\|(a_{ij})\|_M = n \max\limits_{i,j=1,\ldots,n} |a_{ij}|$;

(4) $\|(a_{ij})\|_2 = \max\limits_{i=1,\ldots,n} \nu_i$ is the spectral norm. Here ν_i are singular values of (a_{ij}), i.e., ν_i^2 are the eigenvalues of $(a_{ik}a_{jk})$;

(5) $\|(a_{ij})\|_E = \left(\sum\limits_{i,j=1}^{n} |a_{ij}|^2 \right)^{1/2}$ is the Euclidean or ℓ_2 norm;

(6) $\|(a_{ij})\|_{\ell_1} = \sum\limits_{i,j=1}^{n} |a_{ij}|$ is the ℓ_1 norm.

The reader should verify that every norm from this list satisfies the norm properties (i)–(iii). Clearly one can construct many matrix norms. For example, $\max(\|(a_{ij})\|_\alpha, \|(a_{ij})\|_\beta)$ is also a matrix norm if α, β stand for any of the symbols $1, 2, \infty, M, E, \ell_1$. Why, if all norms are equivalent on a finite dimensional space, does one need so many definitions? The answer lies in the simplicity of checking certain proofs and calculating certain norms numerically. Matrix norms for the matrix of components of a second-order tensor in some basis are not invariant under change of basis, in general. But for a fixed basis or a Cartesian frame we may use a matrix norm as a tensor norm, too.

Exercise 2.42. Find $\|(\delta_{ij})\|$ for the norms (1)–(6) in (a) $\mathbb{R}^3$ and (b) $\mathbb{R}^n$.

Considering a tensor as a linear operator acting in $\mathbb{R}^n$, we introduce the *operator norm* as follows:

$$\|\mathbf{A}\| = \max_{\|\mathbf{x}\|=1} \|\mathbf{A} \cdot \mathbf{x}\|.$$

So the operator norm is induced by some vector norm.

Exercise 2.43. Show that the operator norm of $\mathbf{E}$ is 1 (regardless of the form taken by the vector norm). Hence the Euclidean norm $(\mathbf{A} \bullet \mathbf{A})^{1/2}$, for instance, cannot be an operator norm for any vector norm in $\mathbb{R}^3$ because $(\mathbf{E} \bullet \mathbf{E})^{1/2} = \sqrt{n} \neq 1$.

The spectral norm plays an important role in the theory of matrices. The operator norm coincides with the spectral norm if we consider the former with respect to the Euclidean vector norm induced by the dot product for vectors. Indeed, $\|\mathbf{A} \cdot \mathbf{x}\|^2 = (\mathbf{A} \cdot \mathbf{x}) \cdot (\mathbf{A} \cdot \mathbf{x}) = \mathbf{x} \cdot \mathbf{A}^T \cdot \mathbf{A} \cdot \mathbf{x}$. The matrix $\mathbf{A}^T \cdot \mathbf{A}$ is symmetric and positive, so there exists a spectral decomposition

$$\mathbf{A}^T \cdot \mathbf{A} = \sum_{i=1}^{n} \nu_i^2 \mathbf{e}_i \mathbf{e}_i$$

where ν_i^2 and $\mathbf{e}_i$ constitute the ith eigenpair of $\mathbf{A}^T \cdot \mathbf{A}$ and $\mathbf{i}_k$ forms an orthonormal basis of $\mathbb{R}^n$. Hence

$$\max_{\|\mathbf{x}\|=1} \|\mathbf{A} \cdot \mathbf{x}\| = \max_{i=1\ldots n} \nu_i$$

and the maximum is attained when $\mathbf{x}$ is equal to the eigenvector corresponding to maximal ν_i.

For a symmetric tensor $\mathbf{A}$, the spectral norm is the maximal absolute eigenvalue of $\mathbf{A}$.

Exercise 2.44. Prove that the operator norm does not change under orthogonal transformation of a tensor, i.e., that $\|\mathbf{A}\| = \|\mathbf{Q} \cdot \mathbf{A} \cdot \mathbf{Q}^T\|$ for any orthogonal tensor $\mathbf{Q}$.

2.7 Analysis of a Vector Function of a Single Real Variable

The availability of a norm permits us to apply familiar notions from calculus to a function $\mathbf{f}\colon (a,b) \to \mathbb{R}^3$, where (a,b) is an open interval along the real line (e.g., a time interval). We say that $\mathbf{f}(t)$ has limit $\mathbf{a}$ at $t = t_0 \in (a,b)$, and write

$$\lim_{t \to t_0} \mathbf{f}(t) = \mathbf{a} \qquad \text{or} \qquad \mathbf{f}(t) \to \mathbf{a} \text{ as } t \to t_0$$

if to each $\varepsilon > 0$ there corresponds $\delta = \delta(\varepsilon) > 0$ such that $\|\mathbf{f}(t) - \mathbf{a}\| < \varepsilon$ whenever $0 < |t - t_0| < \delta$. If $\mathbf{f}(t_0) = \lim_{t \to t_0} \mathbf{f}(t)$, we call $\mathbf{f}(t)$ continuous at $t = t_0$; we call it continuous on (a,b) if it is continuous at each $t \in (a,b)$. The definitions of derivative and integral — themselves limiting operations

— can be adapted to $\mathbf{f}(t)$ by replacing the ordinary absolute value with a norm.

Various norms can be imposed on $\mathbb{R}^3$. However, in linear algebra it is shown that on a finite-dimensional space all norms are equivalent. Equivalence of two norms $\|\cdot\|_1$ and $\|\cdot\|_2$ means that there exist positive constants c_1 and c_2 such that for any $\mathbf{x}$

$$c_1\|\mathbf{x}\|_2 \leq \|\mathbf{x}\|_1 \leq c_2\|\mathbf{x}\|_2.$$

Consequently, either norm can be used in the definition of limit: the limit will exist or not, independently of the form taken by the norm.

The derivative of $\mathbf{f}(t)$, defined by

$$\mathbf{f}'(t) = \frac{d\mathbf{f}}{dt} = \lim_{h\to 0} \frac{\mathbf{f}(t+h) - \mathbf{f}(t)}{h},$$

has familiar properties such as linearity,

$$[\alpha\mathbf{f}(t) + \mathbf{g}(t)]' = \alpha\mathbf{f}'(t) + \mathbf{g}'(t),$$

and the product rule

$$[\mathbf{f}(t)\cdot\mathbf{g}(t)]' = \mathbf{f}(t)\cdot\mathbf{g}'(t) + \mathbf{f}'(t)\cdot\mathbf{g}(t).$$

Hence if $\mathbf{f}(t)$ is expanded in a time-varying basis as $\mathbf{f}(t) = f^k(t)\mathbf{e}_k(t)$ then

$$\frac{d\mathbf{f}}{dt} = \frac{df^k}{dt}\mathbf{e}_k + f^k\frac{d\mathbf{e}_k}{dt}.$$

Of course, if the $\mathbf{e}_k$ do not depend on time then

$$\frac{d\mathbf{f}}{dt} = \frac{df^k}{dt}\mathbf{e}_k.$$

If the $\mathbf{e}_k$ are independent of time, then the definite integral can also be done componentwise,

$$\int_a^b \mathbf{f}(t)\,dt = \mathbf{e}_k \int_a^b f^k(t)\,dt,$$

and has all the properties familiar from calculus.

2.8 Differentiation of Tensorial Functions

By the term *tensorial function* we mean a tensor-valued function of a tensor argument. Since a scalar is a tensor of order zero, this includes the relatively simple case of a function taking scalars into second-order tensors. This situation can be handled using the approach of the preceding section: if $\mathbf{F}(t)$

takes values $t \in (a, b)$ into second-order tensor values, we can expand $\mathbf{F}(t)$ in a fixed, time-independent basis $\mathbf{e}_k$ as $\mathbf{F}(t) = f^{mn}(t)\mathbf{e}_m\mathbf{e}_n$ and differentiate componentwise:

$$\frac{d\mathbf{F}(t)}{dt} = \frac{df^{mn}(t)}{dt}\mathbf{e}_m\mathbf{e}_n.$$

We could directly differentiate $\mathbf{F}$ even for the case when the $\mathbf{e}_k$ also depend on t (using the rule for differentiating a product of three functions). But having in mind the problems of objectivity of the results, analogy with the first differential, etc., we need to introduce the notion of Gâteaux derivative.

Scalar-valued function of a tensor argument

Let $f = f(\mathbf{X})$ map from the set of second-order tensors $\mathbf{X}$ to scalars. If there exists a second-order tensor, denoted $f_{,\mathbf{X}}$, such that for any second-order tensor $\mathbf{B}$ the relation

$$f_{,\mathbf{X}} \odot \mathbf{B} = \frac{d}{d\varepsilon} f(\mathbf{X} + \varepsilon\mathbf{B})\bigg|_{\varepsilon=0} \tag{2.62}$$

holds, then $f_{,\mathbf{X}}$ is called the derivative of f with respect to its tensor argument $\mathbf{X}$. We note that

$$\frac{d}{d\varepsilon} f(\mathbf{X} + \varepsilon\mathbf{B})\bigg|_{\varepsilon=0} = \lim_{\varepsilon\to 0}\frac{f(\mathbf{X} + \varepsilon\mathbf{B}) - f(\mathbf{X})}{\varepsilon},$$

the quantity on the right being known as the *Gâteaux differential* of f. It can be introduced in the same manner for a tensor-valued function of tensor argument $\mathbf{F}$:

$$\frac{d}{d\varepsilon}\mathbf{F}(\mathbf{X} + \varepsilon\mathbf{B})\bigg|_{\varepsilon=0} = \lim_{\varepsilon\to 0}\frac{\mathbf{F}(\mathbf{X} + \varepsilon\mathbf{B}) - \mathbf{F}(\mathbf{X})}{\varepsilon} = \mathbf{F}_{,\mathbf{X}} \odot \mathbf{B}.$$

But we continue with a scalar-valued function f.

Temporarily denoting $\mathbf{B}$ by $d\mathbf{X}$ and $\frac{d}{d\varepsilon} f(\mathbf{X} + \varepsilon\mathbf{B})\big|_{\varepsilon=0}$ by df, we have

$$df = f_{,\mathbf{X}} \odot d\mathbf{X}$$

and can see the analogy with the first differential from ordinary calculus:

$$df = f'(x)\,dx.$$

Let us fix a basis $\mathbf{e}_k$. In this basis, $\mathbf{X} = x^{km}\mathbf{e}_k\mathbf{e}_m$. Suppose $f(\mathbf{X})$ is a given function. We will write interchangeably $f(\mathbf{X})$ and $f(x^{km})$ (remembering that $f(\mathbf{X})$ does not depend on a basis, but that $f(x^{km}) = f(\mathbf{X})$ only if the basis is $\mathbf{e}_k$). We then apply the definition (2.62) to find $f_{,\mathbf{X}}$. The chain rule gives

$$\begin{aligned}\frac{d}{d\varepsilon} f(\mathbf{X}+\varepsilon\mathbf{B})\bigg|_{\varepsilon=0} &= \frac{d}{d\varepsilon} f(x^{mn}+\varepsilon b^{mn})\bigg|_{\varepsilon=0} = \frac{\partial f}{\partial x^{mn}}\, b^{mn} \\ &= \frac{\partial f}{\partial x^{mn}}\, \delta_u^m \delta_v^n\, b^{uv} = \frac{\partial f}{\partial x^{mn}}\, (\mathbf{e}^m \cdot \mathbf{e}_u)(\mathbf{e}^n \cdot \mathbf{e}_v)\, b^{uv} \\ &= \frac{\partial f}{\partial x^{mn}}\, \mathbf{e}^m\mathbf{e}^n \odot b^{uv}\mathbf{e}_u\mathbf{e}_v = \frac{\partial f}{\partial x^{mn}}\, \mathbf{e}^m\mathbf{e}^n \odot \mathbf{B}\end{aligned}$$

and hence by (2.62) the derivative is[4]

$$f_{,\mathbf{X}} = \frac{\partial f}{\partial x^{mn}}\, \mathbf{e}^m\mathbf{e}^n. \tag{2.63}$$

In a Cartesian basis $\mathbf{i}_k$ this can be expressed as

$$f_{,\mathbf{X}} = \frac{\partial f}{\partial x_{mn}}\, \mathbf{i}_m\mathbf{i}_n, \tag{2.64}$$

where the x_{mn} are the Cartesian components of $\mathbf{X}$. The notation $df/d\mathbf{X}$ is sometimes used instead of $f_{,\mathbf{X}}$.

As an example of computing $f_{,\mathbf{X}}$ we may take $f(\mathbf{X}) = \mathbf{X} \cdot\cdot\, \mathbf{X}$. In this case the definition (2.62) can be applied directly and compared with the use of (2.64). By writing

$$\begin{aligned}\frac{d}{d\varepsilon} f(\mathbf{X}+\varepsilon\mathbf{B})\bigg|_{\varepsilon=0} &= \frac{d}{d\varepsilon}[(\mathbf{X}+\varepsilon\mathbf{B}) \cdot\cdot\, (\mathbf{X}+\varepsilon\mathbf{B})]\bigg|_{\varepsilon=0} \\ &= \frac{d}{d\varepsilon}[\mathbf{X} \cdot\cdot\, \mathbf{X} + \varepsilon(\mathbf{B} \cdot\cdot\, \mathbf{X} + \mathbf{X} \cdot\cdot\, \mathbf{B}) + \varepsilon^2 \mathbf{B} \cdot\cdot\, \mathbf{B}]\bigg|_{\varepsilon=0} \\ &= \mathbf{B} \cdot\cdot\, \mathbf{X} + \mathbf{X} \cdot\cdot\, \mathbf{B} = \mathbf{X} \cdot\cdot\, \mathbf{B}^T + \mathbf{X}^T \cdot\cdot\, \mathbf{B}^T \\ &= (\mathbf{X}+\mathbf{X}^T) \cdot\cdot\, \mathbf{B}^T = (\mathbf{X}+\mathbf{X}^T) \odot \mathbf{B}\end{aligned}$$

we see that $f_{,\mathbf{X}} = \mathbf{X} + \mathbf{X}^T$. On the other hand, in Cartesian component form we have

$$f(\mathbf{X}) = x_{ij}\mathbf{i}_i\mathbf{i}_j \cdot\cdot\, x_{kl}\mathbf{i}_k\mathbf{i}_l = x_{ij}x_{kl}\delta_{jk}\delta_{il} = x_{ik}x_{ki}$$

and (2.64) yields

$$\begin{aligned}f_{,\mathbf{X}} &= \frac{\partial(x_{ik}x_{ki})}{\partial x_{mn}}\, \mathbf{i}_m\mathbf{i}_n = \left(x_{ik}\frac{\partial x_{ki}}{\partial x_{mn}} + x_{ki}\frac{\partial x_{ik}}{\partial x_{mn}}\right)\mathbf{i}_m\mathbf{i}_n \\ &= (x_{ik}\delta_{km}\delta_{in} + x_{ki}\delta_{im}\delta_{kn})\mathbf{i}_m\mathbf{i}_n = (x_{nm} + x_{mn})\mathbf{i}_m\mathbf{i}_n\end{aligned}$$

[4]Note that here $\mathbf{X}$ and even $\mathbf{e}_k$ may depend on some parameters like t or coordinates, but the resulting formula does not change.

so we get the same result.

Exercise 2.45. Find $f_{,\mathbf{X}}$ if $f(\mathbf{X})$ is (a) $I_1(\mathbf{X})$, (b) $\operatorname{tr}\mathbf{X}^2$, (c) $\operatorname{tr}\mathbf{X}^3$, (d) $I_2(\mathbf{X})$, (e) $I_3(\mathbf{X})$.

The formulas for differentiating the invariants of the strain tensor are used to write down the constitutive equation for a nonlinear elastic isotropic material under finite deformation.

Exercise 2.46. The strain energy of an isotropic elastic medium is a function $f = f(I_1, I_2, I_3)$ of the invariants of the strain tensor $\mathbf{X}$. Use the results of Exercise 2.45 to show that

$$f_{,\mathbf{X}} = \left(\frac{\partial f}{\partial I_1} + I_1\frac{\partial f}{\partial I_2} + I_2\frac{\partial f}{\partial I_3}\right)\mathbf{E} - \left(\frac{\partial f}{\partial I_2} + I_1\frac{\partial f}{\partial I_3}\right)\mathbf{X}^T + \frac{\partial f}{\partial I_3}(\mathbf{X}^T)^2.$$

Vector-valued function of a vector argument

Let $\mathbf{f}(\mathbf{x})$ be a vector-valued function of the vector argument $\mathbf{x}$. If there exists a second-order tensor $\mathbf{f}_{,\mathbf{x}}$ such that the relation

$$\mathbf{f}_{,\mathbf{x}} \odot \mathbf{b} = \frac{d}{d\varepsilon}\mathbf{f}(\mathbf{x}+\varepsilon\mathbf{b})\bigg|_{\varepsilon=0} \tag{2.65}$$

holds for every vector $\mathbf{b}$, then $\mathbf{f}_{,\mathbf{x}}$ is the derivative of $\mathbf{f}(\mathbf{x})$ with respect to $\mathbf{x}$. Fixing a basis $\mathbf{e}_k$, expanding $\mathbf{x} = x^k\mathbf{e}_k$ and agreeing (as in the previous subsection) to identify $f^r(\mathbf{x})$ and $f^r(x^k)$ for this particular choice of basis, we have

$$\begin{aligned}\frac{d}{d\varepsilon}\mathbf{f}(\mathbf{x}+\varepsilon\mathbf{b})\bigg|_{\varepsilon=0} &= \frac{\partial f^r}{\partial x^m}b^m\mathbf{e}_r = \frac{\partial f^r}{\partial x^m}\mathbf{e}_r\delta^m_u b^u = \frac{\partial f^r}{\partial x^m}\mathbf{e}_r(\mathbf{e}^m\cdot\mathbf{e}_u)\,b^u \\ &= \frac{\partial f^r}{\partial x^m}\mathbf{r}_r\mathbf{e}^m\odot b^u\mathbf{e}_u = \frac{\partial f^r}{\partial x^m}\mathbf{e}_r\mathbf{e}^m\odot\mathbf{b}\end{aligned}$$

so that

$$\mathbf{f}_{,\mathbf{x}} = \frac{\partial f^r}{\partial x^m}\mathbf{e}_r\mathbf{e}^m. \tag{2.66}$$

Other choices of component representation of $\mathbf{x}$ and $\mathbf{f}(\mathbf{x})$ yield the expressions

$$\frac{\partial f_r}{\partial x_m}\mathbf{e}^r\mathbf{e}_m, \qquad \frac{\partial f_r}{\partial x^m}\mathbf{e}^r\mathbf{e}^m, \qquad \frac{\partial f^r}{\partial x_m}\mathbf{e}_r\mathbf{e}_m,$$

although these can also be obtained directly from (2.66) through raising and lowering processes with the metric coefficients:

$$\frac{\partial f^r}{\partial x^m}\mathbf{e}_r\mathbf{e}^m = \frac{\partial f_i}{\partial x_j}g^{ir}g_{jm}\mathbf{e}_r\mathbf{e}^m = \frac{\partial f_i}{\partial x_j}\mathbf{e}^i\mathbf{e}_j,$$

for instance. Note that a superscript in the denominator of an expression behaves for this purpose like a subscript in the numerator. Of course, differentiation is often carried out in a Cartesian basis as

$$\mathbf{f}_{,\mathbf{x}} = \frac{\partial f_r}{\partial x_m}\,\mathbf{i}_r\mathbf{i}_m.$$

The notation $d\mathbf{f}/d\mathbf{x}$ is also used.

Tensor-valued function of a tensor argument

Here we consider a function $\mathbf{F}(\mathbf{X})$ acting in the set of second-order tensors. If there is a fourth-order tensor $\mathbf{F}_{,\mathbf{X}}$ satisfying

$$\mathbf{F}_{,\mathbf{X}} \odot \mathbf{B} = \frac{d}{d\varepsilon}\,\mathbf{F}(\mathbf{X} + \varepsilon\mathbf{B})\bigg|_{\varepsilon=0}$$

for every second-order tensor $\mathbf{B}$, then $\mathbf{F}_{,\mathbf{X}}$ is the derivative of $\mathbf{F}(\mathbf{X})$ with respect to its second-order tensor argument. In this case the expression is

$$\mathbf{F}_{,\mathbf{X}} = \frac{\partial f^{rs}}{\partial x^{mn}}\,\mathbf{e}_r\mathbf{e}_s\mathbf{e}^m\mathbf{e}^n. \tag{2.67}$$

Exercise 2.47. Verify (2.67).

Some important examples are as follows:

(a) If $\mathbf{F}(\mathbf{X}) = \mathbf{C} \cdot\cdot\, \mathbf{X}^T$, then $\mathbf{F}_{,\mathbf{X}} = \mathbf{C}$.
(b) If $\mathbf{F}(\mathbf{X}) = \mathbf{X}$, then $\mathbf{F}_{,\mathbf{X}} = \mathbf{I}$.
(c) If $\mathbf{F}(\mathbf{X}) = \mathbf{X}^T$, then $\mathbf{F}_{,\mathbf{X}} = \mathbf{e}_k\mathbf{E}\mathbf{e}^k$.

The notation $d\mathbf{F}/d\mathbf{X}$ is sometimes used instead of $\mathbf{F}_{,\mathbf{X}}$.

On symmetric tensor functions

The principal tensors of linear elasticity, the stress tensor $\boldsymbol{\sigma}$ and strain tensor $\boldsymbol{\varepsilon}$, are symmetric. So we consider the problem of differentiating a tensor-valued function of a symmetric tensorial argument, as this specific case has some peculiarities.

Let us consider the derivative of a scalar-valued function $f(\mathbf{X})$ of a symmetric second-order tensor $\mathbf{X}$. We modify the definition as follows. The derivative $f_{,\mathbf{X}}$ is a second-order symmetric tensor that satisfies the condition

$$\frac{d}{d\varepsilon}\,f(\mathbf{X} + \varepsilon\, d\mathbf{X})\bigg|_{\varepsilon=0} = f_{,\mathbf{X}} \cdot\cdot\, d\mathbf{X} \tag{2.68}$$

for any symmetric tensor $d\mathbf{X}$. The component representation of $f_{,\mathbf{X}}$ in a Cartesian basis is

$$f_{,\mathbf{X}} = \frac{\partial f}{\partial x_{mn}} \mathbf{i}_m \mathbf{i}_n.$$

So we have brought the symmetry of $f_{,\mathbf{X}}$ into the definition:

$$(f_{,\mathbf{X}})^T = f_{,\mathbf{X}}.$$

Why do we require $f_{,\mathbf{X}}$ to be symmetric? The sets of symmetric and antisymmetric second-order tensors are subspaces of the space of all second-order tensors. Let $\mathbf{A}$ be a symmetric tensor and $\mathbf{B}$ an antisymmetric second-order tensor. It is easy to see that $\mathbf{A} \cdot\cdot \mathbf{B} = 0$. So the subspaces of symmetric and antisymmetric tensors are mutually orthogonal. In (2.68), $d\mathbf{X}$ is arbitrary but symmetric. If in this definition we do not require $f_{,\mathbf{X}}$ to be symmetric, then (2.68), holding only for all the symmetric tensors $d\mathbf{X}$, defines $f_{,\mathbf{X}}$ non-uniquely up to an additive term $\mathbf{B}$ that can be any antisymmetric second-order tensor.

Note that the definition of derivative for a function with respect to a symmetric tensor argument is closely related to the problem of representing a linear function acting on the subspace of symmetric tensors, which differs slightly from the general case. For example, when $f(\mathbf{X}) = \mathbf{X} \cdot\cdot \mathbf{X}$ and $\mathbf{X} = \mathbf{X}^T$ we get $f_{,\mathbf{X}} = 2\mathbf{X} \odot \mathbf{B}$. See Problems 2.32, 2.33, and 2.34.

As an example, we find the derivative of the function

$$f(\mathbf{X}) = \mathbf{C} \cdot\cdot \mathbf{X},$$

where $\mathbf{C}$ is a second-order tensor and $\mathbf{X}^T = \mathbf{X}$. By definition,

$$f_{,\mathbf{X}} \cdot\cdot d\mathbf{X} = \mathbf{C} \cdot\cdot d\mathbf{X}$$

for all $d\mathbf{X}$ such that $d\mathbf{X} = d\mathbf{X}^T$. At first glance it seems that $f_{,\mathbf{X}} = \mathbf{C}$, and this does hold for an arbitrary argument $\mathbf{X}$. But for a symmetric argument $\mathbf{X}$ the answer changes. Indeed the derivative must be a symmetric tensor, so $f_{,\mathbf{X}} = \mathbf{C}$ if and only if $\mathbf{C} = \mathbf{C}^T$. Now we consider the case when $\mathbf{C}$ is not symmetric. We said that $f_{,\mathbf{X}}$ must be symmetric. A consequence is that $f_{,\mathbf{X}}$ becomes

$$f_{,\mathbf{X}} = \tfrac{1}{2}(\mathbf{C} + \mathbf{C}^T).$$

This expression is uniquely defined. Indeed, we can represent $\mathbf{C}$ as a sum of symmetric and antisymmetric tensors:

$$\mathbf{C} = \tfrac{1}{2}(\mathbf{C} + \mathbf{C}^T) + \tfrac{1}{2}(\mathbf{C} - \mathbf{C}^T).$$

But $\frac{1}{2}(\mathbf{C} - \mathbf{C}^T) \cdot\cdot\, d\mathbf{X} = \frac{1}{2}(\mathbf{C} - \mathbf{C}) \cdot\cdot\, d\mathbf{X} = 0$ and thus

$$\mathbf{C} \cdot\cdot\, d\mathbf{X} = \tfrac{1}{2}(\mathbf{C} + \mathbf{C}^T) \cdot\cdot\, d\mathbf{X}$$

as claimed.

In a similar fashion, we should modify the definition of the derivative of a tensor-valued function of a second-order symmetric argument. If $\mathbf{F}$ takes values in the set of tensors of order n, the derivative with respect to $\mathbf{X}$ is denoted by $\mathbf{F}_{,\mathbf{X}}$; it takes values in the subset of tensors of order $n+2$ which have to be symmetric in the last two indices and satisfy the equality

$$\mathbf{F}_{,\mathbf{X}} \cdot\cdot\, d\mathbf{X} = \frac{d}{d\varepsilon} \mathbf{F}(\mathbf{X} + \varepsilon\, d\mathbf{X}) \bigg|_{\varepsilon=0}$$

for all symmetric second-order tensors $d\mathbf{X}$. The symmetry of $\mathbf{F}_{,\mathbf{X}}$ in the last two indices means that for the components $F_{ij\ldots pt}$ of $\mathbf{F}_{,\mathbf{X}}$ we have

$$F_{ij\ldots pt} = F_{ij\ldots tp}.$$

Similar changes permit the definition of partial derivative for tensorial functions in many symmetric tensorial arguments.

Exercise 2.48. Let $\mathbf{F}(\mathbf{X}) = \mathbf{C} \cdot\cdot\, \mathbf{X}$, where $\mathbf{C} = c_{mnpt}\mathbf{i}_m\mathbf{i}_n\mathbf{i}_p\mathbf{i}_t$ is a fourth-order tensor and $\mathbf{X}$ is a second-order symmetric tensor. Show that

$$\mathbf{F}_{,\mathbf{X}} = \tfrac{1}{2}(\mathbf{C} + \mathbf{C}'),$$

where $\mathbf{C}'$ is derived from $\mathbf{C}$ by transposing the last two indices: $\mathbf{C}' = c_{mnpt}\mathbf{i}_m\mathbf{i}_n\mathbf{i}_t\mathbf{i}_p$.

Exercise 2.49. Let $f(\mathbf{X}) = \frac{1}{2}\mathbf{X} \cdot\cdot\, \mathbf{C} \cdot\cdot\, \mathbf{X}$ where $\mathbf{C} = c_{mnpt}\mathbf{i}_m\mathbf{i}_n\mathbf{i}_p\mathbf{i}_t$ and $\mathbf{X}$ is a second-order symmetric tensor. Show that

$$f_{,\mathbf{X}} = \tfrac{1}{4}(\mathbf{C} \cdot\cdot\, \mathbf{X} + \mathbf{C}'' \cdot\cdot\, \mathbf{X} + \mathbf{X} \cdot\cdot\, \mathbf{C} + \mathbf{X} \cdot\cdot\, \mathbf{C}'),$$

where $\mathbf{C}'$ is derived from $\mathbf{C}$ by transposing the last two indices of the components of $\mathbf{C}$, i.e., $\mathbf{C}' = c_{mnpt}\mathbf{i}_m\mathbf{i}_n\mathbf{i}_t\mathbf{i}_p$, and $\mathbf{C}''$ by transposing the first two indices of the components, i.e., $\mathbf{C}'' = c_{mnpt}\mathbf{i}_n\mathbf{i}_m\mathbf{i}_p\mathbf{i}_t$.

Exercise 2.50. Let $f(\mathbf{X}) = \frac{1}{2}\mathbf{X} \cdot\cdot\, \mathbf{C} \cdot\cdot\, \mathbf{X}$ for some fourth-order tensor $\mathbf{C}$ and some second-order symmetric tensor $\mathbf{X}$. Suppose $\mathbf{C} = \mathbf{C}' = \mathbf{C}''$ and $\mathbf{C} \cdot\cdot\, \mathbf{X} = \mathbf{X} \cdot\cdot\, \mathbf{C}$ for any symmetric tensor $\mathbf{X}$; in terms of components, this means the equalities $c_{mnpt} = c_{nmpt}$ and $c_{mnpt} = c_{ptmn}$ hold for any sets of indices. Using the solution of Exercise 2.49, show that $f_{,\mathbf{X}} = \mathbf{C} \cdot\cdot\, \mathbf{X}$. Note these are the conditions that can be found for the elastic moduli tensor in any textbook on linear elasticity.

2.9 Tensor Fields

General coordinates

In a general coordinate system the position of a point in space is determined uniquely by three numbers q^1, q^2, q^3. These are referred to as *curvilinear coordinates* if the frame is not Cartesian. See Fig. 2.2.

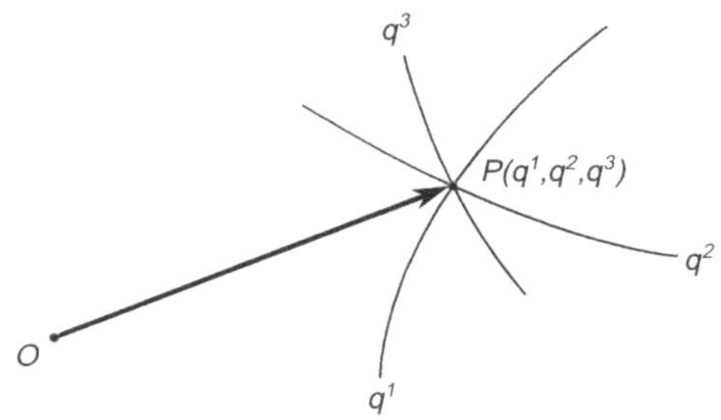

Fig. 2.2 Curvilinear coordinates.

By fixing two of the coordinates and changing the third, we get a space curve called a *coordinate line*. The Cartesian coordinates x^1, x^2, x^3 of a point can be determined through the general curvilinear coordinates by relations of the form

$$x^i = x^i(q^1, q^2, q^3) \qquad (i = 1, 2, 3). \tag{2.69}$$

Except for some set of singular points in space, the correspondence (2.69) is one-to-one. We suppose the functions $x^i = x^i(q^1, q^2, q^3)$ are smooth (continuously differentiable). In this case the local one-to-one correspondence is provided by the requirement that the Jacobian

$$\left| \frac{\partial x^i}{\partial q^j} \right| \neq 0. \tag{2.70}$$

We choose an origin O and locate a point $P(q^1, q^2, q^3)$ with a position vector $\overrightarrow{OP}$ denoted by $\mathbf{r} = \mathbf{r}(q^1, q^2, q^3)$.

When a point moves along a coordinate line, along the q^1 line for instance, the end of its position vector moves along this line and the difference vector

$$\Delta\mathbf{r} = \mathbf{r}(q^1 + \Delta q^1, q^2, q^3) - \mathbf{r}(q^1, q^2, q^3)$$

is directed along the chord (Fig. 2.3). The smaller the value of Δq^1, the closer $\Delta\mathbf{r}$ is to being tangent to the q^1 line. The limit as $\Delta q^1 \to 0$ of the

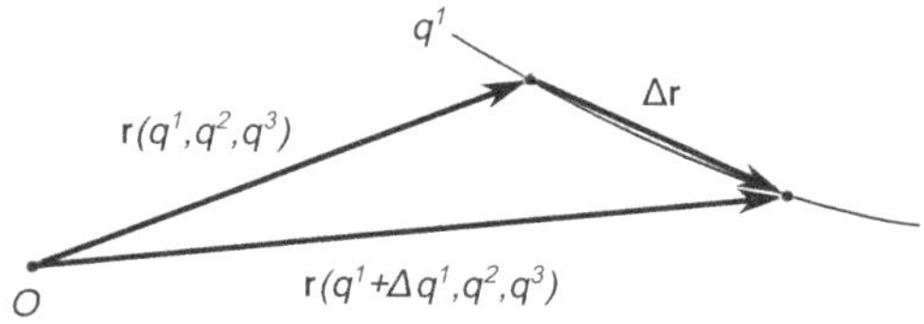

Fig. 2.3 Generation of a local tangent vector to a coordinate line.

ratio $\Delta\mathbf{r}/\Delta q^1$ is a vector

$$\frac{\partial\mathbf{r}(q^1,q^2,q^3)}{\partial q^1} = \mathbf{r}_1$$

tangent to the q^1 coordinate line. So we construct a set

$$\mathbf{r}_i = \frac{\partial\mathbf{r}}{\partial q^i} \qquad (i = 1,2,3)$$

of vectors tangent to the coordinate lines. If the coordinate point (q^1,q^2,q^3) is not singular, the vectors $\mathbf{r}_i$ are non-coplanar and therefore constitute a frame triad of basis vectors associated with the coordinates q^i. Supposing its positiveness or renumbering the vectors if necessary, we get the mixed product $\mathbf{r}_1 \cdot \mathbf{r}_2 \times \mathbf{r}_3 > 0$, the absolute value of which is the volume of the frame parallelepiped. This is also the expression for the Jacobian (2.70). Let us denote it

$$\sqrt{g} = \mathbf{r}_1 \cdot \mathbf{r}_2 \times \mathbf{r}_3 = \left|\frac{\partial x^i}{\partial q^j}\right|.$$

Unlike the previous discussion in which all frame vectors were constant, we now deal with frame vectors that change from point to point. However, at each point we can repeat our prior reasoning. In particular, let us introduce the reciprocal basis $\mathbf{r}^i$ by the relation

$$\mathbf{r}^i \cdot \mathbf{r}_j = \delta^i_j.$$

The metric coefficients

$$g_{ij} = \mathbf{r}_i \cdot \mathbf{r}_j, \qquad g^{ij} = \mathbf{r}^i \cdot \mathbf{r}^j, \qquad g^j_i = \mathbf{r}_i \cdot \mathbf{r}^j = \delta^j_i,$$

are the components of the metric tensor $\mathbf{E}$ at each point in space. Dotting both sides of the identity $\mathbf{r}_i = (\mathbf{r}_j \cdot \mathbf{r}_i)\mathbf{r}^j$, with $\mathbf{r}^k$, we get

$$g_{ij}g^{jk} = \delta^k_i.$$

See Table 2.6 for a summary of the notation introduced above.

Table 2.6 Quantities associated with curvilinear coordinate system.

curvilinear coordinates	(q^1, q^2, q^3), $\frac{\partial q^i}{\partial q^k} = \delta^i_k$
position vector of a point in space	$\mathbf{r}(q^1, q^2, q^3)$
vector tangent to q^i coordinate line	$\mathbf{r}_i = \partial \mathbf{r}/\partial q^i$
frame triad at a point	$(\mathbf{r}_1, \mathbf{r}_2, \mathbf{r}_3)$
volume of frame parallelepiped	$\sqrt{g} = \mathbf{r}_1 \cdot \mathbf{r}_2 \times \mathbf{r}_3$
definition of reciprocal basis at a point	$\mathbf{r}^i \cdot \mathbf{r}_j = \delta^i_j$
metric coefficients	$g_{ij} = \mathbf{r}_i \cdot \mathbf{r}_j$, $g^{ij} = \mathbf{r}^i \cdot \mathbf{r}^j$, $g^j_i = \mathbf{r}_i \cdot \mathbf{r}^j = \delta^j_i$

Components of a vector; physical components

A fixed vector $\mathbf{f}$ at a point (q^1, q^2, q^3) may be resolved into components relative to either of the frames $\{\mathbf{r}_i\}$ or $\{\mathbf{r}^i\}$:

$$\mathbf{f} = f^i \mathbf{r}_i = f_i \mathbf{r}^i$$

where, of course,

$$f^i = \mathbf{f} \cdot \mathbf{r}^i, \qquad f_i = \mathbf{f} \cdot \mathbf{r}_i.$$

It is worth noting carefully that both the magnitudes and the physical dimensions of the frame vectors $\mathbf{r}_i$ and $\mathbf{r}^i$ do depend on the nature of the coordinates q^i. This means that the contravariant components f^i and covariant components f_i of $\mathbf{f}$ may not represent the projections of $\mathbf{f}$ along their corresponding basis vectors or even carry the physical units of $\mathbf{f}$.[5]

It is possible to normalize the frame $\{\mathbf{r}_i\}$ and obtain a related frame $\{\mathbf{r}_{(i)}\}$ as

$$\mathbf{r}_{(1)} = \frac{\mathbf{r}_1}{\sqrt{\mathbf{r}_1 \cdot \mathbf{r}_1}} = \frac{\mathbf{r}_1}{\sqrt{g_{11}}}, \quad \mathbf{r}_{(2)} = \frac{\mathbf{r}_2}{\sqrt{g_{22}}}, \quad \mathbf{r}_{(3)} = \frac{\mathbf{r}_3}{\sqrt{g_{33}}},$$

expanding $\mathbf{f}$ in this new frame at point (q^1, q^2, q^3):

$$\mathbf{f} = f^{(i)} \mathbf{r}_{(i)}, \qquad f^{(1)} = \sqrt{g_{11}}\, f^1, \text{ etc.}$$

Analogously, we can normalize the frame $\{\mathbf{r}^i\}$ and obtain a related frame

[5]The theory of tensors (at least the portion devoted to the basis vectors in space, which are used for all other quantities) is formulated in unitless form; otherwise we can get spatial basis vectors $\mathbf{r}_k$ with disparate units so they cannot belong to the same vector space. This happens, for example, in the polar coordinate system where $\mathbf{r}_1 = \partial \mathbf{r}/\partial r$ is unitless and $\mathbf{r}_2 = \partial \mathbf{r}/\partial \phi$ has units of length. The other vector quantities can have physical units, although classical tensor theory operates only with unitless quantities.

$\{\mathbf{r}^{(i)}\}$ as

$$\mathbf{r}^{(1)} = \frac{\mathbf{r}^1}{\sqrt{\mathbf{r}^1 \cdot \mathbf{r}^1}} = \frac{\mathbf{r}^1}{\sqrt{g^{11}}}, \quad \mathbf{r}^{(2)} = \frac{\mathbf{r}^2}{\sqrt{g^{22}}}, \quad \mathbf{r}^{(3)} = \frac{\mathbf{r}^3}{\sqrt{g^{33}}},$$

and then expand $\mathbf{f}$ as

$$\mathbf{f} = f_{(i)}\mathbf{r}^{(i)}, \qquad f_{(1)} = \sqrt{g^{11}}\, f_1, \text{ etc.}$$

Coordinate transformation

Consider a change of curvilinear coordinates q^k to other coordinates $q^{j'}$:

$$q^{j'} = q^{j'}(q^1, q^2, q^3), \quad q^k = q^k(q^{1'}, q^{2'}, q^{3'}) \qquad (j, k = 1, 2, 3),$$

where the dependence must be one-to-one except for some isolated singular points or lines. Because

$$\mathbf{r}_i = \frac{\partial \mathbf{r}}{\partial q^i} = \frac{\partial \mathbf{r}}{\partial q^{j'}} \frac{\partial q^{j'}}{\partial q^i} = \mathbf{r}_{j'} \frac{\partial q^{j'}}{\partial q^i}$$

we can write

$$\mathbf{r}_i = A_i^{j'} \mathbf{r}_{j'}, \qquad A_i^{j'} = \frac{\partial q^{j'}}{\partial q^i}, \tag{2.71}$$

to describe the change of frame. For the inverse transformation

$$\mathbf{r}_{i'} = A_{i'}^j \mathbf{r}_j, \qquad A_{i'}^j = \frac{\partial q^j}{\partial q^{i'}}. \tag{2.72}$$

Hence for the components of a vector field $\mathbf{f}$,

$$f^{i'} = A_j^{i'} f^j, \qquad f_{i'} = A_{i'}^j f_j, \qquad f^i = A_{j'}^i f^{j'}, \qquad f_i = A_i^{j'} f_{j'}.$$

The $f^{i'}$ are still termed contravariant components of $\mathbf{f}$, whereas the $f_{i'}$ are still termed covariant components — the main difference is that now $A_i^{j'}$ and $A_{i'}^j$ can change from point to point in space. The transformation laws for the components of higher-order tensor fields are written similarly (cf., Table 2.2 on page 17).

Derivative operations

Differentials and the nabla operator

Let us consider the infinitesimal vector extending from point (q^1, q^2, q^3) to point $(q^1 + dq^1, q^2 + dq^2, q^3 + dq^3)$, denoted by $d\mathbf{r}$:

$$d\mathbf{r} = \frac{\partial \mathbf{r}}{\partial q^i}\, dq^i = \mathbf{r}_i\, dq^i.$$

Hence we can get dq^i as

$$dq^i = \mathbf{r}^i \cdot d\mathbf{r} = d\mathbf{r} \cdot \mathbf{r}^i. \tag{2.73}$$

(Here there is no summation over i. Note that we only introduce spatial coordinates having superscripts. However, for Cartesian frames it is common to see subscripts used exclusively.) The length of this infinitesimal vector is defined by

$$(ds)^2 = d\mathbf{r} \cdot d\mathbf{r} = \mathbf{r}_i \, dq^i \cdot \mathbf{r}_j \, dq^j = g_{ij} \, dq^i \, dq^j .$$

On the right we have a quadratic form with respect to the variables dq^i; the coefficients of this quadratic form are the covariant components of the metric tensor.

Exercise 2.51. Write out an expression for the volume element in terms of the dq^i.

Let $f(q^1, q^2, q^3)$ be a scalar differentiable function of the variables q^i. Its differential is

$$df(q^1, q^2, q^3) = \frac{\partial f(q^1, q^2, q^3)}{\partial q^i} \, dq^i .$$

Here and in similar situations the i in the denominator stands as if it were in the lower position, and so we must sum over i. Using (2.73) we can write

$$df = \frac{\partial f}{\partial q^i} \, dq^i = \mathbf{r}^i \frac{\partial f}{\partial q^i} \cdot d\mathbf{r} = \nabla f \cdot d\mathbf{r}$$

where

$$\nabla f = \mathbf{r}^i \frac{\partial f}{\partial q^i}$$

and ∇ is *nabla* or *gradient operator* given by

$$\nabla = \mathbf{r}^i \frac{\partial}{\partial q^i} .$$

Observe that the partial derivatives $\partial f / \partial q^i$ are the covariant components of ∇f.

Exercise 2.52. (a) Show that the unit normal to the surface $\varphi(q^1, q^2, q^3) = c =$ constant is given by

$$\mathbf{n} = \frac{g^{ij} \dfrac{\partial \varphi}{\partial q^i}}{\sqrt{g^{mn} \dfrac{\partial \varphi}{\partial q^m} \dfrac{\partial \varphi}{\partial q^n}}} \, \mathbf{r}_j .$$

(b) Show that the angle at a point of intersection of the surfaces $\varphi(q^1, q^2, q^3) = c_1$ and $\psi(q^1, q^2, q^3) = c_2$ is given by

$$\cos\theta = \frac{g^{ij}\dfrac{\partial\varphi}{\partial q^i}\dfrac{\partial\psi}{\partial q^j}}{\sqrt{g^{mn}\dfrac{\partial\varphi}{\partial q^m}\dfrac{\partial\varphi}{\partial q^n}g^{rt}\dfrac{\partial\psi}{\partial q^r}\dfrac{\partial\psi}{\partial q^t}}}$$

(c) Show that the angle between the coordinate surfaces $q^1 = c_1, q^2 = c_2$ is given by

$$\cos\theta_{12} = \frac{g^{12}}{\sqrt{g^{11}g^{22}}}.$$

(d) Derive the condition for orthogonality of surfaces

$$g^{ij}\frac{\partial\varphi}{\partial q^i}\frac{\partial\psi}{\partial q^j} = 0.$$

Gradient of a vector field

For the differential of a vector function $\mathbf{f} = \mathbf{f}(q^1, q^2, q^3)$, we have

$$d\mathbf{f} = \frac{\partial\mathbf{f}}{\partial q^i}\,dq^i = \frac{\partial\mathbf{f}}{\partial q^i}\,(\mathbf{r}^i \cdot d\mathbf{r}) = (d\mathbf{r}\cdot\mathbf{r}^i)\,\frac{\partial\mathbf{f}}{\partial q^i} = d\mathbf{r}\cdot\mathbf{r}^i\,\frac{\partial\mathbf{f}}{\partial q^i}$$

which we can represent as

$$d\mathbf{f} = d\mathbf{r}\cdot\nabla\mathbf{f} = \nabla\mathbf{f}^T\cdot d\mathbf{r}, \qquad \nabla\mathbf{f} = \mathbf{r}^i\frac{\partial\mathbf{f}}{\partial q^i}.$$

We observe that $\mathbf{r}^i\partial\mathbf{f}/\partial q^i$ is the sum of the tensor products of $\mathbf{r}^i$ by $\partial\mathbf{f}/\partial q^i$ and hence it is a second-order tensor. The quantity $\nabla\mathbf{f}$ is known as the *gradient* of $\mathbf{f}$. Sometimes $\nabla\mathbf{f}^T$ is called the gradient of $\mathbf{f}$; it is also called the derivative of $\mathbf{f}$ in the direction of $\mathbf{r}$ and denoted as

$$\frac{d\mathbf{f}}{d\mathbf{r}} = \nabla\mathbf{f}^T.$$

An application of the gradient to a tensor $\mathbf{A}$ of any order yields a new tensor $\nabla\mathbf{A}$ whose order is one higher than that of the original tensor $\mathbf{A}$.

Divergence and rotation

For a pair of vectors, we introduced two types of multiplication: the dot and cross product operations. We can apply these to the pair consisting of the nabla operator (which is regarded as a formal vector) and a vector function $\mathbf{f} = \mathbf{f}(q^1, q^2, q^3)$. In this way we get two operations: the *divergence*

$$\operatorname{div} \mathbf{f} = \nabla \cdot \mathbf{f} = \mathbf{r}^i \frac{\partial}{\partial q^i} \cdot \mathbf{f} = \mathbf{r}^i \cdot \frac{\partial \mathbf{f}}{\partial q^i},$$

and the *rotation*

$$\operatorname{rot} \mathbf{f} = \nabla \times \mathbf{f} = \mathbf{r}^i \frac{\partial}{\partial q^i} \times \mathbf{f} = \mathbf{r}^i \times \frac{\partial \mathbf{f}}{\partial q^i}.$$

Some authors write $\mathbf{f} = \nabla \times \mathbf{f}$ instead of $\operatorname{rot} \mathbf{f}$ and refer to *curl* instead of rotation.

The vector

$$\boldsymbol{\omega} = \tfrac{1}{2} \operatorname{rot} \mathbf{f}$$

is called the linear *vector of rotation* of $\mathbf{f}$. In terms of this we can introduce a tensor $\boldsymbol{\Omega}$ as

$$\boldsymbol{\Omega} = \mathbf{E} \times \boldsymbol{\omega}.$$

This tensor is antisymmetric (the reader should verify this) and is called the *tensor of spin*. It can be shown that

$$\boldsymbol{\Omega} = \tfrac{1}{2} \left(\nabla \mathbf{f}^T - \nabla \mathbf{f} \right).$$

Similarly we can introduce the divergence and rotation operations for a second-order tensor $\mathbf{A}$:

$$\nabla \cdot \mathbf{A} = \mathbf{r}^i \cdot \frac{\partial \mathbf{A}}{\partial q^i}, \qquad \nabla \times \mathbf{A} = \mathbf{r}^i \times \frac{\partial \mathbf{A}}{\partial q^i}.$$

Applying the nabla operator to the position vector, we obtain

$$\nabla \mathbf{r} = \mathbf{r}^i \frac{\partial \mathbf{r}}{\partial q^i} = \mathbf{r}^i \mathbf{r}_i = \mathbf{E},$$

$$\nabla \cdot \mathbf{r} = \mathbf{r}^i \cdot \frac{\partial \mathbf{r}}{\partial q^i} = \mathbf{r}^i \cdot \mathbf{r}_i = 3,$$

$$\nabla \times \mathbf{r} = \mathbf{r}^i \times \frac{\partial \mathbf{r}}{\partial q^i} = \mathbf{r}^i \times \mathbf{r}_i = \mathbf{0}.$$

Exercise 2.53. Calculate $\nabla \mathbf{E}$, $\nabla \cdot \mathbf{E}$, $\nabla \times \mathbf{E}$, $\nabla \boldsymbol{\mathcal{E}}$, $\nabla \cdot \boldsymbol{\mathcal{E}}$, and $\nabla \times \boldsymbol{\mathcal{E}}$.

Covariant differentiation

To differentiate a vector function $\mathbf{f}(x_1, x_2, x_3) = f^k(x_1, x_2, x_3)\mathbf{i}_k$ with respect to a Cartesian variable x_i, component-wise differentiation is sufficient:

$$\frac{\partial \mathbf{f}}{\partial x_i} = \frac{\partial f^k}{\partial x_i}\mathbf{i}_k. \tag{2.74}$$

However, this is not true for a function

$$\mathbf{f}(q^1, q^2, q^3) = f^i(q^1, q^2, q^3)\,\mathbf{r}_i$$

expressed in curvilinear coordinates.

Exercise 2.54. Discuss the case of a constant function $\mathbf{f}(q^1, q^2, q^3) = \mathbf{c}$.

So we must account for the fact that the frame vectors $\mathbf{r}_i$ depend on the q^k and the local basis $\{\mathbf{r}_i\}$ can change as we move along the q^k coordinate curve. The product rule gives

$$\frac{\partial \mathbf{f}}{\partial q^k} = \frac{\partial (f^i \mathbf{r}_i)}{\partial q^k} = \frac{\partial f^i}{\partial q^k}\mathbf{r}_i + f^i \frac{\partial \mathbf{r}_i}{\partial q^k}. \tag{2.75}$$

It is convenient to express this in a form reminiscent of (2.74). We will write that form as

$$\frac{\partial \mathbf{f}}{\partial q^k} = \mathbf{r}_i \nabla_k f^i$$

and call the quantities $\nabla_k f^i$ the *covariant derivatives* of the f^i with respect to the coordinate q^k.

The second term in (2.75) requires us to investigate the derivatives of the frame vectors $\mathbf{r}_i$. First note that

$$\frac{\partial \mathbf{r}_i}{\partial q^j} = \frac{\partial}{\partial q^j}\left(\frac{\partial \mathbf{r}}{\partial q^i}\right) = \frac{\partial^2 \mathbf{r}}{\partial q^j \partial q^i} = \frac{\partial^2 \mathbf{r}}{\partial q^i \partial q^j} = \frac{\partial}{\partial q^i}\left(\frac{\partial \mathbf{r}}{\partial q^j}\right) = \frac{\partial \mathbf{r}_j}{\partial q^i}. \tag{2.76}$$

Now since the partial derivative of a frame vector is also a vector, it can be expanded in the same frame $\{\mathbf{r}_i\}$. The kth contravariant component of $\partial \mathbf{r}_i / \partial q^j$, denoted Γ^k_{ij}, is called the *Christoffel symbol of the second kind* and we write

$$\frac{\partial \mathbf{r}_i}{\partial q^j} = \Gamma^k_{ij}\mathbf{r}_k. \tag{2.77}$$

By (2.76) we have symmetry in the subscripts:

$$\Gamma^k_{ij} = \Gamma^k_{ji}. \tag{2.78}$$

Hence

$$\frac{\partial \mathbf{f}}{\partial q^k} = \frac{\partial f^i}{\partial q^k}\mathbf{r}_i + \Gamma^j_{ki} f^i \mathbf{r}_j = \frac{\partial f^i}{\partial q^k}\mathbf{r}_i + \Gamma^i_{kt} f^t \mathbf{r}_i = \left(\frac{\partial f^i}{\partial q^k} + \Gamma^i_{kt} f^t\right)\mathbf{r}_i$$

and we conclude that

$$\nabla_k f^i = \frac{\partial f^i}{\partial q^k} + \Gamma^i_{kt} f^t.$$

Exercise 2.55. How many distinct values of the Γ^k_{ij} can exist in three space dimensions?

In Cartesian frames the Christoffel symbols are zero, but in curvilinear frames we will have to compute them. Let us denote

$$\mathbf{r}_{ij} = \frac{\partial \mathbf{r}_i}{\partial q^j} = \frac{\partial^2 \mathbf{r}}{\partial q^i \partial q^j}.$$

Relation (2.77) can be written as

$$\mathbf{r}_{ij} = \Gamma^k_{ij}\mathbf{r}_k \tag{2.79}$$

from which it follows that

$$\mathbf{r}_{ij}\cdot\mathbf{r}_t = \Gamma^k_{ij}\mathbf{r}_k\cdot\mathbf{r}_t = \Gamma^k_{ij} g_{kt}.$$

The left-hand side of this can be expressed in terms of the components of the metric tensor. Indeed

$$\frac{\partial g_{it}}{\partial q^j} = \frac{\partial(\mathbf{r}_i\cdot\mathbf{r}_t)}{\partial q^j} = \mathbf{r}_{ij}\cdot\mathbf{r}_t + \mathbf{r}_i\cdot\mathbf{r}_{tj}. \tag{2.80}$$

Similarly

$$\frac{\partial g_{ji}}{\partial q^t} = \frac{\partial(\mathbf{r}_j\cdot\mathbf{r}_i)}{\partial q^t} = \mathbf{r}_{jt}\cdot\mathbf{r}_i + \mathbf{r}_j\cdot\mathbf{r}_{it}, \tag{2.81}$$

$$\frac{\partial g_{tj}}{\partial q^i} = \frac{\partial(\mathbf{r}_t\cdot\mathbf{r}_j)}{\partial q^i} = \mathbf{r}_{ti}\cdot\mathbf{r}_j + \mathbf{r}_t\cdot\mathbf{r}_{ji}. \tag{2.82}$$

We obtain $\mathbf{r}_{ij}\cdot\mathbf{r}_t$ by subtracting (2.81) from the sum of (2.80) and (2.82) and dividing by 2:

$$\mathbf{r}_{ij}\cdot\mathbf{r}_t = \frac{1}{2}\left(\frac{\partial g_{it}}{\partial q^j} + \frac{\partial g_{tj}}{\partial q^i} - \frac{\partial g_{ji}}{\partial q^t}\right). \tag{2.83}$$

We now introduce a new symbol Γ_{ijt} for $\mathbf{r}_{ij} \cdot \mathbf{r}_t$, known as the *Christoffel symbol of the first kind.* Clearly $\Gamma_{ijk} = \Gamma_{jik}$. Thus we have[6]

$$\Gamma^k_{ij} g_{kt} = \Gamma_{ijt} \tag{2.84}$$

and by applying g^{st} to both sides

$$\Gamma^k_{ij} = g^{kt}\Gamma_{ijt}. \tag{2.85}$$

Finally

$$0 = \frac{\partial \delta^i_j}{\partial q^t} = \frac{\partial(\mathbf{r}^i \cdot \mathbf{r}_j)}{\partial q^t} = \frac{\partial \mathbf{r}^i}{\partial q^t} \cdot \mathbf{r}_j + \mathbf{r}^i \cdot \mathbf{r}_{jt} = \frac{\partial \mathbf{r}^i}{\partial q^t} \cdot \mathbf{r}_j + \mathbf{r}^i \cdot \Gamma^k_{jt}\mathbf{r}_k,$$

so

$$\frac{\partial \mathbf{r}^i}{\partial q^t} \cdot \mathbf{r}_j = -\mathbf{r}^i \cdot \Gamma^k_{jt}\mathbf{r}_k = -\Gamma^k_{jt}\delta^i_k = -\Gamma^i_{jt}$$

and we have

$$\frac{\partial \mathbf{r}^i}{\partial q^t} = -\Gamma^i_{jt}\mathbf{r}^j. \tag{2.86}$$

A convenient summary of the Christoffel symbols is given in Fig. 2.4.

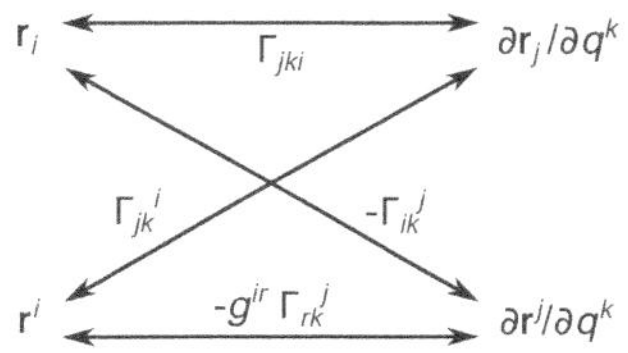

Fig. 2.4 The Christoffel symbols. The values near the arrows equal the respective dot products of the vectors connected by the arrows. (After [Young (1993)].)

Let us examine how the Christoffel symbols behave under coordinate transformation. Suppose the new coordinates $q^{i'}$ are determined by the old coordinates q^i through relations of the form $q^{i'} = q^{i'}(q^1, q^2, q^3)$, and refer back to (2.71)–(2.72). Now (2.79) applied in the new system gives

$$\Gamma^{k'}_{i'j'}\mathbf{r}_{k'} = \mathbf{r}_{i'j'}. \tag{2.87}$$

[6]The appearance of index lowering, while a convenient manipulative aid, can increase the risk that the Christoffel symbols will be mistaken for the components of a tensor. Hence alternate notation for them is often preferred. Some authors use the symbol $\left\{ {k \atop ij} \right\}$ instead of Γ^k_{ij}, and the symbol $[ij, k]$ instead of Γ_{ijk}.

But

$$\mathbf{r}_{i'j'} = \frac{\partial \mathbf{r}_{i'}}{\partial q^{j'}} = \frac{\partial q^m}{\partial q^{j'}} \frac{\partial \mathbf{r}_{i'}}{\partial q^m} = A^m_{j'} \frac{\partial (A^n_{i'} \mathbf{r}_n)}{\partial q^m},$$

and the product rule gives

$$\mathbf{r}_{i'j'} = A^m_{j'} \left(\frac{\partial A^n_{i'}}{\partial q^m} \mathbf{r}_n + A^n_{i'} \frac{\partial \mathbf{r}_n}{\partial q^m} \right) = A^m_{j'} \frac{\partial A^n_{i'}}{\partial q^m} \mathbf{r}_n + A^m_{j'} A^n_{i'} \Gamma^p_{nm} \mathbf{r}_p$$

$$= A^m_{j'} \frac{\partial A^n_{i'}}{\partial q^m} A^{k'}_n \mathbf{r}_{k'} + A^m_{j'} A^n_{i'} \Gamma^p_{nm} A^{k'}_p \mathbf{r}_{k'}.$$

Comparison with (2.87) shows that

$$\Gamma^{k'}_{i'j'} = A^m_{j'} \frac{\partial A^n_{i'}}{\partial q^m} A^{k'}_n + A^n_{i'} A^m_{j'} A^{k'}_p \Gamma^p_{nm}.$$

The presence of the first term on the right means that the Christoffel symbols do not constitute the components of a third-order tensor.

Exercise 2.56. Find the Christoffel symbols for the system of plane elliptic coordinates (u, v) defined by $x = c \cosh u \cos v$ and $y = c \sinh u \sin v$ for a constant c.

We can also introduce covariant differentiation in terms of the covariant components of a vector field. The product rule yields

$$\frac{\partial \mathbf{f}}{\partial q^k} = \frac{\partial (f_i \mathbf{r}^i)}{\partial q^k} = \frac{\partial f_i}{\partial q^k} \mathbf{r}^i + f_i \frac{\partial \mathbf{r}^i}{\partial q^k}$$

so that by (2.86)

$$\frac{\partial \mathbf{f}}{\partial q^k} = \frac{\partial f_i}{\partial q^k} \mathbf{r}^i - f_i \Gamma^i_{kt} \mathbf{r}^t = \left(\frac{\partial f_i}{\partial q^k} - \Gamma^j_{ki} f_j \right) \mathbf{r}^i \equiv \nabla_k f_i$$

where $\nabla_k f_i$ is the covariant derivative[7] of the covariant component f_i of $\mathbf{f}$. Because

$$\nabla \mathbf{f} = \mathbf{r}^i \frac{\partial (f_j \mathbf{r}^j)}{\partial q^i} = \mathbf{r}^i (\nabla_i f_j) \mathbf{r}^j = \mathbf{r}^i \mathbf{r}^j \nabla_i f_j,$$

(note that we place $\nabla_i f_j$ to the right to emphasize that ∇_i should not be applied to the basis vectors) we see that $\nabla_i f_j$ is a covariant component of $\nabla \mathbf{f}$. Similarly,

$$\nabla \mathbf{f} = \mathbf{r}^i \frac{\partial (f^j \mathbf{r}_j)}{\partial q^i} = \mathbf{r}^i \mathbf{r}_j \nabla_i f^j$$

[7]Some authors use a comma notation for covariant differentiation, denoting $\nabla_k f^i$ and $\nabla_k f_i$ by $f^i_{,k}$ and $f_{i,k}$ respectively.

shows that $\nabla_i f^j$ is a mixed component of $\nabla\mathbf{f}$. See Table 2.7 for a summary of some important relationships.

Table 2.7 Covariant differentiation and the Christoffel symbols.

covariant differentiation of scalar f	$\nabla_k f \equiv \frac{\partial f}{\partial q^k}$
covariant differentiation of vector $\mathbf{f}$	$\frac{\partial \mathbf{f}}{\partial q^k} = \mathbf{r}_i \nabla_k f^i = \mathbf{r}^i \nabla_k f_i$
covariant derivative, contravariant component	$\nabla_k f^i = \frac{\partial f^i}{\partial q^k} + \Gamma^i_{kt} f^t$
covariant derivative, covariant component	$\nabla_k f_i = \frac{\partial f_i}{\partial q^k} - \Gamma^j_{ki} f_j$
Christoffel symbol of second kind	$\frac{\partial \mathbf{r}_i}{\partial q^j} = \Gamma^k_{ij}\mathbf{r}_k, \quad \frac{\partial \mathbf{r}^i}{\partial q^k} = -\Gamma^i_{jk}\mathbf{r}^j$
Christoffel symbol of first kind	$\Gamma_{ijt} = \frac{\partial \mathbf{r}_i}{\partial q^j} \cdot \mathbf{r}_t$
symmetry	$\Gamma^k_{ij} = \Gamma^k_{ji}, \quad \Gamma_{ijk} = \Gamma_{jik}$
"index raising/lowering"	$\Gamma^k_{ij} g_{kt} = \Gamma_{ijt}, \quad \Gamma^k_{ij} = g^{kt}\Gamma_{ijt}$
Γ_{ijt} in terms of metric coefficients	$\Gamma_{ijt} = \frac{1}{2}\left(\frac{\partial g_{it}}{\partial q^j} + \frac{\partial g_{tj}}{\partial q^i} - \frac{\partial g_{ji}}{\partial q^t}\right)$
Γ^k_{ij} in terms of metric coefficients	$\Gamma^k_{ij} = \frac{g^{kl}}{2}\left(\frac{\partial g_{il}}{\partial x^j} + \frac{\partial g_{jl}}{\partial x^i} - \frac{\partial g_{ij}}{\partial x^l}\right)$

Exercise 2.57. Express the strain tensor $\frac{1}{2}[\nabla\mathbf{f} + (\nabla\mathbf{f})^T]$ in curvilinear coordinates.

Now let us consider the covariant derivative of a second-order tensor. The partial derivative of $\mathbf{A}$ is

$$\begin{aligned}\frac{\partial \mathbf{A}}{\partial q^k} = \frac{\partial(a^{ij}\mathbf{r}_i\mathbf{r}_j)}{\partial q^k} &= \frac{\partial a^{ij}}{\partial q^k}\mathbf{r}_i\mathbf{r}_j + a^{ij}(\mathbf{r}_{ik}\mathbf{r}_j + \mathbf{r}_i\mathbf{r}_{jk}) \\ &= \frac{\partial a^{ij}}{\partial q^k}\mathbf{r}_i\mathbf{r}_j + a^{ij}(\Gamma^t_{ik}\mathbf{r}_t\mathbf{r}_j + \mathbf{r}_i\Gamma^t_{jk}\mathbf{r}_t).\end{aligned}$$

Changing dummy indices we get

$$\frac{\partial \mathbf{A}}{\partial q^k} = \left(\frac{\partial a^{ij}}{\partial q^k} + \Gamma^i_{ks}a^{sj} + \Gamma^j_{ks}a^{is}\right)\mathbf{r}_i\mathbf{r}_j.$$

The parenthetical expression is designated as a covariant derivative:

$$\nabla_k a^{ij} = \frac{\partial a^{ij}}{\partial q^k} + \Gamma^i_{ks}a^{sj} + \Gamma^j_{ks}a^{is}.$$

Similarly

$$\begin{aligned}\frac{\partial \mathbf{A}}{\partial q^k} &= \frac{\partial(a_{ij}\mathbf{r}^i\mathbf{r}^j)}{\partial q^k} = \frac{\partial a_{ij}}{\partial q^k}\,\mathbf{r}^i\mathbf{r}^j + a_{ij}\left(\frac{\partial \mathbf{r}^i}{\partial q^k}\,\mathbf{r}^j + \mathbf{r}^i\,\frac{\partial \mathbf{r}^j}{\partial q^k}\right)\\ &= \frac{\partial a_{ij}}{\partial q^k}\,\mathbf{r}^i\mathbf{r}^j - a_{ij}(\Gamma^i_{kt}\mathbf{r}^t\mathbf{r}^j + \mathbf{r}^i\Gamma^j_{kt}\mathbf{r}^t)\\ &= \left(\frac{\partial a_{ij}}{\partial q^k} - \Gamma^s_{ki}a_{sj} - \Gamma^s_{kj}a_{is}\right)\mathbf{r}^i\mathbf{r}^j.\end{aligned}$$

As before we denote

$$\nabla_k a_{ij} = \frac{\partial a_{ij}}{\partial q^k} - \Gamma^s_{ki}a_{sj} - \Gamma^s_{kj}a_{is}.$$

Exercise 2.58. Show that

$$\frac{\partial \mathbf{A}}{\partial q^k} = \mathbf{r}^i\mathbf{r}_j\,\nabla_k a_i^{\ j}, \qquad \nabla_k a_i^{\ j} = \frac{\partial a_i^{\ j}}{\partial q^k} - \Gamma^s_{ki}a_s^{\ j} + \Gamma^j_{ks}a_i^{\ s},$$

$$\frac{\partial \mathbf{A}}{\partial q^k} = \mathbf{r}_i\mathbf{r}^j\,\nabla_k a^i_{\ j}, \qquad \nabla_k a^i_{\ j} = \frac{\partial a^i_{\ j}}{\partial q^k} + \Gamma^i_{ks}a^s_{\ j} - \Gamma^s_{kj}a^i_{\ s}.$$

Exercise 2.59. Show that any of the covariant derivatives of any component of the metric tensor is equal to zero.

Exercise 2.60. Show that the components of the metric tensor behave as constants under covariant differentiation of components:

$$\nabla_k g^{st}a_t = g^{st}\nabla_k a_t, \qquad \nabla_k g_{st}a^t = g_{st}\nabla_k a^t.$$

Differential operations

Here we look further at various differential operations that may be performed on vector and tensor fields. We begin with the rotation of a vector:

$$\nabla \times \mathbf{f} = \mathbf{r}^i \times \frac{\partial \mathbf{f}}{\partial q^i} = \mathbf{r}^i \times \mathbf{r}^j\nabla_i f_j = \epsilon^{ijk}\mathbf{r}_k\left(\frac{\partial f_j}{\partial q^i} - \Gamma^n_{ij}f_n\right). \tag{2.88}$$

Considering the second term in parentheses we note that

$$\epsilon^{ijk}\Gamma^n_{ij} = \epsilon^{ijk}\Gamma^n_{ji} = \epsilon^{jik}\Gamma^n_{ij}$$

by (2.78) and a subsequent renaming of dummy indices. But $\epsilon^{jik} = -\epsilon^{ijk}$, hence $\epsilon^{ijk}\Gamma^n_{ij} = -\epsilon^{ijk}\Gamma^n_{ij}$ so that $\epsilon^{ijk}\Gamma^n_{ij} = 0$. Equation (2.88) is thereby reduced to

$$\nabla \times \mathbf{f} = \mathbf{r}_k\epsilon^{ijk}\frac{\partial f_j}{\partial q^i}.$$

For the divergence of $\mathbf{f}$, we start by writing

$$\nabla \cdot \mathbf{f} = \mathbf{r}^i \cdot \frac{\partial \mathbf{f}}{\partial q^i} = \mathbf{r}^i \cdot \mathbf{r}^j \nabla_i f_j = g^{ij} \nabla_i f_j.$$

By the result of Exercise 2.60 this can be written as

$$\nabla \cdot \mathbf{f} = \nabla_i g^{ij} f_j = \nabla_i f^i = \frac{\partial f^i}{\partial q^i} + \Gamma^i_{in} f^n. \tag{2.89}$$

As was the case with (2.88) this can be simplified; we must first develop a useful identity for Γ^i_{in}. We begin by writing

$$\frac{\partial \sqrt{g}}{\partial q^n} = \frac{\partial}{\partial q^n}[\mathbf{r}_1 \cdot (\mathbf{r}_2 \times \mathbf{r}_3)] = \frac{\partial \mathbf{r}_1}{\partial q^n} \cdot (\mathbf{r}_2 \times \mathbf{r}_3) + \mathbf{r}_1 \cdot \frac{\partial (\mathbf{r}_2 \times \mathbf{r}_3)}{\partial q^n},$$

where

$$\begin{aligned} \mathbf{r}_1 \cdot \frac{\partial (\mathbf{r}_2 \times \mathbf{r}_3)}{\partial q^n} &= \mathbf{r}_1 \cdot \mathbf{r}_2 \times \frac{\partial \mathbf{r}_3}{\partial q^n} + \mathbf{r}_1 \cdot \frac{\partial \mathbf{r}_2}{\partial q^n} \times \mathbf{r}_3 \\ &= \frac{\partial \mathbf{r}_3}{\partial q^n} \cdot (\mathbf{r}_1 \times \mathbf{r}_2) + \frac{\partial \mathbf{r}_2}{\partial q^n} \cdot (\mathbf{r}_3 \times \mathbf{r}_1), \end{aligned}$$

so that

$$\frac{\partial \sqrt{g}}{\partial q^n} = \frac{\partial \mathbf{r}_1}{\partial q^n} \cdot (\mathbf{r}_2 \times \mathbf{r}_3) + \frac{\partial \mathbf{r}_2}{\partial q^n} \cdot (\mathbf{r}_3 \times \mathbf{r}_1) + \frac{\partial \mathbf{r}_3}{\partial q^n} \cdot (\mathbf{r}_1 \times \mathbf{r}_2).$$

Continuing to rewrite this we have

$$\begin{aligned} \frac{\partial \sqrt{g}}{\partial q^n} &= \mathbf{r}_{1n} \cdot (\mathbf{r}_2 \times \mathbf{r}_3) + \mathbf{r}_{2n} \cdot (\mathbf{r}_3 \times \mathbf{r}_1) + \mathbf{r}_{3n} \cdot (\mathbf{r}_1 \times \mathbf{r}_2) \\ &= \Gamma^i_{1n} \mathbf{r}_i \cdot (\mathbf{r}_2 \times \mathbf{r}_3) + \Gamma^i_{2n} \mathbf{r}_i \cdot (\mathbf{r}_3 \times \mathbf{r}_1) + \Gamma^i_{3n} \mathbf{r}_i \cdot (\mathbf{r}_1 \times \mathbf{r}_2) \\ &= \Gamma^1_{1n} \sqrt{g} + \Gamma^2_{2n} \sqrt{g} + \Gamma^3_{3n} \sqrt{g} = \sqrt{g}\, \Gamma^i_{in}. \end{aligned}$$

Hence

$$\Gamma^i_{in} = \frac{1}{\sqrt{g}} \frac{\partial \sqrt{g}}{\partial q^n}.$$

This is the needed identity, and with it (2.89) may be written as

$$\nabla \cdot \mathbf{f} = \frac{1}{\sqrt{g}} \frac{\partial}{\partial q^i} \left(\sqrt{g} f^i \right). \tag{2.90}$$

The curl of a tensor field $\mathbf{A}$ may be computed as follows. We start with

$$\begin{aligned} \nabla \times \mathbf{A} &= \mathbf{r}^k \times \frac{\partial \mathbf{A}}{\partial q^k} = \mathbf{r}^k \times \mathbf{r}^i \mathbf{r}^j \nabla_k a_{ij} = \epsilon^{kin} \mathbf{r}_n \mathbf{r}^j \nabla_k a_{ij} \\ &= \epsilon^{kin} \mathbf{r}_n \mathbf{r}^j \left(\frac{\partial a_{ij}}{\partial q^k} - \Gamma^s_{ki} a_{sj} - \Gamma^s_{kj} a_{is} \right). \end{aligned}$$

We then use

$$\epsilon^{kin}\Gamma^s_{ki} = 0 \quad \text{and} \quad -\mathbf{r}^j\Gamma^s_{kj} = \frac{\partial \mathbf{r}^s}{\partial q^k},$$

to get

$$\nabla \times \mathbf{A} = \epsilon^{kin}\mathbf{r}_n \left(\mathbf{r}^j \frac{\partial a_{ij}}{\partial q^k} + \frac{\partial \mathbf{r}^s}{\partial q^k} a_{is} \right)$$

and hence

$$\nabla \times \mathbf{A} = \epsilon^{kin}\mathbf{r}_n \frac{\partial}{\partial q^k} \left(\mathbf{r}^j a_{ij} \right).$$

For the divergence of a tensor field $\mathbf{A}$, we write

$$\begin{aligned}\nabla \cdot \mathbf{A} &= \mathbf{r}^k \cdot \frac{\partial \mathbf{A}}{\partial q^k} = \mathbf{r}^k \cdot \mathbf{r}_i \mathbf{r}_j \nabla_k a^{ij} = \mathbf{r}_j \nabla_i a^{ij} \\ &= \mathbf{r}_j \left(\frac{\partial a^{ij}}{\partial q^i} + \Gamma^i_{is} a^{sj} + \Gamma^j_{is} a^{is} \right).\end{aligned}$$

But

$$\mathbf{r}_j \Gamma^j_{is} = \frac{\partial \mathbf{r}_s}{\partial q^i}$$

so we have

$$\begin{aligned}\nabla \cdot \mathbf{A} &= \mathbf{r}_j \frac{\partial a^{ij}}{\partial q^i} + \mathbf{r}_j \frac{1}{\sqrt{g}} \frac{\partial \sqrt{g}}{\partial q^s} a^{sj} + \frac{\partial \mathbf{r}_s}{\partial q^i} a^{is} \\ &= \frac{\partial}{\partial q^i} \left(\mathbf{r}_j a^{ij} \right) + \mathbf{r}_j \frac{1}{\sqrt{g}} \frac{\partial \sqrt{g}}{\partial q^i} a^{ij}.\end{aligned}$$

Finally then,

$$\nabla \cdot \mathbf{A} = \frac{1}{\sqrt{g}} \frac{\partial}{\partial q^i} \left(\sqrt{g} a^{ij} \mathbf{r}_j \right). \tag{2.91}$$

We now obtain some formulas related to the second-order tensor $\nabla\nabla f$. We have

$$\nabla\nabla f = \mathbf{r}^i \frac{\partial}{\partial q^i} \left(\mathbf{r}^j \frac{\partial}{\partial q^j} f \right) = \mathbf{r}^i \left(\frac{\partial}{\partial q^i} \frac{\partial f}{\partial q^j} - \Gamma^k_{ij} \frac{\partial f}{\partial q^k} \right) \mathbf{r}^j,$$

hence

$$\nabla\nabla f = \mathbf{r}^i \mathbf{r}^j \left(\frac{\partial^2 f}{\partial q^i \partial q^j} - \Gamma^k_{ij} \frac{\partial f}{\partial q^k} \right). \tag{2.92}$$

From this we see that

$$(\nabla\nabla f)^T = \nabla\nabla f \tag{2.93}$$

which is obvious by symmetry (in i and j) of the Christoffel symbol and the rest of the expression on the right side of (2.92). A formal insertion[8] of the dot product operation between the vectors of (2.92) allows us to generate an expression for the Laplacian $\nabla^2 f \equiv \nabla \cdot \nabla f$:

$$\nabla^2 f = \mathbf{r}^i \cdot \mathbf{r}^j \left(\frac{\partial^2 f}{\partial q^i \partial q^j} - \Gamma^k_{ij} \frac{\partial f}{\partial q^k} \right) = g^{ij} \left(\frac{\partial^2 f}{\partial q^i \partial q^j} - \Gamma^k_{ij} \frac{\partial f}{\partial q^k} \right). \qquad (2.94)$$

Exercise 2.61. Show that a formal insertion of the cross product operation between the vectors of (2.92) yields the identity $\nabla \times \nabla f = 0$ for any scalar field f.

For a scalar function f we have $\frac{df}{dq^j} = \nabla_j f$ and so

$$\nabla\nabla f = \mathbf{r}^i \frac{\partial}{\partial q^i} \mathbf{r}^j \frac{\partial}{\partial q^j} f = \mathbf{r}^i \frac{\partial}{\partial q^i} \left(\mathbf{r}^j \nabla_j f \right) = \mathbf{r}^i \mathbf{r}^j \nabla_i \nabla_j f = \mathbf{r}^j \mathbf{r}^i \nabla_i \nabla_j f$$

where in the last step we used (2.93). The corresponding result for the Laplacian is

$$\nabla^2 f = \mathbf{r}^i \cdot \mathbf{r}^j \nabla_i \nabla_j f = g^{ij} \nabla_i \nabla_j f = \nabla^j \nabla_j f$$

where

$$\nabla^j \equiv g^{ij} \nabla_i.$$

For the Laplacian of a vector, we have

$$\nabla^2 \mathbf{f} = \nabla \cdot \nabla \mathbf{f} = \mathbf{r}^k \cdot \frac{\partial}{\partial q^k} \mathbf{r}^i \mathbf{r}_j \nabla_i f^j = \mathbf{r}^k \cdot \mathbf{r}^i \mathbf{r}_j \nabla_k \nabla_i f^j = g^{ki} \mathbf{r}_j \nabla_k \nabla_i f^j$$

hence

$$\nabla^2 \mathbf{f} = \mathbf{r}_j \nabla^i \nabla_i f^j.$$

Also

$$\nabla\nabla \cdot \mathbf{f} = \mathbf{r}^i \frac{\partial}{\partial q^i} \left[\frac{1}{\sqrt{g}} \frac{\partial}{\partial q^j} \left(\sqrt{g} f^j \right) \right] = \mathbf{r}^i \nabla_i \nabla_j f^j.$$

Exercise 2.62. Write out $\nabla^2 \mathbf{f}$ in detail, using the covariant components of $\mathbf{f}$.

See Table 2.8 for a summary.

[8]This sort of operation can be done with any dyad of vectors; we can generate a scalar $\mathbf{a} \cdot \mathbf{b}$ from $\mathbf{ab}$ by inserting a dot. This operation can have additional meaning when done with a tensor $\mathbf{A}$: if we write the tensor in mixed form $a_i^{\ j} \mathbf{r}^i \mathbf{r}_j$ we obtain $a_i^{\ j} \mathbf{r}^i \cdot \mathbf{r}_j = a_i^{\ i}$. This is the first invariant of $\mathbf{A}$, known as the trace of $\mathbf{A}$.

Table 2.8 Differential operations in general coordinates.

covariant derivative	$\nabla^k = g^{kj}\nabla_j$
rotation of vector field	$\nabla \times \mathbf{f} = \mathbf{r}^i \times \frac{\partial \mathbf{f}}{\partial q^i} = \mathbf{r}_k \epsilon^{ijk} \frac{\partial f_j}{\partial q^i}$
divergence of vector field	$\nabla \cdot \mathbf{f} = \mathbf{r}^i \cdot \frac{\partial \mathbf{f}}{\partial q^i} = \nabla_i f^i = \nabla^i f_i = \frac{1}{\sqrt{g}} \frac{\partial}{\partial q^i} \left(\sqrt{g} f^i\right)$
rotation of tensor field	$\nabla \times \mathbf{A} = \epsilon^{kin} \mathbf{r}_n \frac{\partial}{\partial q^k} \left(\mathbf{r}^j a_{ij}\right)$
divergence of tensor field	$\nabla \cdot \mathbf{A} = \frac{1}{\sqrt{g}} \frac{\partial}{\partial q^i} \left(\sqrt{g} a^{ij} \mathbf{r}_j\right)$
"Hessian" of scalar field	$\nabla\nabla f = \mathbf{r}^j \mathbf{r}^i \nabla_i \nabla_j f = \mathbf{r}^i \mathbf{r}^j \left(\frac{\partial^2 f}{\partial q^i \partial q^j} - \Gamma^k_{ij} \frac{\partial f}{\partial q^k}\right)$
Laplacian of scalar field	$\nabla^2 f = \nabla \cdot \nabla f = \nabla^j \nabla_j f = g^{ij} \left(\frac{\partial^2 f}{\partial q^i \partial q^j} - \Gamma^k_{ij} \frac{\partial f}{\partial q^k}\right)$
Laplacian of vector field	$\nabla^2 \mathbf{f} = \mathbf{r}_j \nabla^i \nabla_i f^j$
gradient of divergence	$\nabla\nabla \cdot \mathbf{f} = \mathbf{r}^i \nabla_i \nabla_j f^j$

Differentiation in orthogonal coordinate systems

Orthogonality of the frame vectors results in simplification of many of the formulas discussed above. Moreover, the physical components of a vector are conveniently displayed in orthogonal frames with use of the Lamé coefficients. Important examples of orthogonal systems are the familiar cylindrical and spherical systems of Fig. 2.5. In the cylindrical coordinate system, for instance, we have $(q^1, q^2, q^3) = (\rho, \phi, z)$.

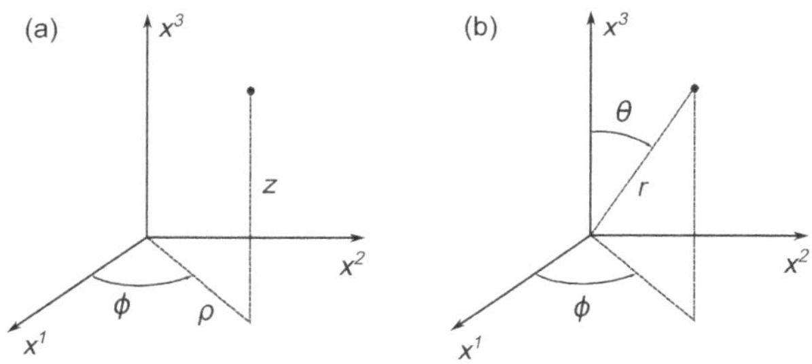

Fig. 2.5 Cylindrical and spherical coordinate systems.

Therefore suppose that at each (q^1, q^2, q^3) we have

$$\mathbf{r}_i \cdot \mathbf{r}_j = 0 \qquad (i \neq j).$$

The Lamé coefficients H_i are

$$(H_i)^2 = g_{ii} = \mathbf{r}_i \cdot \mathbf{r}_i \qquad (i = 1, 2, 3)$$

so that H_i is the magnitude of the frame vector $\mathbf{r}_i$. By construction, the reciprocal basis vectors $\mathbf{r}^i$ are co-directed with $\mathbf{r}_i$ and the product of their lengths is 1. Hence

$$\mathbf{r}^i = \mathbf{r}_i/(H_i)^2 \qquad (i = 1, 2, 3).$$

Let us introduce the frame whose vectors are co-directed with the basis vectors but have unit length:

$$\hat{\mathbf{r}}_i = \mathbf{r}_i/H_i \qquad (i = 1, 2, 3).$$

Hence

$$\mathbf{r}_i = H_i\hat{\mathbf{r}}_i, \qquad \mathbf{r}^i = \hat{\mathbf{r}}_i/H_i \qquad (i = 1, 2, 3)$$

and the ∇-operator becomes

$$\nabla = \frac{\hat{\mathbf{r}}_i}{H_i}\frac{\partial}{\partial q^i}$$

with summation over i understood (note that the frame under consideration here is essentially Cartesian at each point). By (2.83),

$$\begin{aligned}\mathbf{r}_{ij} \cdot \mathbf{r}_t &= \frac{1}{2}\left(\frac{\partial g_{it}}{\partial q^j} + \frac{\partial g_{tj}}{\partial q^i} - \frac{\partial g_{ji}}{\partial q^t}\right) \\ &= H_t\frac{\partial H_i}{\partial q^j}\delta_{it} + H_t\frac{\partial H_j}{\partial q^i}\delta_{jt} - H_i\frac{\partial H_j}{\partial q^t}\delta_{ij}\end{aligned}$$

since $g_{ij} = 0$ for $i \neq j$. On the other hand

$$\begin{aligned}\mathbf{r}_{ij} \cdot \mathbf{r}_t &= \frac{\partial}{\partial q^j}(H_i\hat{\mathbf{r}}_i) \cdot H_t\hat{\mathbf{r}}_t = \frac{\partial H_i}{\partial q^j}\hat{\mathbf{r}}_i \cdot H_t\hat{\mathbf{r}}_t + H_i\frac{\partial \hat{\mathbf{r}}_i}{\partial q^j} \cdot H_t\hat{\mathbf{r}}_t \\ &= \frac{\partial H_i}{\partial q^j}H_t\delta_{it} + H_iH_t\frac{\partial \hat{\mathbf{r}}_i}{\partial q^j} \cdot \hat{\mathbf{r}}_t.\end{aligned}$$

Hence

$$\frac{\partial \hat{\mathbf{r}}_i}{\partial q^j} \cdot \hat{\mathbf{r}}_t = \frac{1}{H_i}\frac{\partial H_t}{\partial q^i}\delta_{jt} - \frac{1}{H_t}\frac{\partial H_i}{\partial q^t}\delta_{ij}.$$

and

$$\frac{\partial \hat{\mathbf{r}}_i}{\partial q^j} = \sum_{t=1}^{3}\left(\frac{1}{H_i}\frac{\partial H_t}{\partial q^i}\delta_{jt} - \frac{1}{H_t}\frac{\partial H_i}{\partial q^t}\delta_{ij}\right)\hat{\mathbf{r}}_t.$$

(Note that the components of vectors are given in the frame of the same $\hat{\mathbf{r}}_i$.) Using this we derive

$$\nabla\mathbf{f} = \sum_{i,j=1}^{3}\frac{\hat{\mathbf{r}}_j}{H_j}\frac{\partial}{\partial q^j}\left(f_i\hat{\mathbf{r}}_i\right).$$

The gradient, divergence, and rotation of a vector field $\mathbf{f}$ are given by

$$\nabla \mathbf{f} = \hat{\mathbf{r}}_i \hat{\mathbf{r}}_j \left(\frac{1}{H_i} \frac{\partial f_j}{\partial q^i} - \frac{f_i}{H_i H_j} \frac{\partial H_i}{\partial q^j} + \delta_{ij} \frac{f_k}{H_k} \frac{1}{H_i} \frac{\partial H_i}{\partial q^k} \right)$$

$$\nabla \cdot \mathbf{f} = \frac{1}{H_1 H_2 H_3} \left(\frac{\partial}{\partial q^1} (H_2 H_3 f_1) + \frac{\partial}{\partial q^2} (H_3 H_1 f_2) + \frac{\partial}{\partial q^3} (H_1 H_2 f_3) \right)$$

and

$$\nabla \times \mathbf{f} = \frac{1}{2} \frac{\hat{\mathbf{r}}_i \times \hat{\mathbf{r}}_j}{H_i H_j} \left(\frac{\partial}{\partial q^i} (H_j f_j) - \frac{\partial}{\partial q^j} (H_i f_i) \right).$$

The Laplacian of a scalar field f is

$$\nabla^2 f = \frac{1}{H_1 H_2 H_3} \left[\frac{\partial}{\partial q^1} \left(\frac{H_2 H_3}{H_1} \frac{\partial f}{\partial q^1} \right) + \frac{\partial}{\partial q^2} \left(\frac{H_3 H_1}{H_2} \frac{\partial f}{\partial q^2} \right) + \frac{\partial}{\partial q^3} \left(\frac{H_1 H_2}{H_3} \frac{\partial f}{\partial q^3} \right) \right].$$

Using the gradient we can write out the tensor of small strainsfor a displacement vector $\mathbf{u} = u_i \hat{\mathbf{r}}_i$, which is the main object of linear elasticity:

$$\begin{aligned} \boldsymbol{\varepsilon} &= \frac{1}{2} \left(\nabla \mathbf{u} + \nabla \mathbf{u}^T \right) \\ &= \frac{1}{2} \hat{\mathbf{r}}_i \hat{\mathbf{r}}_j \left(\frac{1}{H_i} \frac{\partial u_i}{\partial q^i} + \frac{1}{H_j} \frac{\partial u_j}{\partial q^j} - \frac{u_i}{H_i H_j} \frac{\partial H_i}{\partial q^j} \right. \\ &\qquad \left. - \frac{u_j}{H_i H_j} \frac{\partial H_j}{\partial q^i} + 2\delta_{ij} \frac{u_t}{H_i H_j} \frac{\partial H_j}{\partial q^t} \right). \end{aligned} \tag{2.95}$$

Some Formulas of Integration

Let $f(x_1, x_2, x_3)$ be a continuous function of the Cartesian coordinates x_1, x_2, x_3 in a compact volume V. Let q^1, q^2, q^3 be curvilinear coordinates in the same volume, in one-to-one continuously differentiable correspondence with the Cartesian coordinates, so after transformation of the coordinates we shall write out the same function as $f(q^1, q^2, q^3)$. The transformation is

$$\int_V f(x_1, x_2, x_3)\, dx_1\, dx_2\, dx_3 = \int_V f(q^1, q^2, q^3) J\, dq^1\, dq^2\, dq^3,$$

where J is the Jacobian[9]

$$J = \sqrt{g} = \left| \frac{\partial x_i}{\partial q^j} \right|.$$

[9]We suppose $J \geq 0$.

Exercise 2.63. Two successive coordinate transformations are given by $x^i = x^i(q^j)$ and $q^i = q^i(\tilde{q}^j)$. Show that the Jacobian determinant of the composite transformation $x^i = x^i(\tilde{q}^j)$ is given by

$$\left|\frac{\partial(x^1, x^2, x^3)}{\partial(\tilde{q}^1, \tilde{q}^2, \tilde{q}^3)}\right| = \left|\frac{\partial(x^1, x^2, x^3)}{\partial(q^1, q^2, q^3)}\right| \left|\frac{\partial(q^1, q^2, q^3)}{\partial(\tilde{q}^1, \tilde{q}^2, \tilde{q}^3)}\right|.$$

For functions $f(x_1, x_2, x_3)$ and $g(x_1, x_2, x_3)$ that are continuously differentiable on a compact volume V with piecewise smooth boundary S, there is the well known Gauss–Ostrogradsky formula for integration by parts:

$$\int_V \frac{\partial f}{\partial x_k}\, g\, dV = -\int_V \frac{\partial g}{\partial x_k}\, f\, dV + \int_S f g n_k\, dS,$$

where $dV = dx_1\, dx_2\, dx_3$ is the differential volume element, dS is the differential element of area on S, and n_k is the projection of the outward unit normal $\mathbf{n}$ from S onto the axis $\mathbf{i}_k$. In the particular case $g \equiv 1$ we have

$$\int_V \frac{\partial f}{\partial x_k}\, dV = \int_S f n_k\, dS. \tag{2.96}$$

Let us use (2.96) to derive some formulas involving the nabla operator that are frequently used in applications. These formulas will hold in curvilinear coordinates, despite the fact that the intermediate transformations will be done in Cartesian coordinates, because the final results will be written in non-coordinate form. We begin with the integral of ∇f:

$$\int_V \nabla f\, dV = \int_V \mathbf{i}_k \frac{\partial f}{\partial x_k}\, dV = \mathbf{i}_k \int_S f n_k\, dS = \int_S f \mathbf{n}\, dS. \tag{2.97}$$

Now consider an analogous formula for a vector function $\mathbf{f}$:

$$\begin{aligned}\int_V \nabla \mathbf{f}\, dV &= \int_V \mathbf{i}_k \frac{\partial (f_t \mathbf{i}_t)}{\partial x_k}\, dV = \mathbf{i}_k \mathbf{i}_t \int_S n_k f_t\, dS \\ &= \int_S n_k \mathbf{i}_k\, f_t \mathbf{i}_t\, dS = \int_S \mathbf{n}\mathbf{f}\, dS.\end{aligned} \tag{2.98}$$

Since the left- and right-hand sides are written in non-coordinate form we can use this formula with any coordinate frame with $dV = \sqrt{g}\, dq^1\, dq^2\, dq^3$. This is the formula for a tensor $\nabla \mathbf{f}$; for its trace we have

$$\begin{aligned}\int_V \nabla \cdot \mathbf{f}\, dV &= \int_V \mathbf{i}_k \cdot \frac{\partial (f_t \mathbf{i}_t)}{\partial x_k}\, dV = \mathbf{i}_k \cdot \mathbf{i}_t \int_S n_k f_t\, dS \\ &= \int_S n_k \mathbf{i}_k \cdot f_t \mathbf{i}_t\, dS = \int_S \mathbf{n} \cdot \mathbf{f}\, dS.\end{aligned} \tag{2.99}$$

In a similar fashion we can derive

$$\int_V \nabla \times \mathbf{f} \, dV = \int_S \mathbf{n} \times \mathbf{f} \, dS. \tag{2.100}$$

It is easily seen that

$$\int_V (\nabla \mathbf{f})^T \, dV = \int_S (\mathbf{n}\mathbf{f})^T \, dS = \int_S \mathbf{f}\mathbf{n} \, dS. \tag{2.101}$$

The reader can also use (2.96) to derive the formulas

$$\int_V \nabla \mathbf{A} \, dV = \int_S \mathbf{n}\mathbf{A} \, dS, \tag{2.102}$$

$$\int_V \nabla \cdot \mathbf{A} \, dV = \int_S \mathbf{n} \cdot \mathbf{A} \, dS, \tag{2.103}$$

$$\int_V \nabla \times \mathbf{A} \, dV = \int_S \mathbf{n} \times \mathbf{A} \, dS, \tag{2.104}$$

for a tensor field $\mathbf{A}$.

Finally we write out Stokes's formula in non-coordinate form. Recall that this relates a vector function $\mathbf{f}$ given on a piecewise smooth oriented surface S to its circulation over the piecewise smooth boundary contour Γ:

$$\oint_\Gamma \mathbf{f} \cdot d\mathbf{r} = \int_S (\mathbf{n} \times \nabla) \cdot \mathbf{f} \, dS, \tag{2.105}$$

where $d\mathbf{r}$ is such that for an observer following ∂S along $d\mathbf{r}$, (1) the surface S is to the observer's left, and (2) the surface normal $\mathbf{n}$ is directed upward. For a tensor $\mathbf{A}$ of second-order, this extends to the two formulas

$$\oint_\Gamma d\mathbf{r} \cdot \mathbf{A} = \int_S (\mathbf{n} \times \nabla) \cdot \mathbf{A} \, dS, \tag{2.106}$$

$$\oint_\Gamma \mathbf{A} \cdot d\mathbf{r} = \int_S (\mathbf{n} \times \nabla) \cdot \mathbf{A}^T \, dS. \tag{2.107}$$

2.10 Pseudovectors, Pseudoscalars, and Pseudotensors

In physics there are many quantities which are of vector nature: forces, displacements, velocities, accelerations. But there are also "vectors" that behave differently from these under certain changes of basis.

When we apply to a force vector $\mathbf{F}$ a mirror reflection with respect to the plane orthogonal to the direction of $\mathbf{F}$, the vector reverses direction: it transforms to $-\mathbf{F}$. This idea is illustrated in Fig. 2.6. The same thing happens with vectors of displacement, velocity, acceleration, and many others. Such vectors are said to be *polar*. Other quantities, however, can be regarded as vectorial but behave differently under such transformations. For example, let us consider the rotation of a rigid body through an angle α about a fixed axis. For convenience we associate this rotation with an arrow of magnitude α along the rotation axis, and use the familiar right-hand rule to select from among the two possible directions. The result is a "vector" $\boldsymbol{\varphi}$. What happens when we reflect the picture of this rotation through a planar mirror orthogonal to the rotation axis? The direction of rotation does *not* change, and so we should associate with this mirrored rotation the same vector $\boldsymbol{\varphi}$: not $-\boldsymbol{\varphi}$ as was the case with $\mathbf{F}$. The "vector" $\boldsymbol{\varphi}$ is representative of a class of quantities, many related in some way to rotation and invoking the right-hand rule in a similar fashion.

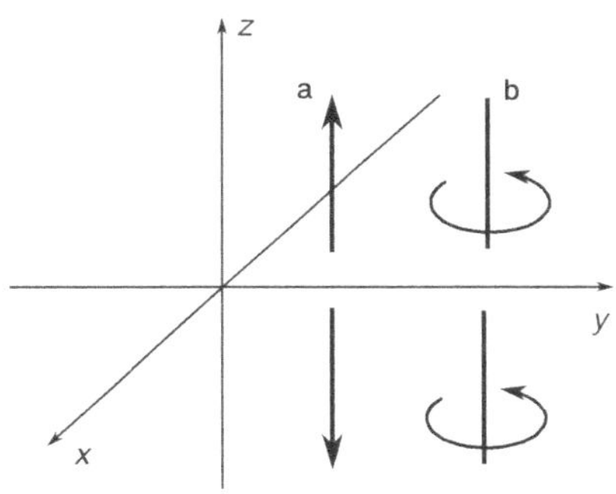

Fig. 2.6 Polar and axial vectors and their mirror image reflections.

Quantities of "vectorial" nature which under mirror reflections with respect to planes perpendicular to their directional axes do not reverse are called *axial* vectors or *pseudovectors*.

The pseudovectors also behave "strangely" compared with polar vectors when we simultaneously invert all the frame vectors (note that this maintains the handedness of the frame). Another way to distinguish vectors from pseudovectors is to watch how a quantity changes when the handedness of the frame changes. In this case polar vectors are unchanged but pseudovectors change sign.

If mirror reflections, basis inversions, and changes of frame orientation are not permitted in a given discussion where vectorial quantities are involved, then the distinction between vectors and pseudovectors becomes largely unimportant (although a vector cannot be added to a pseudovector in any circumstance).

A common source of pseudovectors in practice is the ordinary cross product. Since the magnitude of $\mathbf{a} \times \mathbf{b}$ is the area of the parallelogram generated by $\mathbf{a}$ and $\mathbf{b}$, it is immediately clear that $\mathbf{a} \times \mathbf{b}$ differs in nature from either of its factors. But for convenience we naturally regard $\mathbf{a} \times \mathbf{b}$ as a "vector" orthogonal to $\mathbf{a}$ and $\mathbf{b}$. Here again — as was the case for rotations — one must select from two possible directions. The quantity $\mathbf{a} \times \mathbf{b}$ is actually a pseudovector; it is only for convenience in elementary physics that we use it alongside ordinary vectors without belaboring this point.

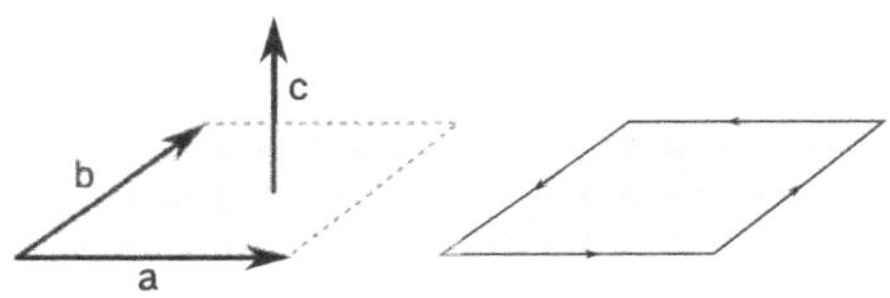

Fig. 2.7 Cross product of two vectors and related oriented area.

Pseudoscalars

In a similar way, there are scalar quantities which can be split into two groups: true scalars, and axial or pseudoscalars. The latter change sign upon reversal of frame handedness. The scalars obtained from physical experiments — such as temperature, mass, and density — are usually true scalars. But mathematical operations involving cross products may produce pseudoscalars. The vector triple product is an example. Indeed, for noncoplanar polar vectors $\mathbf{a}, \mathbf{b}, \mathbf{c}$ the triple product satisfies

$$(-\mathbf{E} \cdot \mathbf{a}) \cdot [(-\mathbf{E} \cdot \mathbf{b}) \times (-\mathbf{E} \cdot \mathbf{c})] = -\mathbf{a} \cdot (\mathbf{b} \times \mathbf{c}).$$

Hence the scalar $\mathbf{a} \cdot (\mathbf{b} \times \mathbf{c})$ changes sign upon coordinate inversion. Another example of a pseudoscalar is the signed curvature of a plane curve. We say that the curvature k of a plane curve is positive if the curve is concave up and negative if concave down. Obviously, the sign of the curvature depends on our sign convention. Consider, for example, the two curves given in Fig. 2.8. The curvature of curve 1 is positive whereas that of curve 2 is negative. But if we consider the mirror image reflection of these curves, the curvatures change signs (dashed lines in the figure).

The theory of mechanical beams provides an example of the use of pseudoscalars. Indeed, in beam theory we use a sign convention for positive

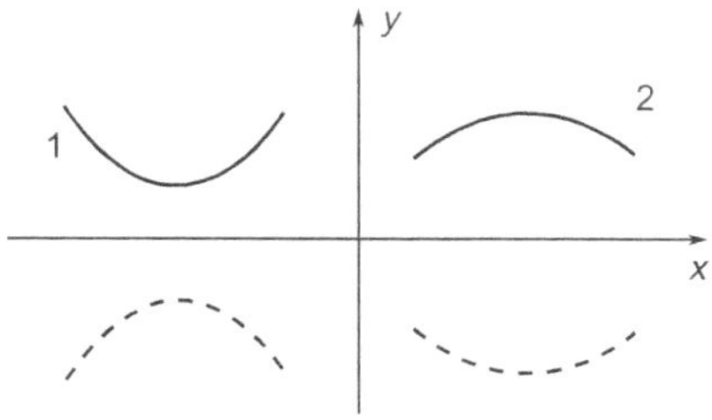

Fig. 2.8 Curves with (1) positive and (2) negative curvatures and their mirror image reflections (dashed).

bending moment which is quite similar to the convention for curvature. Here we have constitutive relations of the form $M = D\kappa$ where M is the moment, κ is the beam curvature, and D is the stiffness. It is clear that since κ is pseudoscalar then so is M, as D is a true scalar.

Pseudotensors

Just as there are polar and axial vectors, there are polar and axial tensors (or pseudotensors). For the components of an axial second-order tensor, the transformation formulae are

$$\tilde{a}^{ij} = \pm A^i_k A^j_l a^{kl}, \tag{2.108}$$

where sign choice is determined by the sign of the determinant of the A^j_i. An example of an axial second-order tensor is the skew tensor $\mathbf{A} = \mathbf{a} \times \mathbf{E}$ with polar vector $\mathbf{a}$. The permutation tensor $-\mathbf{E} \times \mathbf{E}$ is an axial third-order tensor.

The stress and strain tensors of elasticity are polar tensors, and the couple stress tensor and wryness tensors used in micropolar mechanics are axial. To axial tensors belong also moment (couple) tensors and the tensor of change of curvature used in the theory of plates and shells.

When the distinction between axial and polar quantities does not affect mechanical or physical theories, it is typically ignored. This is nearly the case with linear elasticity. However, sometimes the various properties with respect to proper and non-proper orthogonal transformations become important. This occurs if we consider various crystallographic groups; it also occurs with piezoelectric and piezomagnetic solids, with liquid crystals, with the electrodynamics of deformable media, etc. (see [Nye (1957); Feynman et al. (1977); Spencer (1971); Boehler (1987); Eringen and Maugin (1990); Maugin (1988); de Gennes and Prost (1993);

Chandrasekhar (1977)]). As an example, let us consider two scalar functions

$$W = W_1(\mathbf{S}, \mathbf{m}), \quad W = W_2(\mathbf{S}, \mathbf{mm}),$$

where $\mathbf{S}$ is a symmetric second-order tensor, $\mathbf{m}$ is a polar vector, and $\mathbf{mm}$ is the dyadic product of two vectors $\mathbf{m}$. Such functions are used to construct the constitutive equations of transversely isotropic materials. They also appear in the theory of nematic and cholesteric liquid crystals. Assuming the values of W to be true scalars (e.g., strain energies), we consider the invariance of W under orthogonal transformations. It is clear that the mirror image reflection invariance involves the constraint $W_1(\mathbf{S}, \mathbf{m}) = W_1(\mathbf{S}, -\mathbf{m})$ for the first function. If $\mathbf{m}$ is an axial vector, which occurs for the magnetic field vector in magnetoelasticity, there is no such restriction.

Note that in the mechanics of a Cosserat continuum we are obliged to distinguish between axial and polar vectors and tensors.

The mathematical and physical literature display more general formulas for the transformations of tensors [Korn and Korn (1968)] such as

$$\tilde{a}^{ij} = \det(A^n_m)^\eta A^i_k A^j_l a^{kl}, \tag{2.109}$$

where $\eta = 0$ for true or polar tensors (also called absolute tensors) and $\eta \neq 0$ for so-called tensor densities of weight η.

Two more notes on units and pseudovectors

In many theories pertaining to continuum media, we employ vectors (and tensors) having differing units; examples are displacements and force vectors. In coordinate form both vectors are written in the same basis, which means that the basis vectors should be unitless. The basis vectors are introduced through the derivatives $\partial\mathbf{r}/\,dq^i$. But the q^i have different units in the most common curvilinear coordinate systems — cylindrical and spherical — where some coordinates are in radians and other carry units of meters. One approach is to have the position vector $\mathbf{r}$ and the q^i unitless; the other vectors and tensors may have components with physical units, typically called physical components when the basis vectors are normalized. Many textbooks and papers on continuum media present material in unitless form.

On pseudotensors. In the coordinate-free equations of continuum theory we often encounter tensors and pseudotensors simultaneously. To write out such equations in coordinate form, we should present all pseudovectors and pseudotensors using, in the basis, only polar vectors and not axial vectors like $\mathbf{i}\times\mathbf{j}$. Only the components should reflect their "pseudo" nature, with certain factors dependent on what happens to the components on the inversion of the basis.

2.11 More on Invariants of Tensors and Vectors

Earlier for a second-order symmetric tensor we introduced the principal invariants by formulas (2.24) or (2.25). We now recall that the principal invariants of second-order symmetric tensor $\mathbf{S}$ given by

$$I_1(\mathbf{S}) = \operatorname{tr}\mathbf{S}, \quad I_2(\mathbf{S}) = \tfrac{1}{2}[\operatorname{tr}^2\mathbf{S} - \operatorname{tr}\mathbf{S}^2], \quad I_3(\mathbf{S}) = \det\mathbf{S} \tag{2.110}$$

are invariant under orthogonal transformation of $\mathbf{S}$, i.e., they do not change under the replacement $\mathbf{S} \to \mathbf{Q}\cdot\mathbf{S}\cdot\mathbf{Q}^T$ for any orthogonal tensor $\mathbf{Q}$, which is defined by the equality $\mathbf{Q}^{-1} = \mathbf{Q}^T$:

$$I_k(\mathbf{S}) = I_k(\mathbf{Q}\cdot\mathbf{S}\cdot\mathbf{Q}^T).$$

This also means that I_k is an isotropic function of $\mathbf{S}$. Moreover, any isotropic scalar-valued function of a second-order symmetric tensor is a function of its invariants. Instead of I_k one may use another set of invariants $\operatorname{tr}\mathbf{S}$, $\operatorname{tr}\mathbf{S}^2$, $\operatorname{tr}\mathbf{S}^3$, equivalent to the set of I_k. Both sets are called *polynomial invariants* since they are polynomials of $\mathbf{S}$.

This is an example of application of the *theory of invariants*. The theory plays an important role in many branches of mechanics, physics, and mathematics. In mechanics, the theory of invariants is used to derive a simplified version of the constitutive equation. Suppose a set of invariants is given for some sets of tensors, vectors, and a transformation group. A central problem of invariant theory [Spencer (1971)] is to determine a so-called *basis* from which all other invariants can be generated. Frequently employed are polynomial invariants which constitute the *integrity basis*. The integrity basis is a set of polynomial invariants such that any polynomial invariant can be expressed as a polynomial in terms of the integrity basis. A polynomial invariant is called *reducible* if it can be expressed as a polynomial of other invariants; otherwise it is *irreducible*. Obviously, one is interested in finding a *minimal integrity basis* which contains the smallest possible number of members. Non-polynomial dependencies between invariants are called *syzygies*. Existence of an integrity basis follows from Hilbert's theorem [Spencer (1971); Zheng (1994)]:

For any finite system of vector and tensors there exists an integrity basis which consists of a finite number of invariants.

Usually they consider the invariance under the action of the full orthogonal group, the proper orthogonal group (rotation group) or its subgroups. Let us recall that for the transformations from the proper orthogonal group there is no difference in the transformations of the axial or polar vector and tensors. But there exists a difference in case of the transformations from the full orthogonal group. Following [Spencer (1971)], we distinguish the *absolute invariants* which are invariant under all the orthogonal transformations and the *relative invariants* which are invariant under rotations but change sign under the non-proper orthogonal transformations. For example, the trace $\operatorname{tr}\boldsymbol{\Omega}$ of a pseudotensor $\boldsymbol{\Omega}$ is a relative invariant whereas $\operatorname{tr}\boldsymbol{\Omega}^2$ is an absolute invariant. For the theory of invariants and its application in mechanics we refer to [Boehler (1987); Spencer (1971)], and to the review [Zheng (1994)]. Here we restrict ourselves to some applications of the theory of invariants. Below we use the following nomenclature: $\mathbf{A}$, $\mathbf{B}$, etc., denote second-order polar (true or absolute) tensors, $\mathbf{S}$, $\mathbf{S}_k$ denote symmetric second-order polar (true or absolute) tensors, $\boldsymbol{\Omega}$, $\boldsymbol{\Omega}_k$, $\boldsymbol{\Gamma}$ denote second-order axial tensors (pseudotensors), $\mathbf{W}$, $\mathbf{W}_k$, $\mathbf{V}$ are skew tensors, $\mathbf{a}$, $\mathbf{b}$, $\mathbf{c}$, ... denote polar vectors, $\boldsymbol{\omega}$, $\boldsymbol{\varphi}$, ... denote axial vectors (pseudovectors), and indices S and A stand for symmetric and skew-symmetric parts of a tensor, respectively. For example, $\mathbf{A}_S = \frac{1}{2}(\mathbf{A} + \mathbf{A}^T)$ and $\mathbf{A}_A = \frac{1}{2}(\mathbf{A} - \mathbf{A}^T)$.

Earlier we introduced isotropic functions as the functions invariant under all orthogonal transformations. In the theory of invariants they consider also scalar-, vector-, and tensor-valued functions which are invariant under certain orthogonal transformations. The set of invariant transformations forms the *symmetry group*. First, let us recall that scalar- and tensor-valued functions are called *isotropic* if for any orthogonal transformations $(\mathbf{Q}^T = \mathbf{Q}^{-1})$

$$
\begin{aligned}
&f(\mathbf{Q}\cdot\mathbf{A}\cdot\mathbf{Q}^T,\ldots,(\det\mathbf{Q})\mathbf{Q}\cdot\boldsymbol{\Omega}\cdot\mathbf{Q}^T,\ldots,\mathbf{Q}\cdot\mathbf{a},\ldots,(\det\mathbf{Q})\mathbf{Q}\cdot\boldsymbol{\omega},\ldots)\\
&\quad = f(\mathbf{A},\ldots,\boldsymbol{\Omega},\ldots,\mathbf{a},\ldots,\boldsymbol{\omega},\ldots)
\end{aligned}
\tag{2.111}
$$

$$
\begin{aligned}
&\mathbf{F}(\mathbf{Q}\cdot\mathbf{A}\cdot\mathbf{Q}^T,\ldots,(\det\mathbf{Q})\mathbf{Q}\cdot\boldsymbol{\Omega}\cdot\mathbf{Q}^T,\ldots,\mathbf{Q}\cdot\mathbf{a},\ldots,(\det\mathbf{Q})\mathbf{Q}\cdot\boldsymbol{\omega},\ldots)\\
&\quad = \mathbf{Q}\cdot\mathbf{F}(\mathbf{A},\ldots,\boldsymbol{\Omega},\ldots,\mathbf{a},\ldots,\boldsymbol{\omega},\ldots)\cdot\mathbf{Q}^T,
\end{aligned}
\tag{2.112}
$$

$$\boldsymbol{\Gamma}(\mathbf{Q}\cdot\mathbf{A}\cdot\mathbf{Q}^T,\ldots,(\det\mathbf{Q})\mathbf{Q}\cdot\boldsymbol{\Omega}\cdot\mathbf{Q}^T,\ldots,\mathbf{Q}\cdot\mathbf{a},\ldots,(\det\mathbf{Q})\mathbf{Q}\cdot\boldsymbol{\omega},\ldots)$$
$$=(\det\mathbf{Q})\mathbf{Q}\cdot\boldsymbol{\Gamma}(\mathbf{A},\ldots,\boldsymbol{\Omega},\ldots,\mathbf{a},\ldots,\boldsymbol{\omega},\ldots)\cdot\mathbf{Q}^T. \tag{2.113}$$

Here we consider an isotropic scalar-valued function f (a true scalar), a polar tensor-valued function $\mathbf{F}$, and an axial tensor-valued function $\boldsymbol{\Gamma}$. The symbols "..." in the argument lists for these functions stand for the arguments of the same nature as the preceding ones.

For isotropic functions the symmetry group is the full orthogonal group Orth. To determine the difference between the proper and improper orthogonal tensors, we introduce hemitropic functions which satisfy (2.111) and (2.112) only for the rotation tensor. So for the hemitropic functions we get

$$h(\mathbf{Q}\cdot\mathbf{A}\cdot\mathbf{Q}^T,\ldots,\mathbf{Q}\cdot\boldsymbol{\Omega}\cdot\mathbf{Q}^T,\ldots,\mathbf{Q}\cdot\mathbf{a},\ldots,\mathbf{Q}\cdot\boldsymbol{\omega},\ldots)$$
$$=h(\mathbf{A},\ldots,\boldsymbol{\Omega},\ldots,\mathbf{a},\ldots,\boldsymbol{\omega},\ldots), \tag{2.114}$$

$$\mathbf{H}(\mathbf{Q}\cdot\mathbf{A}\cdot\mathbf{Q}^T,\ldots,\mathbf{Q}\cdot\boldsymbol{\Omega}\cdot\mathbf{Q}^T,\ldots,\mathbf{Q}\cdot\mathbf{a},\ldots,\mathbf{Q}\cdot\boldsymbol{\omega},\ldots)$$
$$=\mathbf{Q}\cdot\mathbf{H}(\mathbf{A},\ldots,\boldsymbol{\Omega},\ldots,\mathbf{a},\ldots,\boldsymbol{\omega},\ldots)\cdot\mathbf{Q}^T, \tag{2.115}$$

with $\mathbf{Q}^T=\mathbf{Q}^{-1}$ and $\det\mathbf{Q}=1$. In other words, the symmetry group for a hemitropic function is Orth_+ and the formulas do not depend on the nature of tensors or vectors, polar or axial. In most cases of classical mechanics, hemitropic and isotropic functions do not differ as their arguments are only polar vectors and tensors.

As an example of isotropic function, let us present isotropic scalar-valued functions of a symmetric tensor:

$$f=f(I_1,I_2,I_3)=\tilde{f}(\operatorname{tr}\mathbf{S},\operatorname{tr}\mathbf{S}^2,\operatorname{tr}\mathbf{S}^3) \tag{2.116}$$
$$\mathbf{F}=f_0\mathbf{E}+f_1\mathbf{S}+f_2\mathbf{S}^2,\quad f_k=f_k(I_1,I_2,I_3). \tag{2.117}$$

In other words, any isotropic scalar function of a symmetric tensor is a function of its principal invariants. Equation (2.116) gives us significant reduction in the number of possible arguments. Indeed, $\mathbf{S}$ contains six components as a symmetric tensor, so $f(\mathbf{S})$ is a function of six scalar variables, in general. But an isotropic function of $\mathbf{S}$ depends only on three independent variables. With use of the formulas for differentiating tensor invariants (Exercises 2.45 and 2.46), the derivative of an isotropic function $df/d\mathbf{S}$ takes the form (2.117). However, there are isotropic tensor functions $\mathbf{F}(\mathbf{S})$ having no potential function $f(\mathbf{S})$ such that $\mathbf{F}=df/d\mathbf{S}$.

Representations (2.116) and (2.117) are widely used in nonlinear elasticity [Lurie (1990); Ogden (1997)] and hydromechanics [Astarita and Marrucci (1974); Truesdell and Noll (2004)].

Following [Spencer (1971); Zheng (1994)], where joint invariants of up to six vectors or tensors are given, in Table 2.9 we present invariants for some sets of vectors and tensors. It shows that the reduction of possible arguments of isotropic scalar functions is more for a smaller number of arguments. For example, for an isotropic function of a symmetric tensor, the reduction is from six to three variables, whereas in the case of a function of two symmetric tensors the number only drops from 12 to 10. We can also observe similarity in skew tensors and vectors in three-dimensional space. Indeed, a skew tensor has only three independent components like a vector, so it has the same number of invariants as a vector.

Table 2.9 Invariants for some sets of tensors and vectors.

Agencies	Isotropy	Hemitropy
$\mathbf{a}$	$\mathbf{a}\cdot\mathbf{a}$	$\mathbf{a}\cdot\mathbf{a}$
$\mathbf{a}$, $\mathbf{b}$	$\mathbf{a}\cdot\mathbf{a}$, $\mathbf{b}\cdot\mathbf{b}$, $\mathbf{a}\cdot\mathbf{b}$	$\mathbf{a}\cdot\mathbf{a}$, $\mathbf{b}\cdot\mathbf{b}$, $\mathbf{a}\cdot\mathbf{b}$
$\mathbf{S}$	$\operatorname{tr}\mathbf{S}$, $\operatorname{tr}\mathbf{S}^2$, $\operatorname{tr}\mathbf{S}^3$	$\operatorname{tr}\mathbf{S}$, $\operatorname{tr}\mathbf{S}^2$, $\operatorname{tr}\mathbf{S}^3$
$\mathbf{S}_1$, $\mathbf{S}_2$	$\operatorname{tr}\mathbf{S}_1$, $\operatorname{tr}\mathbf{S}_1^2$, $\operatorname{tr}\mathbf{S}_1^3$,	$\operatorname{tr}\mathbf{S}_1$, $\operatorname{tr}\mathbf{S}_1^2$, $\operatorname{tr}\mathbf{S}_1^3$,
	$\operatorname{tr}\mathbf{S}_2$, $\operatorname{tr}\mathbf{S}_2^2$, $\operatorname{tr}\mathbf{S}_2^3$,	$\operatorname{tr}\mathbf{S}_2$, $\operatorname{tr}\mathbf{S}_2^2$, $\operatorname{tr}\mathbf{S}_2^3$,
	$\operatorname{tr}\mathbf{S}_1\cdot\mathbf{S}_2$, $\operatorname{tr}\mathbf{S}_1^2\cdot\mathbf{S}_2$,	$\operatorname{tr}\mathbf{S}_1\cdot\mathbf{S}_2$, $\operatorname{tr}\mathbf{S}_1^2\cdot\mathbf{S}_2$,
	$\operatorname{tr}\mathbf{S}_1\cdot\mathbf{S}_2^2$, $\operatorname{tr}\mathbf{S}_1^2\cdot\mathbf{S}_2^2$	$\operatorname{tr}\mathbf{S}_1\cdot\mathbf{S}_2^2$, $\operatorname{tr}\mathbf{S}_1^2\cdot\mathbf{S}_2^2$
$\mathbf{a}$, $\mathbf{S}$	$\mathbf{a}\cdot\mathbf{a}$, $\operatorname{tr}\mathbf{S}$, $\operatorname{tr}\mathbf{S}^2$, $\operatorname{tr}\mathbf{S}^3$,	$\mathbf{a}\cdot\mathbf{a}$, $\operatorname{tr}\mathbf{S}$, $\operatorname{tr}\mathbf{S}^2$, $\operatorname{tr}\mathbf{S}^3$,
	$\mathbf{a}\cdot\mathbf{S}\cdot\mathbf{a}$, $\mathbf{a}\cdot\mathbf{S}^2\cdot\mathbf{a}$	$\mathbf{a}\cdot\mathbf{S}\cdot\mathbf{a}$, $\mathbf{a}\cdot\mathbf{S}^2\cdot\mathbf{a}$, $\mathbf{a}\cdot[(\mathbf{S}\cdot\mathbf{a})\times(\mathbf{S}^2\cdot\mathbf{a})]$
$\mathbf{W}$	$\operatorname{tr}\mathbf{W}^2$	$\operatorname{tr}\mathbf{W}^2$
$\mathbf{W}_1$, $\mathbf{W}_2$	$\operatorname{tr}\mathbf{W}_1^2$, $\operatorname{tr}\mathbf{W}_2^2$, $\operatorname{tr}\mathbf{W}_1\cdot\mathbf{W}_2$	$\operatorname{tr}\mathbf{W}_1^2$, $\operatorname{tr}\mathbf{W}_2^2$, $\operatorname{tr}\mathbf{W}_1\cdot\mathbf{W}_2$
$\mathbf{W}$, $\mathbf{S}$	$\operatorname{tr}\mathbf{W}^2$, $\operatorname{tr}\mathbf{S}$, $\operatorname{tr}\mathbf{S}^2$, $\operatorname{tr}\mathbf{S}^3$,	$\operatorname{tr}\mathbf{W}^2$, $\operatorname{tr}\mathbf{S}$, $\operatorname{tr}\mathbf{S}^2$, $\operatorname{tr}\mathbf{S}^3$,
	$\operatorname{tr}\mathbf{S}\cdot\mathbf{W}^2$, $\operatorname{tr}\mathbf{S}^2\cdot\mathbf{W}^2$,	$\operatorname{tr}\mathbf{S}\cdot\mathbf{W}^2$, $\operatorname{tr}\mathbf{S}^2\cdot\mathbf{W}^2$,
	$\operatorname{tr}\mathbf{S}^2\cdot\mathbf{W}^2\cdot\mathbf{S}\cdot\mathbf{W}$	$\operatorname{tr}\mathbf{S}^2\cdot\mathbf{W}^2\cdot\mathbf{S}\cdot\mathbf{W}$

Exercise 2.64. Find the invariants of (a) two dyads $\mathbf{aa}$ and $\mathbf{bb}$; (b) dyad $\mathbf{aa}$ and symmetric tensor $\mathbf{S}$; (c) dyad $\mathbf{aa}$ and skew tensor $\mathbf{W}$; (d) two tensors $\mathbf{E}\times\mathbf{a}$ and $\mathbf{E}\times\mathbf{b}$; (e) tensor $\mathbf{ab}+\mathbf{ba}$; (f) tensor $\mathbf{ab}-\mathbf{ba}$.

In Table 2.9 both relative and absolute invariants are listed. For example, $\mathbf{a}\cdot\mathbf{a}$ and $\boldsymbol{\omega}\cdot\boldsymbol{\omega}$ are absolute invariants independently for polar vector

$\mathbf{a}$ and pseudovector $\boldsymbol{\omega}$, and $\mathbf{a} \cdot \boldsymbol{\omega}$ is a relative invariant. Here the difference between polar and axial tensors may lead to some consequences in the representations of isotropic functions. As mentioned, an isotropic function f of a polar symmetric tensor $\mathbf{S}$ has a form $f(\operatorname{tr}\mathbf{S}, \operatorname{tr}\mathbf{S}^2, \operatorname{tr}\mathbf{S}^3)$. Let us now see what happens if we consider an isotropic function of an axial symmetric tensor $\boldsymbol{\Omega}$. At first glance we see no difference, as the list of invariants is the same as in case of $\mathbf{S}$. So we have

$$f(\boldsymbol{\Omega}) = f(\operatorname{tr}\boldsymbol{\Omega}, \operatorname{tr}\boldsymbol{\Omega}^2, \operatorname{tr}\boldsymbol{\Omega}^3).$$

But according to definition (2.111), $f(\boldsymbol{\Omega}) = f(-\boldsymbol{\Omega})$ and our function of three invariants must satisfy the relation

$$f(\operatorname{tr}\boldsymbol{\Omega}, \operatorname{tr}\boldsymbol{\Omega}^2, \operatorname{tr}\boldsymbol{\Omega}^3) = f(-\operatorname{tr}\boldsymbol{\Omega}, \operatorname{tr}\boldsymbol{\Omega}^2, -\operatorname{tr}\boldsymbol{\Omega}^3).$$

So for an isotropic function of an axial tensor we get some additional constraints.

In addition to invariants listed in Table 2.9, we present the invariants of tensor $\mathbf{A}$ which is neither symmetric nor antisymmetric. These invariants can be obtained considering the invariants of two tensors $\mathbf{S} = \frac{1}{2}(\mathbf{A} + \mathbf{A}^T)$ and $\mathbf{W} = \frac{1}{2}(\mathbf{A} - \mathbf{A}^T)$. Omitting the calculations we present the final result. The list of six invariants of a general second-order tensor $\mathbf{A}$ is

$$\begin{aligned} &J_1(\mathbf{A}) = \operatorname{tr}\mathbf{A}, \quad J_2(\mathbf{A}) = \operatorname{tr}\mathbf{A}^2, \quad J_3(\mathbf{A}) = \operatorname{tr}\mathbf{A}^3, \\ &J_4(\mathbf{A}) = \operatorname{tr}\mathbf{A} \cdot \mathbf{A}^T, \quad J_5(\mathbf{A}) = \operatorname{tr}\mathbf{A}^2 \cdot \mathbf{A}^T, \quad J_6(\mathbf{A}) = \operatorname{tr}\mathbf{A}^2 \cdot \mathbf{A}^{T2}. \end{aligned} \tag{2.118}$$

Among these, the invariant $\operatorname{tr}\mathbf{A} \cdot \mathbf{A}^T$ is one of the most important from the mathematical point of view as it is a norm of $\mathbf{A}$. So an isotropic scalar function f of a second-order $\mathbf{A}$ is a function of the six invariants (2.118):

$$f(\mathbf{A}) = f(J_1, J_2, J_3, J_4, J_5, J_6).$$

For an isotropic scalar function of a pseudotensor $\boldsymbol{\Omega}$ we obtain a similar representation

$$f(\boldsymbol{\Omega}) = f(J_1, J_2, J_3, J_4, J_5, J_6), \quad J_k = J_k(\boldsymbol{\Omega})$$

with an additional requirement following from the relation $f(\boldsymbol{\Omega}) = f(-\boldsymbol{\Omega})$:

$$f(J_1, J_2, J_3, J_4, J_5, J_6) = f(-J_1, J_2, -J_3, J_4, -J_5, J_6).$$

The invariants of two tensors $\mathbf{A}$ and $\mathbf{B}$ are

$$\begin{aligned} &J_1(\mathbf{A}), \quad J_2(\mathbf{A}), \quad J_3(\mathbf{A}), \quad J_4(\mathbf{A}), \quad J_5(\mathbf{A}), \quad J_6(\mathbf{A}), \\ &J_1(\mathbf{B}), \quad J_2(\mathbf{B}), \quad J_3(\mathbf{B}), \quad J_4(\mathbf{B}), \quad J_5(\mathbf{B}), \quad J_6(\mathbf{B}), \\ &\operatorname{tr}\mathbf{A} \cdot \mathbf{B}, \quad \operatorname{tr}\mathbf{A}^2 \cdot \mathbf{B}, \quad \operatorname{tr}\mathbf{A} \cdot \mathbf{B}^2. \end{aligned} \tag{2.119}$$

So an isotropic function $f(\mathbf{A})$ depends on six invariants instead of nine components of $\mathbf{A}$, whereas an isotropic function $f(\mathbf{A}, \mathbf{B})$ depends on 15 invariants instead of 18 components of tensors $\mathbf{A}$ and $\mathbf{B}$. The invariants of a single tensors and two tensors which are neither symmetric nor antisymmetric, were used for example, to derive constitutive equations of elastic micropolar media [Kafadar and Eringen (1971)].

On the basis of invariants from Table 2.9, we can determine the form of such functions as a strain energy density or a dissipative potential function of an isotropic material.

Exercise 2.65. Find the forms for the presentation of isotropic scalar quadratic functions depending on (a) symmetric polar tensor $\mathbf{S}$; (b) symmetric axial tensor $\boldsymbol{\Omega}$; (c) two symmetric polar tensor $\mathbf{S}_1$ and $\mathbf{S}_2$; (d) symmetric polar tensor $\mathbf{S}$ and axial symmetric tensor $\boldsymbol{\Omega}$; (e) two symmetric axial tensors $\boldsymbol{\Omega}_1$ and $\boldsymbol{\Omega}_2$.

Exercise 2.66. Find the forms for the presentation of isotropic scalar cubic functions depending on (a) symmetric polar tensor $\mathbf{S}$; (b) symmetric axial tensor $\boldsymbol{\Omega}$.

Exercise 2.67. Find the invariants of $\mathbf{S}$ and $\mathbf{W}$ if $\mathbf{S} = \frac{1}{2}(\mathbf{A} + \mathbf{A}^T)$ and $\mathbf{W} = \frac{1}{2}(\mathbf{A} - \mathbf{A}^T)$.

Unfortunately in mechanics, isotropic or hemitropic scalar functions are not the only ones needed. Sometimes one needs isotropic tensor-valued functions such as (2.117). We are also interested in symmetric tensor-valued functions such as the stress-strain dependence, so in Table 2.10 we present some symmetric tensor-valued functions.

According to Table 2.10, a symmetric tensor-valued isotropic function of two symmetric tensors takes the form

$$\begin{aligned} \mathbf{F}(\mathbf{S}_1, \mathbf{S}_2) = {} & f_0\mathbf{E} + f_1\mathbf{S}_1 + f_2\mathbf{S}_1^2 + f_3\mathbf{S}_2 + f_4\mathbf{S}_2^2 \\ & + f_5(\mathbf{S}_1 \cdot \mathbf{S}_2 + \mathbf{S}_2 \cdot \mathbf{S}_1) + f_6(\mathbf{S}_1^2 \cdot \mathbf{S}_2 + \mathbf{S}_2 \cdot \mathbf{S}_1^2) \\ & + f_7(\mathbf{S}_1 \cdot \mathbf{S}_2^2 + \mathbf{S}_2^2 \cdot \mathbf{S}_1), \end{aligned} \tag{2.120}$$

where f_k $(k = 0, 1, \ldots, 7)$ are functions of invariants of $\mathbf{S}_1$ and $\mathbf{S}_2$ from Table 2.9. A symmetric tensor-valued isotropic function of a vector and symmetric tensor takes the form

$$\begin{aligned} \mathbf{F}(\mathbf{a}, \mathbf{S}) = {} & f_0\mathbf{E} + f_1\mathbf{S} + f_2\mathbf{S}^2 + f_3\mathbf{aa} + f_4(\mathbf{aS} \cdot \mathbf{a} + \mathbf{a} \cdot \mathbf{Sa}) \\ & + f_5(\mathbf{aS}^2 \cdot \mathbf{a} + \mathbf{a} \cdot \mathbf{S}^2\mathbf{a}), \end{aligned} \tag{2.121}$$

Table 2.10 Symmetric tensor-valued functions of tensors and vectors.

Agencies	Isotropic	Hemitropic
$\mathbf{a}$	$\mathbf{aa}$	$\mathbf{aa}$
$\mathbf{S}$	$\mathbf{E}$, $\mathbf{S}$, $\mathbf{S}^2$	$\mathbf{E}$, $\mathbf{S}$, $\mathbf{S}^2$
$\mathbf{S}_1$, $\mathbf{S}_2$	$\mathbf{E}$, $\mathbf{S}_1$, $\mathbf{S}_1^2$, $\mathbf{S}_2$, $\mathbf{S}_2^2$,	$\mathbf{E}$, $\mathbf{S}_1$, $\mathbf{S}_1^2$, $\mathbf{S}_2$, $\mathbf{S}_2^2$,
	$\mathbf{S}_1 \cdot \mathbf{S}_2 + \mathbf{S}_2 \cdot \mathbf{S}_1$,	$\mathbf{S}_1 \cdot \mathbf{S}_2 + \mathbf{S}_2 \cdot \mathbf{S}_1$,
	$\mathbf{S}_1^2 \cdot \mathbf{S}_2 + \mathbf{S}_2 \cdot \mathbf{S}_1^2$,	$\mathbf{S}_1^2 \cdot \mathbf{S}_2 + \mathbf{S}_2 \cdot \mathbf{S}_1^2$,
	$\mathbf{S}_1 \cdot \mathbf{S}_2^2 + \mathbf{S}_2^2 \cdot \mathbf{S}_1$	$\mathbf{S}_1 \cdot \mathbf{S}_2^2 + \mathbf{S}_2^2 \cdot \mathbf{S}_1$
$\mathbf{a}$, $\mathbf{S}$	$\mathbf{E}$, $\mathbf{S}$, $\mathbf{S}^2$, $\mathbf{aa}$,	$\mathbf{E}$, $\mathbf{S}$, $\mathbf{S}^2$, $\mathbf{aa}$,
	$(\mathbf{aS} \cdot \mathbf{a})_S$, $(\mathbf{aS}^2 \cdot \mathbf{a})_S$	$(\mathbf{aS} \cdot \mathbf{a})_S$, $(\mathbf{S} \times \mathbf{a})_S$,
		$(\mathbf{S}^2 \times \mathbf{a})_S$, $(\mathbf{aa} \times \mathbf{S} \cdot \mathbf{a})_S$
$\mathbf{W}$	$\mathbf{W}^2$	$\mathbf{W}^2$
$\mathbf{S}$, $\mathbf{W}$	$\mathbf{E}$, $\mathbf{S}$, $\mathbf{S}^2$, $\mathbf{W}^2$,	$\mathbf{E}$, $\mathbf{S}$, $\mathbf{S}^2$, $\mathbf{W}^2$,
	$\mathbf{S} \cdot \mathbf{W} - \mathbf{W} \cdot \mathbf{S}$, $\mathbf{S} \cdot \mathbf{W}^2 + \mathbf{W}^2 \cdot \mathbf{S}$,	$\mathbf{S} \cdot \mathbf{W} - \mathbf{W} \cdot \mathbf{S}$, $\mathbf{S} \cdot \mathbf{W}^2 + \mathbf{W}^2 \cdot \mathbf{S}$,
	$\mathbf{W} \cdot \mathbf{S} \cdot \mathbf{W}^2 - \mathbf{W}^2 \cdot \mathbf{S} \cdot \mathbf{W}$,	$\mathbf{W} \cdot \mathbf{S} \cdot \mathbf{W}^2 - \mathbf{W}^2 \cdot \mathbf{S} \cdot \mathbf{W}$,
	$\mathbf{S}^2 \cdot \mathbf{W} - \mathbf{W} \cdot \mathbf{S}^2$	$\mathbf{S}^2 \cdot \mathbf{W} - \mathbf{W} \cdot \mathbf{S}^2$

where f_k $(k = 0, 1, \ldots, 5)$ are functions of invariants of $\mathbf{a}$ and $\mathbf{S}$ from Table 2.9.

Table 2.11 Antisymmetric tensor functions of tensors and vectors.

Agencies	Isotropic	Hemitropic
$\mathbf{a}$	–	$\mathbf{E} \times \mathbf{a}$
$\mathbf{W}$	$\mathbf{W}$	$\mathbf{W}$
$\mathbf{S}$, $\mathbf{W}$	$\mathbf{W}$, $\mathbf{S} \cdot \mathbf{W} + \mathbf{W} \cdot \mathbf{S}$,	$\mathbf{W}$, $\mathbf{S} \cdot \mathbf{W} + \mathbf{W} \cdot \mathbf{S}$,
	$\mathbf{S} \cdot \mathbf{W}^2 - \mathbf{W}^2 \cdot \mathbf{S}$,	$\mathbf{S} \cdot \mathbf{W}^2 - \mathbf{W}^2 \cdot \mathbf{S}$,
$\mathbf{a}$, $\mathbf{S}$	$(\mathbf{aS} \cdot \mathbf{a})_A$, $(\mathbf{aS}^2 \cdot \mathbf{a})_A$, $(\mathbf{S} \cdot \mathbf{aS}^2 \cdot \mathbf{a})_A$	$\mathbf{E} \times \mathbf{a}$, $(\mathbf{aS} \cdot \mathbf{a})_A$ $(\mathbf{S} \times \mathbf{a})_A$

For consistency, in Table 2.11 we also present antisymmetric tensor-valued functions. The functions from both Tables 2.10 and 2.11 may be combined to find various functional dependencies. For example, a tensor function of a second-order tensor $\mathbf{A}$ can be treated as a sum of functions from the two tables, which depend on symmetric and antisymmetric tensors. Indeed, we represent $\mathbf{F}(\mathbf{A})$ as

$$\mathbf{F}(\mathbf{A}) = \mathbf{F}_S(\mathbf{S}, \mathbf{W}) + \mathbf{F}_A(\mathbf{S}, \mathbf{W}),$$

where

$$\mathbf{F}_S = \tfrac{1}{2}(\mathbf{F}(\mathbf{A}) + \mathbf{F}^T(\mathbf{A})), \qquad \mathbf{S} = \tfrac{1}{2}(\mathbf{A} + \mathbf{A}^T),$$
$$\mathbf{F}_A = \tfrac{1}{2}(\mathbf{F}(\mathbf{A}) - \mathbf{F}^T(\mathbf{A})), \qquad \mathbf{W} = \tfrac{1}{2}(\mathbf{A} - \mathbf{A}^T).$$

For $\mathbf{F}_S$ and $\mathbf{F}_A$ we get

$$\begin{aligned}\mathbf{F}_S(\mathbf{S},\mathbf{W}) &= s_0\mathbf{E} + s_1\mathbf{S} + s_2\mathbf{S}^2 + s_3\mathbf{W}^2 + s_4(\mathbf{S}\cdot\mathbf{W} - \mathbf{W}\cdot\mathbf{S})\\ &\quad + s_5(\mathbf{S}\cdot\mathbf{W}^2 + \mathbf{W}^2\cdot\mathbf{S}) + s_6(\mathbf{W}\cdot\mathbf{S}\cdot\mathbf{W}^2 - \mathbf{W}^2\cdot\mathbf{S}\cdot\mathbf{W})\\ &\quad + s_7(\mathbf{S}^2\cdot\mathbf{W} - \mathbf{W}\cdot\mathbf{S}^2),\\ \mathbf{F}_A(\mathbf{S},\mathbf{W}) &= a_0\mathbf{W} + a_1(\mathbf{S}\cdot\mathbf{W} + \mathbf{W}\cdot\mathbf{S}) + a_3(\mathbf{S}\cdot\mathbf{W}^2 - \mathbf{W}^2\cdot\mathbf{S}),\end{aligned}$$

where s_k and a_k are functions of invariants of $\mathbf{A}$ given in (2.118). Replacing $\mathbf{S}$ and $\mathbf{W}$ by their expressions through $\mathbf{A}$ and $\mathbf{A}^T$, we can obtain the form of an isotropic tensor function of non-symmetric tensor $\mathbf{A}$.

In addition to Orth and Orth_+, we can introduce other symmetry groups. For example, the *transverse symmetry group* includes the rotational symmetry; it contains all rotation tensors $\mathbf{R}(\omega\mathbf{e})$, where ω and $\mathbf{e}$ are the angle and fixed axis of rotation respectively. In general, there are five transverse isotropy groups, mostly differing by the inclusion of different reflection transformations [Zheng (1994)], see Table 2.12. Here $\mathbf{i}$ is a unit vector orthogonal to $\mathbf{e}$ and

$$\mathbf{R}_\mathbf{i} = \mathbf{E} - 2\mathbf{i}\mathbf{i}.$$

The simplest group, called the rotational symmetry group $\mathcal{C}_\infty$, contains only the rotations about $\mathbf{e}$.

Table 2.12 Transverse symmetry groups

Group	Generators	Structural tensors
$\mathcal{C}_\infty$	$\mathbf{R}(\omega\mathbf{e})$	$\mathbf{e}$, $\boldsymbol{\mathcal{E}}$
$\mathcal{C}_{\infty\nu}$	$\mathbf{R}(\omega\mathbf{e})$, $\mathbf{R}_\mathbf{i}$	$\mathbf{e}$
$\mathcal{C}_{\infty h}$	$\mathbf{R}(\omega\mathbf{e})$, $-\mathbf{E}$	$\mathbf{E}\times\mathbf{e}$
$\mathcal{D}_{\infty h}$	$\mathbf{R}(\omega\mathbf{e})$, $-\mathbf{E}$, $\mathbf{R}_\mathbf{i}$	$\mathbf{e}\mathbf{e}$
$\mathcal{D}_\infty$	$\mathbf{R}(\omega\mathbf{e})$, $-\mathbf{R}_\mathbf{i}$	$\mathbf{e}\mathbf{e}$, $\boldsymbol{\mathcal{E}}$

Table 2.13 presents the invariants of tensors with respect to transverse symmetry groups $D_{\infty h}$, D_∞, and $C_{\infty\nu}$, which can be used in the representations of transversely isotropic functions. So a transversely isotropic

Table 2.13 Invariants with respect to $\mathcal{D}_{\infty h}$, $\mathcal{D}_{\infty}$ and $\mathcal{C}_{\infty\nu}$.

Agencies	Invariants
$\mathbf{S}$	$\operatorname{tr}\mathbf{S}$, $\operatorname{tr}\mathbf{S}^2$, $\operatorname{tr}\mathbf{S}^3$, $\mathbf{e}\cdot\mathbf{S}\cdot\mathbf{e}$, $\mathbf{e}\cdot\mathbf{S}^2\cdot\mathbf{e}$
$\mathbf{W}$	$\operatorname{tr}\mathbf{W}^2$, $\mathbf{e}\cdot\mathbf{W}^2\cdot\mathbf{e}$
$\mathbf{S}$, $\mathbf{W}$	$\operatorname{tr}\mathbf{S}$, $\operatorname{tr}\mathbf{S}^2$, $\operatorname{tr}\mathbf{S}^3$, $\mathbf{e}\cdot\mathbf{S}\cdot\mathbf{e}$, $\mathbf{e}\cdot\mathbf{S}^2\cdot\mathbf{S}$, $\operatorname{tr}\mathbf{W}^2$, $\mathbf{e}\cdot\mathbf{W}^2\cdot\mathbf{e}$, $\operatorname{tr}\mathbf{S}\cdot\mathbf{W}^2$, $\operatorname{tr}\mathbf{S}^2\cdot\mathbf{W}^2$, $\operatorname{tr}\mathbf{S}^2\cdot\mathbf{W}^2\cdot\mathbf{S}\cdot\mathbf{W}$, $\mathbf{e}\cdot\mathbf{S}\cdot\mathbf{W}\cdot\mathbf{e}$, $\mathbf{e}\cdot\mathbf{S}^2\cdot\mathbf{W}\cdot\mathbf{e}$, $\mathbf{e}\cdot\mathbf{W}\cdot\mathbf{S}\cdot\mathbf{W}^2\cdot\mathbf{e}$

function $f(\mathbf{S})$ is a function of five invariants from Table 2.13:

$$f(\mathbf{S}) = f(\operatorname{tr}\mathbf{S}, \operatorname{tr}\mathbf{S}^2, \operatorname{tr}\mathbf{S}^3, \mathbf{e}\cdot\mathbf{S}\cdot\mathbf{e}, \mathbf{e}\cdot\mathbf{S}^2\cdot\mathbf{e}). \tag{2.122}$$

Let us note that if we consider the function defined in (2.122) as a function of $\mathbf{S}$ and unit vector $\mathbf{e}$, the new function $f(\mathbf{S}, \mathbf{e})$ is isotropic. So the type of isotropy depends on the set of explicitly chosen arguments. Moreover, the set of the invariants with respect to a considered symmetry group can be obtained as follows. First we introduce a *structural tensor* which has exactly the same symmetry as the considered symmetry group. For example, the dyad $\mathbf{ee}$ is invariant under any rotation about $\mathbf{e}$, under the mirror reflection $-\mathbf{E}$, and under reflection $\mathbf{R_i}$ with $\mathbf{i}$ orthogonal to $\mathbf{e}$. Thus, $\mathbf{ee}$ can be treated as the structural tensor for $\mathcal{D}_{\infty h}$. Then we should find the set of isotropic invariants for the initial set of vectors and tensors together with chosen structural tensor. In other words, instead of finding a function invariant under certain symmetry group transformations, we should find an isotropic function of the initial set of vectors and tensors supplemented by the structural tensor. For example, the transverse symmetric function $f = f(\mathbf{S})$ with respect to $\mathcal{D}_{\infty h}$ can be constructed as an isotropic function of two arguments:

$$f = f(\mathbf{S}, \mathbf{ee}).$$

So we transfer the difficulties in finding invariants under transformations from certain symmetry group to the problem of finding of isotropic invariants but for the extended set of tensors. This makes the problem of construction of isotropic functions even more important as isotropic functions can also be used to derive constitutive equations of anisotropic materials and generalized media. Sometimes one should use vector, second- or higher-order tensor or even a collection of such vectors and tensors as structural tensors which correspond to a considered symmetry group. Structural tensors for transverse symmetry groups are listed in Table 2.12.

Table 2.14 Invariants with respect to $\mathcal{D}_{\infty h}$, $\mathcal{D}_\infty$ and $\mathcal{C}_{\infty\nu}$.

Agencies	$\mathcal{D}_{\infty h}$	$\mathcal{C}_{\infty\nu}$	$\mathcal{D}_\infty$
$\mathbf{a}$	$\mathbf{a}\cdot\mathbf{a}$, $(\mathbf{e}\cdot\mathbf{a})^2$	$\mathbf{a}\cdot\mathbf{a}$, $\mathbf{e}\cdot\mathbf{a}$	$\mathbf{a}\cdot\mathbf{a}$, $(\mathbf{e}\cdot\mathbf{a})^2$
$\mathbf{a}$, $\mathbf{S}$	$\mathbf{a}\cdot\mathbf{a}$, $(\mathbf{e}\cdot\mathbf{a})^2$, $\operatorname{tr}\mathbf{S}$, $\operatorname{tr}\mathbf{S}^2$, $\operatorname{tr}\mathbf{S}^3$, $\mathbf{e}\cdot\mathbf{S}\cdot\mathbf{e}$, $\mathbf{e}\cdot\mathbf{S}^2\cdot\mathbf{S}$, $\mathbf{a}\cdot\mathbf{S}\cdot\mathbf{a}$, $\mathbf{a}\cdot\mathbf{S}^2\cdot\mathbf{a}$, $(\mathbf{e}\cdot\mathbf{a})(\mathbf{e}\cdot\mathbf{S}\cdot\mathbf{a})$	$\mathbf{a}\cdot\mathbf{a}$, $\mathbf{e}\cdot\mathbf{a}$, $\operatorname{tr}\mathbf{S}$, $\operatorname{tr}\mathbf{S}^2$, $\operatorname{tr}\mathbf{S}^3$, $\mathbf{e}\cdot\mathbf{S}\cdot\mathbf{e}$, $\mathbf{e}\cdot\mathbf{S}^2\cdot\mathbf{S}$, $\mathbf{a}\cdot\mathbf{S}\cdot\mathbf{a}$, $\mathbf{e}\cdot\mathbf{S}\cdot\mathbf{a}$, $\mathbf{e}\cdot\mathbf{S}^2\cdot\mathbf{a}$	$\mathbf{a}\cdot\mathbf{a}$, $(\mathbf{e}\cdot\mathbf{a})^2$, $\operatorname{tr}\mathbf{S}$, $\operatorname{tr}\mathbf{S}^2$, $\operatorname{tr}\mathbf{S}^3$, $\mathbf{e}\cdot\mathbf{S}\cdot\mathbf{e}$, $\mathbf{e}\cdot\mathbf{S}^2\cdot\mathbf{S}$, $\mathbf{a}\cdot\mathbf{S}\cdot\mathbf{a}$, $\mathbf{a}\cdot\mathbf{S}^2\cdot\mathbf{a}$, $\mathbf{a}\cdot(\mathbf{a}\cdot\mathbf{S}\times\mathbf{S}^2\cdot\mathbf{a})$, $\mathbf{e}\cdot(\mathbf{a}\times\mathbf{S}\cdot\mathbf{e})$, $\mathbf{e}\cdot(\mathbf{a}\times\mathbf{S}^2\cdot\mathbf{e})$, $(\mathbf{e}\cdot\mathbf{a})\mathbf{e}\cdot(\mathbf{a}\times\mathbf{S}\cdot\mathbf{a})$
$\mathbf{a}$, $\mathbf{W}$	$\mathbf{a}\cdot\mathbf{a}$, $(\mathbf{e}\cdot\mathbf{a})^2$, $\operatorname{tr}\mathbf{W}^2$, $\mathbf{e}\cdot\mathbf{W}^2\cdot\mathbf{e}$, $\mathbf{a}\cdot\mathbf{W}^2\cdot\mathbf{a}$, $(\mathbf{e}\cdot\mathbf{a})(\mathbf{e}\cdot\mathbf{W}\cdot\mathbf{a})$, $(\mathbf{e}\cdot\mathbf{W}\cdot\mathbf{a})(\mathbf{e}\cdot\mathbf{W}^2\cdot\mathbf{a})$	$\mathbf{a}\cdot\mathbf{a}$, $\mathbf{e}\cdot\mathbf{a}$, $\operatorname{tr}\mathbf{W}^2$, $\mathbf{e}\cdot\mathbf{W}^2\cdot\mathbf{e}$, $\mathbf{e}\cdot\mathbf{W}\cdot\mathbf{a}$, $\mathbf{e}\cdot\mathbf{W}^2\cdot\mathbf{a}$	$\mathbf{a}\cdot\mathbf{a}$, $(\mathbf{e}\cdot\mathbf{a})^2$, $\operatorname{tr}\mathbf{W}^2$, $\mathbf{e}\cdot\mathbf{W}^2\cdot\mathbf{e}$ $\mathbf{a}\cdot\mathbf{W}_\times$, $\mathbf{e}\cdot(\mathbf{a}\times\mathbf{W}\cdot\mathbf{e})$, $\mathbf{e}\cdot(\mathbf{a}\times\mathbf{W}^2\cdot\mathbf{e})$, $(\mathbf{e}\cdot\mathbf{a})(\mathbf{e}\cdot\mathbf{W}\cdot\mathbf{a})$

Observing that invariants from Table 2.13 coincide for the groups $D_{\infty h}$, D_∞, and $C_{\infty\nu}$, one may conclude that there is no difference for these transversely isotropic groups in case of the invariants of a single tensor or of two tensors. However, the difference arises in more general cases, as demonstrated by the invariants of vector and tensor given in Table 2.14 for the same symmetry groups. For complete analysis of invariants for these and other transverse isotropic groups $\mathcal{C}_\infty$ and $\mathcal{C}_{\infty h}$ including the structure of vector- and tensor-valued functions, we refer to [Zheng (1994)].

An efficient way to calculate a number of invariants and invariants themselves with respect to arbitrary rotations is based on the theory of first-order differential equations. The invariance of f under rotations about axis $\mathbf{e}$ means that

$$f(\mathbf{S}) = f(\mathbf{Q}\cdot\mathbf{S}\cdot\mathbf{Q}^T), \quad \mathbf{Q} = \mathbf{R}(\omega\mathbf{e}) \tag{2.123}$$

for all ω. In other words, $f(\mathbf{R}(\omega\mathbf{e})\cdot\mathbf{S}\cdot\mathbf{R}^T(\omega\mathbf{e}))$ does not depend on ω.

Differentiating (2.123) with respect to ω we get

$$
\begin{aligned}
0 &= \frac{d}{d\omega} f(\mathbf{Q}\cdot\mathbf{S}\cdot\mathbf{Q}^T) \\
&= \frac{d}{d\omega} f(\mathbf{R}(\omega\mathbf{e})\cdot\mathbf{S}\cdot\mathbf{R}^T(\omega\mathbf{e})) \\
&= f_{,(\mathbf{R}\cdot\mathbf{S}\cdot\mathbf{R}^T)} \bullet \left[\frac{d}{d\omega}\mathbf{R}(\omega\mathbf{e})\cdot\mathbf{S}\cdot\mathbf{R}^T(\omega\mathbf{e}) + \mathbf{R}(\omega\mathbf{e})\cdot\mathbf{S}\cdot\frac{d}{d\omega}\mathbf{R}^T(\omega\mathbf{e})\right] \\
&= f_{,\mathbf{S}} \bullet \left\{\mathbf{R}^T\cdot\left[\frac{d}{d\omega}\mathbf{R}\cdot\mathbf{S}\cdot\mathbf{R}^T + \mathbf{R}\cdot\mathbf{S}\cdot\frac{d}{d\omega}\mathbf{R}^T\right]\cdot\mathbf{R}\right\} \\
&= f_{,\mathbf{S}} \bullet \left[\mathbf{R}^T\cdot\frac{d\mathbf{R}}{d\omega}\cdot\mathbf{S} + \mathbf{S}\cdot\frac{d\mathbf{R}^T}{d\omega}\cdot\mathbf{R}\right].
\end{aligned}
$$

As $\mathbf{R}^T\cdot\mathbf{R} = \mathbf{E}$, we have

$$
\mathbf{R}^T\cdot\frac{d\mathbf{R}}{d\omega} + \frac{d\mathbf{R}^T}{d\omega}\cdot\mathbf{R} = \mathbf{0}.
$$

So $\mathbf{R}^T\cdot\frac{d\mathbf{R}}{d\omega}$ is a skew tensor and may be represented as $\mathbf{E}\times\boldsymbol{\omega}$ with a conjugate vector $\boldsymbol{\omega}$. Calculating explicitly the derivative of the rotation tensor

$$
\begin{aligned}
\frac{d}{d\omega}\mathbf{R}(\omega\mathbf{e}) &= \frac{d}{d\omega}\left[\mathbf{E}\cos\omega + (1-\cos\omega)\mathbf{e}\mathbf{e} - \mathbf{e}\times\mathbf{E}\sin\omega\right] \\
&= -\mathbf{E}\sin\omega + \sin\omega\,\mathbf{e}\mathbf{e} - \mathbf{e}\times\mathbf{E}\cos\omega \\
&= -(\mathbf{E}-\mathbf{e}\mathbf{e})\sin\omega + \mathbf{e}\times\mathbf{E}\cos\omega, \\
&= -\mathbf{R}(\omega\mathbf{e})\times\mathbf{e},
\end{aligned}
$$

$$
\frac{d}{d\omega}\mathbf{R}^T(\omega\mathbf{e}) = \mathbf{e}\times\mathbf{R}^T(\omega\mathbf{e})
$$

we find that the conjugate vector $\boldsymbol{\omega} = \mathbf{e}$. So the invariance of f transforms to

$$
\frac{df}{d\omega} = f_{,\mathbf{S}} \bullet [-\mathbf{e}\times\mathbf{S} + \mathbf{S}\times\mathbf{e}] = 0. \tag{2.124}
$$

As a result, we obtain the first-order linear homogeneous differential equation for $f(\mathbf{S})$. The methods for solution of such equations are well-known, see for example [Arnold (1992)] or the classic book [Courant and Hilbert (1989)]. In particular, (2.124) reduces to an ordinary differential equation called the *characteristic equation*

$$
\dot{\mathbf{S}} = -\mathbf{e}\times\mathbf{S} + \mathbf{S}\times\mathbf{e}. \tag{2.125}
$$

Function f is a solution of (2.124) if and only if it is a first integral of the characteristic equation (2.125). Note that in the general case of n simultaneous ordinary equations, there exist $n-1$ functionally independent first integrals [Arnold (1992)]. Equation (2.125) is a system of six scalar differential equations for the components of symmetric tensor $\mathbf{S}$, so it has five functionally independent first integrals which are invariant with respect to rotational symmetry.

The method can be extended for analysis of invariance under rotational symmetry of functions of several vector and tensor arguments.

Exercise 2.68. Prove that $I_6(\mathbf{S}) = \mathbf{e}\cdot\mathbf{S}^2\cdot(\mathbf{e}\times\mathbf{S}\cdot\mathbf{e})$ is also an invariant of $\mathbf{S}$ with respect to the rotational symmetry. Demonstrate that I_6 is a relative invariant.

Exercise 2.69. Prove that for $I_6(\mathbf{S}) \equiv \mathbf{e}\cdot\mathbf{S}^2\cdot(\mathbf{e}\times\mathbf{S}\cdot\mathbf{e})$ there holds [Naumenko and Altenbach (2016)]

$$I_6^2(\mathbf{S}) = \begin{vmatrix} 1 & \mathbf{e}\cdot\mathbf{S}\cdot\mathbf{e} & \mathbf{e}\cdot\mathbf{S}^2\cdot\mathbf{e} \\ \mathbf{e}\cdot\mathbf{S}\cdot\mathbf{e} & \mathbf{e}\cdot\mathbf{S}^2\cdot\mathbf{e} & \mathbf{e}\cdot\mathbf{S}^3\cdot\mathbf{e} \\ \mathbf{e}\cdot\mathbf{S}^2\cdot\mathbf{e} & \mathbf{e}\cdot\mathbf{S}^3\cdot\mathbf{e} & \mathbf{e}\cdot\mathbf{S}^4\cdot\mathbf{e} \end{vmatrix}.$$

So I_6 depends on the other five transverse invariants of $\mathbf{S}$.

The theory of invariants answer questions on the number of invariants and the invariants themselves for other symmetry groups, including the discrete groups for the representations of scalar-, vector-, and tensor-valued functions with respect to corresponding symmetry. Additionally, let us present the invariants of a symmetric tensor with respect to the orthotropic group. The *orthotropic symmetry group* contains the rotations with angle π about two orthogonal axes with unit vectors $\mathbf{i}_1$ and $\mathbf{i}_2$. So the following proper orthogonal tensors can be considered as the generators:

$$\mathbf{E}, \quad \mathbf{Q}_1 \equiv \mathbf{R}(\pi\mathbf{i}_1) = 2\mathbf{i}_1\mathbf{i}_1 - \mathbf{E}, \quad \mathbf{Q}_2 \equiv \mathbf{R}(\pi\mathbf{i}_2) = 2\mathbf{i}_2\mathbf{i}_2 - \mathbf{E}.$$

As by definition $\mathbf{Q}_3 \equiv \mathbf{Q}_1\cdot\mathbf{Q}_2 = 2\mathbf{i}_3\mathbf{i}_3 - \mathbf{E}$ also belongs to the group, note that the rotations with angle π about vector $\mathbf{i}_3$ orthogonal to $\mathbf{i}_1$ and $\mathbf{i}_2$ are also included in the orthotropic symmetry group. The invariants of a symmetric tensor $\mathbf{S}$ are [Lurie (1990)]

$$\mathbf{i}_1\cdot\mathbf{S}\cdot\mathbf{i}_1, \quad \mathbf{i}_2\cdot\mathbf{S}\cdot\mathbf{i}_2, \quad \mathbf{i}_3\cdot\mathbf{S}\cdot\mathbf{i}_3, \quad (\mathbf{i}_1\cdot\mathbf{S}\cdot\mathbf{i}_2)^2, \quad (\mathbf{i}_1\cdot\mathbf{S}\cdot\mathbf{i}_3)^2, \quad (\mathbf{i}_2\cdot\mathbf{S}\cdot\mathbf{i}_3)^2, \quad \det\mathbf{S}. \tag{2.126}$$

Exercise 2.70. Show that the following also qualifies as a list of principal invariants [Naumenko and Altenbach (2016)]: $\mathbf{i}_1 \cdot \mathbf{S} \cdot \mathbf{i}_1$, $\mathbf{i}_2 \cdot \mathbf{S} \cdot \mathbf{i}_2$, $\mathbf{i}_3 \cdot \mathbf{S} \cdot \mathbf{i}_3$, $\mathbf{i}_1 \cdot \mathbf{S}^2 \cdot \mathbf{i}_1$, $\mathbf{i}_2 \cdot \mathbf{S}^2 \cdot \mathbf{i}_2$, $\mathbf{i}_3 \cdot \mathbf{S}^2 \cdot \mathbf{i}_3$, and $\operatorname{tr} \mathbf{S}^3$.

Finally, we should emphasize that this paragraph presents some, mostly classic, results from the theory of invariants. There remain many unsolved problems in this area. Here we mention only the problem of deriving the invariants of higher-order tensors. A novice could refer to [Spencer (1971); Boehler (1987); Zheng (1994)].

2.12 Problems

Here, unless otherwise stated, we use $\mathbf{A}, \mathbf{B}, \mathbf{C}, \mathbf{X}$ to denote second-order tensors, λ a scalar, $\boldsymbol{\Omega}$ an antisymmetric second-order tensor, and $\mathbf{Q}$ an orthogonal second-order tensor.

2.1 Let $\mathbf{A}$ and $\mathbf{B}$ be second-order tensors. Show that if $\mathbf{A}$ is symmetric and $\mathbf{B}$ is antisymmetric, then $\mathbf{B} \cdot \mathbf{A} \cdot \mathbf{B}$ is symmetric.

2.2 Show that if $\mathbf{A}$ and $\mathbf{B}$ are both symmetric or both antisymmetric, then the tensor $\mathbf{C} = \mathbf{A} \cdot \mathbf{B} - \mathbf{B} \cdot \mathbf{A}$ is antisymmetric.

2.3 Show that

(a) $\mathbf{A} \bullet \mathbf{B} = \operatorname{tr}(\mathbf{A} \cdot \mathbf{B}^T)$;
(b) $\operatorname{tr}(\mathbf{A} \cdot \mathbf{B}) = \operatorname{tr}(\mathbf{B} \cdot \mathbf{A})$;
(c) $\operatorname{tr}(\mathbf{A} \cdot \mathbf{B}) = \mathbf{A} \cdot\cdot\, \mathbf{B} = \mathbf{A}^T \cdot\cdot\, \mathbf{B}^T$;
(d) $\operatorname{tr}(\mathbf{a} \times \mathbf{E} \times \mathbf{b}) = -2\mathbf{a} \cdot \mathbf{b}$;
(e) $\operatorname{tr}[\mathbf{u} \times (\mathbf{v} \times \mathbf{A})] = \mathbf{u} \cdot \mathbf{A} \cdot \mathbf{v} - \mathbf{u} \cdot \mathbf{v} \operatorname{tr} \mathbf{A}$.

2.4 Show that $\operatorname{tr}(\mathbf{A} \times \mathbf{v}) = \operatorname{tr}(\mathbf{v} \times \mathbf{A})$, and that these quantities vanish for $\mathbf{A}$ symmetric.

2.5 Find the ball and deviator parts of the tensors (a) $\mathbf{i}_1\mathbf{i}_2$ and (b) $\mathbf{aa}$.

2.6 Prove (2.25).

2.7 Let $\mathbf{A}$ be a symmetric, invertible tensor. Show that

$$I_1(\mathbf{A}^{-1}) = \frac{I_2(\mathbf{A})}{I_3(\mathbf{A})}, \qquad I_2(\mathbf{A}^{-1}) = \frac{I_1(\mathbf{A})}{I_3(\mathbf{A})}, \qquad I_3(\mathbf{A}^{-1}) = I_3^{-1}(\mathbf{A}).$$

2.8 Find the principal invariants of the tensors (a) $\mathbf{aa}$ and (b) $\lambda\mathbf{E}$.

2.9 Find the left and right polar decompositions of the tensors (a) $\lambda\mathbf{E}$, (b) $\mathbf{aa}+\mathbf{bb}+\mathbf{cc}$ for $\mathbf{a},\mathbf{b},\mathbf{c}$ mutually orthogonal, (c) $\mathbf{E}+\gamma\mathbf{i}_1\mathbf{i}_2$, and (d) $\lambda_1\mathbf{i}_1\mathbf{i}_1+\lambda_2\mathbf{i}_2\mathbf{i}_2+\lambda_3\mathbf{i}_3\mathbf{i}_3+\gamma\mathbf{i}_1\mathbf{i}_2$.

2.10 Show that $I_2(\operatorname{dev}\mathbf{A})\le 0$ for any symmetric tensor $\mathbf{A}$.

2.11 For a second-order tensor $\mathbf{A}=a_{mn}\mathbf{e}^m\mathbf{e}^n$, the *vectorial invariant* was introduced by J.W. Gibbs (1839–1903) as $\mathbf{A}_\times=a_{mn}\mathbf{e}^m\times\mathbf{e}^n$. Find the vectorial invariants of (a) $\mathbf{ab}$, (b) $\mathbf{aa}$, (c) $\mathbf{E}\times\mathbf{v}$, (d) $\mathbf{ab}-\mathbf{ba}$, (e) a symmetric tensor $\mathbf{A}$, and (f) $\mathbf{E}$.

2.12 Prove:

(a) $\mathbf{A}_\times=\frac{1}{2}(\mathbf{A}-\mathbf{A}^T)_\times$;
(b) $(\mathbf{v}\times\mathbf{A})_\times=\mathbf{A}\cdot\mathbf{v}-\mathbf{v}\operatorname{tr}\mathbf{A}$;
(c) $(\mathbf{v}\times\mathbf{E})_\times=-2\mathbf{v}$.

2.13 Show that if $\mathbf{A}$ and $\mathbf{B}$ are symmetric, then $(\mathbf{A}\cdot\mathbf{B})_\times+(\mathbf{B}\cdot\mathbf{A})_\times=\mathbf{0}$.

2.14 Show that $(\mathbf{E}\times\mathbf{v})\cdot\mathbf{A}=\mathbf{v}\times\mathbf{A}$.

2.15 Show that $\mathbf{A}\times\mathbf{v}=-(\mathbf{v}\times\mathbf{A}^T)^T$ and $\mathbf{v}\times\mathbf{A}=-(\mathbf{A}^T\times\mathbf{v})^T$.

2.16 Show that $\mathbf{Q}=\mathbf{E}-2\mathbf{ee}$ is orthogonal if $\mathbf{e}\cdot\mathbf{e}=1$.

2.17 Show that $\mathbf{v}\times\mathbf{E}\times\mathbf{v}=\mathbf{vv}-\mathbf{E}$ if $\mathbf{v}\cdot\mathbf{v}=1$.

2.18 Show that $\mathbf{Q}=(\mathbf{E}+\boldsymbol{\Omega})\cdot(\mathbf{E}-\boldsymbol{\Omega})^{-1}$ is orthogonal if $\boldsymbol{\Omega}$ is antisymmetric.

2.19 Show that if $\mathbf{Q}$ is orthogonal then $(\mathbf{Q}-\mathbf{E})\cdot\cdot(\mathbf{Q}^T-\mathbf{E})=6-2\operatorname{tr}\mathbf{Q}$.

2.20 Show that if $\mathbf{Q}$ is proper orthogonal tensor then $\mathbf{Q}_\times=2\sin\omega\mathbf{e}$ where ω and $\mathbf{e}$ are the rotation angle and axis, respectively.

2.21 Show that $\det\mathbf{A}=\frac{1}{6}(\operatorname{tr}^3\mathbf{A}-3\operatorname{tr}\mathbf{A}\operatorname{tr}\mathbf{A}^2+2\operatorname{tr}\mathbf{A}^3)$.

2.22 Show that $\det e^{\mathbf{A}}=e^{\operatorname{tr}\mathbf{A}}$ (*Liouville's formula*) for symmetric $\mathbf{A}$.

2.23 Show that $e^{\mathbf{A}}$ is nonsingular.

2.24 Show that $e^{\boldsymbol{\Omega}}$ is orthogonal if $\boldsymbol{\Omega}$ is antisymmetric.

2.25 Find $f_{,\mathbf{X}}$ if $f(\mathbf{X})=\mathbf{a}\cdot\mathbf{X}\cdot\mathbf{b}$.

2.26 Let $f(x)$ be an analytic function in an open ball $x < r$ centered at $x = 0$; that is, suppose the representation

$$f(x) = \sum_{k=0}^{\infty} \frac{f^{(k)}(0)}{k!} x^k$$

holds and the series converges for $x < r$. In the ball, the relation

$$f'(x) = \sum_{k=1}^{\infty} \frac{f^{(k)}(0)}{(k-1)!} x^{k-1}$$

holds. Let

$$\mathbf{F}(\mathbf{X}) = \sum_{k=0}^{\infty} \frac{f^{(k)}(0)}{k!} \mathbf{X}^k.$$

Demonstrate that for $\|\mathbf{X}\| < r$

$$[\text{tr}(\mathbf{F}(\mathbf{X})]_{,\mathbf{X}} = \sum_{k=1}^{\infty} \frac{f^{(k)}(0)}{(k-1)!} \mathbf{X}^{T^{k-1}} = f'(\mathbf{X}^T).$$

2.27 Let $f(\mathbf{X})$ be a scalar-valued function. Show that the derivative of $f(\mathbf{X})$ with respect to $\mathbf{X}^T$ is equal to the transpose of its derivative with respect to $\mathbf{X}$, i.e., $f(\mathbf{X})_{,\mathbf{X}^T} = (f(\mathbf{X})_{,\mathbf{X}})^T$.

2.28 Differentiate the following tensor-valued functions: (a) $\text{tr}(\mathbf{A} \cdot \mathbf{X}^T)\mathbf{E}$, (b) $\text{tr}(\mathbf{A} \cdot \mathbf{X})\mathbf{E}$, and (c) $\mathbf{a} \cdot \mathbf{X} \cdot \mathbf{bcd}$ with respect to $\mathbf{X}$.

2.29 Show that if $\mathbf{X}$ is nonsingular then $(\det \mathbf{X})_{,\mathbf{X}} = (\det \mathbf{X})\mathbf{X}^{-T}$.

2.30 Show that

$$\frac{df}{d\mathbf{X}^{-1}} = -\mathbf{X}^T \cdot f_{,\mathbf{X}} \cdot \mathbf{X}^T.$$

2.31 Show that the set S of symmetric second-order tensors is a subspace of the linear space of second-order tensors. On S the inner product $\mathbf{A} \cdot\cdot \mathbf{B}^T$ of the whole space takes the form $\mathbf{A} \cdot\cdot \mathbf{B}$.

2.32 Let f be a scalar-valued function of a symmetric second-order tensor $\mathbf{A}$. Show that there is a unique symmetric second-order tensor $\mathbf{B}$ such that $f(\mathbf{A}) = \text{tr}(\mathbf{B} \cdot \mathbf{A})$ for any symmetric $\mathbf{A}$.

2.33 Let $\mathbf{F} = \mathbf{F}(\mathbf{A})$ be a linear function from the set of second-order symmetric tensors $\mathbf{A}$ to the set of second-order tensors. Show that there is a unique fourth-order tensor $\mathbf{C}$ such that $\mathbf{F}(\mathbf{A}) = \mathbf{C} \cdot\cdot \mathbf{A}$, $\mathbf{C} = c_{mnpt}\mathbf{i}_m\mathbf{i}_n\mathbf{i}_p\mathbf{i}_t$, and $c_{mnpt} = c_{mntp}$.

2.34 Let $\mathbf{F} = \mathbf{F}(\mathbf{A})$ be a linear function from the set of second-order symmetric tensors $\mathbf{A}$ to the same set of symmetric second-order tensors $(\mathbf{F}(\mathbf{A}) = \mathbf{F}(\mathbf{A})^T)$. Show that there is a unique fourth-order tensor $\mathbf{C}$ such that $\mathbf{F}(\mathbf{A}) = \mathbf{C} \cdot\cdot \mathbf{A}$, $\mathbf{C} = c_{mnpt}\mathbf{i}_m\mathbf{i}_n\mathbf{i}_p\mathbf{i}_t$, $c_{mnpt} = c_{mntp}$, and $c_{mnpt} = c_{nmpt}$.

2.35 Let Q be the set of all antisymmetric second-order tensors. Show that Q is a linear subspace of the linear space of all second-order tensors. On Q, the inner product $\mathbf{A}\cdot\cdot\mathbf{B}^T$ of the whole space takes the form $\mathbf{A}\cdot\cdot\mathbf{B}^T = -\mathbf{A} \cdot\cdot \mathbf{B} = -\mathbf{A}^T \cdot\cdot \mathbf{B}^T$.

2.36 Let $\mathbf{C}$ be a tensor of order n that is antisymmetric in its last two indices: $c_{mn\ldots pt} = -c_{mn\ldots tp}$. Let $\mathbf{A}$ be an arbitrary symmetric second-order tensor. Show that $\mathbf{C} \cdot\cdot \mathbf{A} = \mathbf{0}$, where $\mathbf{0}$ is the zero tensor of order $n-2$.

In the remaining problems, we let $\mathbf{u}$, $\mathbf{v}$, $\boldsymbol{\omega}$ denote vector fields; f, g, h smooth functions; $\mathbf{a}, \mathbf{b}$ arbitrary constant vectors; a, b, c arbitrary constants; $\mathbf{A}$ a second-order tensor; $\mathbf{r}$ the position vector of a body point; and $\mathbf{n}$ the unit external normal to the body boundary.

2.37 Demonstrate:

(a) $\nabla \cdot \nabla\mathbf{u} = \nabla(\nabla \cdot \mathbf{u}) - \nabla \times (\nabla \times \mathbf{u})$;
(b) $\nabla(\mathbf{u} \times \mathbf{v}) = (\nabla\mathbf{u}) \times \mathbf{v} - (\nabla\mathbf{v}) \times \mathbf{u}$;
(c) $\nabla \cdot (\mathbf{u} \times \mathbf{v}) = \mathbf{v} \cdot (\nabla \times \mathbf{u}) - \mathbf{u} \cdot (\nabla \times \mathbf{v})$;
(d) $\nabla \cdot (\mathbf{u}\mathbf{v}) = (\nabla \cdot \mathbf{u})\mathbf{v} + \mathbf{u} \cdot (\nabla\mathbf{v})$;
(e) $\nabla \times (\mathbf{u}\mathbf{v}) = (\nabla \times \mathbf{u})\mathbf{v} - \mathbf{u} \times (\nabla\mathbf{v})$;
(f) $\nabla \times (\mathbf{u} \times \mathbf{v}) = \mathbf{v} \cdot \nabla\mathbf{u} - \mathbf{v}\nabla \cdot \mathbf{u} - \mathbf{u} \cdot \nabla\mathbf{v} + \mathbf{u}\nabla \cdot \mathbf{v}$;
(g) $\mathbf{u} \cdot \nabla\mathbf{u} = \frac{1}{2}\nabla(\mathbf{u} \cdot \mathbf{u}) + (\nabla \times \mathbf{u}) \times \mathbf{u}$.

2.38 Demonstrate:

(a) $\nabla \cdot (\mathbf{A} \cdot \mathbf{f}) = (\nabla \cdot \mathbf{A}) \cdot \mathbf{f} + \mathbf{A}^T \cdot\cdot \nabla\mathbf{f}$;
(b) $\nabla \cdot (\mathbf{A} \cdot \mathbf{B}) = (\nabla \cdot \mathbf{A}) \cdot \mathbf{B} + \mathbf{A}^T \cdot\cdot \nabla\mathbf{B}$;
(c) $\nabla \cdot (\mathbf{A} \times \mathbf{r}) = (\nabla \cdot \mathbf{A}) \times \mathbf{r}$, if $\mathbf{A}$ is a symmetric tensor: $\mathbf{A} = \mathbf{A}^T$;
(d) $\nabla \cdot [(\mathbf{E} \times \boldsymbol{\omega}) \times \mathbf{r}] = 2\boldsymbol{\omega} + (\nabla \times \boldsymbol{\omega}) \times \mathbf{r}$;
(e) $\nabla \times (\nabla \times \mathbf{A}) = \nabla(\nabla \cdot \mathbf{A}) - \nabla \cdot (\nabla\mathbf{A})$;
(f) $\nabla \times (\mathbf{f} \times \mathbf{r}) = \mathbf{r} \cdot \nabla\mathbf{f} - \mathbf{r}(\nabla \cdot \mathbf{f}) + 2\mathbf{f}$;
(g) $\operatorname{tr}[\nabla \times (\mathbf{E} \times \boldsymbol{\omega})] = -2\nabla \cdot \boldsymbol{\omega}$;
(h) $\nabla \cdot [(\nabla\mathbf{f})^T - (\nabla \cdot \mathbf{f})\mathbf{E}] = \mathbf{0}$;
(i) $\nabla \cdot (\nabla f \times \nabla g) = 0$;
(j) $\nabla \times (\nabla \times \mathbf{A})^T$ is a symmetric tensor if $\mathbf{A}$ is symmetric;
(k) $\nabla \cdot (\mathbf{f}\,\mathbf{f}) = \mathbf{f}\nabla \cdot \mathbf{f} + (\nabla\mathbf{f}) \cdot \mathbf{f} - \mathbf{f} \times (\nabla \times \mathbf{f})$;

(l) $\operatorname{tr}[\nabla \times (\mathbf{A} \cdot \mathbf{B})] = \mathbf{B} \cdot\cdot (\nabla \times \mathbf{A}) - \mathbf{A}^T \cdot\cdot (\nabla \times \mathbf{B}^T)$;
(m) $(\nabla \mathbf{f})_\times = \nabla \times \mathbf{f}$;
(n) $\nabla \times (\mathbf{E} \times \boldsymbol{\omega}) = (\nabla \boldsymbol{\omega})^T - \mathbf{E}\nabla \cdot \boldsymbol{\omega}$;
(o) $\nabla \cdot \mathbf{A} = \nabla \cdot \mathbf{S} + \operatorname{rot} \boldsymbol{\omega}$, where $\mathbf{S} = \frac{1}{2}(\mathbf{A} + \mathbf{A}^T)$ and $\boldsymbol{\omega}$ is the vector conjugate to $\boldsymbol{\Omega} = \frac{1}{2}(\mathbf{A} - \mathbf{A}^T)$;
(p) $(\nabla \times \mathbf{A})_\times = \nabla \cdot (\mathbf{A}^T - \mathbf{E} \operatorname{tr} \mathbf{A})$;
(q) $\nabla \cdot \mathbf{A} = [\nabla \times (\mathbf{A}^T - \frac{1}{2}\mathbf{E} \operatorname{tr} \mathbf{A})]_\times$;
(r) $\nabla |\mathbf{a} \times \mathbf{r}|^2 = 2\mathbf{r}(\mathbf{a} \cdot \mathbf{a}) - 2\mathbf{a}(\mathbf{a} \cdot \mathbf{r})$;
(s) $\nabla [r(\mathbf{a} \times \mathbf{r})] = \mathbf{0}$;
(t) $\nabla \cdot [\mathbf{a} \times (\mathbf{r} \times \mathbf{b})] = 2\mathbf{a} \cdot \mathbf{b}$;
(u) $\nabla \times (f(r)\mathbf{r}) = \mathbf{0}$;
(v) $\nabla \times [\mathbf{b}(\mathbf{r} \cdot \mathbf{a})] = \mathbf{a} \times \mathbf{b}$;
(w) $\nabla \times (r\mathbf{a}) = \frac{\mathbf{r} \times \mathbf{a}}{r}$;
(x) $\mathbf{a} \times \nabla f = -\nabla \times (f\mathbf{a})$;
(y) $r(\mathbf{a} \times \mathbf{r}) = -\nabla \times (\frac{1}{3} r^3 \mathbf{a})$;
(z) $\nabla \times (\nabla \times \mathbf{A}) = \nabla\nabla \cdot \mathbf{A} - \nabla \cdot \nabla \mathbf{A}$.

2.39 Find

(a) $\nabla \cdot (\mathbf{E}\mathbf{r})$;
(b) $\nabla \times [\nabla \times (\mathbf{r} \times \mathbf{A} \times \mathbf{r})]^T$, if $\mathbf{A} = \mathbf{A}^T = \text{constant}$;
(c) $\nabla \cdot (\mathbf{A} \cdot \mathbf{r})$ if $\mathbf{A} = \text{constant}$;
(d) $\nabla \cdot (f\mathbf{E})$;
(e) $\nabla \cdot (\mathbf{r}\mathbf{E})$;
(f) $\nabla \cdot (\mathbf{r}\mathbf{r})$.

2.40 Let f, g, h be arbitrary smooth functions and $r^2 = \mathbf{r} \cdot \mathbf{r}$. Find the divergence of the tensors $\mathbf{A}$ given by the following formulas.

(a) $\mathbf{A} = f(r)\mathbf{e}_r\mathbf{e}_r + g(r)\mathbf{e}_\phi\mathbf{e}_\phi + h(z)\mathbf{e}_z\mathbf{e}_z$;
(b) $\mathbf{A} = f(r)\mathbf{e}_r\mathbf{e}_r + g(r)\mathbf{e}_\phi\mathbf{e}_\phi + g(r)\mathbf{e}_\theta\mathbf{e}_\theta$;
(c) $\mathbf{A} = f(r)\mathbf{e}_r\mathbf{e}_r + g(r)\mathbf{e}_\phi\mathbf{e}_\phi + h(r)\mathbf{e}_z\mathbf{e}_z$;
(d) $\mathbf{A} = f(x_1)\mathbf{i}_1\mathbf{i}_1 + g(x_2)\mathbf{i}_2\mathbf{i}_2 + h(x_3)\mathbf{i}_3\mathbf{i}_3$;
(e) $\mathbf{A} = f(x_2)\mathbf{i}_1\mathbf{i}_1 + g(x_3)\mathbf{i}_2\mathbf{i}_2 + h(x_1)\mathbf{i}_3\mathbf{i}_3$;
(f) $\mathbf{A} = f(r)\mathbf{e}_r\mathbf{e}_\phi + g(r)\mathbf{e}_\phi\mathbf{e}_r + h(r)\mathbf{e}_r\mathbf{e}_z$;
(g) $\mathbf{A} = f(z)\mathbf{e}_r\mathbf{e}_z + g(z)\mathbf{e}_\phi\mathbf{e}_r + h(z)\mathbf{e}_z\mathbf{e}_r$.

2.41 Let $\mathbf{A}$ be a symmetric second-order tensor depending on the coordinates. Denote $\mathbf{f} = \nabla \cdot \mathbf{A}$. Find $\nabla \cdot (\mathbf{A} \times \mathbf{r})$ in terms of $\mathbf{f}$.

2.42 The following describe coordinate systems and specify their relations to Cartesian coordinates x_1, x_2, x_3. In each case, demonstrate that the coordinates are orthogonal and find the Lamé coefficients. Take a to be a positive parameter.

(a) elliptic cylindrical coordinates σ, τ, z:

$$x_1 = a\sigma\tau, \quad x_2 = \pm a\sqrt{(\sigma^2 - 1)(1 - \tau^2)}, \quad x_3 = z,$$

where $\sigma \geq 1$ and $|\tau| \leq 1$.

(b) parabolic coordinates σ, τ, ϕ:

$$x_1 = \sigma\tau\cos\phi, \quad x_2 = \sigma\tau\sin\phi, \quad x_3 = \tfrac{1}{2}(\tau^2 - \sigma^2).$$

(c) bipolar cylindrical coordinates σ, τ, z:

$$x_1 = \frac{a\sinh\tau}{\cosh\tau - \cos\sigma}, \quad x_2 = \frac{a\sin\sigma}{\cosh\tau - \cos\sigma}, \quad x_3 = z.$$

(d) bipolar coordinates σ, τ, ϕ:

$$x_1 = \frac{a\sin\sigma\cos\phi}{\cosh\tau - \cos\sigma}, \quad x_2 = \frac{a\sin\sigma\sin\phi}{\cosh\tau - \cos\sigma}, \quad x_3 = \frac{a\sinh\tau}{\cosh\tau - \cos\sigma},$$

where $0 \leq \sigma < \pi$ and $0 \leq \phi < 2\pi$.

(e) toroidal coordinates σ, τ, ϕ:

$$x_1 = \frac{a\sinh\tau\cos\phi}{\cosh\tau - \cos\sigma}, \quad x_2 = \frac{a\sinh\tau\sin\phi}{\cosh\tau - \cos\sigma}, \quad x_3 = \frac{a\sin\sigma}{\cosh\tau - \cos\sigma},$$

where $-\pi \leq \sigma \leq \pi$, $0 \leq \tau$, and $0 \leq \phi < 2\pi$.

2.43 Prove (2.100), (2.101), (2.102), (2.103), and (2.104).

2.44 Use the Gauss–Ostrogradsky theorem to show that

$$\int_S \mathbf{nr}\, dS = V\mathbf{E}$$

where V is the volume of the domain bounded by surface S.

2.45 Show that the volume V of the body bounded by surface S is given by the formulas.

$$V = \frac{1}{3}\int_S \mathbf{n}\cdot\mathbf{r}\, dS \qquad \text{and} \qquad V = \frac{1}{6}\int_S \mathbf{n}\cdot\nabla r^2\, dS \quad (r^2 = \mathbf{r}\cdot\mathbf{r}).$$

2.46 Let $\mathbf{a}$ be a vector field with $\nabla\cdot\mathbf{a} = 0$. Using the Gauss–Ostrogradsky formula, show that

$$\int_S f\mathbf{n}\cdot\mathbf{a}\, dS = \int_V (\nabla f)\cdot\mathbf{a}\, dV.$$

2.47 Show that for a closed surface S we have

$$\int_S \mathbf{n}\, dS = \mathbf{0}.$$

2.48 Show that if $\mathbf{A}$ is a second-order symmetric tensor, then

$$\int_S \mathbf{r} \times (\mathbf{n} \cdot \mathbf{A})\, dS = \int_V \mathbf{r} \times (\nabla \cdot \mathbf{A})\, dV.$$

2.49 Show that

$$\int_V \nabla^2 \mathbf{A}\, dV = \int_S \mathbf{n} \cdot \nabla \mathbf{A}\, dS.$$

2.50 Given a second-order tensor $\mathbf{A}$, denote $\mathbf{f} = \nabla \cdot \mathbf{A}$ in a volume V and $\mathbf{n} \cdot \mathbf{A}|_S = \mathbf{g}$ over the boundary S of V. Find $\int_V \mathbf{A}\, dV$.

2.51 Prove the identity $(\mathbf{n} \times \nabla) \cdot \mathbf{A} = \mathbf{n} \cdot \operatorname{rot} \mathbf{A}$.

Chapter 3

Elements of Differential Geometry

Differential geometry elucidates the properties of curves and surfaces. It is a highly developed subject with many deep and useful results. We cover the main technical formulas of the theory of curves and surfaces which are used in applications.

3.1 Basic Theory of Curves

Analytic representation of a space curve

In differential geometry, a curve C is described by a position vector

$$\mathbf{r} = \mathbf{r}(t) \tag{3.1}$$

whose tip touches the points of C as the real parameter t runs through an interval. In Cartesian coordinates this representation takes the form

$$\mathbf{r}(t) = \mathbf{i}_1 x_1(t) + \mathbf{i}_2 x_2(t) + \mathbf{i}_3 x_3(t) \qquad (t_1 \le t \le t_2).$$

Some curves are given in Table 3.1.

Table 3.1 Parametric representations of familiar curves.

curve	$x_1(t)$	$x_2(t)$	$x_3(t)$	parameter range
straight line	$a_1 + b_1 t$	$a_2 + b_2 t$	$a_3 + b_3 t$	$-\infty < t < \infty$
circle	$a \cos t$	$a \sin t$	0	$0 \le t < 2\pi$
ellipse	$a \cos t$	$b \sin t$	0	$0 \le t < 2\pi$
cycloid	$a(t - \sin t)$	$a(1 - \cos t)$	0	$0 \le t < 2\pi$
trochoid	$at - h \sin t$	$a - h \cos t$	0	$0 \le t < 2\pi$
helix	$\cos t$	$\sin t$	t	$-\infty < t < \infty$

Unless otherwise stated, we take the dependence of $\mathbf{r}$ on t to be sufficiently smooth (e.g., $\mathbf{r}'(t)$ is continuous and so is $\mathbf{r}''(t)$ if necessary). The notion of norm is required here, of course, but in this chapter we denote this norm as simply $|\cdot|$.

Tangent vector and arc length

In Chapter 2 we introduced frame vectors $\mathbf{r}_i$ tangent to the coordinate lines. We now introduce the tangent vector to an arbitrary curve C at a point t:

$$\mathbf{r}'(t) = \frac{d\mathbf{r}(t)}{dt}.$$

The differential of the position vector for the curve (3.1) is

$$d\mathbf{r}(t) = \mathbf{r}'(t)\,dt.$$

The differential arc length along C is $ds = |\mathbf{r}'(t)|\,dt$, and the length of the portion of C for $a \le t \le b$ is

$$s = \int_a^b |\mathbf{r}'(t)|\,dt.$$

For the helix of Table 3.1, for instance,

$$\mathbf{r}'(t) = -\mathbf{i}_1 \sin t + \mathbf{i}_2 \cos t + \mathbf{i}_3, \qquad |\mathbf{r}'(t)| = \sqrt{2},$$

and a single turn has length

$$s = \int_0^{2\pi} \sqrt{2}\,dt = 2\pi\sqrt{2}.$$

More convenient is the *natural parametrization* where the parameter represents the length of C calculated from a reference point:

$$\mathbf{r} = \mathbf{r}(s).$$

Here $ds = |\mathbf{r}'(s)|\,ds$ so that $|\mathbf{r}'(s)| = 1$. Hence when C is parametrized naturally $\boldsymbol{\tau}(s) = \mathbf{r}'(s)$ is the unit tangent at point s. Let us treat the helix in terms of its natural length parameter s. We have

$$s(t) = \int_0^t \sqrt{2}\,dt = t\sqrt{2}$$

and

$$\mathbf{r}(s) = \mathbf{i}_1 \cos\left(\frac{s}{\sqrt{2}}\right) + \mathbf{i}_2 \sin\left(\frac{s}{\sqrt{2}}\right) + \mathbf{i}_3\left(\frac{s}{\sqrt{2}}\right),$$

so the unit tangent is

$$\mathbf{r}'(s) = \frac{1}{\sqrt{2}}\left[-\mathbf{i}_1 \sin\left(\frac{s}{\sqrt{2}}\right) + \mathbf{i}_2 \cos\left(\frac{s}{\sqrt{2}}\right) + \mathbf{i}_3\right].$$

In this section s denotes the length parameter. If another parametrization t of C is related to the natural parametrization as $s = s(t)$, then the vector $d\mathbf{r}(s(t))/dt$ is also tangent to C at the point $s = s(t)$. The derivative of a vector function is given by the chain rule

$$\frac{d\mathbf{r}(s(t))}{dt} = \left.\frac{d\mathbf{r}(s)}{ds}\right|_{s=s(t)} \frac{ds(t)}{dt}.$$

As seen in Table 3.1, a straight line has parametric representation of the form

$$\begin{aligned}\mathbf{r}(t) &= \mathbf{i}_1 x_1(t) + \mathbf{i}_2 x_2(t) + \mathbf{i}_3 x_3(t)\\ &= \mathbf{i}_1(a_1 + b_1 t) + \mathbf{i}_2(a_2 + b_2 t) + \mathbf{i}_3(a_3 + b_3 t)\\ &= \mathbf{a} + t\mathbf{b}\end{aligned}$$

where $\mathbf{a}$ fixes one point of the line ($t = 0$) and $\mathbf{b}$ specifies the direction. This gives the vector equation of the line tangent to a curve C at point $\mathbf{r}(t_0)$:

$$\mathbf{r} = \mathbf{r}(t_0) + t\mathbf{r}'(t_0).$$

In Cartesian coordinates (x, y, z) this is equivalent to the three scalar equations

$$\begin{aligned}x &= x(t_0) + tx'(t_0),\\ y &= y(t_0) + ty'(t_0),\\ z &= z(t_0) + tz'(t_0),\end{aligned}$$

and also to the nonparametric form

$$\frac{x - x(t_0)}{x'(t_0)} = \frac{y - y(t_0)}{y'(t_0)} = \frac{z - z(t_0)}{z'(t_0)}.$$

Note that the construction of a tangent vector is possible if $\mathbf{r}'(t_0) \neq \mathbf{0}$. A point t_0 where $\mathbf{r}'(t_0) = \mathbf{0}$ is called *singular*. In the case of a singular point it is also possible to construct a tangent vector if there is a higher derivative $\mathbf{r}^{(n)}(t_0) \neq \mathbf{0}$.

Curvature

Curvature expresses the rate of change of the tangent as a point moves along a curve. For a plane curve, it is defined as the limit

$$k = \lim_{\Delta s \to 0} \frac{\Delta \varphi}{\Delta s}$$

where $\Delta\varphi$ is the change in angle of the tangent at point s. We can say that a circle of radius R has $\Delta s = R\Delta\varphi$ and therefore $k = 1/R$. However, with plane curves $\Delta\varphi$ can be reckoned positive or negative to indicate a sense of convexity (e.g., concave up or concave down).

For a more general space curve C, we define curvature as the nonnegative quantity

$$k = \lim_{\Delta s \to 0} \frac{|\Delta \varphi|}{\Delta s}.$$

Fig. 3.1 shows a point moving through an element of C corresponding to Δs as the unit tangent $\boldsymbol{\tau}(s)$ turns through an angle $\Delta\varphi$. Noting that

$$|\boldsymbol{\tau}(s + \Delta s) - \boldsymbol{\tau}(s)| = 2\sin(\Delta\varphi/2),$$

and using the classic limit

$$\lim_{\Delta\varphi \to 0} \frac{\sin(\Delta\varphi/2)}{(\Delta\varphi/2)} = 1,$$

we obtain

$$k = \lim_{\Delta s \to 0} \frac{|\boldsymbol{\tau}(s + \Delta s) - \boldsymbol{\tau}(s)|}{\Delta s} = \lim_{\Delta s \to 0} \frac{|\mathbf{r}'(s + \Delta s) - \mathbf{r}'(s)|}{\Delta s} = |\mathbf{r}''(s)|.$$

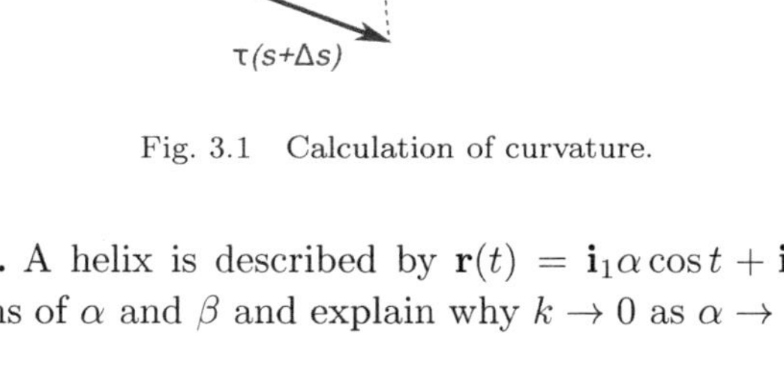

Fig. 3.1 Calculation of curvature.

Exercise 3.1. A helix is described by $\mathbf{r}(t) = \mathbf{i}_1\alpha\cos t + \mathbf{i}_2\alpha\sin t + \mathbf{i}_3\beta t$. Find k in terms of α and β and explain why $k \to 0$ as $\alpha \to 0$ and $k \to 1/\alpha$ as $\beta \to 0$.

Moving trihedron

Suppose $\mathbf{e}(s)$ is any unit vector dependent on a natural curve parameter s. On one hand we have $\mathbf{e}(s)\cdot\mathbf{e}(s) \equiv 1$ for all s. On the other hand, the product rule gives $[\mathbf{e}(s)\cdot\mathbf{e}(s)]' = 2\mathbf{e}(s)\cdot\mathbf{e}'(s)$ and we conclude that $\mathbf{e}(s)\cdot\mathbf{e}'(s) = 0$. An important example is the fact that the vector $\boldsymbol{\tau}'(s) = \mathbf{r}''(s)$ is orthogonal to $\boldsymbol{\tau}(s)$ at any point on a curve C.

We define the *principal normal* at point s of C as the vector

$$\boldsymbol{\nu}(s) = \frac{\mathbf{r}''(s)}{k} \tag{3.2}$$

if $k \neq 0$. If C lies in a plane then clearly $\boldsymbol{\nu}$ lies in the same plane. Any plane through a point s and containing $\boldsymbol{\nu}(s)$ can be called a tangent plane. However, one such plane contains both $\boldsymbol{\nu}(s)$ and $\boldsymbol{\tau}(s)$ and is regarded as "the most locally tangent" plane at $\mathbf{r}(s)$. This is the *osculating plane* at $\mathbf{r}(s)$; it hugs the curve better than any other plane passing through a given point.

Exercise 3.2. Find the plane osculating to the curve $\mathbf{r}(t) = \mathbf{i}_1 t + \mathbf{i}_2 2t^2 + \mathbf{i}_3 t^3$ at $t = 1$.

A unit normal to the osculating plane at s is defined by

$$\boldsymbol{\beta}(s) = \boldsymbol{\tau}(s) \times \boldsymbol{\nu}(s).$$

The triplet $(\boldsymbol{\tau}, \boldsymbol{\nu}, \boldsymbol{\beta})$, known as the *moving trihedron*, forms a local basis at each point s of C. Three mutually perpendicular planes are associated with the moving trihedron; besides the osculating plane mentioned above, we have the *normal plane* and the *rectifying plane*. The characteristics of these *fundamental planes* are summarized in Table 3.2. For general parametrization of the curve $\mathbf{r} = \mathbf{r}(t)$, the osculating plane is described in Cartesian coordinates by

$$\begin{vmatrix} x - x(t_0) & y - y(t_0) & z - z(t_0) \\ x'(t_0) & y'(t_0) & z'(t_0) \\ x''(t_0) & y''(t_0) & z''(t_0) \end{vmatrix} = 0.$$

Exercise 3.3. Calculate $\boldsymbol{\nu}$ and $\boldsymbol{\beta}$ for the helix of Exercise 3.1. Then find the equation of the rectifying plane at $t = t_0$.

Using the chain rule we can present the expression for k for an arbitrary parametrization of a curve:

$$k^2 = \frac{[\mathbf{r}'(t) \times \mathbf{r}''(t)]^2}{[\mathbf{r}'(t) \cdot \mathbf{r}'(t)]^3}. \tag{3.3}$$

Table 3.2 Fundamental planes through a point s_0 of a space curve.

name	vectors contained	describing equation
osculating plane	$\boldsymbol{\tau}$ and $\boldsymbol{\nu}$	$[\mathbf{r} - \mathbf{r}(s_0)] \cdot \boldsymbol{\beta}(s_0) = 0$
normal plane	$\boldsymbol{\nu}$ and $\boldsymbol{\beta}$	$[\mathbf{r} - \mathbf{r}(s_0)] \cdot \boldsymbol{\tau}(s_0) = 0$
rectifying plane	$\boldsymbol{\beta}$ and $\boldsymbol{\tau}$	$[\mathbf{r} - \mathbf{r}(s_0)] \cdot \boldsymbol{\nu}(s_0) = 0$

Its derivation is left to the reader. We shall now denote the curvature by k_1, because there is another quantity that characterizes how a curve differs from a straight line. In terms of k_1 we may define the *radius of curvature* as the number $R = 1/k_1$.

For a space curve, $\boldsymbol{\nu}$ and $\boldsymbol{\beta}$ are not defined uniquely at *points of inflection* where $\boldsymbol{\tau}'(s) = \mathbf{0}$. Unless otherwise stated we assume $\boldsymbol{\tau}'(s) \neq \mathbf{0}$ for all s.

Torsion

As we move along a curve C, the rate of rotation of the osculating plane indicates the tendency of C to be "non-planar." Since $\boldsymbol{\beta}$ is normal to the osculating plane, this tendency can be measured by the rate of turn of $\boldsymbol{\beta}$. The *torsion* or *second curvature* of the curve C at a point is the number

$$k_2 = \lim_{\Delta s \to 0} \frac{\Delta \vartheta}{\Delta s}$$

where $\Delta\vartheta$ is the angle of turn of the binormal $\boldsymbol{\beta}$ over a change Δs in the natural parameter. The torsion is the magnitude of the rate of change of the binormal as we move along C; it measures the tendency of C to twist out of the osculating plane. The algebraic sign of k_2 will be discussed below.

Theorem 3.1. *Let* $\mathbf{r} = \mathbf{r}(s)$ *be a three times continuously differentiable vector function of* s. *At any point where* $k_1 \neq 0$, *the torsion of the curve is*

$$k_2 = -\frac{[\mathbf{r}'(s) \times \mathbf{r}''(s)] \cdot \mathbf{r}'''(s)}{k_1^2}.$$

Proof. At a point s where $k_1 \neq 0$, both $\boldsymbol{\beta}$ and $\boldsymbol{\nu}$ are uniquely defined. Noting that $\boldsymbol{\beta}$ is a unit vector, we have $|\boldsymbol{\beta}(s+\Delta s) - \boldsymbol{\beta}(s)| = 2\,|\sin \Delta\vartheta/2|$ and therefore

$$\begin{aligned} |k_2| &= \lim_{\Delta s \to 0} \left| \frac{\Delta\vartheta}{\Delta s} \right| = \lim_{\Delta s \to 0} \left| \frac{\Delta\vartheta}{2\sin(\Delta\vartheta/2)} \right| \left| \frac{2\sin(\Delta\vartheta/2)}{\Delta s} \right| \\ &= \lim_{\Delta s \to 0} \left| \frac{\boldsymbol{\beta}(s+\Delta s) - \boldsymbol{\beta}(s)}{\Delta s} \right| = \left| \boldsymbol{\beta}'(s) \right| . \end{aligned}$$

Furthermore, $\boldsymbol{\beta}'$ is orthogonal to $\boldsymbol{\beta}$ and hence lies in the osculating plane. Since $\boldsymbol{\tau}'$ is parallel to $\boldsymbol{\nu}$, we have

$$\boldsymbol{\beta}' = (\boldsymbol{\tau} \times \boldsymbol{\nu})' = \boldsymbol{\tau}' \times \boldsymbol{\nu} + \boldsymbol{\tau} \times \boldsymbol{\nu}' = \boldsymbol{\tau} \times \boldsymbol{\nu}'.$$

Hence $\boldsymbol{\beta}'$ and $\boldsymbol{\tau}$ are orthogonal so that $\boldsymbol{\beta}'$ and $\boldsymbol{\nu}$ are parallel. But $|\boldsymbol{\nu}| = 1$, so $|\boldsymbol{\beta}' \cdot \boldsymbol{\nu}| = |\boldsymbol{\beta}'|$ and therefore

$$|k_2| = |\boldsymbol{\beta}' \cdot \boldsymbol{\nu}|.$$

Let us use the fact that $\boldsymbol{\nu} = \mathbf{r}''/k_1$, and thus

$$\boldsymbol{\nu}' = \mathbf{r}'''/k_1 - \mathbf{r}''k_1{}'/k_1^2, \qquad \boldsymbol{\beta} = (\mathbf{r}' \times \mathbf{r}'')/k_1.$$

By the properties of the scalar triple product,

$$|k_2| = |(\boldsymbol{\tau} \times \boldsymbol{\nu}') \cdot \boldsymbol{\nu}| = \frac{|(\mathbf{r}' \times \mathbf{r}'') \cdot \mathbf{r}'''|}{{k_1}^2}.$$

We define

$$k_2 = -\frac{(\mathbf{r}' \times \mathbf{r}'') \cdot \mathbf{r}'''}{{k_1}^2}$$

with the agreement that if $\boldsymbol{\beta}$ turns toward $\boldsymbol{\nu}$ then the sign should be positive. (Intuitively, if C has positive torsion then it will twist in the manner of a right-hand screw thread with increasing s.) That is,

$$\boldsymbol{\beta}' = k_2 \boldsymbol{\nu}. \tag{3.4}$$

This completes the proof. □

For a parameter t other than the natural parameter s, the chain rule gives

$$k_2 = -\frac{[\mathbf{r}'(t) \times \mathbf{r}''(t)] \cdot \mathbf{r}'''(t)}{[\mathbf{r}'(t) \times \mathbf{r}''(t)]^2}. \tag{3.5}$$

To see that k_2 indicates the rate at which C distinguishes itself from a plane curve, suppose $k_2 = 0$ for all s. Then $0 = |k_2| = |\boldsymbol{\beta}'(s) \cdot \boldsymbol{\nu}(s)|$ so that $|\boldsymbol{\beta}'(s)| = 0$ (since $\boldsymbol{\nu}$ is a unit vector) and thus $\boldsymbol{\beta}(s) = \boldsymbol{\beta}^0 = \text{constant}$. The tangent vector $\boldsymbol{\tau}$ is orthogonal to $\boldsymbol{\beta}^0$, so $0 = \boldsymbol{\tau} \cdot \boldsymbol{\beta}^0 = \mathbf{r}'(s) \cdot \boldsymbol{\beta}^0$ and integration yields $[\mathbf{r}(s) - \mathbf{r}(s_0)] \cdot \boldsymbol{\beta}^0 = 0$. The curve lies in a plane.

By formula (3.4), $k_2 = |d\boldsymbol{\beta}/ds|$. So $k_2 = \lim_{\Delta s \to 0} |\Delta\boldsymbol{\beta}/\Delta s|$. But up to small quantities of the second order of Δs, the change $|\Delta\mathbf{e}|$ of a unit vector $\mathbf{e}$ is equal to the angle of rotation of $\mathbf{e}$ when moved through a distance Δs. Hence $|\Delta\boldsymbol{\beta}|$ is approximately the rotation angle of the binormal as the point shifts through Δs, and k_2 measures the rate of rotation of $\boldsymbol{\beta}$ when a point moves along the curve.

Exercise 3.4. Show that a smooth curve lies in a plane only if $k_2 = 0$.

Exercise 3.5. Show that in Cartesian coordinates

$$k_1 = \frac{\left[(y'z'' - z'y'')^2 + (z'x'' - x'z'')^2 + (x'y'' - y'x'')^2\right]^{1/2}}{\left[(x')^2 + (y')^2 + (z')^2\right]^{3/2}}$$

and

$$k_2 = \frac{\begin{vmatrix} x' & y' & z' \\ x'' & y'' & z'' \\ x''' & y''' & z''' \end{vmatrix}}{(y'z'' - z'y'')^2 + (z'x'' - x'z'')^2 + (x'y'' - y'x'')^2}$$

where the prime denotes d/dt.

Exercise 3.6. What happens to k_1 and k_2 (and the moving trihedron of a curve) if the direction of change of the parameter is reversed ($t \mapsto -t$)?

Exercise 3.7. Calculate k_2 for the helix of Exercise 3.1.

Frenet–Serret equations

Frenet established a system of three ordinary differential equations that govern the right-handed orthogonal triad $(\boldsymbol{\tau}, \boldsymbol{\nu}, \boldsymbol{\beta})$. Two of these are (3.2) and (3.4). The former will be written as $\boldsymbol{\tau}' = k_1\boldsymbol{\nu}$. The third equation of the Frenet–Serret system is obtained as

$$\begin{aligned} \boldsymbol{\nu}' &= (\boldsymbol{\beta} \times \boldsymbol{\tau})' = \boldsymbol{\beta}' \times \boldsymbol{\tau} + \boldsymbol{\beta} \times \boldsymbol{\tau}' = k_2\boldsymbol{\nu} \times \boldsymbol{\tau} + \boldsymbol{\beta} \times (k_1\boldsymbol{\nu}) \\ &= -k_1\boldsymbol{\tau} - k_2\boldsymbol{\beta}. \end{aligned}$$

Hence the system is

$$\boldsymbol{\tau}' = k_1\boldsymbol{\nu}, \qquad \boldsymbol{\nu}' = -k_1\boldsymbol{\tau} - k_2\boldsymbol{\beta}, \qquad \boldsymbol{\beta}' = k_2\boldsymbol{\nu}. \tag{3.6}$$

We stress that these equations are written in the natural parametrization, with the length parameter s.

If the functions $k_1(s)$ and $k_2(s)$ are known, (3.6) constitutes a linear system of ordinary vector differential equations (nine scalar equations in nine unknowns). Fixing some point of the curve in space and an orthonormal triad $\boldsymbol{\tau}, \boldsymbol{\nu}, \boldsymbol{\beta}$ at the point, by this system we can define $\boldsymbol{\tau}(s)$, $\boldsymbol{\nu}(s)$, and $\boldsymbol{\beta}(s)$ uniquely; then, by the equation $\mathbf{r}'(s) = \boldsymbol{\tau}(s)$, we define the curve $\mathbf{r} = \mathbf{r}(s)$

uniquely as well. Thus $k_1(s)$ and $k_2(s)$ define the curve up to a motion as a rigid body in space. That is why the pair of equations for $k_1(s), k_2(s)$ is called the set of natural equations of the curve.

Let us demonstrate how to use the Frenet–Serret equations to characterize the curve locally. We employ the Taylor expansion of the radius vector of a curve at point s:

$$\mathbf{r}(s+\Delta s) = \mathbf{r}(s) + \Delta s\mathbf{r}'(s) + \frac{(\Delta s)^2}{2}\mathbf{r}''(s) + \frac{(\Delta s)^3}{6}\mathbf{r}'''(s) + o(|\Delta s|^3).$$

By the definition $\boldsymbol{\nu} = \mathbf{r}''/k_1$ and by the second of equations (3.6) we get

$$\mathbf{r}''' = (k_1\boldsymbol{\nu})' = k_1'\boldsymbol{\nu} + k_1\boldsymbol{\nu}' = k_1'\boldsymbol{\nu} + k_1(-k_1\boldsymbol{\tau} - k_2\boldsymbol{\beta}).$$

Substitution yields

$$\mathbf{r}(s+\Delta s) = \mathbf{r}(s) + \Delta s\boldsymbol{\tau} + \frac{(\Delta s)^2}{2}k_1\boldsymbol{\nu} + \frac{(\Delta s)^3}{6}(k_1'\boldsymbol{\nu} - k_1^2\boldsymbol{\tau} - k_1k_2\boldsymbol{\beta}) + o(|\Delta s|^3).$$

This shows that, to the order of $(\Delta s)^2$, the curve lies in the osculating plane at point s. At a point s the triad $\boldsymbol{\tau}, \boldsymbol{\nu}, \boldsymbol{\beta}$ is orthonormal. Let us fix this frame and place its origin at $\mathbf{r}(s) = 0$. Defining x, y, z as the components of $\mathbf{r}(s+\Delta s)$ in this frame, we obtain the approximate representation for the curve

$$x = \Delta s - \frac{k_1^2(\Delta s)^3}{6} + o(|\Delta s|^3),$$
$$y = \frac{k_1(\Delta s)^2}{2} + \frac{k_1'(\Delta s)^3}{6} + o(|\Delta s|^3),$$
$$z = -\frac{k_1k_2(\Delta s)^3}{6} + o(|\Delta s|^3).$$

As another application of the Frenet–Serret equations, we find the velocity and acceleration of a point moving along the curve in space. Let s be the length parameter of the trajectory of the point, and let t denote time. We shall use the triad $(\boldsymbol{\tau}, \boldsymbol{\nu}, \boldsymbol{\beta})$ of the trajectory $\mathbf{r} = \mathbf{r}(s)$. The position of the point is given by the equation

$$\mathbf{r} = \mathbf{r}(s(t)).$$

The velocity of the point is

$$\mathbf{v} = \frac{d}{dt}\mathbf{r}(s(t)) = \frac{d\mathbf{r}}{ds}\frac{ds}{dt} = \frac{ds}{dt}\boldsymbol{\tau}.$$

Denoting $v = ds/dt$, the particle speed at each point along its path, we can write $\mathbf{v} = v\boldsymbol{\tau}$. Now let us find the acceleration of the point in the same frame, which is

$$\mathbf{a} = \frac{d^2}{dt^2}\mathbf{r}(s(t)) = \frac{d}{dt}\mathbf{v}(s(t)) = \frac{d}{dt}\left(\frac{ds}{dt}\boldsymbol{\tau}\right)$$
$$= \frac{d^2s}{dt^2}\boldsymbol{\tau} + \frac{ds}{dt}\frac{d\boldsymbol{\tau}}{ds}\frac{ds}{dt} = s''(t)\boldsymbol{\tau} + v^2\frac{d\boldsymbol{\tau}}{ds}.$$

Using the Frenet–Serret equations we get

$$\mathbf{a} = s''(t)\boldsymbol{\tau} + k_1 v^2 \boldsymbol{\nu}$$

where k_1 is the principal curvature of the trajectory. In mechanics this is commonly written as

$$\mathbf{a} = s''(t)\boldsymbol{\tau} + (v^2/\rho)\boldsymbol{\nu}$$

where $\rho = 1/k_1$ is the radius of curvature of the curve. Hence $\mathbf{a}$ always lies in the osculating plane.

Exercise 3.8. Express $d^3\mathbf{r}/ds^3$ in terms of the moving trihedron.

Exercise 3.9. Show that the Frenet–Serret equations can be written as

$$\boldsymbol{\tau}' = \boldsymbol{\delta} \times \boldsymbol{\tau}, \qquad \boldsymbol{\nu}' = \boldsymbol{\delta} \times \boldsymbol{\nu}, \qquad \boldsymbol{\beta}' = \boldsymbol{\delta} \times \boldsymbol{\beta},$$

where $\boldsymbol{\delta} = k_1\boldsymbol{\beta} - k_2\boldsymbol{\tau}$ is called the *Darboux vector*.

3.2 Basic Theory of Surfaces

Here we show how to describe complicated surfaces using a position vector representation. This will enable us to calculate distances and angles on a general surface, as well as study curvature properties. The result will be a full local description of the surface. We shall study other notions as well, concluding our treatment by adapting tensor analysis to two dimensions and applying the resulting tools to the study of surfaces.

Analytic representation of a surface

For a surface S, the parametric position vector representation takes the form

$$\mathbf{r} = \mathbf{r}(u^1, u^2), \tag{3.7}$$

where each parameter u^i $(i = 1, 2)$ runs through its own real set. In a Cartesian frame

$$\mathbf{r}(u^1, u^2) = \mathbf{i}_1\, x(u^1, u^2) + \mathbf{i}_2\, y(u^1, u^2) + \mathbf{i}_3\, z(u^1, u^2)$$

for functions $x(u^1,u^2)$, $y(u^1,u^2)$, and $z(u^1,u^2)$. A sphere of radius a centered at the origin, for instance, can be described as

$$\mathbf{r}(\theta,\phi) = \mathbf{i}_1\, a\sin\theta\cos\phi + \mathbf{i}_2\, a\sin\theta\sin\phi + \mathbf{i}_3\, a\cos\theta \qquad \begin{cases} 0 \le \theta \le \pi \\ 0 \le \phi < 2\pi \end{cases}$$

where θ and ϕ are the spherical coordinate angles.

In general, the pair (u^1,u^2) can be regarded as the coordinates of a point in the u^1u^2-plane, and a portion of S may be taken as the image of a domain in this plane. It is in this sense that even very complicated, nonintuitive surfaces can safely be called "two-dimensional." We can even lay out a coordinate system on S. If we fix u^2, the locus of $\mathbf{r}$ is a space curve along which only u^1 varies. For each such u^2 a space curve exists, and if we let u^2 vary, we obtain a locus of curves which forms S. Holding u^2 constant, we get a u^1*-curve*. Holding u^1 constant, we get a u^2*-curve*. These curves are called *coordinate lines*. A net of such parametric curves covers S. On a sphere, the latitude and longitude lines are parametric curves of the angular coordinates θ and ϕ.

The term *intrinsic coordinates* is often applied to the u^i. This emphasizes that the coordinates refer to the surface; they are taken *in the surface itself*, not from the surrounding space somehow.

Osculating plane and local bases

So we begin to study a surface S described by (3.7). For simplicity we take S to be smooth everywhere except possibly at certain points (e.g., the vertex of a cone) or along certain curves. Along each u^i-curve we can introduce a tangent vector

$$\mathbf{r}_i = \frac{\partial \mathbf{r}(u^1,u^2)}{\partial u^i} \qquad (i = 1,2).$$

We assume that, except at certain singular lines or poles in the (u^1,u^2) plane, the $\mathbf{r}_i$ are continuously differentiable functions of the q^i. We also assume that, except on singularities, the $\mathbf{r}_i$ are not collinear and hence

$$\mathbf{N} = \mathbf{r}_1 \times \mathbf{r}_2 \neq \mathbf{0}.$$

The resulting unit normal to S,

$$\mathbf{n} = \frac{\mathbf{r}_1 \times \mathbf{r}_2}{|\mathbf{r}_1 \times \mathbf{r}_2|},$$

is also normal to the plane that osculates to S at point (u^1, u^2). The latter plane (Fig. 3.2) is the plane for which

$$\lim_{\Delta s \to 0} \frac{h}{\Delta s} = 0$$

where

$$\Delta s = \left| \mathbf{r}(u^1 + \Delta u^1, u^2 + \Delta u^2) - \mathbf{r}(u^1, u^2) \right|$$

and h is the distance from the point with coordinates $(u^1 + \Delta u^1, u^2 + \Delta u^2)$ to the plane. Note that this must hold for any curve on the surface through the point. Take the plane orthogonal to $\mathbf{n}$. Then

$$h = \left| \left[\mathbf{r}(u^1 + \Delta u^1, u^2 + \Delta u^2) - \mathbf{r}(u^1, u^2) \right] \cdot \mathbf{n} \right| .$$

By definition of the differential we have

$$\begin{aligned} \frac{h}{\Delta s} &= \frac{\left| [\mathbf{r}(u^1 + \Delta u^1, u^2 + \Delta u^2) - \mathbf{r}(u^1, u^2)] \cdot \mathbf{n} \right|}{\left| \mathbf{r}(u^1 + \Delta u^1, u^2 + \Delta u^2) - \mathbf{r}(u^1, u^2) \right|} \\ &= \frac{\left| [\mathbf{r}_i \Delta u^i + o(|\Delta u^1| + |\Delta u^2|)] \cdot \mathbf{n} \right|}{\left| \mathbf{r}_i \Delta u^i + o(|\Delta u^1| + |\Delta u^2|) \right|} \\ &= \frac{o(|\Delta u^1| + |\Delta u^2|)}{\left| \mathbf{r}_i \Delta u^i + o(|\Delta u^1| + |\Delta u^2|) \right|} . \end{aligned}$$

This means that the plane we chose is osculating.

Much of what was said about the $\mathbf{r}_i$ in Chapter 2 can be simply reformulated for the present case if we use the vectors $\mathbf{r}_1$, $\mathbf{r}_2$, and $\mathbf{r}^3 = \mathbf{n}$ as a local, three-dimensional basis on S. First we can introduce the reciprocal basis in space. The third vector is still[1] $\mathbf{n}$, while the two others are defined by

$$\mathbf{r}^i \cdot \mathbf{r}_j = \delta^i_j \quad (i = 1, 2), \qquad \mathbf{r}^i \cdot \mathbf{n} = 0.$$

The frame $(\mathbf{r}^1, \mathbf{r}^2)$ is reciprocal to the frame $(\mathbf{r}_1, \mathbf{r}_2)$ in the osculating plane at (u^1, u^2).

[1] As a result of a cross product, $\mathbf{n}$ is a pseudovector. But just as for $\mathbf{k}$ from the trihedron $(\mathbf{i}, \mathbf{j}, \mathbf{k})$, as we wish to use $(\mathbf{r}_1, \mathbf{r}_2, \mathbf{n})$ for a local basis at a point, we change the pseudovector $\mathbf{n}$ to the corresponding polar vector $\mathbf{n}$ and then "forget" about its pseudo-nature.

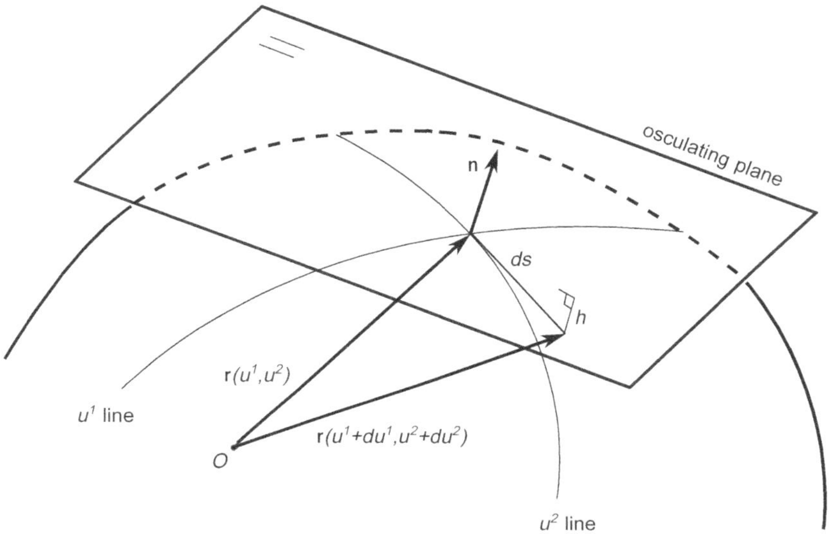

Fig. 3.2 Osculating plane to a surface.

Arc length on a surface; first fundamental form

In this section, summation indices take the values 1, 2. The differential

$$d\mathbf{r} = \mathbf{r}_i \, du^i \tag{3.8}$$

is the main linear part in du^i of the increment

$$\Delta\mathbf{r} = \mathbf{r}(u^1 + du^1, u^2 + du^2) - \mathbf{r}(u^1, u^2).$$

The element of arc between nearby points $\mathbf{r}(u^1+du^1, u^2+du^2)$ and $\mathbf{r}(u^1, u^2)$ is given by

$$(ds)^2 = d\mathbf{r} \cdot d\mathbf{r} = \mathbf{r}_i \, du^i \cdot \mathbf{r}_j \, du^j = g_{ij} \, du^i \, du^j$$

where $g_{ij} = \mathbf{r}_i \cdot \mathbf{r}_j$ is the metric tensor of S. The quadratic form on the right-hand side of the equation

$$(ds)^2 = g_{ij} \, du^i \, du^j$$

is called the *first fundamental form* of S. Beginning with Gauss, the metric coefficients were denoted by

$$E = g_{11}, \qquad F = g_{12} = g_{21}, \qquad G = g_{22},$$

and we write

$$(ds)^2 = E(du^1)^2 + 2F\,du^1\,du^2 + G(du^2)^2$$

where E, F, G are the *first fundamental coefficients* of the surface. Along a u^1-curve we have $du^2 = 0$ and hence $ds = \sqrt{E}\,du^1$. Along a u^2-curve, $ds = \sqrt{G}\,du^2$. Since the $\partial\mathbf{r}/\partial u^i$ are tangent vectors to the u^i-curves, the parametric curves form an orthogonal system if and only if $F = 0$.

Exercise 3.10. Find the first fundamental form of a sphere of radius a.

Besides the length of an elementary curve on the surface, formula (3.8) allows us to find an angle between two elementary curves $d\mathbf{r} = \mathbf{r}_i\,du^i$ and $d\tilde{\mathbf{r}} = \mathbf{r}_i\,d\tilde{u}^i$. Indeed, by the definition of the dot product we have

$$\begin{aligned}\cos\varphi &= \frac{d\mathbf{r}\cdot d\tilde{\mathbf{r}}}{|d\mathbf{r}|\,|d\tilde{\mathbf{r}}|}\\ &= \frac{E\,du^1\,d\tilde{u}^1 + F(du^1\,d\tilde{u}^2 + du^2\,d\tilde{u}^1) + G\,du^2\,d\tilde{u}^2}{\sqrt{E(du^1)^2 + 2F\,du^1\,du^2 + G(du^2)^2}\sqrt{E(d\tilde{u}^1)^2 + 2F\,d\tilde{u}^1\,d\tilde{u}^2 + G(d\tilde{u}^2)^2}}.\end{aligned} \tag{3.9}$$

Exercise 3.11. (a) Find the expression for the angle between coordinate lines of a surface. (b) Calculate the angle at which the curve $\theta = \phi$ crosses the equator of a sphere.

The area of the parallelogram that is based on the elementary vectors $\mathbf{r}_1\,du^1$ and $\mathbf{r}_2\,du^2$ which are tangent to the coordinate lines is

$$dS = \left|\mathbf{r}_1\,du^1 \times \mathbf{r}_2\,du^2\right|.$$

This approximates up to higher order terms the area of the corresponding curvilinear figure bordered by the coordinate lines

$$u^1 = \text{const}, \quad u^2 = \text{const}, \quad u^1 + du^1 = \text{const}, \quad u^2 + du^2 = \text{const}.$$

(See Fig. 3.3.) Let us find dS in terms of the first fundamental form. We need to calculate $|\mathbf{r}_1 \times \mathbf{r}_2|$. Let us demonstrate that

$$|\mathbf{r}_1 \times \mathbf{r}_2| = \sqrt{EG - F^2}. \tag{3.10}$$

Indeed, by definition of the dot and cross product we have

$$|\mathbf{r}_1 \times \mathbf{r}_2|^2 + (\mathbf{r}_1\cdot\mathbf{r}_2)^2 = |\mathbf{r}_1|^2\,|\mathbf{r}_2|^2\sin^2\varphi + |\mathbf{r}_1|^2\,|\mathbf{r}_2|^2\cos^2\varphi = |\mathbf{r}_1|^2\,|\mathbf{r}_2|^2.$$

Thus

$$|\mathbf{r}_1 \times \mathbf{r}_2|^2 = |\mathbf{r}_1|^2\,|\mathbf{r}_2|^2 - (\mathbf{r}_1\cdot\mathbf{r}_2)^2 = EG - F^2$$

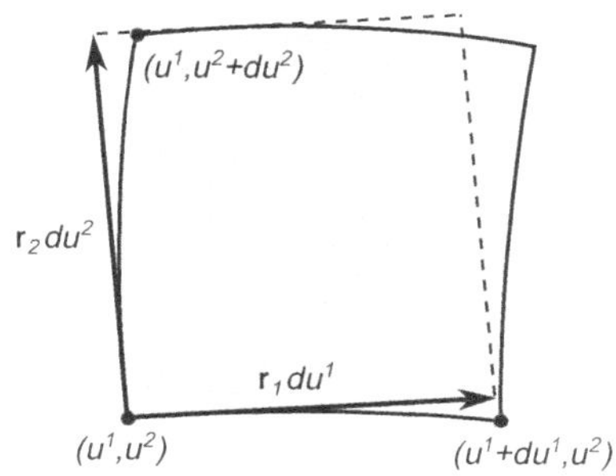

Fig. 3.3 Approximating a curvilinear patch with a parallelogram generated by tangent vectors.

and (3.10) follows.

Summing the elementary areas corresponding to a domain A in the (u^1, u^2) plane and doing the limit passage we get the value

$$S = \int_A \sqrt{EG - F^2}\, du^1\, du^2.$$

It can be shown that for a smooth surface this limit S does not depend on the parametrization and hence is the area of the portion A of the surface.

Exercise 3.12. A cone is described by the position vector

$$\mathbf{r}(u^1, u^2) = \mathbf{i}_1 u^1 \cos u^2 + \mathbf{i}_2 u^1 \sin u^2 + \mathbf{i}_3 u^1,$$

where $0 \le u^1 \le a$ and $0 \le u^2 < 2\pi$. Find the surface area of the cone.

Exercise 3.13. Let two smooth surfaces be parameterized in such a way that the coefficients of their first fundamental forms are proportional at any point with the same coordinates (so this defines a map of one surface onto the other). Show that the map preserves the angles between corresponding directions on the surfaces.

Second fundamental form and curvature

We have seen that the first fundamental form carries the metrical properties of a surface (i.e., its coefficients play roles in the calculation of arc length, angles, and surface area). The second fundamental form, in contrast, carries the curvature properties.

Using the Taylor expansion we can find the change of $\mathbf{r}(u^1, u^2)$ to a higher degree of approximation:

$$\mathbf{r}(u^1 + \Delta u^1, u^2 + \Delta u^2) = \mathbf{r}(u^1, u^2) + \frac{\partial \mathbf{r}(u^1, u^2)}{\partial u^i}\Delta u^i + \\ + \frac{1}{2}\frac{\partial^2 \mathbf{r}(u^1, u^2)}{\partial u^i \partial u^j}\Delta u^i \Delta u^j + o\left((\Delta u^1)^2 + (\Delta u^2)^2\right).$$

Here we apply the o-notation to vector quantities, indicating that the norm of the remainder has a higher order of smallness. We seek the deviation of S from the osculating plane in a neighborhood of a point (u^1, u^2). This value is given through the dot product by the normal vector $\mathbf{n}$:

$$\begin{aligned}
&(\mathbf{r}(u^1 + \Delta u^1, u^2 + \Delta u^2) - \mathbf{r}(u^1, u^2)) \cdot \mathbf{n} \\
&= \frac{\partial \mathbf{r}(u^1, u^2)}{\partial u^i} \cdot \mathbf{n}\Delta u^i + \frac{1}{2}\frac{\partial^2 \mathbf{r}(u^1, u^2)}{\partial u^i \partial u^j} \cdot \mathbf{n}\Delta u^i \Delta u^j + o\left((\Delta u^1)^2 + (\Delta u^2)^2\right) \\
&= \frac{1}{2}\frac{\partial^2 \mathbf{r}(u^1, u^2)}{\partial u^i \partial u^j} \cdot \mathbf{n}\Delta u^i \Delta u^j + o\left((\Delta u^1)^2 + (\Delta u^2)^2\right).
\end{aligned}$$

We have used the fact that $\mathbf{r}_i \cdot \mathbf{n} = 0$. The terms of the second order of smallness in Δu^i constitute the *second fundamental form.* Denoting

$$\frac{\partial^2 \mathbf{r}(u^1, u^2)}{(\partial u^1)^2} \cdot \mathbf{n} = L(u^1, u^2) = L,$$

$$\frac{\partial^2 \mathbf{r}(u^1, u^2)}{\partial u^1 \partial u^2} \cdot \mathbf{n} = M(u^1, u^2) = M,$$

$$\frac{\partial^2 \mathbf{r}(u^1, u^2)}{(\partial u^2)^2} \cdot \mathbf{n} = N(u^1, u^2) = N,$$

we introduce the second fundamental form by

$$d^2\mathbf{r} \cdot \mathbf{n} = L(du^1)^2 + 2M du^1 du^2 + N(du^2)^2 \tag{3.11}$$

where L, M, N are the *second fundamental coefficients.* The deviation of S from the osculating plane in a small neighborhood of (u^1, u^2) is approximately

$$z = \tfrac{1}{2}\left(L(du^1)^2 + 2M du^1 du^2 + N(du^2)^2\right), \tag{3.12}$$

and so (3.12) determines the qualitative nature of S in this neighborhood. This formula defines a paraboloid

$$z = \tfrac{1}{2}\left(L(v^1)^2 + 2M v^1 v^2 + N(v^2)^2\right) \tag{3.13}$$

written in the triad $\left(\mathbf{r}_1(u^1, u^2), \mathbf{r}_2(u^1, u^2), \mathbf{n}(u^1, u^2)\right)$ when the origin is the point of the surface with coordinates (u^1, u^2). Depending on the coefficients

L, M, N, this *osculating paraboloid* can be elliptic, hyperbolic, or parabolic (Table 3.3). If $L = M = N = 0$ then the corresponding point is called a planar point.

Table 3.3 Osculating paraboloid at a point P of a surface S.

classification of P	**condition**	**shape of S at P**
elliptic point	$LN - M^2 > 0$	elliptic paraboloid
hyperbolic point	$LN - M^2 < 0$	hyperbolic paraboloid
parabolic point	$LN - M^2 = 0$ but $L^2 + M^2 + N^2 \neq 0$	parabolic cylinder
planar point	L, M, N all 0	plane

For the second fundamental form of the surface there is a representation different from (3.11). Differentiating the identity $\mathbf{r}_i \cdot \mathbf{n} = 0$ we get

$$\frac{\partial \mathbf{r}_i}{\partial u^j} \cdot \mathbf{n} = -\mathbf{r}_i \cdot \frac{\partial \mathbf{n}}{\partial u^j}. \tag{3.14}$$

Dot multiplying $d\mathbf{r} \cdot d\mathbf{n}$ we get

$$\mathbf{r}_i\, du^i \cdot \frac{\partial \mathbf{n}}{\partial u^j}\, du^j = -\left(\frac{\partial \mathbf{r}_i}{\partial u^j}\, du^i\, du^j\right) \cdot \mathbf{n} = -\frac{\partial}{\partial u^j}\left(\frac{\partial \mathbf{r}}{\partial u^i}\right) du^i\, du^j = -d^2\mathbf{r} \cdot \mathbf{n},$$

which means that

$$-d\mathbf{r} \cdot d\mathbf{n} = L(du^1)^2 + 2M\, du^1\, du^2 + N(du^2)^2.$$

Exercise 3.14. Let S be described in Cartesian components by

$$\mathbf{r} = x(u,v)\mathbf{i}_1 + y(u,v)\mathbf{i}_2 + z(u,v)\mathbf{i}_3.$$

Let the subscripts u and v denote corresponding partial derivatives with respect to u and v. Show that

$$E = \mathbf{r}_u^2 = x_u^2 + y_u^2 + z_u^2,$$
$$F = \mathbf{r}_u \cdot \mathbf{r}_v = x_u x_v + y_u y_v + z_u z_v,$$
$$G = \mathbf{r}_v^2 = x_v^2 + y_v^2 + z_v^2,$$

and

$$L = \frac{\begin{vmatrix} x_{uu} & y_{uu} & z_{uu} \\ x_u & y_u & z_u \\ x_v & y_v & z_v \end{vmatrix}}{\sqrt{EG - F^2}}, \qquad M = \frac{\begin{vmatrix} x_{uv} & y_{uv} & z_{uv} \\ x_u & y_u & z_u \\ x_v & y_v & z_v \end{vmatrix}}{\sqrt{EG - F^2}}, \qquad N = \frac{\begin{vmatrix} x_{vv} & y_{vv} & z_{vv} \\ x_u & y_u & z_u \\ x_v & y_v & z_v \end{vmatrix}}{\sqrt{EG - F^2}}.$$

Let us consider a curve lying on the surface. The curve can be uniquely defined by a parameter s in such a way that $u^1 = u^1(s)$, $u^2 = u^2(s)$; for simplicity we assume s to be the length parameter. Then the equation of the curve is

$$\mathbf{r} = \mathbf{r}(u^1(s), u^2(s)) = \mathbf{r}(s)$$

and a tangential vector to the curve at (u^1, u^2) is $(\mathbf{r}_i du^i/ds)$. Consider the curvature of this curve, which can be found using $\mathbf{r}''(s)$: the normal $\boldsymbol{\nu}$ to the curve has the same direction and k_1 equals $|\mathbf{r}''(s)|$. It follows that

$$\mathbf{r}''(s) \cdot \mathbf{n} = k_1 \cos\vartheta \tag{3.15}$$

where ϑ is the angle between $\boldsymbol{\nu}$ and $\mathbf{n}$. Let us rewrite (3.15) in the form

$$\begin{aligned} k_1 \cos\vartheta \\ &= \frac{\frac{\partial^2\mathbf{r}}{(\partial u^1)^2}\left(\frac{du^1}{ds}\right)^2 (ds)^2 + 2\frac{\partial^2\mathbf{r}}{\partial u^1 \partial u^2}\frac{du^1}{ds}\frac{du^2}{ds}(ds)^2 + \frac{\partial^2\mathbf{r}}{(\partial u^2)^2}\left(\frac{du^2}{ds}\right)^2 (ds)^2}{(ds)^2} \cdot \mathbf{n} \\ &= \frac{L(du^1(s))^2 + 2M\,du^1(s)\,du^2(s) + N(du^2(s))^2}{E(du^1(s))^2 + 2F\,du^1(s)\,du^2(s) + G(du^2(s))^2}. \end{aligned} \tag{3.16}$$

For all the curves with the same direction defined by the constant ratio $du^1(s) : du^2(s)$, the right-hand side of (3.16) is the same as the ratio of the second to the first fundamental forms. We define this value k_0 as the *normal curvature* of the surface in the direction $du^1(s) : du^2(s)$. Geometrically, this is the curvature of the curve formed by intersecting the surface with the plane through $(u^1(s), u^2(s))$ that is parallel to $\mathbf{n}$ and $\mathbf{r}_i\, du^i(s)$. This normal curvature satisfies the equality

$$k_0 = k_1 \cos\vartheta$$

which is the main part of *Meusnier's theorem*. From this formula for curvatures it follows that the normal curvature of the surface in the direction $du^1 : du^2$ has minimal absolute value among all the curvatures of the curves on the surface through $(u^1(s), u^2(s))$ having the same direction $du^1 : du^2$.

Let us note that the curvature of the osculating paraboloid is characterized by the same equation. The normal curvature depends on the direction $du^1 : du^2 = x : y$ only. Let us rewrite its expression as

$$k_0 = \frac{Lx^2 + 2Mxy + Ny^2}{Ex^2 + 2Fxy + Gy^2}. \tag{3.17}$$

The right-hand side of (3.17) is a homogeneous form with respect to x, y of zero order; it is easy to see that the minimum and maximum of k_0 can be found as the minimum and maximum of the quadratic form

$$Lx^2 + 2Mxy + Ny^2$$

when

$$Ex^2 + 2Fxy + Gy^2 = 1. \tag{3.18}$$

This problem can be solved with use of the Lagrange multiplier. On the curve (3.18) denote

$$\min(Lx^2 + 2Mxy + Ny^2) = k_{\min}, \qquad \max(Lx^2 + 2Mxy + Ny^2) = k_{\max}.$$

Let the corresponding values for x, y be (x_1, y_1) and (x_2, y_2), respectively. If $k_{\min} \neq k_{\max}$ these values define only two directions $x_j : y_j$ $(j = 1, 2)$ and it can be shown that

$$Ex_1x_2 + F(x_1y_2 + x_2y_1) + Gy_1y_2 = 0,$$

which, by (3.9), means that the directions corresponding to the extreme values of the normal curvature are mutually orthogonal. In the case when $k_{\min} = k_{\max}$ the normal curvature is the same in all directions (the osculating paraboloid corresponds to a paraboloid of revolution). These values $k_{\min}, k_{\max}$ are the extreme normal curvatures of the surface at a point; they characterize the surface at a point and hence are invariant with respect to the change of coordinates of the surface. So are two other characteristics of the surface at a point: the *mean curvature*

$$H = \tfrac{1}{2}(k_{\min} + k_{\max}),$$

and the *Gaussian* or *complete curvature*

$$K = k_{\min}k_{\max}.$$

We present without proof the equations in terms of the fundamental forms of the surface:

$$H = \frac{1}{2}\frac{LG - 2MF + NE}{EG - F^2}, \qquad K = \frac{LN - M^2}{EG - F^2}.$$

Since $EG - F^2 > 0$, the sign of K is the sign of $LN - M^2$. When $K = 0$ everywhere the surface is developable.[2] These and many other facts on the

[2]Roughly speaking, a *developable surface* is one that can be flattened into a portion of a plane without compressing or stretching any part of it. Examples include cones and cylinders, but not spheres. A developable surface can be generated by sweeping a straight line (*generator*) along a curve through space. See page 139 for a more precise definition.

properties of surfaces with given H and K can be found in any textbook on differential geometry (e.g., [Pogorelov (1957)]).

Exercise 3.15. Find the mean and Gaussian curvatures for the right helicoid $x = u\cos\phi$, $y = u\sin\phi$, $z = c\phi$.

Exercise 3.16. A surface is described by $z = f(x,y)$ so that $\mathbf{r} = \mathbf{i}_1 x + \mathbf{i}_2 y + \mathbf{i}_3 f(x,y)$. Show that (a)

$$E = \mathbf{r}_x^2 = 1 + f_x^2, \qquad F = \mathbf{r}_x \cdot \mathbf{r}_y = f_x f_y, \qquad G = \mathbf{r}_y^2 = 1 + f_y^2;$$

(b)

$$EG - F^2 = 1 + f_x^2 + f_y^2;$$

(c)

$$L = \frac{f_{xx}}{\sqrt{1 + f_x^2 + f_y^2}}, \qquad M = \frac{f_{xy}}{\sqrt{1 + f_x^2 + f_y^2}}, \qquad N = \frac{f_{yy}}{\sqrt{1 + f_x^2 + f_y^2}};$$

(d) the area of a portion of the surface is

$$S = \int_D \sqrt{1 + f_x^2 + f_y^2}\, dx\, dy;$$

(e) the Gaussian curvature is

$$K = \frac{LN - M^2}{EG - F^2} = \frac{f_{xx} f_{yy} - f_{xy}^2}{\left(1 + f_x^2 + f_y^2\right)^2}.$$

Derivation formulas

At each point of a surface there is defined the frame triad

$$\mathbf{r}_1(u^1, u^2), \quad \mathbf{r}_2(u^1, u^2), \quad \mathbf{n}(u^1, u^2).$$

In the theory of shells they introduce the curvilinear coordinates in a neighborhood of the surface in such a way that on the surface the triad $(\mathbf{r}_1, \mathbf{r}_2, \mathbf{n})$ is preserved. Since we seek different characteristics of fields given on the surface and outside of it, we must find the derivatives of the triad with respect to the coordinates. The goal of this section is to present these in terms of the surface we have introduced: i.e., in terms of the first and second fundamental forms of the surface.

We start with the representation for the derivatives through the Christoffel notation. This is valid because, by assumption, $(\mathbf{r}_1, \mathbf{r}_2, \mathbf{n})$ is a basis of the space of vectors. So

$$\mathbf{r}_{ij} = \frac{\partial^2 \mathbf{r}}{\partial u^i \partial u^j} = \Gamma^t_{ij} \mathbf{r}_t + \lambda_{ij} \mathbf{n}. \tag{3.19}$$

We recall that we use indices i, j, t taking values from the set $\{1, 2\}$, which is why we introduce the notation λ_{ij} for the coefficients of $\mathbf{n}$. Similarly let us introduce the expansion for the derivatives of $\mathbf{n}$. To derive the coefficients λ_{ij} we dot multiply (3.19) by $\mathbf{n}$. Using the expressions for the second fundamental form we get

$$\lambda_{11} = L, \qquad \lambda_{12} = \lambda_{21} = M, \qquad \lambda_{22} = N.$$

Next, dot multiplying (3.19) first by $\mathbf{r}_1$ and then by $\mathbf{r}_2$, we get six equations in the six unknown Christoffel symbols:

$$\begin{aligned}
\frac{1}{2}\frac{\partial E}{\partial u^1} &= \Gamma^1_{11} E + \Gamma^2_{11} F, & \frac{1}{2}\frac{\partial G}{\partial u^1} &= \Gamma^1_{12} F + \Gamma^2_{12} G, \\
\frac{\partial F}{\partial u^1} - \frac{1}{2}\frac{\partial E}{\partial u^2} &= \Gamma^1_{11} F + \Gamma^2_{11} G, & \frac{\partial F}{\partial u^2} - \frac{1}{2}\frac{\partial G}{\partial u^1} &= \Gamma^1_{22} E + \Gamma^2_{22} F, \\
\frac{1}{2}\frac{\partial E}{\partial u^2} &= \Gamma^1_{12} E + \Gamma^2_{12} F, & \frac{1}{2}\frac{\partial G}{\partial u^2} &= \Gamma^1_{22} F + \Gamma^2_{22} G.
\end{aligned}$$

This system can be easily solved for the Christoffel coefficients since the system splits into pairs of equations with respect to the pairs of Christoffel symbols. The first pair yields, for instance,

$$\Gamma^1_{11} = \frac{1}{EG - F^2}\left(\frac{G}{2}\frac{\partial E}{\partial u^1} - F\frac{\partial F}{\partial u^1} + \frac{F}{2}\frac{\partial E}{\partial u^2}\right),$$

and a similar expression for Γ^2_{11}. We leave the rest to the reader, noting that the expressions for the Christoffel symbols involve only the coefficients of the first fundamental form. Next we consider the expressions for the derivatives of the normal $\mathbf{n}$:

$$\mathbf{n}_i \equiv \frac{\partial \mathbf{n}}{\partial u^i} = \mu_i{}^t \mathbf{r}_t + \mu_i{}^3 \mathbf{n}. \tag{3.20}$$

Let us find the coefficients of these expansions through the coefficients of the first and second fundamental forms. Since

$$0 = \frac{\partial}{\partial u^i} 1 = \frac{\partial}{\partial u^i}(\mathbf{n} \cdot \mathbf{n}) = 2\mathbf{n}_i \cdot \mathbf{n}$$

the third coefficient is zero:

$$\mu_i{}^3 = 0$$

Let us dot multiply (3.20) when $i = 1$ by $\mathbf{r}_1$ and $\mathbf{r}_2$ successively. Using the definitions for the coefficients of the first and second fundamental forms we have the equations

$$-L = \mu_1{}^1 E + \mu_1{}^2 F, \qquad -M = \mu_1{}^1 F + \mu_1{}^2 G,$$

from which we get

$$\mu_1{}^1 = \frac{-LG + MF}{EG - F^2}, \qquad \mu_1{}^2 = \frac{LF - ME}{EG - F^2}.$$

Repeating this procedure for (3.20) when $i = 2$ we similarly obtain

$$\mu_2{}^1 = \frac{NF - MG}{EG - F^2}, \qquad \mu_2{}^2 = \frac{-NE + MF}{EG - F^2}.$$

Exercise 3.17. Derive the Christoffel symbols for the case when the first fundamental form is $(du^1)^2 + G(du^2)^2$.

Some useful formulas

Certain identities are frequently employed. Their derivation is lengthy so we omit it here. Let us redenote the coefficients of the second fundamental form using index notations:

$$b_{11} = L, \qquad b_{12} = b_{21} = M, \qquad b_{22} = N.$$

These are the covariant components of a corresponding symmetric tensor of order two. The first formula is due to Gauss:

$$b_{11}b_{22} - b_{12}^2 = \frac{\partial^2 g_{12}}{\partial u^1 \partial u^2} - \frac{1}{2}\left(\frac{\partial^2 g_{11}}{(\partial u^2)^2} + \frac{\partial^2 g_{22}}{(\partial u^1)^2}\right) + \left(\Gamma^i_{12}\Gamma^j_{12} - \Gamma^i_{11}\Gamma^j_{22}\right) g_{ij},$$

from which (and the form of K, see page 129) it follows that the Gaussian curvature of the surface can be expressed purely in terms of the first fundamental form. The next formulas are due to Peterson and Codazzi:

$$\frac{\partial b_{i1}}{\partial u^2} - \frac{\partial b_{i2}}{\partial u^1} = \Gamma^t_{i2} b_{t1} - \Gamma^t_{i1} b_{t2} \qquad (i = 1, 2).$$

Using the formulas of this paragraph, we can derive the formulas for differentiating a spatial vector

$$\mathbf{f} = f^i \mathbf{r}_i + f_3 \mathbf{n} = f_i \mathbf{r}^i + f_3 \mathbf{n}$$

given on the surface. Denoting $\mathbf{r}_3 = \mathbf{r}^3 = \mathbf{n}$, we can derive the following formulas that mimic the formulas of Chapter 4:

$$\frac{\partial \mathbf{f}}{\partial u^i} = \nabla_i f^t \mathbf{r}_t = \nabla_i f_t \mathbf{r}^t.$$

In these, the covariant derivatives are

$$\nabla_i f^j = \frac{\partial f^j}{\partial u^i} + \Gamma^j_{it} f^t - b^j_i f_3,$$
$$\nabla_i f_j = \frac{\partial f_j}{\partial u^i} + \Gamma^t_{ij} f_t - b_{ij} f_3,$$
$$\nabla_i f_3 = \frac{\partial f_3}{\partial u^i} + b_{it} f^t = \frac{\partial f_3}{\partial u^i} + b^t_i f_t.$$

Here i, j take the values $1, 2$, while t takes the values $1, 2, 3$. Summation is carried out over i, j, and t in the respective ranges. When $\mathbf{f} = \mathbf{n}$, i.e., when $f^1 = f^2 = 0$ and $f^3 = 1$, we get the following formula for the derivatives of $\mathbf{n}$:

$$\frac{\partial \mathbf{n}}{\partial u^i} = -b^t_i \mathbf{r}_t = -b_{it} \mathbf{r}^t. \tag{3.21}$$

Introducing the *surface gradient* or *surface nabla operator* by the formula

$$\widetilde{\nabla} = \mathbf{r}^i \frac{\partial}{\partial u^i} \qquad (i = 1, 2),$$

we find that

$$\widetilde{\nabla} \mathbf{f} = \nabla_i f^t \mathbf{r}^i \mathbf{r}_t = \nabla_i f_t \mathbf{r}^i \mathbf{r}^t, \quad \widetilde{\nabla} \mathbf{n} = -\mathbf{B},$$

where $\mathbf{B}$ is the *surface curvature tensor* given by

$$\mathbf{B} = b^j_i \mathbf{r}^i \mathbf{r}_j = b_{ij} \mathbf{r}^i \mathbf{r}^j = b^{ij} \mathbf{r}_i \mathbf{r}_j.$$

One may introduce the surface divergence and rotation operations

$$\operatorname{div} \mathbf{f} = \widetilde{\nabla} \cdot \mathbf{f} = \mathbf{r}^i \cdot \frac{\partial \mathbf{f}}{\partial u^i}, \quad \operatorname{rot} \mathbf{f} = \widetilde{\nabla} \times \mathbf{f} = \mathbf{r}^i \times \frac{\partial \mathbf{f}}{\partial u^i}.$$

The applications of the differential calculus on a surface are presented in Chapter 7, where shell theories are considered.

Osculating paraboloid

When considering the structure of a surface at a point, it is often helpful to approximate the surface using another surface whose behavior is more easily visualized. A spherical surface would be insufficient for this purpose because it has the same normal curvature in all directions. We can, however, use the osculating paraboloid introduced in (3.13). Let us reconsider this paraboloid from the point of view of local approximation.

Let O be a fixed point of a surface, and assume the surface is sufficiently smooth at O. At O we determine the osculating plane and introduce a

Cartesian frame $(\mathbf{i}_1, \mathbf{i}_2, \mathbf{n})$, where $(\mathbf{i}_1, \mathbf{i}_2)$ is a Cartesian frame with origin O *on the osculating plane* and $\mathbf{n}$ is normal to both the surface and the osculating plane. For a smooth surface, the Cartesian coordinates (x, y) can play the role of surface coordinates since they uniquely define any point of the surface at O. The surface in the vicinity of O can be described by the equation $z = z(x, y)$. We suppose that $z(x, y)$ is twice continuously differentiable near O. In the neighborhood of $(0, 0)$ we can use the Taylor expansion of $z(x, y)$:

$$\begin{aligned} z(x, y) &= z(0,0) + z_x(0,0)x + z_y(0,0)y \\ &\quad + \tfrac{1}{2}\left(z_{xx}x^2 + 2z_{xy}xy + z_{yy}y^2\right) + o(x^2 + y^2), \end{aligned}$$

where the indices x, y indicate that we take partial derivatives with respect to x, y, respectively (and in this section, evaluate at the point $(0,0)$). Since the coordinates (x, y) are in the osculating plane, the partial derivatives

$$z_x(0,0) = 0, \qquad z_y(0,0) = 0.$$

By choice of the origin, $z(0,0) = 0$. Thus the Taylor expansion is

$$z(x, y) = \tfrac{1}{2}\left(z_{xx}x^2 + 2z_{xy}xy + z_{yy}y^2\right) + o(x^2 + y^2). \tag{3.22}$$

Let us consider the paraboloid

$$z(x, y) = \tfrac{1}{2}\left(z_{xx}x^2 + 2z_{xy}xy + z_{yy}y^2\right). \tag{3.23}$$

The difference between the surfaces described by (3.22) and (3.23) is small (as indicated by the o term). This allows us to show that it is an osculating paraboloid that approximates the behavior of the surface at point O. A change of coordinate frames can show that the paraboloid (3.23) coincides with (3.13). Its coefficients are L, M, N at the point O.

Let us note that the second fundamental form of the initial surface at a point coincides with that of the osculating paraboloid, and the situation is the same with the normal curvatures. We denote

$$r = z_{xx}(0,0), \qquad s = z_{xy}(0,0), \qquad t = z_{yy}(0,0),$$

so that the equation of the osculating paraboloid is

$$z = \tfrac{1}{2}\left(rx^2 + 2sxy + ty^2\right).$$

Let us draw the projection onto the xy-coordinate plane of the cross-sections of the osculating paraboloid by the two planes $z = \pm h$, $h > 0$. This is the curve defined by

$$\tfrac{1}{2}\left|rx^2 + 2sxy + ty^2\right| = h. \tag{3.24}$$

The curve is an ellipse when the paraboloid is elliptical ($rt - s^2 > 0$) or a hyperbola when it is hyperbolic ($rt - s^2 < 0$), or a family of straight lines when it is parabolic ($rt - s^2 = 0$). It can be shown that the radius of normal curvature of the surface in the direction $x : y$ is proportional to the squared distance from the origin to the point of the curve (3.24) taken in the same direction. This curve is called the *Dupin indicatrix.*

The Dupin indicatrix is a curve of second order on the plane, and thus it has some special directions (axes) and special values characterizing the curve. The special directions are known as *principal directions.* In the next section we consider this from another vantage point.

Principal curvatures of a surface

Let us discuss in more detail the properties of a surface connected with its second fundamental form. We now denote the coordinates without indices, using $u = u^1, v = u^2$. We also use subscripts u, v to indicate partial differentiation:

$$\mathbf{r}_1 = \mathbf{r}_u = \frac{\partial \mathbf{r}}{\partial u}, \qquad \mathbf{r}_2 = \mathbf{r}_v = \frac{\partial \mathbf{r}}{\partial v}, \qquad \mathbf{n}_u = \frac{\partial \mathbf{n}}{\partial u}, \qquad \mathbf{n}_v = \frac{\partial \mathbf{n}}{\partial v}.$$

The second fundamental form is

$$L\,du^2 + 2M\,du\,dv + N\,dv^2 = -d\mathbf{r}\cdot d\mathbf{n},$$

where

$$L = \mathbf{r}_{uu}\cdot\mathbf{n}, \qquad M = \mathbf{r}_{uv}\cdot\mathbf{n}, \qquad N = \mathbf{r}_{vv}\cdot\mathbf{n}.$$

Consider the first differential of $\mathbf{n}$ at a point P:

$$d\mathbf{n} = \mathbf{n}_u\,du + \mathbf{n}_v\,dv.$$

Since $\mathbf{n}$ is a unit vector, its differential $d\mathbf{n}$ is orthogonal to $\mathbf{n}$ and thus lies in the osculating plane to the surface at P. We know that the differential $d\mathbf{r} = \mathbf{r}_u\,du + \mathbf{r}_v\,dv$ also lies in the plane osculating at P. Let us consider the relation between the differentials $d\mathbf{r}$ and $d\mathbf{n}$ with respect to the variables du, dv now considered as independent variables. It is clearly a linear correspondence $d\mathbf{r} \mapsto d\mathbf{n}$. Thus it defines a tensor $\mathbf{A}$ in two-dimensional space such that

$$d\mathbf{n} = \mathbf{A}\cdot d\mathbf{r}.$$

This tensor is completely defined by its values $\mathbf{n}_u = \mathbf{A}\cdot\mathbf{r}_u$ and $\mathbf{n}_v = \mathbf{A}\cdot\mathbf{r}_v$.

Lemma 3.1. *The tensor* $\mathbf{A}$ *is symmetric.*

Proof. It is enough to establish the equality

$$\mathbf{x}_1 \cdot (\mathbf{A} \cdot \mathbf{x}_2) = \mathbf{x}_2 \cdot (\mathbf{A} \cdot \mathbf{x}_1)$$

for a pair of linearly independent vectors $(\mathbf{x}_1, \mathbf{x}_2)$. To show symmetry of $\mathbf{A}$ consider

$$\mathbf{r}_u \cdot (\mathbf{A} \cdot \mathbf{r}_v) = \mathbf{r}_u \cdot \mathbf{n}_v.$$

Similarly

$$\mathbf{r}_v \cdot (\mathbf{A} \cdot \mathbf{r}_u) = \mathbf{r}_v \cdot \mathbf{n}_u.$$

The symmetry of $\mathbf{A}$ follows from the identity

$$\mathbf{r}_u \cdot \mathbf{n}_v = \mathbf{r}_v \cdot \mathbf{n}_u;$$

this is derived by differentiating the identity $\mathbf{n} \cdot \mathbf{r}_u = 0$ with respect to v, then differentiating $\mathbf{n} \cdot \mathbf{r}_v = 0$ with respect to u and eliminating the term containing $\mathbf{r}_{uv}$. □

We know that a symmetric tensor has real eigenvalues; in this case there are no more than two eigenvalues λ_1 and λ_2 to which there correspond mutually orthogonal eigenvectors $\mathbf{x}_1, \mathbf{x}_2$, respectively (if $\lambda_1 \neq \lambda_2$).

Let us find the equation for the eigenvalues. A vector in the osculating plane can be represented as $x\mathbf{r}_u + y\mathbf{r}_v$ since $(\mathbf{r}_u, \mathbf{r}_v)$ is a basis in it. By definition of $\mathbf{A}$ we have

$$\mathbf{A} \cdot (x\mathbf{r}_u + y\mathbf{r}_v) = x\mathbf{n}_u + y\mathbf{n}_v.$$

On the other hand, $x\mathbf{r}_u + y\mathbf{r}_v \neq 0$ is an eigenvector if there exists λ such that

$$\mathbf{A} \cdot (x\mathbf{r}_u + y\mathbf{r}_v) = \lambda (x\mathbf{r}_u + y\mathbf{r}_v).$$

Thus, for the same vector we get

$$x\mathbf{n}_u + y\mathbf{n}_v = \lambda (x\mathbf{r}_u + y\mathbf{r}_v). \tag{3.25}$$

We see that an eigenvector is defined by the same condition as the principal directions of the previous section, since this equation means that we find the direction in which $d\mathbf{r}$ and $d\mathbf{n}$ are parallel.

From the last vector equation, let us derive scalar equations. For this, dot multiply (3.25) first by $\mathbf{r}_u$, then by $\mathbf{r}_v$. By (3.14) we have

$$L = -\mathbf{n}_u \cdot \mathbf{r}_u, \qquad M = -\mathbf{n}_u \cdot \mathbf{r}_v, \qquad N = -\mathbf{n}_v \cdot \mathbf{r}_v,$$

hence

$$-Lx - My = \lambda(Ex + Fy),$$
$$-Mx - Ny = \lambda(Fx + Gy).$$

Rewriting this as

$$\begin{aligned}(L + \lambda E)x + (M + \lambda F)y = 0,\\ (M + \lambda F)x + (N + \lambda G)y = 0,\end{aligned} \tag{3.26}$$

we have a homogeneous linear system of algebraic equations. This system has nontrivial solutions when its determinant vanishes:

$$(EG - F^2)\lambda^2 - (2MF - EN - LG)\lambda + (LN - M^2) = 0. \tag{3.27}$$

In the general case, there are two roots λ_1 and λ_2, to which there correspond two directions previously called the principal directions. They can be found by elimination of λ from (3.26):

$$\begin{vmatrix} -x^2 & xy & -y^2 \\ E & F & G \\ L & M & N \end{vmatrix} = 0.$$

By this we define two mutually orthogonal directions $x : y$. If $\lambda_1 = \lambda_2$, then all the directions are principal so we can choose any two mutually orthogonal directions and regard them as principal.

Now we would like to consider another question that will bring us to the same equations. It is the question of finding the extreme normal curvatures at a point of a surface. We have established that in the direction $x : y$ the normal curvature of a surface is given by the formula

$$k = \frac{Lx^2 + 2Mxy + Ny^2}{Ex^2 + 2Fxy + Gy^2}.$$

Because of the second-order homogeneity in x, y of the numerator and denominator we can reformulate the problem of finding extreme curvatures as the problem of finding extremal points of the function

$$Lx^2 + 2Mxy + Ny^2$$

under the restriction

$$Ex^2 + 2Fxy + Gy^2 = 1.$$

Applying the theory of Lagrange multipliers to this, we should find the extreme points of the function

$$Lx^2 + 2Mxy + Ny^2 - k(Ex^2 + 2Fxy + Gy^2),$$

which leads to the equations

$$\begin{aligned}(L-kE)x+(M-kF)y&=0,\\(M-kF)x+(N-kG)y&=0.\end{aligned}\tag{3.28}$$

The systems (3.26) and (3.28) coincide if we put $k=-\lambda$, which means that the principal directions and the directions found here are the same. Moreover, it is seen that λ_1 and λ_2 found as the eigenvalues of the tensor $\mathbf{A}$ give the extreme curvatures of the surface at a point, which are $k_1=-\lambda_1$ and $k_2=-\lambda_2$. Remembering that they are the roots of the polynomial (3.27) and using the Viète theorem, we get

$$k_1+k_2=\frac{EN-2FM+GL}{EG-F^2},\qquad k_1k_2=\frac{LN-M^2}{EG-F^2}.$$

We have met these expressions in the Gaussian curvature $K=k_1k_2$ and the mean curvature $H=(k_1+k_2)/2$.

We see that both are expressed through the coefficients of the first and second fundamental forms of the surface. There is a famous theorem due to Gauss that K can be expressed only in terms of the coefficients of the first fundamental form.

When the principal curvatures and their directions are known, the Euler formula gives the normal curvature at any direction of the curve which composes the angle ϕ with the first principal direction at the same point, corresponding to k_1:

$$k_\phi=k_1\cos\phi+k_2\sin\phi.$$

The principal directions of a surface define the *lines of curvature* of the surface. A line in the surface is called a line of curvature if at each point its tangent is directed along one of the principal directions at this point of the surface.

Two lines of curvature pass through each point. Denoting their directions by $du:dv$ and $\delta u:\delta v$ we have two equations, the first of which means orthogonality of the directions and the second their conjugation:

$$E\,du\,\delta u+F(du\,\delta v+dv\,\delta u)+G\,dv\,\delta v=0,$$
$$L\,du\,\delta u+M(du\,\delta v+dv\,\delta u)+N\,dv\,\delta v=0.$$

It is convenient to take the family of the lines of curvature of the surface as the coordinate lines of the surface. For these curvilinear coordinates $F=M=0$.

Two surfaces are called *isometric* if there is a one-to-one correspondence between the points of the surfaces such that corresponding curves in the surfaces have the same length.

A planar surface and a cylindrical surface provide an example of isometric surfaces since we can develop the cylindrical surface over the plane. The correspondence between points is defined by coincidence of the points in this developing.

We can refer to local isometry of a surface at point A_1 to another surface at point B_2 if there are neighborhoods of points on the surfaces that are isometric.

Two smooth surfaces are locally isometric if and only if there are parametrizations of the surfaces at corresponding points such that the coefficients of the first principal forms of the surfaces coincide:

$$E_1 = E_2, \qquad F_1 = F_2, \qquad G_1 = G_2.$$

A surface is *developable* if it is locally isometric to the plane at each point. It turns out that a surface is developable if and only if its Gaussian curvature is zero at each point.

Surfaces with zero Gaussian curvature appear once more in the problem of finding spatial surfaces of minimal area that have given boundaries. This popular physics problem describes the shape assumed by a soap film on a wire frame. The energy of such a film is proportional to its area, and thus the form actually taken by such a film is the surface of minimum area.

3.3 Problems

3.1 Find the parametrization and singular points of the plane curve

$$|x|^{\frac{2}{3}} + |y|^{\frac{2}{3}} = |a|^{\frac{2}{3}},$$

where a is a parameter.

3.2 Find the singular points of the plane curve given by the equations

$$x = a(t - \sin t), \qquad y = a(1 - \cos t),$$

where a is a parameter.

3.3 Find the equation of the tangent to the curve $x^2 + y^2 + z^2 = 1$, $x^2 + y^2 = x$ at the point $(0, 0, 1)$.

3.4 Find the length of the curve $\mathbf{r}(t) = \mathbf{i}_1 \cosh 2t + \mathbf{i}_2 \sinh 2t + \mathbf{i}_3 2t$ for $0 \leq t \leq \pi$.

3.5 Find the lengths of the following curves for $0 \le t < 2\pi$: (a) the astroid $x = a\cos^3 t$, $y = a\sin^3 t$; (b) the cycloid $x = a(t - \sin t)$, $y = a(1 - \cos t)$.

3.6 Find the curvature of the curve $x = t - \sin t$, $y = 1 - \cos t$, $z = 4\sin(t/2)$.

3.7 Find the curvature and torsion of the curve $x = a\cosh t$, $y = a\sinh t$, $z = at$ at an arbitrary point.

3.8 Find the torsion of the curve $x = a\cosh t\cos t$, $y = a\cosh t\sin t$, $z = at$ at an arbitrary point.

3.9 Let a triad of orthonormal vectors $\mathbf{e}_1(s)$, $\mathbf{e}_2(s)$, $\mathbf{e}_3(s)$ (i.e., vectors satisfying $\mathbf{e}_i \cdot \mathbf{e}_j = \delta_{ij}$) be given along a curve. Show that

$$\frac{d}{ds}\mathbf{e}_i(s) = \mathbf{d} \times \mathbf{e}_i(s)$$

where $\mathbf{d} = -\frac{1}{2}(\mathbf{e}_i' \times \mathbf{e}_i)$. Note that these formulas are analogous to the Frenet–Serret equations with the Darboux vector $\boldsymbol{\delta}$.

3.10 Let $\mathbf{Q}(s)$ be an orthogonal tensor given along a curve. Verify the formulas $\mathbf{Q}' = \mathbf{d} \times \mathbf{Q}$ and $\mathbf{d} = -\frac{1}{2}(\mathbf{Q}' \times \mathbf{Q}^T)_\times$.

3.11 Find the second quadratic form of the surface given by the equations $x = u\cos v$, $y = u\sin v$, and $z = v$.

3.12 A surface is defined by the equation $z = f(x, y)$. Show that the coefficients of its second principal form are

$$b_{11} = \frac{f_{xx}}{\sqrt{1 + f_x^2 + f_y^2}}, \qquad b_{12} = b_{21} = \frac{f_{xy}}{\sqrt{1 + f_x^2 + f_y^2}}, \qquad b_{22} = \frac{f_{yy}}{\sqrt{1 + f_x^2 + f_y^2}}.$$

3.13 A surface is defined by the equation $z = f(x, y)$, where f satisfies Laplace's equation $\nabla^2 f = 0$. Demonstrate that its Gaussian curvature satisfies $K \le 0$.

3.14 Show that the mean curvature of the surface $z = f(x, y)$ is given by

$$H = \operatorname{div}\left(\frac{\operatorname{grad} f}{\sqrt{1 + |\operatorname{grad} f|^2}}\right) \quad \text{where} \quad \operatorname{grad} = \mathbf{i}_1\frac{\partial}{\partial x} + \mathbf{i}_2\frac{\partial}{\partial y}.$$

3.15 Show that the mean curvature of a surface is given by $H = -\widetilde{\nabla} \cdot \mathbf{n}/2$.

3.16 Demonstrate that $\widetilde{\nabla} \cdot \mathbf{A} = 2H\mathbf{n}$ where $\mathbf{A} = \mathbf{E} - \mathbf{nn}$.

3.17 Let $\mathbf{X}$ be a second-order tensor. Show that $\mathbf{n}\cdot(\widetilde{\nabla}\times\mathbf{X}) = -\widetilde{\nabla}\cdot(\mathbf{n}\times\mathbf{X})$.

3.18 Find the Gaussian curvature of a surface given by the relation

$$z = f(x) + g(y).$$

3.19 The first principal form of a surface is $A^2\,du^2 + B^2\,dv^2$. Determine its Gaussian curvature.

3.20 Let $\mathbf{X}$ be a second-order tensor. Prove the following analog of the Gauss–Ostrogradsky theorem on a surface S having boundary contour Γ:

$$\int_S \left(\widetilde{\nabla}\cdot\mathbf{X} + 2H\mathbf{n}\cdot\mathbf{X}\right) dS = \oint_\Gamma \boldsymbol{\nu}\cdot\mathbf{X}\,ds, \tag{3.29}$$

where $\boldsymbol{\nu}$ is the outward unit normal to Γ lying in the tangent plane, i.e., $\boldsymbol{\nu}\cdot\mathbf{n} = 0$. Note that when S is a closed surface,

$$\int_S \widetilde{\nabla}\cdot\mathbf{X}\,dS = -\int_S 2H\mathbf{n}\cdot\mathbf{X}\,dS.$$

3.21 Prove that (3.29) holds for a tensor field $\mathbf{X}$ of any order.

3.22 Use the solution of the previous problem to prove

$$\int_S \left(\widetilde{\nabla}\mathbf{X} + 2H\mathbf{n}\mathbf{X}\right) dS = \oint_\Gamma \boldsymbol{\nu}\mathbf{X}\,ds,$$

$$\int_S \left(\widetilde{\nabla}\times\mathbf{X} + 2H\mathbf{n}\times\mathbf{X}\right) dS = \oint_\Gamma \boldsymbol{\nu}\times\mathbf{X}\,ds,$$

$$\int_S \widetilde{\nabla}\times(\mathbf{n}\mathbf{X})\,dS = \oint_\Gamma \boldsymbol{\tau}\mathbf{X}\,ds,$$

where $\boldsymbol{\tau} = \boldsymbol{\nu}\times\mathbf{n}$ is the unit tangent vector to Γ.

Chapter 4

Linear Elasticity

Linear elasticity uses tensor tools to build on the ideas of classical mechanics. It treats small deformations of elastic bodies described by linear constitutive equations relating stress to strain. This chapter introduces the main concepts of linear elasticity, starting with Cauchy's stress tensor. The main types of boundary value problems are formulated and some of their properties are studied. Finally, the role of variational principles is explored.

4.1 Stress Tensor

The notion of stress is introduced as the ratio of the value of a force distributed over a surface element, inside a body or on its boundary, to the element area. The difference between stress in elasticity and pressure in ideal fluids is that in fluids the forces are always normal to the surface element. In elastic bodies they can have any direction and the magnitude depends on the direction.

Consider a long thin cylindrical bar having cross-sectional area S and stretched by a force f uniformly distributed over the end caps (Fig. 4.1). Suppose the force is equidistributed over any normal cross-section of the bar, and the stress σ is defined as $\sigma = f/S$. But this example, while instructive initially, is too elementary to be representative; the force situation internal to a solid deformable body can be much more complex. At a chosen point in such a body, the contact force may not be directed normal to a given area element. Furthermore, with a change of element orientation the force density acting across the element may change in both magnitude and direction.

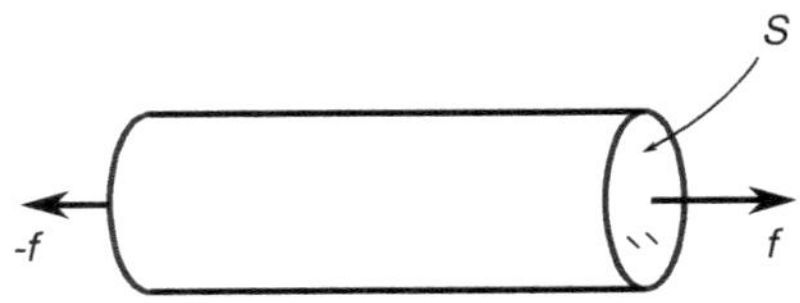

Fig. 4.1 The stress $\sigma = f/S$ in a bar having cross-sectional area S and stretched by a force f.

Forces

Let us discuss the conditions for equilibrium of a rigid body, on the basis of which we will construct the equilibrium conditions in a deformable body. These consist of the two vector equations

$$\sum_k \mathbf{f}_k = \mathbf{0}, \qquad \sum_k (\mathbf{r}_k - \mathbf{r}_0) \times \mathbf{f}_k = \mathbf{0}, \tag{4.1}$$

where $\mathbf{f}_k$ denotes a force applied to a point located by position vector $\mathbf{r}_k$, and $\mathbf{r}_0$ locates an arbitrary but fixed point with respect to which the moments are taken. So both the net (or *resultant*) force and the net moment acting on the body must vanish. Here we do not include the lumped couples that are needed in the more complex models presented in Chapter 6.

To define the equilibrium of a deformable body, we apply the equilibrium equations for a *rigid* body to any portion of the deformable body. We take this as an axiom and call it the

Solidification principle. *In equilibrium, any part of a deformable body obeys the equilibrium equations as if it were a rigid body under the action of (1) all the external forces, and (2) the force reaction imposed by the remainder of the body on the part under consideration.*

This principle allows us to apply the results of classical mechanics — obtained for non-deformable objects — to deformable objects. We will use the following terminology. A solid body occupies a certain volume in three-dimensional space. The mapping which takes its material points into the spatial points is a *configuration* of the body; roughly, it describes the geometry of the body in space. If the body is not under load, the configuration is the *initial* or *reference configuration.* For a deformed body in equilibrium[1] under a given load, the configuration is the *actual configuration.* Let us apply the solidification principle in the actual configuration of a deformable

[1]Note that the body may be prestressed in the actual configuration. However, the methods of linear elasticity neglect this possible effect and study only additional stresses-deformations.

body in equilibrium. Take an arbitrary portion $\mathcal{P}$ as shown in Fig. 4.2. Two types of forces act on $\mathcal{P}$. *Body forces* (e.g., gravitational forces) act on the interior of $\mathcal{P}$ and do not depend on the conditions over the boundary surface of $\mathcal{P}$. *Contact forces* act over the boundary of $\mathcal{P}$ and represent the reaction forces imposed on $\mathcal{P}$ by the rest of the body. To see how the latter arise, imagine that we isolate $\mathcal{P}$ and replace the effects of the rest of the body by some forces. These reactions are applied only to that part of the boundary of $\mathcal{P}$ that lies in contact with the rest of the body (hence the term "contact forces"). Note that in classical theories of elasticity, the contact forces are the only forces that can act on $\mathcal{P}$ as reaction forces from the remainder of the body. In addition, surface forces may act over the external boundary of the entire body. We will refer to these, together with the body forces, as the *external forces.* By the solidification principle, $\mathcal{P}$ must be in equilibrium under the action of all forces as though it were a rigid body in classic mechanics.

So the total force acting on $\mathcal{P}$ is

$$\mathbf{f}(\mathcal{P}) = \mathbf{f}_B(\mathcal{P}) + \mathbf{f}_C(\mathcal{P}),$$

where the subscripts B and C denote the body and contact forces, respectively. To characterize the spatial distribution of the forces, we introduce force (specific) densities $\mathbf{f}$ and $\mathbf{t}$. Let B be a cube with side δa centered at $\mathbf{x}$, and let $\delta\mathbf{F}$ be the net force of some field (like the gravitational field) acting on the material volume B with specific density ρ. Then the density of the external force field, the volume force density $\mathbf{f}$, at $\mathbf{x}$ is

$$\mathbf{f} = \lim_{\delta a \to 0} \frac{\delta\mathbf{F}}{\text{volume of } B} = \lim_{\delta a \to 0} \frac{\delta\mathbf{F}}{\delta a^3}$$

and, as we require existence of the limit,

$$\mathbf{f}_B(\mathcal{P}) = \int_{V_{\mathcal{P}}} \rho\mathbf{f}\, dV,$$

where $V_{\mathcal{P}}$ is the space volume of $\mathcal{P}$.

To introduce the specific contact density on an elementary surface element at point $\mathbf{x}$ with the orientation $\mathbf{n}$, we take a small square δS with side δa centered at $\mathbf{x}$ and denote by $\delta\mathbf{T}$ the net contact force acting on δS from the connected part of the body or from the external surface forces if $\mathbf{x}$ is on the boundary of the body. The contact force density we obtain through the limit procedure is

$$\mathbf{t} = \lim_{\delta a \to 0} \frac{\delta\mathbf{T}}{\text{area of } \delta S} = \lim_{\delta a \to 0} \frac{\delta\mathbf{T}}{\delta a^2}.$$

The quantity $\mathbf{t}$ is called the *stress vector* or *traction.* Note that $\mathbf{t}$ depends on the point $\mathbf{x}$ and the orientation of the square. Now we can write out the net contact force acting on the boundary $\Sigma_{\mathcal{P}} = \partial V_{\mathcal{P}}$ of $\mathcal{P}$:

$$\mathbf{f}_C(\mathcal{P}) = \int_{\Sigma_{\mathcal{P}}} \mathbf{t}\, d\Sigma.$$

The density $\mathbf{t}$ carries dimensions of force per unit area just as classical pressure does, while $\mathbf{f}$ carries those of force per unit mass. So the units of $\mathbf{f}_B(\mathcal{P})$ and $\mathbf{f}_C(\mathcal{P})$ coincide. However, for consistency we should present all the variables in unitless form. We emphasize that $\mathbf{f}_B(\mathcal{P})$ and $\mathbf{f}_C(\mathcal{P})$ are defined in the actual configuration.

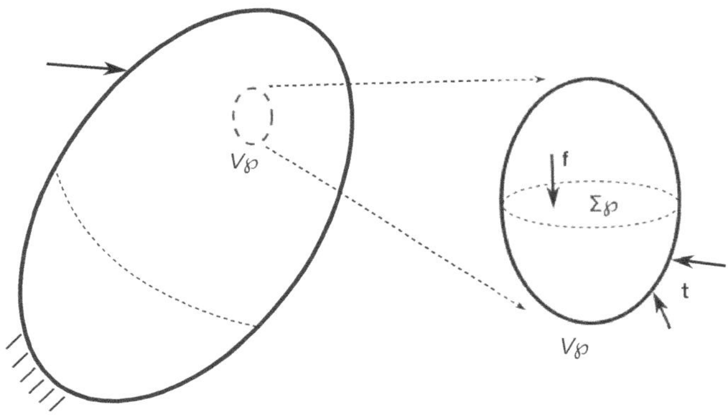

Fig. 4.2 Forces acting on a portion $V_{\mathcal{P}}$ of the body.

Equilibrium equations of a continuum medium

By the solidification principle, the rigid body equilibrium equations (4.1) become the following two conditions for a deformable body.

1. The net force acting on any portion $\mathcal{P}$ is zero:

$$\int_{V_{\mathcal{P}}} \rho \mathbf{f}\, dV + \int_{\Sigma_{\mathcal{P}}} \mathbf{t}\, d\Sigma = \mathbf{0}. \tag{4.2}$$

2. The net moment acting on any portion $\mathcal{P}$ is zero:

$$\int_{V_{\mathcal{P}}} \{(\mathbf{r} - \mathbf{r}_0) \times \rho \mathbf{f}\}\, dV + \int_{\Sigma_{\mathcal{P}}} \{(\mathbf{r} - \mathbf{r}_0) \times \mathbf{t}\}\, d\Sigma = \mathbf{0}, \tag{4.3}$$

where $\mathbf{r}$ locates a material point and $\mathbf{r}_0$ some fixed reference point (one may show, using (4.2), that (4.3) does not depend on the choice of $\mathbf{r}_0$).

The last two equations will define differential equilibrium equations in terms of $\mathbf{t}$ and some its properties.

Note that $\mathcal{P}$ cannot be fully arbitrary, as the integration operation over $\mathcal{P}$ must make sense.

Stress tensor

In general, the stress vector depends on the position $\mathbf{r}$ of a particle and on the normal $\mathbf{n}$ to the area element in the body. We will always take the normal outward from the portion of the body under consideration. Note again that everything is done in the actual configuration.

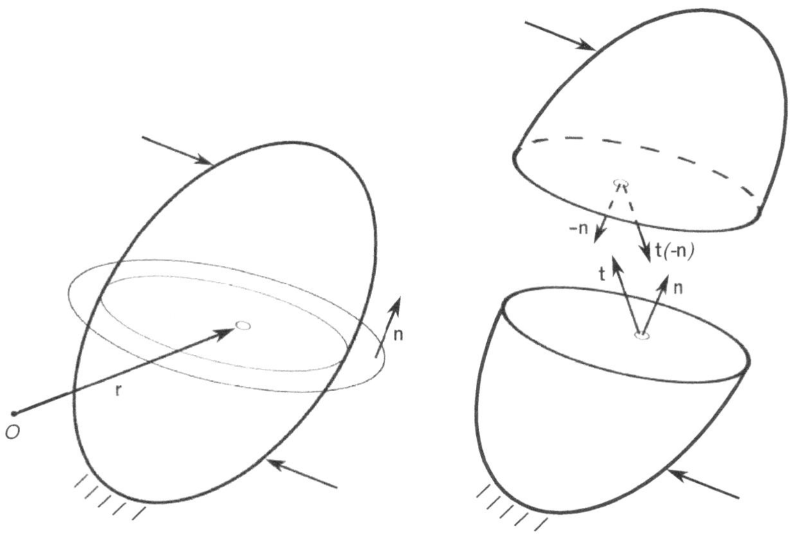

Fig. 4.3 Interaction of two parts of the body.

In continuum mechanics, Newton's third law regarding action-reaction pairs is called *Cauchy's lemma* and is expressed as

$$\mathbf{t}(\mathbf{r}, \mathbf{n}) = -\mathbf{t}(\mathbf{r}, -\mathbf{n}). \tag{4.4}$$

This describes the interaction of the contacting parts of the body shown in Fig. 4.3. So we suppose that on the area element the only forces acting are the contact forces asserted by one part of the body on another part in contact across the area element. Cauchy's lemma allows us to introduce the stress tensor, which describes the dependence of $\mathbf{t}$ on $\mathbf{n}$, the unit normal to the cut. We will see this in Cauchy's theorem below. Note that it is not an absolute axiom; there are, in continuum mechanics, theories where this equality is not valid.

First, however, we will use (4.2) to obtain the equilibrium equation in differential form. We fix an arbitrary point P in V, the volume of the body, and use it as a vertex of an arbitrary small parallelepiped Π having faces parallel to the Cartesian coordinate planes as shown in Fig. 4.4. Hence the normals to the faces lie along the orthonormal basis vectors $\mathbf{i}_1, \mathbf{i}_2, \mathbf{i}_3$.

Let us expand the stress vector $\mathbf{t}(\mathbf{i}_k)$ on the face having normal $\mathbf{i}_k$:

$$\mathbf{t}(\mathbf{i}_k) = t_{ks}\mathbf{i}_s.$$

That is, the t_{ks} are the components of $\mathbf{t}(\mathbf{i}_k)$ in the basis $\mathbf{i}_s$. For brevity we omit $\mathbf{r}$ in the arguments of $\mathbf{t}(\mathbf{r}, \mathbf{n})$ here.

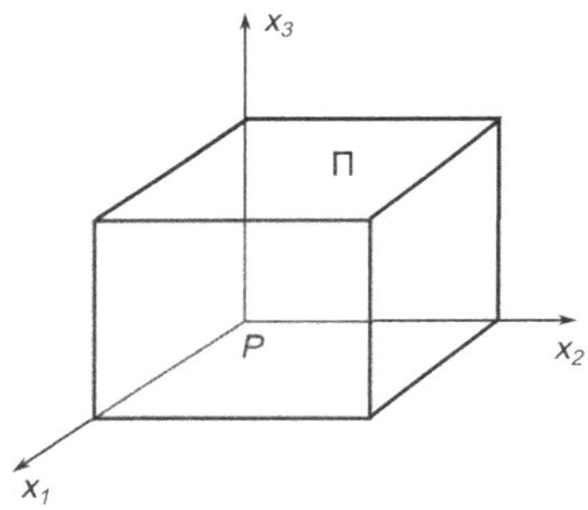

Fig. 4.4 Parallelepiped Π.

Theorem 4.1. *The following* ***equilibrium equation*** *holds in* V*:*

$$\rho\mathbf{f} + \frac{\partial t_{ks}}{\partial x_k}\mathbf{i}_s = \mathbf{0}. \tag{4.5}$$

Proof. Equation (4.2) applied to Π is

$$\int_{V_\Pi} \rho\mathbf{f}\, dV + \int_{\Sigma_\Pi} \mathbf{t}\, d\Sigma = \mathbf{0}.$$

The surface integral extends over the faces of Π, each normal to one of the $\mathbf{i}_k$. On the face with normal $\mathbf{n} = \mathbf{i}_k$ we have $\mathbf{t}(\mathbf{i}_k) = n_k t_{1s}\mathbf{i}_s$, where $n_k = 1$ and the other two components of $\mathbf{n}$ vanish. On the opposite face the normal is $-\mathbf{i}_k$, so $n_k = -1$ and the remaining components vanish. By (4.4) we have $\mathbf{t}(-\mathbf{i}_k) = -\mathbf{t}(\mathbf{i}_k)$ so that $\mathbf{t}(-\mathbf{i}_k) = n_k t_{1s}\mathbf{i}_s$. Hence

$$\int_{V_\Pi} \rho \mathbf{f}\, dV + \int_{\Sigma_\Pi} n_k t_{ks}\mathbf{i}_s\, d\Sigma = \mathbf{0}.$$

Applying (2.96) to the surface integral, we have

$$\int_{\Sigma_\Pi} n_k t_{ks}\mathbf{i}_s\, d\Sigma = \int_{V_\Pi} \frac{\partial t_{ks}}{\partial x_k}\mathbf{i}_s\, dV \tag{4.6}$$

and it follows that

$$\int_{V_\Pi} \left(\rho \mathbf{f} + \frac{\partial t_{ks}}{\partial x_k}\mathbf{i}_s\right) dV = \mathbf{0}. \tag{4.7}$$

Suppose the integrand in (4.7) is continuous. As the vertex P of Π is fixed and Π is arbitrarily small, the differential equation (4.5) must hold at P. Since P is arbitrary, (4.5) holds in V. □

Exercise 4.1. Prove (4.6) by direct integration of the right-hand side.

Now we formulate

Theorem 4.2. ***(Cauchy).*** *At any point of a body, the dependence of* $\mathbf{t}$ *on* $\mathbf{n}$*, the normal to an elementary area at a point, is linear:*

$$\mathbf{t} = \mathbf{n} \cdot \boldsymbol{\sigma}.$$

Here $\boldsymbol{\sigma}$ *is a second-order tensor depending on the point; it is the* ***Cauchy stress tensor****.*

Proof. We construct a tetrahedron T with vertex O at an arbitrary but fixed point of V as shown in Fig. 4.5. Applying (4.2) to T, we get

$$\int_{V_T} \rho \mathbf{f}\, dV + \int_{\Sigma_T} n_k t_{ks}\mathbf{i}_s\, d\Sigma + \int_{M_1M_2M_3} \mathbf{t}(\mathbf{n})\, d\Sigma = \mathbf{0}, \tag{4.8}$$

where V_T is the volume of T, Σ_T is the portion of the boundary of T consisting of faces parallel to the coordinate planes, and $M_1M_2M_3$ is the inclined face. On the face normal to $\mathbf{i}_k$ we use the representation $\mathbf{t}(\mathbf{i}_k) = n_k t_{1s}\mathbf{i}_s$ from the proof of Theorem 4.1.

Using (4.5), we write (4.8) as

$$\int_{V_T} \frac{\partial t_{ks}}{\partial x_k}\mathbf{i}_s\, dV = \int_{\Sigma_T} n_k t_{ks}\mathbf{i}_s\, d\Sigma + \int_{M_1M_2M_3} \mathbf{t}(\mathbf{n})\, d\Sigma.$$

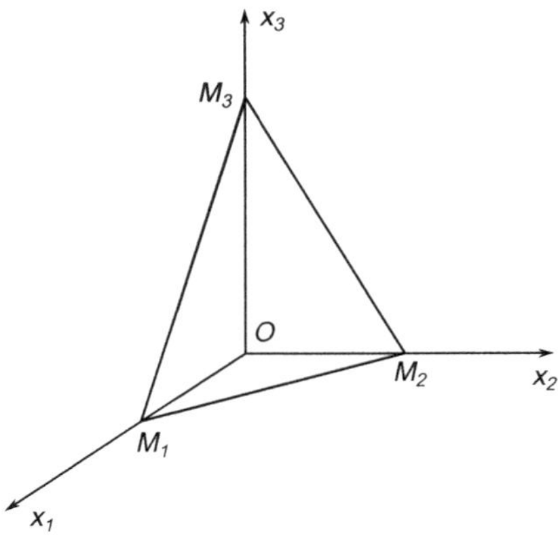

Fig. 4.5 Tetrahedron T.

From (2.96) it follows that

$$\int_{V_T} \frac{\partial t_{ks}}{\partial x_k} \mathbf{i}_s \, dV = \int_{\Sigma_T} n_k t_{ks} \mathbf{i}_s \, d\Sigma + \int_{M_1 M_2 M_3} n_k t_{ks} \mathbf{i}_s \, d\Sigma.$$

These two equations yield

$$\int_{M_1 M_2 M_3} \big(\mathbf{t}(\mathbf{n}) - n_k t_{ks} \mathbf{i}_s\big) \, d\Sigma = \mathbf{0},$$

and since T is small and arbitrary at point O we obtain

$$\mathbf{t}(\mathbf{n}) - n_k t_{ks} \mathbf{i}_s = \mathbf{0}.$$

Because $M_1 M_2 M_3$ can have an arbitrary orientation $\mathbf{n}$, this holds for all $\mathbf{n}$ and, moreover, at all points of V.

So $\mathbf{t}$ depends on $\mathbf{n}$ linearly. A linear dependence between two vectors is given by a second-order tensor that we denote by $\boldsymbol{\sigma}$:

$$\mathbf{t}(\mathbf{n}) = \mathbf{n} \cdot \boldsymbol{\sigma}.$$

The proof of Cauchy's theorem is complete. □

The proof shows that the components of the matrix (t_{sk}) are the components of the stress tensor in a Cartesian frame:

$$\boldsymbol{\sigma} = t_{sk} \mathbf{i}_k \mathbf{i}_s.$$

To maintain the correspondence between the notations for $\boldsymbol{\sigma}$ and its components, we change t_{sk} to σ_{sk} and write

$$\boldsymbol{\sigma} = \sigma_{sk} \mathbf{i}_k \mathbf{i}_s.$$

Each subscript of σ_{sk} has a geometric meaning. The first subscript designates the area element with normal $\mathbf{i}_k$, while the second designates the direction of the projection of the stress vector onto $\mathbf{i}_s$. For example, σ_{31} is the projection of $\mathbf{t}(\mathbf{i}_3)$ onto the x_1-axis, and the stress vector $\mathbf{t}(\mathbf{i}_3)$ acts on the surface element normal to $\mathbf{i}_3$.

Now we return to the equilibrium equation. It is easy to see that (4.5) is the Cartesian form of the equation

$$\nabla \cdot \boldsymbol{\sigma} + \rho \mathbf{f} = \mathbf{0}. \tag{4.9}$$

Hence we have the component-free form of the equilibrium equation for the body and can write out the component equations in any coordinate frame. Note that the equation does not depend on the properties of the material that makes up the body (aside from the dependence on ρ). We recall that (4.9) is the differential form of the condition that the net force acting on an arbitrary part of the body is zero. Having exploited the force equation (4.2), let us formulate the consequences of the moment equation (4.3).

Theorem 4.3. *Equation* (4.3), *holding for any part of the body, implies that the tensor* $\boldsymbol{\sigma}$ *is symmetric.*

Proof. Using (2.96), we change the surface integral in (4.3) to an integral over $V_{\mathcal{P}}$:

$$\begin{aligned}\int_{\Sigma_{\mathcal{P}}} (\mathbf{r} - \mathbf{r}_0) \times \mathbf{t} \, d\Sigma &= \int_{\Sigma_{\mathcal{P}}} (\mathbf{r} - \mathbf{r}_0) \times (\mathbf{n} \cdot \boldsymbol{\sigma}) \, d\Sigma \\ &= -\int_{\Sigma_{\mathcal{P}}} \mathbf{n} \cdot \boldsymbol{\sigma} \times (\mathbf{r} - \mathbf{r}_0) \, d\Sigma \\ &= -\int_{V_{\mathcal{P}}} \nabla \cdot [\boldsymbol{\sigma} \times (\mathbf{r} - \mathbf{r}_0)] \, dV.\end{aligned}$$

Let us transform the integrand of the last integral. Because $\partial \mathbf{r}/\partial x_k = \mathbf{i}_k$ and $\partial \mathbf{r}_0/\partial x_k = \mathbf{0}$, we have

$$\begin{aligned}\nabla \cdot [\boldsymbol{\sigma} \times (\mathbf{r} - \mathbf{r}_0)] &= \nabla \cdot \boldsymbol{\sigma} \times (\mathbf{r} - \mathbf{r}_0) + \mathbf{i}_k \cdot \boldsymbol{\sigma} \times \frac{\partial}{\partial x_k}(\mathbf{r} - \mathbf{r}_0) \\ &= -(\mathbf{r} - \mathbf{r}_0) \times \nabla \cdot \boldsymbol{\sigma} + \mathbf{i}_k \cdot \boldsymbol{\sigma} \times \mathbf{i}_k \\ &= -(\mathbf{r} - \mathbf{r}_0) \times \nabla \cdot \boldsymbol{\sigma} - \sigma_{ks} \mathbf{i}_k \times \mathbf{i}_s.\end{aligned}$$

Thus the condition for moment equilibrium (that the resultant moment of all forces acting on $V_{\mathcal{P}}$ is zero) yields

$$\int_{V_{\mathcal{P}}} (\mathbf{r} - \mathbf{r}_0) \times (\rho \mathbf{f} + \nabla \cdot \boldsymbol{\sigma}) \, dV + \int_{V_{\mathcal{P}}} \sigma_{ks} \mathbf{i}_k \times \mathbf{i}_s \, dV = \mathbf{0}. \tag{4.10}$$

The first integral in (4.10) is zero by (4.9). So the second integral in (4.10) is zero for arbitrary $V_{\mathcal{P}}$, and it follows that

$$\sigma_{ks}\mathbf{i}_k \times \mathbf{i}_s = \mathbf{0} \quad \text{in } V.$$

This holds if and only if $\boldsymbol{\sigma} = \boldsymbol{\sigma}^T$ at each point. Indeed, consider the part of the sum $\sigma_{ks}\mathbf{i}_k \times \mathbf{i}_s$ when k, s are 1 or 2. We have

$$\begin{aligned} \sigma_{ks}\mathbf{i}_k \times \mathbf{i}_s &= \sigma_{11}\mathbf{i}_1 \times \mathbf{i}_1 + \sigma_{22}\mathbf{i}_2 \times \mathbf{i}_2 + \sigma_{12}\mathbf{i}_1 \times \mathbf{i}_2 + \sigma_{21}\mathbf{i}_2 \times \mathbf{i}_1 \\ &= (\sigma_{12} - \sigma_{21})\mathbf{i}_3 = \mathbf{0}, \end{aligned}$$

hence $\sigma_{12} = \sigma_{21}$. Similarly, $\sigma_{23} = \sigma_{32}$ and $\sigma_{13} = \sigma_{31}$. □

It is worth noting that in continuum mechanics other types of stresses, such as couple stresses, can be introduced [Cosserat and Cosserat (1909); Eringen (1999)]. For such models (discussed further in Chapter 6), the Cauchy stress tensor is not symmetric in general.

Principal stresses and principal area elements

In a general basis $\mathbf{e}^k$, Cauchy's stress tensor $\boldsymbol{\sigma}$ takes the form

$$\boldsymbol{\sigma} = \sigma_{sk}\mathbf{e}^s\mathbf{e}^k$$

where only six of the σ_{sk} are independent.

Because $\boldsymbol{\sigma}$ is symmetric, there exists the spectral expansion (2.28):

$$\boldsymbol{\sigma} = \sigma_1\mathbf{i}_1\mathbf{i}_1 + \sigma_2\mathbf{i}_2\mathbf{i}_2 + \sigma_3\mathbf{i}_3\mathbf{i}_3.$$

Here the eigenvalues σ_k of the matrix (σ_{sk}) are the *principal stresses*, and the orthonormal eigenvectors $\mathbf{i}'_k$ of $\boldsymbol{\sigma}$ are the *principal axes* of $\boldsymbol{\sigma}$. On the principal area element having normal $\mathbf{i}_k$, the tangential stresses are absent. When the σ_k are distinct, the frame $\mathbf{i}_1, \mathbf{i}_2, \mathbf{i}_3$ is orthonormal. For the case of repeated σ_k, the frame of $\mathbf{i}_k$ is not unique; even in this case, however, we can select an orthonormal set $\mathbf{i}_1, \mathbf{i}_2, \mathbf{i}_3$.

4.2 Strain Tensor

Bodies change shape under load, and such deformations are described by the strain tensor. We consider only small deformations here.

Let us illustrate the strain idea with a stretched bar. If the bar has length l_0 when undeformed and length l when under load, then the strain is given by

$$\varepsilon = \frac{l - l_0}{l_0} \equiv \frac{\Delta l}{l_0}.$$

Generalizing this to three dimensions is not straightforward, as we must consider changes in shape of a body along all directions.

Let a body initially occupy a volume region V. When deformed under some external load, it occupies region v. Let $\mathbf{r}_0$ and $\mathbf{r}$ be the position vectors of a particle in the initial and deformed states, respectively. The *displacement vector*

$$\mathbf{u} = \mathbf{r} - \mathbf{r}_0$$

describes the displacement of a particle due to deformation (Fig. 4.6) and is the primary field quantity in solid mechanics. Here we restrict ourselves to the case in which $\|\mathbf{u}\| \ll 1$ and all the first derivatives of $\mathbf{u}$ in Cartesian coordinates are small in comparison with 1. So we will omit all terms of the second order of smallness in any expression containing first-order terms. Moreover, in the case of small deformations (e.g., in the case of *linear* elasticity),

the initial and actual states of the body will not be distinguished;

that is, all quantities will be considered to be given in the initial volume V.

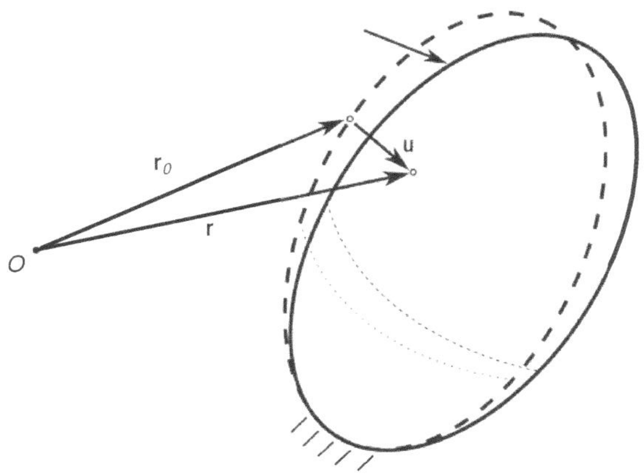

Fig. 4.6 Deformation of a three-dimensional body. The dashed contour indicates the shape of the undeformed body.

Let us consider the change of an infinitesimal vector-segment $d\mathbf{r}_0$ due to deformation. After deformation the segment is given by the vector $d\mathbf{r}$. We have

$$d\mathbf{r} = d\mathbf{r}_0 \cdot \mathbf{F}, \tag{4.11}$$

where $\mathbf{F} = \mathbf{E} + \nabla\mathbf{u}$ is the the *deformation gradient tensor.*[2]

Exercise 4.2. Derive (4.11) from the relation $\mathbf{r} = \mathbf{r}_0 + \mathbf{u}$.

Next we consider the change in length of the segment during deformation. Before deformation, the squared length is

$$dS^2 = d\mathbf{r}_0 \cdot d\mathbf{r}_0.$$

After deformation it is $ds^2 = d\mathbf{r} \cdot d\mathbf{r} = d\mathbf{r}_0 \cdot \mathbf{F} \cdot \mathbf{F}^T \cdot d\mathbf{r}_0$. We have

$$\begin{aligned} ds^2 - dS^2 &= d\mathbf{r}_0 \cdot (\mathbf{F} \cdot \mathbf{F}^T - \mathbf{E}) \cdot d\mathbf{r}_0 \\ &= d\mathbf{r}_0 \cdot \left[(\nabla\mathbf{u} + (\nabla\mathbf{u})^T + (\nabla\mathbf{u}) \cdot \nabla\mathbf{u}^T)\right] \cdot d\mathbf{r}_0. \end{aligned}$$

For small deformations, we may omit all the squared quantities and write

$$ds^2 - dS^2 = d\mathbf{r}_0 \cdot (\nabla\mathbf{u} + (\nabla\mathbf{u})^T) \cdot d\mathbf{r}_0 = 2d\mathbf{r}_0 \cdot \boldsymbol{\varepsilon} \cdot d\mathbf{r}_0,$$

where

$$\boldsymbol{\varepsilon} = \tfrac{1}{2}\left(\nabla\mathbf{u} + (\nabla\mathbf{u})^T\right) \tag{4.12}$$

is the *linear strain tensor*. It is clearly symmetric.

In a Cartesian frame, $\boldsymbol{\varepsilon}$ is given by $\boldsymbol{\varepsilon} = \varepsilon_{mn}\mathbf{i}_m\mathbf{i}_n$ where

$$\begin{aligned} \varepsilon_{11} &= \frac{\partial u_1}{\partial x_1}, & \varepsilon_{12} &= \frac{1}{2}\left(\frac{\partial u_1}{\partial x_2} + \frac{\partial u_2}{\partial x_1}\right), \\ \varepsilon_{22} &= \frac{\partial u_2}{\partial x_2}, & \varepsilon_{13} &= \frac{1}{2}\left(\frac{\partial u_1}{\partial x_3} + \frac{\partial u_3}{\partial x_1}\right), \\ \varepsilon_{33} &= \frac{\partial u_3}{\partial x_3}, & \varepsilon_{23} &= \frac{1}{2}\left(\frac{\partial u_2}{\partial x_3} + \frac{\partial u_3}{\partial x_2}\right). \end{aligned}$$

The diagonal components $\varepsilon_{11}, \varepsilon_{22}, \varepsilon_{33}$ describe the changes in the lengths of elementary segments along the $\mathbf{i}_1$, $\mathbf{i}_2$, $\mathbf{i}_3$ directions, respectively. The other components ε_{mn} $(m \neq n)$ represent skewing of the body; they characterize the deformational changes in the angles between elementary segments lying initially along the axes.

In arbitrary curvilinear coordinates q^1, q^2, q^3 with basis $\mathbf{r}_k$ $(k = 1, 2, 3)$ and reciprocal basis $\mathbf{r}^k$ $(k = 1, 2, 3)$, the tensor $\boldsymbol{\varepsilon}$ is given by

$$\boldsymbol{\varepsilon} = \varepsilon_{st}\mathbf{r}^s\mathbf{r}^t, \qquad \varepsilon_{st} = \frac{1}{2}\left(\frac{\partial u_s}{\partial q^t} + \frac{\partial u_t}{\partial q^s}\right) - \Gamma^r_{st}u_r.$$

In arbitrary orthogonal curvilinear coordinates, $\boldsymbol{\varepsilon}$ was represented in (2.95). See Appendix C for $\boldsymbol{\varepsilon}$ in the cylindrical and spherical systems.

[2]Some books use $\mathbf{F} = \mathbf{E} + \nabla\mathbf{u}^T$, in which case (4.11) takes the form $d\mathbf{r} = \mathbf{F} \cdot d\mathbf{r}_0$.

Equation (4.12) expresses $\boldsymbol{\varepsilon}$ as a tensor-valued function of $\mathbf{u}$. The inverse problem, of finding $\mathbf{u}$ when $\boldsymbol{\varepsilon}$ is given, has a solution if and only if the compatibility condition

$$\nabla \times (\nabla \times \boldsymbol{\varepsilon})^T = \mathbf{0} \tag{4.13}$$

holds. Then $\mathbf{u}$ can be found using *Cesàro's formula*

$$\mathbf{u} = \mathbf{u}_0 + \boldsymbol{\omega}_0 \times (\mathbf{r} - \mathbf{r}_0) + \int_{M_0}^{M} \{\boldsymbol{\varepsilon}(s) + [\mathbf{r}(s) - \mathbf{r}] \times \nabla \times \boldsymbol{\varepsilon}(s)\} \cdot d\mathbf{r}(s), \tag{4.14}$$

where $\mathbf{u}_0$ and $\boldsymbol{\omega}_0$ are arbitrary but fixed vectors, and the integration path M_0M joins the points M_0 and M whose position vectors are $\mathbf{r}_0$ and $\mathbf{r}$, respectively. Here $\mathbf{r}(s)$ locates an arbitrary point on M_0M. See, e.g., [Green and Zerna (1954); Lurie (2005)] for a derivation.

4.3 Equation of Motion

Using the equilibrium equations (4.9) and d'Alembert's principle of mechanics, we can immediately obtain the equation of motion for a body. The technique is to formally add the inertia forces to the body forces:

$$\mathbf{f} \to \mathbf{f} - \rho \frac{\partial^2 \mathbf{u}}{\partial t^2},$$

where t denotes time and ρ the material density. The equation of motion is

$$\nabla \cdot \boldsymbol{\sigma} + \rho \mathbf{f} = \rho \frac{\partial^2 \mathbf{u}}{\partial t^2}. \tag{4.15}$$

The form of (4.15) is simplest in Cartesian coordinates. Putting $\boldsymbol{\sigma} = \sigma_{mn}\mathbf{i}_m\mathbf{i}_n$ and $\mathbf{u} = u_m\mathbf{i}_m$, we get

$$\frac{\partial \sigma_{ij}}{\partial x_i} + \rho f_j = \rho \frac{\partial^2 u_j}{\partial t^2} \qquad (j = 1, 2, 3)$$

or explicitly

$$\frac{\partial \sigma_{11}}{\partial x_1} + \frac{\partial \sigma_{21}}{\partial x_2} + \frac{\partial \sigma_{31}}{\partial x_3} + \rho f_1 = \rho \frac{\partial^2 u_1}{\partial t^2},$$
$$\frac{\partial \sigma_{12}}{\partial x_1} + \frac{\partial \sigma_{22}}{\partial x_2} + \frac{\partial \sigma_{32}}{\partial x_3} + \rho f_2 = \rho \frac{\partial^2 u_2}{\partial t^2},$$
$$\frac{\partial \sigma_{13}}{\partial x_1} + \frac{\partial \sigma_{23}}{\partial x_2} + \frac{\partial \sigma_{33}}{\partial x_3} + \rho f_3 = \rho \frac{\partial^2 u_3}{\partial t^2}.$$

With regard for (2.91), in curvilinear coordinates equation (4.15) takes the form

$$\frac{1}{\sqrt{g}}\frac{\partial}{\partial q^i}\left(\sqrt{g}\sigma^{ij}\mathbf{r}_j\right) + \rho\mathbf{f} = \rho\frac{\partial^2\mathbf{u}}{\partial t^2}. \tag{4.16}$$

In components it is

$$\frac{\partial}{\partial q^i}\left(\sqrt{g}\sigma^{ij}\right) + \Gamma^j_{mn}\sigma^{mn} + \rho\sqrt{g}f^j = \rho\sqrt{g}\frac{\partial^2 u^j}{\partial t^2} \qquad (j = 1, 2, 3).$$

The equations of motion can be simplified in orthogonal coordinates. See Appendix C for expressions in cylindrical and spherical coordinates.

4.4 Hooke's Law

Equations (4.12) and (4.15) apply to any small deformation of a continuous medium. However, they do not uniquely define the deformations and stresses in a body. To study a body under load, we should relate the stresses to the strains through the material properties of the body. These relations are called *constitutive equations.* Here and in the next chapter we consider linearly elastic materials for which the constitutive equation expresses a linear dependence between $\boldsymbol{\sigma}$ and $\boldsymbol{\varepsilon}$. The simplest version is Hooke's law

$$\sigma = E\varepsilon, \tag{4.17}$$

describing the elastic properties of a thin rod under tension or compression. Here E is the elastic modulus of the material from which the rod is made; it is known as *Young's modulus.*

Robert Hooke (1635–1703) was the first to establish the linear dependence $f \sim \Delta l$ between the applied force and the elongation of a bar similar to that in Fig. 4.1. Thomas Young (1773–1829) introduced the elastic modulus E as a quantity that does not depend on the cross-sectional area of the bar; rather, E characterizes the material itself. The linear dependence (4.17) holds only in some range $|\varepsilon| < \varepsilon_0$, where ε_0 depends on the material, temperature, and other factors. Note that for certain materials the linear region is absent. But the importance of Hooke's law in engineering cannot be overestimated; it adequately describes many common construction materials such as steels, aluminium, and various types of wood.

In the general case, a linear dependence between the second-order tensors $\boldsymbol{\sigma}$ and $\boldsymbol{\varepsilon}$ is presented in equation (2.56), known as the *generalized Hooke's law*:

$$\boldsymbol{\sigma} = \mathbf{C} \cdot\cdot\, \boldsymbol{\varepsilon}. \tag{4.18}$$

In a Cartesian basis it is $\sigma_{ij} = c_{ijmn}\varepsilon_{mn}$. The fourth-order *tensor of elastic moduli*

$$\mathbf{C} = c_{ijmn}\mathbf{i}_i\mathbf{i}_j\mathbf{i}_m\mathbf{i}_n$$

has 81 components; only 36 of these are independent, however, as the symmetries of $\boldsymbol{\sigma}$ and $\boldsymbol{\varepsilon}$ lead to the conditions

$$c_{ijmn} = c_{jimn} = c_{ijnm}.$$

In linear elasticity it is shown [Lurie (2005)] that we can introduce the *strain energy*

$$\frac{1}{2}\int_V W\,dV$$

stored in the elastic body by virtue of its deformation. The integrand $W = W(\boldsymbol{\varepsilon})$, the *strain energy function*, is a quadratic form in $\boldsymbol{\varepsilon}$:

$$W = \tfrac{1}{2}\boldsymbol{\varepsilon} \cdot\cdot\, \mathbf{C} \cdot\cdot\, \boldsymbol{\varepsilon} = \tfrac{1}{2}\varepsilon_{ij}c_{ijmn}\varepsilon_{mn}.$$

The fact that W is uniquely defined for any deformation requires $\mathbf{C}$ to have the additional symmetry property

$$c_{ijmn} = c_{mnij}$$

for all i, j, m, n. Indeed, we can represent $\mathbf{C}$ as the sum

$$\mathbf{C} = \mathbf{C}' + \mathbf{C}'',$$

where the components of $\mathbf{C}'$ and $\mathbf{C}''$ satisfy

$$c'_{ijmn} = c'_{mnij} \quad \text{and} \quad c''_{ijmn} = -c''_{mnij}.$$

(This is similar to the representation of a second-order tensor as the sum of symmetric and antisymmetric tensors.) For any symmetric tensors $\boldsymbol{\varepsilon}_k$, we have

$$\boldsymbol{\varepsilon}_1 \cdot\cdot\, \mathbf{C}'' \cdot\cdot\, \boldsymbol{\varepsilon}_2 = -\boldsymbol{\varepsilon}_2 \cdot\cdot\, \mathbf{C}'' \cdot\cdot\, \boldsymbol{\varepsilon}_1,$$

hence for any strain tensor $\boldsymbol{\varepsilon}$,

$$\boldsymbol{\varepsilon} \cdot\cdot\, \mathbf{C}'' \cdot\cdot\, \boldsymbol{\varepsilon} = -\boldsymbol{\varepsilon} \cdot\cdot\, \mathbf{C}'' \cdot\cdot\, \boldsymbol{\varepsilon}$$

and

$$\boldsymbol{\varepsilon} \cdot\cdot\, \mathbf{C}'' \cdot\cdot\, \boldsymbol{\varepsilon} = 0.$$

It follows that

$$\boldsymbol{\varepsilon} \cdot\cdot\, \mathbf{C} \cdot\cdot\, \boldsymbol{\varepsilon} = \boldsymbol{\varepsilon} \cdot\cdot\, \mathbf{C}' \cdot\cdot\, \boldsymbol{\varepsilon}.$$

Thus $\mathbf{C}''$ does not affect the values of W, and the constitutive relations should not include it. Setting $\mathbf{C}''$ to zero, we get $\mathbf{C} = \mathbf{C}'$. Thus the components of $\mathbf{C}$ have the following symmetry properties:

$$c_{ijmn} = c_{jimn} = c_{ijnm}, \qquad c_{ijmn} = c_{mnij},$$

for any indices i, j, n, m. Elementary calculation shows that these are 60 equalities. So in Hooke's law, of the 81 components of $\mathbf{C}$ there remain only $81 - 60 = 21$ independent elastic constants.

It follows from Exercise 2.48 that W is a potential for $\boldsymbol{\sigma}$, i.e., that

$$\boldsymbol{\sigma} = W_{,\boldsymbol{\varepsilon}}$$

where $W_{,\boldsymbol{\varepsilon}}$ is the derivative of W with respect to $\boldsymbol{\varepsilon}$. A dual relation

$$\boldsymbol{\varepsilon} = W_{,\boldsymbol{\sigma}}$$

holds when we express W in terms of $\boldsymbol{\sigma}$, i.e., $W = W(\boldsymbol{\sigma})$. We urge the reader to show this as an exercise.

In engineering analysis, the vector and matrix notations are typically used to describe an elastic body. Because of the symmetry properties of the tensors, all the relations are written in terms of formal six-dimensional "vectors" for the components of $\boldsymbol{\sigma}$ and $\boldsymbol{\varepsilon}$, and 6×6 matrices for $\mathbf{C}$. *Voigt's rule* shows one how to transform the tensor notation to the matrix-vector notation. The rule for changing $c_{ijmn} \to C_{pq}$ is as follows. The pairs of indices 11, 22, 33 change to $1, 2, 3$, the pairs 23 and 32 to 4, the pairs 13 and 31 to 5, and the pairs 12 and 21 to 6. For example, $c_{1122} \to C_{12}$ and $c_{1232} \to C_{64}$. The symmetry property $C_{pq} = C_{qp}$ holds. The components of $\boldsymbol{\sigma}$ and $\boldsymbol{\varepsilon}$ are transformed by the formulas

$$\begin{bmatrix} \sigma_{11} \\ \sigma_{22} \\ \sigma_{33} \\ \sigma_{23} \\ \sigma_{31} \\ \sigma_{12} \end{bmatrix} = \begin{bmatrix} \sigma_1 \\ \sigma_2 \\ \sigma_3 \\ \sigma_4 \\ \sigma_5 \\ \sigma_6 \end{bmatrix}, \qquad \begin{bmatrix} \varepsilon_{11} \\ \varepsilon_{22} \\ \varepsilon_{33} \\ 2\varepsilon_{23} \\ 2\varepsilon_{31} \\ 2\varepsilon_{12} \end{bmatrix} = \begin{bmatrix} \varepsilon_1 \\ \varepsilon_2 \\ \varepsilon_3 \\ \varepsilon_4 \\ \varepsilon_5 \\ \varepsilon_6 \end{bmatrix}.$$

In this notation Hooke's law takes the form

$$\begin{bmatrix} \sigma_1 \\ \sigma_2 \\ \sigma_3 \\ \sigma_4 \\ \sigma_5 \\ \sigma_6 \end{bmatrix} = \begin{bmatrix} C_{11} & C_{12} & C_{13} & C_{14} & C_{15} & C_{16} \\ C_{12} & C_{22} & C_{23} & C_{24} & C_{25} & C_{26} \\ C_{13} & C_{23} & C_{33} & C_{34} & C_{35} & C_{36} \\ C_{14} & C_{24} & C_{34} & C_{44} & C_{45} & C_{46} \\ C_{15} & C_{25} & C_{35} & C_{45} & C_{55} & C_{56} \\ C_{16} & C_{26} & C_{36} & C_{46} & C_{56} & C_{66} \end{bmatrix} \begin{bmatrix} \varepsilon_1 \\ \varepsilon_2 \\ \varepsilon_3 \\ \varepsilon_4 \\ \varepsilon_5 \\ \varepsilon_6 \end{bmatrix}. \tag{4.19}$$

The general case in which all 21 elastic constants in (4.19) are independent is not common in applications. In engineering practice, the materials usually possess symmetries that reduce the number of independent constants. Most common are isotropic materials; such a material exhibits complete symmetry of its properties in space so that at each point the material properties do not vary with direction. Mathematically, symmetry is expressed in the language of group theory. Examples of isotropic materials are steel, aluminum, many other metals, polymeric materials, etc.

For an isotropic material, the relation between $\boldsymbol{\sigma}$ and $\boldsymbol{\varepsilon}$ is given by a linear isotropic function taking the form

$$\boldsymbol{\sigma} = \lambda \mathbf{E} \operatorname{tr} \boldsymbol{\varepsilon} + 2\mu \boldsymbol{\varepsilon}, \tag{4.20}$$

where λ and μ are *Lamé's moduli*. We may also refer to μ as the shear modulus. For an isotropic material,

$$\mathbf{C} = \lambda \mathbf{E}\mathbf{E} + \mu(\mathbf{e}_k \mathbf{E} \mathbf{e}^k + \mathbf{I}).$$

In matrix notation, to $\mathbf{C}$ there corresponds the diagonal matrix

$$\begin{bmatrix} \lambda + 2\mu & 0 & 0 & 0 & 0 & 0 \\ 0 & \lambda + 2\mu & 0 & 0 & 0 & 0 \\ 0 & 0 & \lambda + 2\mu & 0 & 0 & 0 \\ 0 & 0 & 0 & 2\mu & 0 & 0 \\ 0 & 0 & 0 & 0 & 2\mu & 0 \\ 0 & 0 & 0 & 0 & 0 & 2\mu \end{bmatrix}.$$

Other elastic constants were used historically in engineering work, and in terms of these, Hooke's law takes other forms. Common engineering constants include *Young's modulus* E, *Poisson's ratio* ν, the *bulk modulus* k, and the *shear modulus* G. For isotropic materials, these are used in pairs: E and ν, and k and G. The modulus E originated in bar stretching problems; it relates the tension applied to the bar with the resulting strain. The modulus k is used to describe uniform volume deformation of a material, say of a ball under uniform pressure. G describes the shear characteristics of a material. Finally, ν describes the lateral shortening of a stretched band. We will derive some relations between the constants.

Let us split $\boldsymbol{\sigma}$ into a ball tensor and deviator:

$$\boldsymbol{\sigma} = \tfrac{1}{3}\sigma \mathbf{E} + \operatorname{dev} \boldsymbol{\sigma}.$$

Then (4.20) takes the form

$$\sigma = \tfrac{1}{3}(3\lambda + 2\mu) \operatorname{tr} \boldsymbol{\varepsilon}, \qquad \operatorname{dev} \boldsymbol{\sigma} = 2\mu \operatorname{dev} \boldsymbol{\varepsilon}.$$

The bulk modulus k relates the mean stress σ with the bulk strain $\operatorname{tr}\boldsymbol{\varepsilon}$ through $\sigma = k \operatorname{tr}\boldsymbol{\varepsilon}$ and so $k = \lambda + 2\mu/3$. Shear stresses are related to shear strains by

$$\sigma_{ij} = 2G\varepsilon_{ij} \qquad (i \neq j).$$

The above deviator equation in components is

$$\sigma_{ij} = 2\mu\varepsilon_{ij} \qquad (i \neq j),$$

which yields $G = \mu$.

Now we turn to Young's modulus E. Let us consider a uniaxial homogeneous deformation of a bar; this occurs when the material is uniformly stretched or compressed. Take the unit vector $\mathbf{i}_1$ along the bar axis. Then $\boldsymbol{\sigma} = \sigma_{11}\mathbf{i}_1\mathbf{i}_1$. Relation (4.20) reduces to the three nontrivial component equations

$$\sigma_{11} = \lambda \operatorname{tr}\boldsymbol{\varepsilon} + 2\mu\varepsilon_{11}, \qquad 0 = \lambda \operatorname{tr}\boldsymbol{\varepsilon} + 2\mu\varepsilon_{22}, \qquad 0 = \lambda \operatorname{tr}\boldsymbol{\varepsilon} + 2\mu\varepsilon_{33}.$$

Eliminating ε_{22} and ε_{33} from these, we get

$$\sigma_{11} = E\varepsilon_{11}, \quad \text{where} \quad E = \frac{\mu(3\lambda + 2\mu)}{\lambda + \mu}.$$

When a bar is uniformly stretched by a force and its strain is ε, experiment shows that its transverse dimensions decrease. The transverse strain for this deformation is proportional to the longitudinal strain $-\nu\varepsilon$, the coefficient of proportionality being Poisson ratio. Its relation to the other constants is defined by the following

Exercise 4.3. For the above uniaxial strain state of the bar, find $\varepsilon_{22}/\varepsilon_{11} = -\nu$. Show that $\nu = \lambda/(2\lambda + 2\mu)$. Note that in the uniaxially stretched bar, ν defines the dependence of the lateral strain ε_{22} on the axial strain ε_{11}.

The relations between various pairs of moduli used in the literature are summarized in Appendix C.

Thermodynamic considerations require the strain energy to be positive:

$$W(\boldsymbol{\varepsilon}) > 0 \quad \text{whenever } \boldsymbol{\varepsilon} \neq \mathbf{0}. \tag{4.21}$$

This puts additional restrictions on the elastic moduli for an isotropic material. We have

$$W(\boldsymbol{\varepsilon}) = \tfrac{1}{2}\boldsymbol{\sigma} \cdot\cdot\, \boldsymbol{\varepsilon} = \tfrac{1}{2}\lambda(\operatorname{tr}\boldsymbol{\varepsilon})^2 + \mu\boldsymbol{\varepsilon} \cdot\cdot\, \boldsymbol{\varepsilon} \geq 0.$$

Let us represent $\boldsymbol{\varepsilon}$ as the sum of its ball and deviator terms:

$$\boldsymbol{\varepsilon} = \tfrac{1}{3}\mathbf{E} \operatorname{tr}\boldsymbol{\varepsilon} + \operatorname{dev}\boldsymbol{\varepsilon}.$$

Because $\varepsilon \cdot\cdot \varepsilon = \frac{1}{3}(\operatorname{tr}\varepsilon)^2 + \operatorname{dev}\varepsilon \cdot\cdot \operatorname{dev}\varepsilon$, we obtain

$$W(\varepsilon) = \tfrac{1}{6}(3\lambda + 2\mu)(\operatorname{tr}\varepsilon)^2 + \mu \operatorname{dev}\varepsilon \cdot\cdot \operatorname{dev}\varepsilon.$$

But $\operatorname{tr}\varepsilon$ and $\operatorname{dev}\varepsilon$ are independent quantities, and from (4.21) it follows that

$$3\lambda + 2\mu > 0, \qquad \mu > 0. \tag{4.22}$$

Hence the bulk modulus $k = \lambda + 2/3\mu$ and shear modulus $G = \mu$ are positive. A consequence of (4.22) is explored in

Exercise 4.4. Show that $E > 0$ and $-1 < \nu \leq 1/2$.

For most engineering materials we have $\nu > 0$. The values of ν for rubbers are close to $1/2$; materials like re-entrant foams can have negative values of ν. Data for various materials are tabulated in technical reference books. For regular steels, the values of E are about 2×10^{11} Pa, and ν lies in the range 0.25–0.33.

Exercise 4.5. For an isotropic body, express ε and W in terms of $\boldsymbol{\sigma}$.

4.5 Equilibrium Equations in Displacements

We derived the equation of motion (4.15) and the equilibrium equation (4.9). Equation (4.9) is written in terms of stresses. In component form, (4.9) contains three equations in six independent unknowns σ_{ij}. Using the definition of strain tensor and Hooke's law, we can transform (4.15) and (4.9) to a system of three simultaneous equations with respect to the components of the displacement vector $\mathbf{u}$. So the systems (4.15) and (4.9) reduce to systems involving three equations in the three unknown components of $\mathbf{u}$. This corresponds to the expectation that a well-posed problem should contain the same number of equations as unknowns.

For simplicity, we derive the equilibrium equations for an isotropic homogeneous material defined by (4.20). Suppose λ and μ are constants. First we derive $\nabla \cdot \boldsymbol{\sigma}$ in terms of $\mathbf{u}$:

$$\nabla \cdot \boldsymbol{\sigma} = \nabla \cdot (\lambda \mathbf{E} \operatorname{tr}\varepsilon + 2\mu\varepsilon) = \lambda \nabla \operatorname{tr}\varepsilon + \mu \nabla \cdot \nabla \mathbf{u} + \mu \nabla \cdot (\nabla \mathbf{u})^T.$$

Because $\operatorname{tr}\varepsilon = \nabla \cdot \mathbf{u}$ and $\nabla \cdot (\nabla \mathbf{u})^T = \nabla\nabla \cdot \mathbf{u}$, we get

$$\nabla \cdot \boldsymbol{\sigma} = (\lambda + \mu)\nabla\nabla \cdot \mathbf{u} + \mu \nabla \cdot \nabla \mathbf{u}.$$

So the equilibrium equation takes the form

$$(\lambda + \mu)\nabla\nabla \cdot \mathbf{u} + \mu \nabla \cdot \nabla \mathbf{u} + \rho \mathbf{f} = \mathbf{0}. \tag{4.23}$$

Exercise 4.6. Show that $\nabla \cdot (\nabla \mathbf{u})^T = \nabla\nabla \cdot \mathbf{u}$.

In Cartesian coordinates, equation (4.23) is

$$(\lambda + \mu)\frac{\partial u_i}{\partial x_k \partial x_i} + \mu \frac{\partial^2 u_k}{\partial x_i \partial x_i} + \rho f_k = 0 \qquad (k = 1, 2, 3). \tag{4.24}$$

That is,

$$(\lambda + \mu)\frac{\partial}{\partial x_1}\left(\frac{\partial u_1}{\partial x_1} + \frac{\partial u_2}{\partial x_2} + \frac{\partial u_3}{\partial x_3}\right) + \mu \Delta u_1 + \rho f_1 = 0,$$
$$(\lambda + \mu)\frac{\partial}{\partial x_2}\left(\frac{\partial u_1}{\partial x_1} + \frac{\partial u_2}{\partial x_2} + \frac{\partial u_3}{\partial x_3}\right) + \mu \Delta u_2 + \rho f_2 = 0,$$
$$(\lambda + \mu)\frac{\partial}{\partial x_1}\left(\frac{\partial u_1}{\partial x_1} + \frac{\partial u_2}{\partial x_2} + \frac{\partial u_3}{\partial x_3}\right) + \mu \Delta u_3 + \rho f_3 = 0,$$

where

$$\Delta = \frac{\partial^2}{\partial x_1^2} + \frac{\partial^2}{\partial x_2^2} + \frac{\partial^2}{\partial x_3^2}.$$

In curvilinear coordinates, the component representation of (4.23) is

$$(\lambda + \mu)\frac{\partial}{\partial q^k}\varepsilon + \mu g^{mn} \nabla_m \nabla_n u_k + \rho f_k = 0 \qquad (k = 1, 2, 3) \tag{4.25}$$

where

$$\varepsilon = \operatorname{tr} \boldsymbol{\varepsilon} = \frac{1}{\sqrt{g}} \frac{\partial(\sqrt{g} g^{mn} u_n)}{\partial q^m}.$$

Exercise 4.7. Derive (4.25).

See Appendix C for the corresponding relations in cylindrical and spherical coordinates.

Exercise 4.8. Write down the equations of motion in displacements.

4.6 Boundary Conditions and Boundary Value Problems

To ensure uniqueness of solution to the equilibrium equations, we must supplement the equations with conditions on the boundary of the domain where the equations hold. We first pose the main equilibrium problems for elastic bodies. Then we touch on the uniqueness question.

Physical intuition tells us that at a boundary point we should appoint either a displacement vector or a stress vector defined by the contact load.

Appointment of both at the same point is impossible; this will be confirmed later. In elasticity there are three main boundary value problems. Two were mentioned above: one with given displacements, one with given contact forces. The third is a mixed problem with some combination of displacements and contact loads given on the boundary.

In *the first boundary value problem* of elasticity, we supplement the equilibrium equations in displacements (4.23) or other forms of the equations (4.24)–(4.25) with a given displacement field $\mathbf{u}^0$ on the whole boundary Σ:

$$\mathbf{u}\big|_{\Sigma} = \mathbf{u}^0. \tag{4.26}$$

This is known as the *kinematic boundary condition.*

In *the second boundary value problem*, we supplement (4.23) with a prescribed contact load $\mathbf{t}^0$ on Σ:

$$\mathbf{n}\cdot\boldsymbol{\sigma}\big|_{\Sigma} = \mathbf{t}^0. \tag{4.27}$$

This is known as a *static boundary condition.* Since no point of the body is fixed, the body can move freely in space. With this condition, the static boundary value problem is well-posed only if the load acting on the body is self-balanced; this means that the resultant force and resultant moment of all the external forces acting on the body must vanish.

The reader familiar with partial differential equations may recognize that these types of boundary conditions are analogous to the Dirichlet and Neumann conditions for Poisson's equation. It is worth noting that as the Dirichlet and Neumann problems for Poisson's equation are the canonical problems for elliptic linear equations of the second order, the above problems of linear elasticity are the canonical problems for nontrivial linear elliptic systems for partial differential equations. These should be studied before any attempt to extend the results to more general boundary value problems for linear elliptic systems of second order equations.

The *third boundary value problem* of elasticity requires us to supplement (4.23) with *mixed boundary conditions.* That is, we must assign kinematic conditions on some portion Σ_1 of the boundary, and static conditions $\mathbf{t}^0$ on the remainder Σ_2:

$$\mathbf{u}\big|_{\Sigma_1} = \mathbf{u}^0, \qquad \mathbf{n}\cdot\boldsymbol{\sigma}\big|_{\Sigma_2} = \mathbf{t}^0. \tag{4.28}$$

For other "mixed problems" of elasticity, at each boundary point we assign three conditions combining the given displacements and loads in some fashion. Moreover, on the boundary we can combine conditions involving contact with other elastic or inelastic bodies or media. As a rule, such problems require consideration of well-posedness, although engineering intuition commonly permits investigators to propose meaningful boundary conditions of that type. An example of a meaningless problem is the problem with simultaneously given normal displacements and normal stresses on the boundary; these cannot be specified independently.

Henceforth we will assume V is a bounded volume whose boundary Σ is sufficiently regular that we can apply the technique of integration by parts.

4.7 Equilibrium Equations in Stresses

A *boundary value problem in displacements* involves finding a displacement field satisfying the equilibrium equations (4.23) and one of the sets of boundary conditions (4.26), (4.27), or (4.28). It is assumed that the stress tensor in (4.27) or (4.28) is expressed in terms of $\mathbf{u}$ by (4.20) or (4.12). When the stresses are prescribed over the whole boundary, we can try finding the unknown $\boldsymbol{\sigma}$ via (4.9) and (4.27). However, the solution for $\boldsymbol{\sigma}$ is not unique. We must invoke other equations that take into account Hooke's law. We can do this as follows. Suppose we have found a solution $\boldsymbol{\sigma}$ of equations (4.9) supplemented by (4.27). By $\boldsymbol{\sigma}$, through Hooke's law (4.20), we define $\boldsymbol{\varepsilon}$. Calculating the trace of equation (4.20), we get

$$\operatorname{tr}\boldsymbol{\varepsilon} = \frac{1}{3\lambda + 2\mu}\operatorname{tr}\boldsymbol{\sigma}.$$

Substituting this into (4.20), we have

$$\boldsymbol{\varepsilon} = \frac{1}{2\mu}\left[\boldsymbol{\sigma} - \frac{\lambda}{3\lambda + 2\mu}\mathbf{E}\operatorname{tr}\boldsymbol{\sigma}\right] = \frac{1}{2\mu}\left[\boldsymbol{\sigma} - \frac{\nu}{1+\nu}\mathbf{E}\operatorname{tr}\boldsymbol{\sigma}\right]. \tag{4.29}$$

But we know that the strain tensor makes sense, i.e., by $\boldsymbol{\varepsilon}$ we can find the displacement field, if and only if the compatibility equation (4.13) holds:

$$\nabla \times (\nabla \times \boldsymbol{\varepsilon})^T = \mathbf{0}.$$

Substituting (4.29) into the compatibility equation, we get the equation in terms of $\boldsymbol{\sigma}$:

$$\nabla \times (\nabla \times \boldsymbol{\varepsilon})^T = \frac{1}{2\mu}\left[\nabla \times (\nabla \times \boldsymbol{\sigma})^T - \frac{\nu}{1+\nu}\nabla \times \left(\nabla \times (\mathbf{E}\operatorname{tr}\boldsymbol{\sigma})^T\right]\right] = \mathbf{0}.$$

Omitting cumbersome calculations, we can reduce this to the *Beltrami–Michell equations*

$$\nabla \cdot \nabla\boldsymbol{\sigma} + \frac{\nu}{1+\nu}\nabla\nabla\operatorname{tr}\boldsymbol{\sigma} + \rho\nabla\mathbf{f} + \rho(\nabla\mathbf{f})^T + \mathbf{E}\frac{\nu}{1-\nu}\rho\nabla \cdot \mathbf{f} = \mathbf{0}. \tag{4.30}$$

These equations, being supplementary to (4.9) and (4.27), present us with the complete setup of the equilibrium problem in stresses. Note that this problem has a solution only if the set of external forces, consisting of the body forces and those acting on the boundary, is self-balanced.

When body forces are absent, (4.30) reduces to

$$\nabla \cdot \nabla \boldsymbol{\sigma} + \frac{\nu}{1+\nu} \nabla \nabla \operatorname{tr} \boldsymbol{\sigma} = \mathbf{0}. \tag{4.31}$$

In Cartesian coordinates, (4.31) implies

$$\begin{aligned}
&\Delta\sigma_{11} + \frac{1}{1+\nu}\frac{\partial^2\sigma}{\partial x_1^2} = 0, &&\Delta\sigma_{12} + \frac{1}{1+\nu}\frac{\partial^2\sigma}{\partial x_1 \partial x_2} = 0,\\
&\Delta\sigma_{22} + \frac{1}{1+\nu}\frac{\partial^2\sigma}{\partial x_2^2} = 0, &&\Delta\sigma_{23} + \frac{1}{1+\nu}\frac{\partial^2\sigma}{\partial x_2 \partial x_3} = 0,\\
&\Delta\sigma_{33} + \frac{1}{1+\nu}\frac{\partial^2\sigma}{\partial x_3^2} = 0, &&\Delta\sigma_{13} + \frac{1}{1+\nu}\frac{\partial^2\sigma}{\partial x_1 \partial x_3} = 0,
\end{aligned}$$

where

$$\sigma = \operatorname{tr} \boldsymbol{\sigma} = \sigma_{11} + \sigma_{22} + \sigma_{33}.$$

To solve the equilibrium problem completely, we first find $\boldsymbol{\sigma}$ by (4.9), (4.27), and the Beltrami–Michell equations. Then we use Hooke's law to find $\boldsymbol{\varepsilon}$. Finally, Cesàro's formula (4.14) yields the displacement field $\mathbf{u}$, which is uniquely defined up to a rigid motion of the body.

We should note that, from the standpoint of classical mathematical physics, the structure of the boundary value problem in stresses is a bit strange. The unknowns are the six components of $\boldsymbol{\sigma}$, satisfying nine equations — three of first order, and six of second order with respect to the components of $\boldsymbol{\sigma}$. These are supplemented with three boundary conditions. By its derivation, this boundary value problem is equivalent to the boundary value problem in displacements. The theory of such "strange" boundary value problems is far from complete.

4.8 Uniqueness of Solution for the Boundary Value Problems of Elasticity

Uniqueness of solution to a boundary value problem is an important aspect of its well-posedness. A uniqueness theorem for elasticity problems was established by Gustav R. Kirchhoff (1824–1887).

We will prove uniqueness of solution for equations (4.23) supplemented with boundary conditions (4.28), which include conditions (4.26) as a special case.

To the contrary, suppose there are two solutions $\mathbf{u}_1$ and $\mathbf{u}_2$ satisfying (4.23) and (4.28). We denote the respective stress tensors by $\boldsymbol{\sigma}_1$ and $\boldsymbol{\sigma}_2$. So the following two sets of equations hold:

$$\nabla \cdot \boldsymbol{\sigma}_1 + \rho \mathbf{f} = \mathbf{0} \text{ in } V, \quad \mathbf{u}_1\big|_{\Sigma_1} = \mathbf{u}^0, \quad \mathbf{n} \cdot \boldsymbol{\sigma}_1\big|_{\Sigma_2} = \mathbf{t}^0,$$

$$\nabla \cdot \boldsymbol{\sigma}_2 + \rho \mathbf{f} = \mathbf{0} \text{ in } V, \quad \mathbf{u}_2\big|_{\Sigma_1} = \mathbf{u}^0, \quad \mathbf{n} \cdot \boldsymbol{\sigma}_2\big|_{\Sigma_2} = \mathbf{t}^0.$$

Consider the difference $\mathbf{u} = \mathbf{u}_1 - \mathbf{u}_2$ and its corresponding stress field $\boldsymbol{\sigma} = \boldsymbol{\sigma}_1 - \boldsymbol{\sigma}_2$. Subtracting the equations for $\mathbf{u}_1$ and $\mathbf{u}_2$, we see that $\mathbf{u}$ satisfies the equilibrium problem

$$\nabla \cdot \boldsymbol{\sigma} = \mathbf{0} \text{ in } V, \quad \mathbf{u}\big|_{\Sigma_1} = \mathbf{0}, \quad \mathbf{n} \cdot \boldsymbol{\sigma}\big|_{\Sigma_2} = \mathbf{0}.$$

Dot-multiplying the equilibrium equation by $\mathbf{u}$ and integrating over V, we get

$$\int_V (\nabla \cdot \boldsymbol{\sigma}) \cdot \mathbf{u} \, dV = 0.$$

The Gauss–Ostrogradsky theorem yields

$$-\int_V \boldsymbol{\sigma} \cdot\cdot (\nabla \mathbf{u})^T \, dV + \int_\Sigma \mathbf{n} \cdot \boldsymbol{\sigma} \cdot \mathbf{u} = 0.$$

At each point of Σ, one of the conditions $\mathbf{u} = \mathbf{0}$ or $\mathbf{n} \cdot \boldsymbol{\sigma} = \mathbf{0}$ holds. So the last surface integral is zero. The tensor $\boldsymbol{\sigma}$ is symmetric. So

$$\boldsymbol{\sigma} \cdot\cdot (\nabla \mathbf{u})^T = \boldsymbol{\sigma} \cdot\cdot (\nabla \mathbf{u}).$$

This allows us to represent the volume integral as

$$-\int_V \boldsymbol{\sigma} \cdot\cdot \boldsymbol{\varepsilon} \, dV = -2 \int_V W(\boldsymbol{\varepsilon}) \, dV = 0.$$

Recall that $W(\boldsymbol{\varepsilon})$ is positive definite. Hence the equality of the integral to zero implies that its integrand vanishes everywhere in V: $W(\boldsymbol{\varepsilon}) = 0$. By positiveness of W, it follows that

$$\boldsymbol{\varepsilon} = \mathbf{0}.$$

By Cesàro's formula (4.14), when $\boldsymbol{\varepsilon}(\mathbf{u}) = \mathbf{0}$, the displacement vector takes the form

$$\mathbf{u} = \mathbf{u}_0 + \boldsymbol{\omega}_0 \times (\mathbf{r} - \mathbf{r}_0), \tag{4.32}$$

where $\mathbf{u}_0, \boldsymbol{\omega}_0$ and $\mathbf{r}_0$ are constant vectors. But it is a small displacement field for the rigid volume V. Because $\mathbf{u}|_{\Sigma_1} = \mathbf{0}$, we have $\mathbf{u} = \mathbf{0}$ in V. Hence $\mathbf{u}_1 = \mathbf{u}_2$ and we have uniqueness of solution for the first and third boundary value problems of elasticity.

For the second problem, the initial steps of the proof remain valid. We find that $\boldsymbol{\varepsilon} = \mathbf{0}$ in V, hence the two solutions to the problem differ by a rigid-body displacement:

$$\mathbf{u}_1 = \mathbf{u}_2 + \mathbf{u}_0 + \boldsymbol{\omega}_0 \times (\mathbf{r} - \mathbf{r}_0).$$

But this time we cannot conclude that $\mathbf{u}_1 = \mathbf{u}_2$. From physics, the situation is clear: the second problem of elasticity describes a body free of geometric restrictions, so its solution must be defined up to a rigid-body displacement. This is what the last formula states.

4.9 Betti's Reciprocity Theorem

The solutions to the equilibrium problems for an elastic body under two different loads obey *Betti's reciprocal work theorem.*

Let us consider a body under the action of contact forces $\mathbf{t}'$ and body forces $\mathbf{f}'$, and the same body under the action of another pair of external forces $\mathbf{t}''$ and $\mathbf{f}''$. The displacements and other quantities for the two corresponding problems will be denoted similarly, using primes and double-primes, respectively. Now we have two solutions of two different second boundary value problems for the elastic body.

Theorem 4.4. *The solutions $\mathbf{u}'$ and $\mathbf{u}''$ of the second boundary value problem of elasticity that correspond to the respective loading pairs $\mathbf{t}', \mathbf{f}'$ and $\mathbf{t}'', \mathbf{f}''$ satisfy the following relation:*

$$\int_V \rho \mathbf{f}' \cdot \mathbf{u}'' \, dV + \int_\Sigma \mathbf{t}' \cdot \mathbf{u}'' \, d\Sigma = \int_V \rho \mathbf{f}'' \cdot \mathbf{u}' \, dV + \int_\Sigma \mathbf{t}'' \cdot \mathbf{u}' \, d\Sigma. \tag{4.33}$$

This equality is Betti's theorem. First derived for beam theory and later extended to many portions of physics, it is used to derive the equations of the boundary finite element methods. Each of the expressions on the left and right represents the work of a load over a displacement. So we can give a mechanical formulation of Betti's theorem.

The work of the first system of forces over the displacements of the body due to the action of the second system of forces is equal to the work of the second system of forces over the displacements due to the action of the first system of forces.

Proof. The vectors $\mathbf{u}'$ and $\mathbf{u}''$ respectively satisfy the following equilibrium equations and boundary conditions:

$$\nabla \cdot \boldsymbol{\sigma}' + \rho \mathbf{f}' = \mathbf{0} \text{ in } V, \quad \mathbf{n} \cdot \boldsymbol{\sigma}'\big|_\Sigma = \mathbf{t}';$$
$$\nabla \cdot \boldsymbol{\sigma}'' + \rho \mathbf{f}'' = \mathbf{0} \text{ in } V, \quad \mathbf{n} \cdot \boldsymbol{\sigma}''\big|_\Sigma = \mathbf{t}''.$$

Let us consider the left-hand side of (4.33). Writing the forces in terms of the stresses, we get

$$\int_V \rho \mathbf{f}' \cdot \mathbf{u}'' \, dV + \int_\Sigma \mathbf{t}' \cdot \mathbf{u}'' \, d\Sigma = -\int_V (\nabla \cdot \boldsymbol{\sigma}') \cdot \mathbf{u}'' \, dV + \int_\Sigma \mathbf{n} \cdot \boldsymbol{\sigma}' \cdot \mathbf{u}'' \, d\Sigma.$$

Applying the Gauss–Ostrogradsky theorem to the surface integral, we get

$$\begin{aligned} -\int_V (\nabla \cdot \boldsymbol{\sigma}') \cdot \mathbf{u}'' \, dV &+ \int_\Sigma \mathbf{n} \cdot \boldsymbol{\sigma}' \cdot \mathbf{u}'' \, d\Sigma \\ &= \int_V [-(\nabla \cdot \boldsymbol{\sigma}') \cdot \mathbf{u}'' + \nabla \cdot (\boldsymbol{\sigma}' \cdot \mathbf{u}'')] \, dV \\ &= \int_V \boldsymbol{\sigma}' \cdot\cdot (\nabla \mathbf{u}'')^T \, dV \\ &= \int_V \boldsymbol{\sigma}' \cdot\cdot \boldsymbol{\varepsilon}'' \, dV. \end{aligned}$$

Similarly,

$$\int_V \rho \mathbf{f}'' \cdot \mathbf{u}' \, dV + \int_\Sigma \mathbf{t}'' \cdot \mathbf{u}' \, d\Sigma = \int_V \boldsymbol{\sigma}'' \cdot\cdot \boldsymbol{\varepsilon}' \, dV.$$

So the proof reduces to verification of the equality

$$\int_V \boldsymbol{\sigma}' \cdot\cdot \boldsymbol{\varepsilon}'' \, dV = \int_V \boldsymbol{\sigma}'' \cdot\cdot \boldsymbol{\varepsilon}' \, dV.$$

Using Hooke's law and recalling the symmetry properties of $\mathbf{C}$, we have

$$\boldsymbol{\sigma}' \cdot\cdot \boldsymbol{\varepsilon}'' = (\mathbf{C} \cdot\cdot \boldsymbol{\varepsilon}') \cdot\cdot \boldsymbol{\varepsilon}'' = \boldsymbol{\varepsilon}'' \cdot\cdot \mathbf{C} \cdot\cdot \boldsymbol{\varepsilon}' = \boldsymbol{\varepsilon}' \cdot\cdot \mathbf{C} \cdot\cdot \boldsymbol{\varepsilon}'' = \boldsymbol{\sigma}'' \cdot\cdot \boldsymbol{\varepsilon}'.$$

This completes the proof. □

Exercise 4.9. Let a portion Σ_1 of the boundary be fixed: $\mathbf{u}|_{\Sigma_1} = \mathbf{0}$. Suppose two boundary value problems include this condition and, moreover, that the body is under the action of one of the two systems of forces $\mathbf{t}', \mathbf{f}'$ or $\mathbf{t}'', \mathbf{f}''$, with $\mathbf{t}', \mathbf{t}''$ given on $\Sigma_2 = \Sigma \setminus \Sigma_1$. Prove that in this case Betti's equality takes the form

$$\int_V \rho \mathbf{f}' \cdot \mathbf{u}'' \, dV + \int_{\Sigma_2} \mathbf{t}' \cdot \mathbf{u}'' \, d\Sigma = \int_V \rho \mathbf{f}'' \cdot \mathbf{u}' \, dV + \int_{\Sigma_2} \mathbf{t}'' \cdot \mathbf{u}' \, d\Sigma.$$

Exercise 4.10. Consider two equilibrium problems for a body under two systems of external forces, as in the previous exercise. This time, however, two distinct displacement fields are prescribed over Σ_1:

$$\mathbf{u}'\big|_{\Sigma_1} = \mathbf{a}', \qquad \mathbf{u}''\big|_{\Sigma_1} = \mathbf{a}''. \tag{4.34}$$

As above, denote the solutions of the corresponding equilibrium problems by $\mathbf{u}'$ and $\mathbf{u}''$. Denote the solutions of the equilibrium problems for a body free of external forces, that satisfy the corresponding conditions (4.34), by $\mathbf{u}_0'$ and $\mathbf{u}_0''$, respectively. Finally, introduce $\tilde{\mathbf{u}}' = \mathbf{u}' - \mathbf{u}_0'$ and $\tilde{\mathbf{u}}'' = \mathbf{u}'' - \mathbf{u}_0''$. Prove that Betti's equality takes the form

$$\int_V \rho \mathbf{f}' \cdot \tilde{\mathbf{u}}''\, dV + \int_{\Sigma_2} \mathbf{t}' \cdot \tilde{\mathbf{u}}''\, d\Sigma = \int_V \rho \mathbf{f}'' \cdot \tilde{\mathbf{u}}'\, dV + \int_{\Sigma_2} \mathbf{t}'' \cdot \tilde{\mathbf{u}}'\, d\Sigma.$$

4.10 Minimum Total Energy Principle

On a curved surface, a ball in equilibrium takes the lowest position; we call the equilibrium *stable*. In elementary physics, it is said that at such an equilibrium point, the potential energy of the ball takes a minimum value. A stationary point of potential energy corresponds to an equilibrium position of the ball as well, but such an equilibrium may be unstable. An equilibrium that corresponds to a non-minimum point normally is unstable. The stability result for the potential energy of a particle was first extended to classical mechanics by Lagrange. The minimum potential energy principle is now one of the most important principles of physics, holding also for distributed systems like elastic bodies. However, it is not always straightforward to formulate, since we must find an expression for the potential energy and prove that it takes a minimum value in equilibrium.

As the expression for the energy of an elastic body under load, we propose the *total potential energy*

$$\mathcal{E} = \int_V W(\boldsymbol{\varepsilon})\, dV - \int_V \rho \mathbf{f} \cdot \mathbf{u}\, dV - \int_{\Sigma_2} \mathbf{t}^0 \cdot \mathbf{u}\, d\Sigma.$$

The first term is the strain energy associated with deformation. If we slowly decrease the external forces acting on the body to zero, the strain energy is the maximum work that the body can produce. The other terms, with negative signs, represent the work of external forces on the displacement field of the body. These terms are analogous to the expression for the work of the gravitational force acting on a moving particle, and are so similar to the gravitational potential that they may be called potential energy terms.[3] Again, we must prove that $\mathcal{E}$ really takes its minimum value when the body is in equilibrium. Clearly, we should explain what is meant by "the minimum value of $\mathcal{E}$" and introduce the tools necessary to establish the minimum principle.

The reader is familiar with the definition of a local minimum point for a function in n variables: there should exist a neighborhood of the point in which all the values of the function are no less than its value at the point. For a global minimum point, this value should be no greater than the values of the function at any point. These definitions can be extended to quantities like $\mathcal{E}$, as they take values in the set of real numbers. We could call $\mathcal{E}$ a function, but because it depends on the vector function $\mathbf{u}$, it is called a *functional*. A functional is a correspondence that takes its argument, which can be a whole function as it is here, to at most one real number.

To extend the idea of minimum to functionals, we should treat the notions of independent variable, the domain of such a variable, and the neighborhood of a "point" (which is now a vector function $\mathbf{u}$ in the domain). For simplicity, we will assume the domain of E consists of vector functions possessing all continuous derivatives in V up to the second order. Moreover, we suppose that for any $\mathbf{u}$, the kinematic boundary conditions hold. Such vector functions will be called *admissible*. To define a neighborhood, we must introduce a metric or norm on the set of admissible vector functions. The reader interested in the formalities should consult standard books on the calculus of variations. Our present approach will be essentially that of the pioneers of the subject, which was to obtain results without worrying too much about formal justification. In fact, we can completely avoid the question about neighborhoods, since for the problems under consideration, in the equilibrium state $\mathcal{E}$ takes its global minimum value.

Now we will describe the technical tools that permit us to seek the minimum points of a functional. We wish to derive for $\mathcal{E}$ some analogue of Fermat's theorem for a differentiable function in one variable: the first derivative must vanish at a minimum point. Fermat's theorem extends to the theory of functions in many variables: all the first partial derivatives of the function must vanish. Let us review how this extension is accomplished.

Let f be an ordinary real-valued function of a vector variable, and suppose f takes its minimum at a point $\mathbf{x} \in \mathbb{R}^n$. If $\mathbf{a}$ is any vector and τ is a real parameter, then $f(\mathbf{x} + \tau\mathbf{a})$ can be regarded as a function of τ that

[3]This analogy is strict when the field of external forces is conservative, i.e., when there exists a potential function for the forces and hence a potential energy.

takes its minimum value at $\tau = 0$. Therefore, at $\tau = 0$ its derivative with respect to τ must vanish:

$$\left.\frac{df(\mathbf{x} + \tau\mathbf{a})}{d\tau}\right|_{\tau=0} = 0. \tag{4.35}$$

At a minimum point $\mathbf{x}$, this must hold for any $\mathbf{a}$. As the reader is aware, the expression on the left is the directional derivative of f at $\mathbf{x}$ along the direction of $\mathbf{a}$. Moreover, if we formally set $\mathbf{a} = d\mathbf{x}$, then the left member takes the form of the first differential df at point $\mathbf{x}$:

$$df = \left.\frac{df(\mathbf{x} + \tau\, d\mathbf{x})}{d\tau}\right|_{\tau=0} = \nabla f\big|_{\mathbf{x}} \cdot d\mathbf{x}.$$

Putting $\mathbf{a} = \mathbf{i}_k$ in (4.35), we obtain Fermat's theorem for a function in n variables: the partial derivative of f with respect to x_k at $\mathbf{x}$ must vanish if $\nabla f|_{\mathbf{x}}$ exists and $\mathbf{x}$ is a minimum point of f. We emphasize that this is merely a necessary condition. If at some point $\mathbf{x}$ equation (4.35) holds, then $\mathbf{x}$ may not be a minimum (or a maximum) point. Any $\mathbf{x}$ satisfying (4.35) is called a *stationary point* of f. It could be a minimum point, a maximum point, or a saddle point.

This procedure extends to functionals in a straightforward manner. Suppose $\mathbf{u}$ is a minimum point of $\mathcal{E}$. Let us fix some $\delta\mathbf{u}$ such that $\mathbf{u} + \tau\,\delta\mathbf{u}$ is an admissible displacement for all small τ. Substitute this into $\mathcal{E}$. Then $\mathcal{E}(\mathbf{u}+\tau\,\delta\mathbf{u})$ is a function of the real variable τ that takes its minimum value at $\tau = 0$. If the derivative of this function with respect to τ exists, then it vanishes at $\tau = 0$:

$$\delta\mathcal{E} = \left.\frac{d}{d\tau}E(\mathbf{u} + \tau\,\delta\mathbf{u})\right|_{\tau=0} = 0$$

for any $\delta\mathbf{u}$ such that $\mathbf{u} + \tau\,\delta\mathbf{u}$ is admissible for any small τ.

We call $\delta\mathcal{E}$ the *first variation* of the functional $\mathcal{E}$. Its zeros are called *stationary points* of $\mathcal{E}$; they are analogous to the stationary points of an ordinary function f. The equation we obtained is a necessary condition for $\mathbf{u}$ to minimize $\mathcal{E}$.

Note that the techniques described here and below belong to the part of mathematics called the calculus of variations. In its infancy it was largely developed by Euler and Lagrange. The reader interested in more detail and generality should consult textbooks on the subject.

We assumed the displacements $\mathbf{u} + \tau\,\delta\mathbf{u}$ were admissible for all small τ. Clearly, if all derivatives of $\mathbf{u}$ and $\delta\mathbf{u}$ up to order two are continuous in

V, the same is true of $\mathbf{u} + \tau\,\delta\mathbf{u}$. We required $\mathbf{u}$ to satisfy the kinematic restrictions on Σ_1. Hence if $\delta\mathbf{u}$ satisfies

$$\delta\mathbf{u}\big|_{\Sigma_1} = \mathbf{0},$$

then $\mathbf{u}+\tau\,\delta\mathbf{u}$ also satisfies the kinematic restrictions for any τ. In this case $\delta\mathbf{u}$ is called a *virtual displacement.*

We note that the definition of $\delta\mathcal{E}$ can be applied to any sufficiently smooth $\mathcal{E}$. (We leave this on an intuitive level, as a formal definition would require a background in functional analysis.) It is called the *Gâteaux derivative* of the functional. In the notation $\delta\mathcal{E}$, we regard the symbol δ as some action on $\mathcal{E}$ given by a formula similar to the one used to find the differential of a function. By $\delta\mathbf{u}$ we denote a virtual displacement; this could be denoted by $\mathbf{v}$ as well, but we would have to keep clarifying that it is a *virtual* displacement. That is, in this case the symbol δ is merely a notation; it is not an operation and appears for historical reasons. When we apply the operation δ to an ordinary function, it coincides with the operation of taking its first differential. The differentials of variables in this case are written with the use of the symbol δ instead of d; for example, we have $\delta(x^2) = 2x\,\delta x$.

Using the definition, let us find $\delta\mathcal{E}$ for our problem:

$$\delta\mathcal{E} = \int_V \delta W(\boldsymbol{\varepsilon})\,dV - \int_V \rho\mathbf{f}\cdot\delta\mathbf{u}\,dV - \int_{\Sigma_2} \mathbf{t}^0\cdot\delta\mathbf{u}\,d\Sigma.$$

Calculating δW in the integrand and recalling that for intermediate steps we can use the formulas for the first differential, we get

$$\delta W(\boldsymbol{\varepsilon}) = \tfrac{1}{2}\delta(\boldsymbol{\varepsilon}\cdot\cdot\,\mathbf{C}\cdot\cdot\,\boldsymbol{\varepsilon}) = \boldsymbol{\varepsilon}\cdot\cdot\,\mathbf{C}\cdot\cdot\,\delta\boldsymbol{\varepsilon} = \boldsymbol{\sigma}\cdot\cdot\,\delta\boldsymbol{\varepsilon},$$

where

$$\delta\boldsymbol{\varepsilon} = \boldsymbol{\varepsilon}(\delta\mathbf{u}) = \tfrac{1}{2}\left(\nabla\delta\mathbf{u} + (\nabla\delta\mathbf{u})^T\right).$$

It follows that

$$\delta\mathcal{E} = \int_V \boldsymbol{\sigma}\cdot\cdot\,\delta\boldsymbol{\varepsilon}\,dV - \int_V \rho\mathbf{f}\cdot\delta\mathbf{u}\,dV - \int_{\Sigma_2} \mathbf{t}^0\cdot\delta\mathbf{u}\,d\Sigma.$$

Now we wish to find a relationship between a solution to the equilibrium problem for an elastic body and the minimum problem for $\mathcal{E}$. We start with the following theorem.

Theorem 4.5. *A stationary point* $\mathbf{u}$ *of* $\mathcal{E}$ *on the set of admissible displacements satisfies the equilibrium equations of the elastic body in the volume* V *and the boundary condition* $\mathbf{n}\cdot\boldsymbol{\sigma}|_{\Sigma_2} = \mathbf{t}^0$, *and conversely.*

Proof. First we prove that $\mathbf{u}$, a solution of the equilibrium problem for an elastic body under load, is a stationary point of $\mathcal{E}$; that is, $\delta\mathcal{E} = 0$ when $\mathbf{u}$ is a solution and $\delta\mathbf{u}$ is an arbitrary virtual displacement. Assume $\mathbf{u}$ satisfies

$$\nabla \cdot \boldsymbol{\sigma} + \rho\mathbf{f} = \mathbf{0} \text{ in } V, \qquad \mathbf{u}\big|_{\Sigma_1} = \mathbf{u}^0, \quad \mathbf{n} \cdot \boldsymbol{\sigma}\big|_{\Sigma_2} = \mathbf{t}^0. \tag{4.36}$$

Dot-multiply the equilibrium equation by an admissible $\delta\mathbf{u}$ and integrate over V. Then apply the Gauss–Ostrogradsky theorem:

$$\begin{aligned} 0 &= \int_V [(\nabla \cdot \boldsymbol{\sigma}) \cdot \delta\mathbf{u} + \rho\mathbf{f} \cdot \delta\mathbf{u}]\, dV \\ &= \int_V \left[-\boldsymbol{\sigma} \cdot\cdot (\nabla\delta\mathbf{u})^T + \rho\mathbf{f} \cdot \delta\mathbf{u}\right] dV + \int_\Sigma \mathbf{n} \cdot \boldsymbol{\sigma} \cdot \delta\mathbf{u}\, d\Sigma \\ &= \int_V [-\boldsymbol{\sigma} \cdot\cdot \delta\boldsymbol{\varepsilon} + \rho\mathbf{f} \cdot \delta\mathbf{u}]\, dV + \int_{\Sigma_2} \mathbf{t}^0 \cdot \delta\mathbf{u}\, d\Sigma \\ &= -\delta\mathcal{E}. \end{aligned}$$

We used the fact that $\delta\mathbf{u} = \mathbf{0}$ on Σ_1 and $\mathbf{n} \cdot \boldsymbol{\sigma} = \mathbf{t}^0$ on Σ_2. Hence $\mathbf{u}$ is a stationary point of $\mathcal{E}$.

Now we prove the converse statement. Let $\mathbf{u}$ be a stationary point of $\mathcal{E}$, so that it satisfies the equation $\delta\mathcal{E} = 0$ for any virtual $\delta\mathbf{u}$ being sufficiently smooth, and let $\mathbf{u}|_{\Sigma_1} = \mathbf{u}^0$. We will show that $\mathbf{u}$ satisfies the equilibrium equations in V as well as the contact condition $\mathbf{n} \cdot \boldsymbol{\sigma} = \mathbf{t}^0$ on Σ_2. Indeed, doing the above calculations in reverse order, we get

$$\begin{aligned} \delta\mathcal{E} &= \int_V \boldsymbol{\sigma} \cdot\cdot \delta\boldsymbol{\varepsilon}\, dV - \int_V \rho\mathbf{f} \cdot \delta\mathbf{u}\, dV - \int_{\Sigma_2} \mathbf{t}^0 \cdot \delta\mathbf{u}\, d\Sigma \\ &= \int_V \boldsymbol{\sigma} \cdot\cdot (\nabla\delta\mathbf{u})^T\, dV - \int_V \rho\mathbf{f} \cdot \delta\mathbf{u}\, dV - \int_{\Sigma_2} \mathbf{t}^0 \cdot \delta\mathbf{u}\, d\Sigma \\ &= -\int_V [(\nabla \cdot \boldsymbol{\sigma}) \cdot \delta\mathbf{u} + \rho\mathbf{f} \cdot \delta\mathbf{u}]\, dV + \int_\Sigma \mathbf{n} \cdot \boldsymbol{\sigma} \cdot \delta\mathbf{u}\, d\Sigma - \int_{\Sigma_2} \mathbf{t}^0 \cdot \delta\mathbf{u}\, d\Sigma \\ &= -\int_V [\nabla \cdot \boldsymbol{\sigma} + \rho\mathbf{f}] \cdot \delta\mathbf{u}\, dV \\ &\quad + \int_{\Sigma_1} \mathbf{n} \cdot \boldsymbol{\sigma} \cdot \delta\mathbf{u}\, d\Sigma - \int_{\Sigma_2} \left[\mathbf{n} \cdot \boldsymbol{\sigma} - \mathbf{t}^0\right] \cdot \delta\mathbf{u}\, d\Sigma. \end{aligned}$$

Because $\delta\mathbf{u} = 0$ on Σ_1, we finally obtain

$$\delta\mathcal{E} = -\int_V [\nabla \cdot \boldsymbol{\sigma} + \rho\mathbf{f}] \cdot \delta\mathbf{u}\, dV - \int_{\Sigma_2} \left[\mathbf{n} \cdot \boldsymbol{\sigma} - \mathbf{t}^0\right] \cdot \delta\mathbf{u}\, d\Sigma = 0. \tag{4.37}$$

We assumed $\mathbf{u}$ to be smooth enough that the above integrands are continuous. As admissible $\delta\mathbf{u}$ is arbitrary, from this equality it follows that

$$\nabla \cdot \boldsymbol{\sigma} + \rho\mathbf{f} = \mathbf{0} \text{ in } V, \qquad \mathbf{n} \cdot \boldsymbol{\sigma}\big|_{\Sigma_2} = \mathbf{t}^0. \tag{4.38}$$

The justification of this last step requires standard material from the calculus of variations. We postpone this discussion until the end of the present section. □

Now we would like to prove

Theorem 4.6. *Let* $\mathbf{u}$ *be a solution of the boundary problem* (4.36). *Then it is a point of global minimum of* $\mathcal{E}$.

Proof. The statement is a consequence of the fact that W is a positive quadratic form with respect to $\boldsymbol{\varepsilon}$. Indeed, let $\mathbf{v}$ be an arbitrary admissible displacement so that it is smooth enough and satisfies the kinematic condition $\mathbf{v}|_{\Sigma_1} = \mathbf{u}^0$. Consider the difference $\Delta\mathcal{E} = \mathcal{E}(\mathbf{v}) - \mathcal{E}(\mathbf{u})$. We have

$$\begin{aligned}
\Delta\mathcal{E} &= \int_V W(\boldsymbol{\varepsilon}(\mathbf{v}))\,dV - \int_V \rho\mathbf{f}\cdot\mathbf{v}\,dV - \int_{\Sigma_2} \mathbf{t}^0\cdot\mathbf{v}\,d\Sigma \\
&\quad - \int_V W(\boldsymbol{\varepsilon}(\mathbf{u}))\,dV + \int_V \rho\mathbf{f}\cdot\mathbf{u}\,dV + \int_{\Sigma_2} \mathbf{t}^0\cdot\mathbf{u}\,d\Sigma \\
&= \int_V \left[W(\boldsymbol{\varepsilon}(\mathbf{v})) - W(\boldsymbol{\varepsilon}(\mathbf{u}))\right] dV \\
&\quad - \int_V \rho\mathbf{f}\cdot(\mathbf{v}-\mathbf{u})\,dV - \int_{\Sigma_2} \mathbf{t}^0\cdot(\mathbf{v}-\mathbf{u})\,d\Sigma.
\end{aligned}$$

Let $\mathbf{w} = \mathbf{v} - \mathbf{u}$. Because $\mathbf{u}$ and $\mathbf{v}$ coincide on Σ_1, we have $\mathbf{w}|_{\Sigma_1} = \mathbf{0}$. Next,

$$\begin{aligned}
2\left[W(\boldsymbol{\varepsilon}(\mathbf{v})) - W(\boldsymbol{\varepsilon}(\mathbf{u}))\right] &= \boldsymbol{\varepsilon}(\mathbf{v}) \cdot\cdot\, \mathbf{C} \cdot\cdot\, \boldsymbol{\varepsilon}(\mathbf{v}) - \boldsymbol{\varepsilon}(\mathbf{u}) \cdot\cdot\, \mathbf{C} \cdot\cdot\, \boldsymbol{\varepsilon}(\mathbf{u}) \\
&= \boldsymbol{\varepsilon}(\mathbf{w}) \cdot\cdot\, \mathbf{C} \cdot\cdot\, \boldsymbol{\varepsilon}(\mathbf{w}) + 2\boldsymbol{\varepsilon}(\mathbf{u}) \cdot\cdot\, \mathbf{C} \cdot\cdot\, \boldsymbol{\varepsilon}(\mathbf{w}).
\end{aligned}$$

Therefore

$$\begin{aligned}
\Delta\mathcal{E} &= \frac{1}{2}\int_V \boldsymbol{\varepsilon}(\mathbf{w}) \cdot\cdot\, \mathbf{C} \cdot\cdot\, \boldsymbol{\varepsilon}(\mathbf{w})\,dV \\
&\quad + \int_V \boldsymbol{\varepsilon}(\mathbf{u}) \cdot\cdot\, \mathbf{C} \cdot\cdot\, \boldsymbol{\varepsilon}(\mathbf{w})\,dV - \int_V \rho\mathbf{f}\cdot\mathbf{w}\,dV - \int_{\Sigma_2} \mathbf{t}^0\cdot\mathbf{w}\,d\Sigma.
\end{aligned}$$

The sum of the terms in the second line constitutes $\delta\mathcal{E}$ at $\mathbf{u}$ with $\delta\mathbf{u} = \mathbf{w}$, which is a virtual displacement, so this sum is zero. Thus we have

$$\Delta\mathcal{E} = \frac{1}{2}\int_V \boldsymbol{\varepsilon}(\mathbf{w}) \cdot\cdot\, \mathbf{C} \cdot\cdot\, \boldsymbol{\varepsilon}(\mathbf{w})\,dV = \int_V W(\boldsymbol{\varepsilon}(\mathbf{w}))\,dV. \tag{4.39}$$

As W is a positive definite form, $\Delta\mathcal{E} \geq 0$ for any admissible $\mathbf{v}$. Hence $\mathbf{u}$ is a global minimizer of $\mathcal{E}$. □

The two preceding theorems can be summarized as *Lagrange's variational principle.*

Theorem 4.7. *The solution of the equilibrium problem for an elastic body is equivalent to the problem of minimizing the total energy functional $\mathcal{E}$ over the set of all kinematically admissible smooth displacement fields.*

To complete the proof of Theorem 4.5, we must show how to obtain (4.38) from (4.37). It will suffice to do this for one of the component equations. Denoting the kth component of the expressions in the brackets in (4.37) by F and f respectively, the kth component of $\delta\mathbf{u}$ by δu, and setting the remaining components of $\delta\mathbf{u}$ to zero, we reduce equation (4.37) to

$$\int_V F\,\delta u\,dV + \int_{\Sigma_2} f\,\delta u\,d\Sigma = 0. \tag{4.40}$$

Equation (4.38) follows from the next theorem.

Theorem 4.8. *Let F be continuous in V and let f be continuous on Σ_2. Suppose* (4.40) *holds for all sufficiently smooth functions δu that vanish on Σ_1. Then $F = 0$ in V and $f = 0$ on Σ_2.*

Proof. The proof is done in two steps. We begin by restricting the set of all admissible δu to those that vanish on the whole boundary Σ. Then (4.40) becomes

$$\int_V F\,\delta u\,dV = 0.$$

Suppose, contrary to the theorem statement, that $F(\mathbf{x}_*) = a \neq 0$ at some point $\mathbf{x}_*$. Without loss of generality we take $\mathbf{x}_*$ to be an interior point of V and $a > 0$. By continuity of F there is an open ball B_r with center $\mathbf{x}_*$ and radius $r > 0$ that lies inside V such that $F(\mathbf{x}) > a/2 > 0$ on B_r. Now if we take a smooth function δu_0 that is positive on B_r and zero outside B_r, we get

$$\int_V F\,\delta u_0\,dV > 0,$$

which contradicts the above equality that must hold for all δu that vanish on the boundary of V. It is clear that such a function δu_0 exists; the reader can find examples of "bell-shaped functions" in textbooks on the calculus of variations, cf. [Lebedev and Cloud (2003)]. So $F = 0$ in V.

The second step is to prove that $f = 0$ on Σ_2. For this we return to (4.40). Because $F = 0$, the first integral is zero for any admissible δu. So we find that

$$\int_{\Sigma_2} f\, \delta u\, d\Sigma = 0$$

holds for all admissible δu that take arbitrary values on Σ_2. But now we are in a position similar to the proof for F. We can repeat that proof, but on a two-dimensional domain Σ_2, so $f = 0$ on Σ_2. □

A reader familiar with the calculus of variations may recognize the above procedure as leading to one version of the "main lemma" of that subject.

It is worth noting that in the theory of elasticity there are several variational principles. Some are of minimax type; others are of stationary type. We can mention principles that carry the names of Castigliano, Reissner, Tonti, Hamilton, etc.

4.11 Ritz's Method

The minimum total energy principle is of great physical importance. It is also the basis for introducing the generalized setup of the equilibrium problems, which in turn defines *weak solutions* in elasticity. These solutions have finite strain energies. In this way one can prove the existence of generalized solutions, although more modern mathematical techniques involving weak derivatives, Sobolev spaces, and some elements of functional analysis are required. The principle is even more important from a practical standpoint. The variational equation $\delta\mathcal{E} = 0$ is the basis for various numerical methods, including the Ritz and Galerkin methods, the variational finite difference methods, and the finite element methods. Minimality of the total energy on the solution warrants stability of the algorithms of these methods. For details, the reader should consult specialized literature. Here we present Ritz's method, which spawned the other methods mentioned above.

The minimum total energy principle states that a solution of the equilibrium problem

$$\nabla \cdot \boldsymbol{\sigma} + \rho \mathbf{f} = \mathbf{0} \text{ in } V, \qquad \mathbf{u}\big|_{\Sigma_1} = \mathbf{u}^0, \quad \mathbf{n} \cdot \boldsymbol{\sigma}\big|_{\Sigma_2} = \mathbf{t}^0,$$

expressed in terms of $\mathbf{u}$, is a point of global minimum of the total energy functional

$$\mathcal{E} = \int_V W(\boldsymbol{\varepsilon})\,dV - \int_V \rho \mathbf{f} \cdot \mathbf{u}\,dV - \int_{\Sigma_2} \mathbf{t}^0 \cdot \mathbf{u}\,d\Sigma$$

on the set of admissible displacements. The converse statement is also valid, hence by the uniqueness theorem this point of minimum is unique. Walter Ritz (1878–1909) thought that by finding a minimum point on some subset of admissible displacements, we can get an approximation to the solution. The decisive idea of the method was how to select this subset in such a way that solving the approximate minimum problem would be relatively easy.

Ritz proposed to minimize $\mathcal{E}$ on the set of displacements of the form

$$\mathbf{u}_N = \sum_{k=1}^{N} u_k \boldsymbol{\varphi}_k + \mathbf{u}^*, \qquad \mathbf{u}^*\big|_{\Sigma_1} = \mathbf{u}^0, \quad \boldsymbol{\varphi}_n\big|_{\Sigma_1} = \mathbf{0} \tag{4.41}$$

with some fixed $\boldsymbol{\varphi}_k$ and $\mathbf{u}^*$, by determining the numerical coefficients u_k. The first step is to find a sufficiently smooth vector function $\mathbf{u}^*$ that satisfies the condition $\mathbf{u}^*|_{\Sigma_1} = \mathbf{u}^0$. Then we choose some set of basis elements $\boldsymbol{\varphi}_k$ that vanish on Σ_1. Clearly, any $\mathbf{u}_N$ satisfies the kinematic condition $\mathbf{u}_N|_{\Sigma_1} = \mathbf{u}^0$.

In Ritz's time, when calculations where done manually, the selection of the basis was extremely important; an engineer needed to find a few coefficients u_k in a reasonable time. Now, with powerful computers, the number of basis elements can be large. One requirement for the basis elements is that they constitute a linearly independent set. Another is that using this basis, we can actually approximate a solution with satisfactory precision. We shall not pursue this issue, as it requires a serious excursion into mathematics. However, we note that extensions of Ritz's method like the finite element method, where the number of basis elements can now be very large, are basic tools for complicated engineering projects.

Let us formulate *Ritz's minimization problem for the Nth approximation*:

Minimize $\mathcal{E}$ over the set of all $\mathbf{u}_N$ satisfying (4.41).

That is, find

$$\tilde{\mathbf{u}}_N = \sum_{k=1}^{N} \tilde{u}_k \boldsymbol{\varphi}_k + \mathbf{u}^*$$

such that

$$\mathcal{E}(\tilde{\mathbf{u}}_N) \le \mathcal{E}(\mathbf{u}_N) \text{ for any } \mathbf{u}_N \text{ from the set (4.41)},$$

where

$$\mathcal{E}(\mathbf{u}_N) = \int_V W(\boldsymbol{\varepsilon}(\mathbf{u}_N))\,dV - \int_V \rho\mathbf{f}\cdot\mathbf{u}_N\,dV - \int_{\Sigma_2} \mathbf{t}^0\cdot\mathbf{u}_N\,d\Sigma.$$

Because $\mathbf{u}^*$ is fixed, we should minimize $\mathcal{E}(\mathbf{u}_N)$ on the N-dimensional linear space spanned by the elements $\boldsymbol{\varphi}_1,\ldots,\boldsymbol{\varphi}_N$. Hence we must find the values of the real coefficients $\tilde{u}_1,\ldots,\tilde{u}_N$. From the minimum theorem on the whole set of admissible displacements, it follows that $\mathcal{E}(\mathbf{u}_N)$ has a minimum point in the spanned space.

Now we need a practical way of finding the coefficients of the minimizer of $\mathcal{E}$. Let us write out the expression for $\mathcal{E}(\mathbf{u}_N)$ in detail. We introduce the following notation:

$$\langle\mathbf{u},\mathbf{v}\rangle = \int_V \boldsymbol{\varepsilon}(\mathbf{u})\cdot\cdot\,\mathbf{C}\cdot\cdot\,\boldsymbol{\varepsilon}(\mathbf{v})\,dV.$$

Note that $\langle\mathbf{u},\mathbf{v}\rangle = \langle\mathbf{v},\mathbf{u}\rangle$, and that $\langle\mathbf{u},\mathbf{u}\rangle$ is twice the strain energy for the displacement field $\mathbf{u}$. This form $\langle\mathbf{u},\mathbf{v}\rangle$ can be considered as an inner product on any linear space of sufficiently smooth $\mathbf{u}$ such that $\mathbf{u}|_{\Sigma_1} = \mathbf{0}$, for which the expression $\langle\mathbf{u},\mathbf{v}\rangle$ makes sense.

Next we introduce

$$f_m = \int_V \rho\mathbf{f}\cdot\boldsymbol{\varphi}_m\,dV + \int_{\Sigma_2} \mathbf{t}^0\cdot\boldsymbol{\varphi}_m\,d\Sigma - \langle\mathbf{u}^*,\boldsymbol{\varphi}_m\rangle$$

and

$$F = \int_V \rho\mathbf{f}\cdot\mathbf{u}^*\,dV + \int_{\Sigma_2} \mathbf{t}^0\cdot\mathbf{u}^*\,d\Sigma - \tfrac{1}{2}\langle\mathbf{u}^*,\mathbf{u}^*\rangle.$$

In this notation

$$\mathcal{E}(\mathbf{u}_N) = \frac{1}{2}\sum_{m=1}^{N}\sum_{n=1}^{N} u_m u_n \langle\boldsymbol{\varphi}_m,\boldsymbol{\varphi}_n\rangle - \sum_{n=1}^{N} f_n u_n - F.$$

So $\mathcal{E}(\mathbf{u}_N)$ is a quadratic function in the N variables u_n, and standard tools from calculus apply.

At the point of minimum of $\mathcal{E}(\mathbf{u}_N)$, the simultaneous equations

$$\frac{\partial\mathcal{E}(\mathbf{u}_N)}{\partial u_m} = 0 \qquad (m = 1,\ldots,N)$$

must hold. Writing these as

$$\sum_{n=1}^{N} u_n\langle\boldsymbol{\varphi}_n,\boldsymbol{\varphi}_m\rangle = f_m \qquad (m = 1,\ldots,N), \tag{4.42}$$

we have a system of linear algebraic equations. The matrix of the system is

$$\mathbf{A} = \begin{pmatrix} \langle \varphi_1, \varphi_1 \rangle & \langle \varphi_2, \varphi_1 \rangle & \langle \varphi_3, \varphi_1 \rangle & \cdots & \langle \varphi_N, \varphi_1 \rangle \\ \langle \varphi_1, \varphi_2 \rangle & \langle \varphi_2, \varphi_2 \rangle & \langle \varphi_3, \varphi_2 \rangle & \cdots & \langle \varphi_N, \varphi_2 \rangle \\ \langle \varphi_1, \varphi_3 \rangle & \langle \varphi_2, \varphi_3 \rangle & \langle \varphi_3, \varphi_3 \rangle & \cdots & \langle \varphi_N, \varphi_3 \rangle \\ \vdots & \vdots & \vdots & \ddots & \vdots \\ \langle \varphi_1, \varphi_N \rangle & \langle \varphi_2, \varphi_N \rangle & \langle \varphi_3, \varphi_N \rangle & \cdots & \langle \varphi_N, \varphi_N \rangle \end{pmatrix}.$$

Its determinant $\det(\mathbf{A})$ is Gram's determinant. In linear algebra, it is shown that for a linearly independent system of elements the Gram determinant is nonzero and conversely. But the elements $\varphi_1, \ldots, \varphi_N$ are linearly independent by assumption, hence (4.42) has a unique solution.

Now let us touch on the question of convergence for Ritz's approximations. Our use of the term "approximations" does not, by itself, answer the question whether $\mathbf{u}_N$ is really close to the solution.

We shall reformulate the problem. Let $\mathbf{u}$ be a solution to the equilibrium problem under consideration. Then the Nth approximation of Ritz's method minimizes the functional $\mathcal{E}(\mathbf{u}_N) - \mathcal{E}(\mathbf{u})$ as well. But by the formula (4.39) in the proof of Theorem 4.6, we have

$$\mathcal{E}(\mathbf{u}_N) - \mathcal{E}(\mathbf{u}) = \int_V W(\boldsymbol{\varepsilon}(\mathbf{u}_N - \mathbf{u}))\, dV.$$

Because W is a positive definite form, the minimization procedure seems to be convergent if we can approximate *any* admissible displacement $\mathbf{u}$ by $\mathbf{u}_N$. The sense in which we should approximate the displacements is given by the above limit $\mathcal{E}(\mathbf{u}_N) - \mathcal{E}(\mathbf{u}) \to 0$ as $N \to \infty$. So to get an approximation, it suffices to have the set φ_m be *complete* in the sense that for any admissible $\mathbf{v}$ such that $\mathbf{v}|_{\Sigma_1} = \mathbf{0}$ and $\varepsilon > 0$, we can find $\tilde{\mathbf{u}}_n$ such that

$$\int_V W(\boldsymbol{\varepsilon}(\tilde{\mathbf{u}}_n - \mathbf{v}))\, dV < \varepsilon.$$

In practice we can find complete systems, but proof of completeness is not easy.

Having a complete basis set of φ_m, it seems we could say the following. Because the Nth Ritz approximation is the best approximation of the solution from all the possible approximations, by completeness of the φ_m we immediately obtain convergence of the Ritz approximations $\mathbf{u}_N$ to $\mathbf{u}$ in the sense that

$$\int_V W(\boldsymbol{\varepsilon}(\mathbf{u}_N - \mathbf{u}))\, dV \to 0 \quad \text{as } N \to \infty.$$

This argument is, unfortunately, merely plausible.

We can show that for Ritz's approximations the following holds:

$$\int_V W(\varepsilon(\mathbf{u}_N) - \varepsilon(\mathbf{u}_M))\,dV \to 0 \quad \text{whenever } N, M \to \infty.$$

So $\{\mathbf{u}_N\}$ seems to be a Cauchy sequence in the integral energy sense. Unfortunately, using only classical calculus we cannot conclude that a limit element exists and determine its properties. These issues are discussed in the theory of Sobolev spaces. Engineers who use Ritz's method should understand that they work with smooth approximations that are ordinary functions, but that the limit functions and convergence questions are functional analytic issues; see, for example, [Lebedev and Cloud (2003)].

Often engineers attempt to compensate for a lack of theoretical justification by means of experiment. However, experiments are quite restrictive and cannot provide real justification in all practical cases. They can only support opinion regarding the applicability of a method.

Korn's inequality

The inequality of this subsection plays a significant role in elasticity. It states that a displacement with finite energy belongs to the space $W^{1,2}(V)$; i.e., its Cartesian components all have square-integrable first derivatives. Korn's inequality can also yield an estimate for the displacement field through the expression for the strain energy.

One form of the inequality is

$$\int_V (\mathbf{u} \cdot \mathbf{u} + \nabla\mathbf{u} \cdot\cdot \nabla\mathbf{u}^T)\,dV \le c \int_V W(\varepsilon(\mathbf{u}))\,dV. \tag{4.43}$$

Here $\mathbf{u}$ is sufficiently smooth, and the constant c depends only on V and the portion Σ_1 of the boundary over which $\mathbf{u}|_{\Sigma_1} = \mathbf{0}$. Because $W(\varepsilon(\mathbf{u}))$ is a positive definite quadratic form in ε and we are not interested in the best value of c, we can change $W(\varepsilon(\mathbf{u}))$ to $\operatorname{tr}(\varepsilon \cdot \varepsilon)$. A general proof for $\Sigma_1 \ne \Sigma$ is difficult [Ciarlet (1988)], so we establish the inequality when $\mathbf{u}|_\Sigma = \mathbf{0}$. We will prove

$$\int_V \operatorname{tr}(\nabla\mathbf{u} \cdot \nabla\mathbf{u}^T)\,dV \le 2 \int_V \operatorname{tr}(\varepsilon \cdot \varepsilon)\,dV, \tag{4.44}$$

from which (4.43) follows. In Cartesian coordinates this is

$$\int_V \sum_{i,j=1}^{3} \left(\frac{\partial u_i}{\partial x_j}\right)^2 dV \le 2 \int_V \left(\sum_{i,j=1}^{3} \varepsilon_{ij}^2\right) dV.$$

The rest of the relation follows from *Friedrich's inequality*

$$\int_V \sum_{i=1}^{3} |u_i|^2 \, dV \le c_1 \int_V \sum_{i,j=1}^{3} \left(\frac{\partial u_i}{\partial x_j} \right)^2 dV,$$

holding with a constant c_1 that is independent of u_i for any smooth u_i that vanishes on the boundary [Lebedev and Cloud (2003)].

So we prove (4.44). We have

$$\begin{aligned} \int_V \operatorname{tr}(\varepsilon \cdot \varepsilon) \, dV &= \frac{1}{2} \int_V \left[\operatorname{tr}(\nabla \mathbf{u} \cdot (\nabla \mathbf{u})^T) + \operatorname{tr}(\nabla \mathbf{u} \cdot \nabla \mathbf{u}) \right] dV \\ &= \frac{1}{2} \|\nabla \mathbf{u}\|^2_{L^2(V)} + \frac{1}{2} \int_V \operatorname{tr}(\nabla \mathbf{u} \cdot \nabla \mathbf{u}) \, dV. \end{aligned}$$

On the boundary, $\mathbf{u} = \mathbf{0}$. The Gauss–Ostrogradsky theorem yields

$$\begin{aligned} \int_V \operatorname{tr}(\nabla \mathbf{u} \cdot \nabla \mathbf{u}) \, dV &= \int_V \frac{\partial u_j}{\partial x_i} \frac{\partial u_i}{\partial x_j} \, dV = - \int_V u_j \frac{\partial^2 u_i}{\partial x_i \partial x_j} \, dV \\ &= \int_V \frac{\partial u_i}{\partial x_i} \frac{\partial u_j}{\partial x_j} \, dV = \int_V (\nabla \cdot \mathbf{u})^2 \, dV. \end{aligned}$$

Hence

$$2 \|\varepsilon\|^2 = \|\nabla \mathbf{u}\|^2 + \int_V (\nabla \cdot \mathbf{u})^2 \, dV,$$

and (4.44) follows from the positivity of the second term.

4.12 Rayleigh's Variational Principle

Lagrange's minimum total energy principle applies specifically to equilibrium problems. But for one important dynamics problem there is a variational principle based on a similar minimization idea. It is the eigenoscillation problem for an elastic body. Here we seek solutions to a dynamic homogeneous problem in displacements in the form

$$\mathbf{u} = \mathbf{u}(\mathbf{r}, t) = \mathbf{w}(\mathbf{r}) e^{i\omega t}. \tag{4.45}$$

The dynamical problem is governed by

$$\nabla \cdot \boldsymbol{\sigma} = \rho \ddot{\mathbf{u}} \text{ in } V, \qquad \mathbf{u}\big|_{\Sigma_1} = \mathbf{0}, \quad \mathbf{n} \cdot \boldsymbol{\sigma}\big|_{\Sigma_2} = \mathbf{0}, \tag{4.46}$$

where

$$\boldsymbol{\sigma} = \mathbf{C} \cdot\cdot\, \varepsilon, \qquad \varepsilon = \varepsilon(\mathbf{w}) = \tfrac{1}{2} \left(\nabla \mathbf{w} + (\nabla \mathbf{w})^T \right).$$

Substituting (4.45) into (4.46) expressed in displacements, and canceling the factor $e^{i\omega t}$, we get

$$\nabla \cdot \boldsymbol{\sigma} = -\rho\omega^2 \mathbf{w} \text{ in } V, \qquad \mathbf{w}\big|_{\Sigma_1} = \mathbf{0}, \quad \mathbf{n} \cdot \boldsymbol{\sigma}\big|_{\Sigma_2} = \mathbf{0}. \tag{4.47}$$

These equations constitute an *eigenvalue problem*: we seek positive *eigenfrequencies* ω of the elastic body, for which (4.47) has a nontrivial solution $\mathbf{w}$. The latter is termed an *eigenoscillation*.

It can be shown that the problem has only nonnegative eigenfrequencies, for which $\mathbf{w}$ has only real components, and that the set of eigenfrequencies is countable. To demonstrate that the set of eigenfrequencies is countable and that a complete set of linearly independent eigenmodes forms a basis in the space of modes having finite energy, we require techniques that fall outside our present scope. The proofs require V to be a compact volume with a sufficiently smooth boundary (e.g., without cusps).

We show that an eigenfrequency ω must be real and nonnegative. Suppose to the contrary that ω is complex and hence so is the corresponding $\mathbf{w}$. Dot-multiplying the first equation of (4.47) by $\overline{\mathbf{w}}$ and integrating over V, we get

$$\int_V (\nabla \cdot \boldsymbol{\sigma}) \cdot \overline{\mathbf{w}}\, dV = -\omega^2 \int_V \rho \mathbf{w} \cdot \overline{\mathbf{w}}\, dV.$$

Applying the Gauss–Ostrogradsky theorem to the left side, we get

$$\int_V \boldsymbol{\sigma} \cdot\cdot\, \overline{\boldsymbol{\varepsilon}}\, dV = \omega^2 \int_V \rho \mathbf{w} \cdot \overline{\mathbf{w}}\, dV. \tag{4.48}$$

Because $\boldsymbol{\sigma} \cdot\cdot\, \overline{\boldsymbol{\varepsilon}} \geq 0$ and $\rho \mathbf{w} \cdot \overline{\mathbf{w}} \geq 0$, from (4.48) it follows that ω^2 is a real nonnegative number, hence so is ω. As the eigenfrequency equation is linear with real coefficients, the real and imaginary parts of an eigensolution are eigensolutions themselves. Hence we can consider only real eigenmodes.

Engineers are mostly interested in some range of eigenfrequencies, say the lowest few, or those belonging to some finite range.

For the first and third boundary value problems, the minimum eigenfrequency is positive. Indeed, for $\omega = 0$ the problem (4.47) is described by the equilibrium equations, and by uniqueness we have only the zero solution $\mathbf{w} = \mathbf{0}$. For the second boundary value problem, where $\Sigma_2 = \Sigma$, to the value $\omega = 0$ there corresponds a nontrivial solution that represents a displacement of the body as a rigid whole:

$$\mathbf{w} = \mathbf{w}_0 + \boldsymbol{\omega}_0 \times (\mathbf{r} - \mathbf{r}_0)$$

with arbitrary but constant vectors $\mathbf{w}_0, \boldsymbol{\omega}_0$, and $\mathbf{r}_0$. Note that the set of eigenvalues of a problem does not contain an accumulation point; and so it is for $\omega = 0$ — it is not an accumulation point of the spectrum. We restrict ourselves to the case of positive ω.

By linearity of the problem, if $\mathbf{w}$ is an eigensolution, then so is $a\mathbf{w}$ for any scalar a. We typically choose a so that $a\mathbf{w}$ has unit L^2 norm. Such an eigensolution is called an *oscillation mode.*

We prove the following theorem.

Theorem 4.9. *For oscillation eigenmodes $\mathbf{w}_1$ and $\mathbf{w}_2$ corresponding to distinct eigenfrequencies ω_1 and ω_2 respectively, the relation*

$$\int_V \rho \mathbf{w}_1 \cdot \mathbf{w}_2 \, dV = 0 \tag{4.49}$$

holds. Moreover,

$$\langle \mathbf{w}_1, \mathbf{w}_2 \rangle = 0. \tag{4.50}$$

Equality (4.49) *is the* ***orthogonality relation****, and* (4.50) *is the* ***generalized orthogonality relation*** *between $\mathbf{w}_1$ and $\mathbf{w}_2$.*

Proof. The proof mimics that of Theorem 2.2. Let $\mathbf{w}_1$ and $\mathbf{w}_2$ satisfy

$$\nabla \cdot \boldsymbol{\sigma}_1 = -\rho\omega_1^2 \mathbf{w}_1 \text{ in } V, \qquad \mathbf{w}_1\big|_{\Sigma_1} = \mathbf{0}, \quad \mathbf{n} \cdot \boldsymbol{\sigma}_1\big|_{\Sigma_2} = \mathbf{0},$$

and

$$\nabla \cdot \boldsymbol{\sigma}_2 = -\rho\omega_2^2 \mathbf{w}_2 \text{ in } V, \qquad \mathbf{w}_2\big|_{\Sigma_1} = \mathbf{0}, \quad \mathbf{n} \cdot \boldsymbol{\sigma}_2\big|_{\Sigma_2} = \mathbf{0},$$

where $\boldsymbol{\sigma}_k = \mathbf{C} \cdot\cdot\, \boldsymbol{\varepsilon}_k$ and $\boldsymbol{\varepsilon}_k = \boldsymbol{\varepsilon}(\mathbf{w}_k)$. Dot-multiply the first equation in V by $\mathbf{w}_2$ and integrate over V. Applying the Gauss–Ostrogradsky theorem, we get

$$-\int_V \boldsymbol{\sigma}_1 \cdot\cdot\, \boldsymbol{\varepsilon}_2 \, dV + \omega_1^2 \int_V \rho \mathbf{w}_1 \cdot \mathbf{w}_2 \, dV = 0. \tag{4.51}$$

Similarly,

$$-\int_V \boldsymbol{\sigma}_2 \cdot\cdot\, \boldsymbol{\varepsilon}_1 \, dV + \omega_2^2 \int_V \rho \mathbf{w}_2 \cdot \mathbf{w}_1 \, dV = 0.$$

Subtracting these two equations and using $\boldsymbol{\sigma}_1 \cdot\cdot\, \boldsymbol{\varepsilon}_2 = \boldsymbol{\sigma}_2 \cdot\cdot\, \boldsymbol{\varepsilon}_1$, we get

$$(\omega_1^2 - \omega_2^2) \int_V \rho \mathbf{w}_1 \cdot \mathbf{w}_2 \, dV = 0.$$

Relation (4.49) follows from the fact that $\omega_1 \neq \omega_2$. Substituting this into the above equality, we obtain (4.50). □

Theorem 4.9 states that the oscillation eigenmodes corresponding to distinct eigenfrequencies are orthogonal in the space $L_2(V)$ with weight ρ. From (4.51) it follows that they are also orthogonal with respect to the energy inner product $\langle \cdot, \cdot \rangle$. If to some ω there correspond a few linearly independent eigenmodes, we can construct an orthonormal set of corresponding eigenmodes using the Gram–Schmidt procedure [Yosida (1980)]. This orthogonality is the basis for practical application of the following variational Rayleigh principle.

Theorem 4.10. *Eigenmodes are stationary points of the energy functional*

$$\mathcal{E}_0(\mathbf{w}) = \frac{1}{2}\int_V W(\boldsymbol{\varepsilon}(\mathbf{w}))\,dV$$

on the set of displacements satisfying the boundary conditions $\mathbf{w}|_{\Sigma_1} = \mathbf{0}$ *and subject to the constraint*

$$\frac{1}{2}\int_V \rho\mathbf{w}\cdot\mathbf{w}\,dV = 1. \tag{4.52}$$

Conversely, all the stationary points of $\mathcal{E}_0(\mathbf{w})$ *on the above set of displacements are eigenmodes of the body that correspond to its eigenfrequencies.*

Proof. Let us write out the stationarity condition for $\mathcal{E}_0(\mathbf{w})$. Mimicing the proof of Theorem 4.5, we get

$$\delta\mathcal{E}_0 = -\int_V (\nabla\cdot\boldsymbol{\sigma})\cdot\delta\mathbf{w}\,dV - \int_{\Sigma_2} \mathbf{n}\cdot\boldsymbol{\sigma}\cdot\delta\mathbf{w}\,d\Sigma = 0. \tag{4.53}$$

This should be considered on the set of displacements with the restrictions described in the theorem statement. So $\delta\mathbf{w}$ is not independent as it was in Theorem 4.5, but must satisfy the condition

$$\delta\left(\frac{1}{2}\int_V \rho\mathbf{w}\cdot\mathbf{w}\,dV - 1\right) = \int_V \rho\mathbf{w}\cdot\delta\mathbf{w}\,dV = 0.$$

In the spirit of minimization via Lagrange multipliers, to $\mathcal{E}_0(\mathbf{w})$ we add

$$-\lambda\left(\frac{1}{2}\int_V \rho\mathbf{w}\cdot\mathbf{w}\,dV - 1\right)$$

where λ is an indefinite multiplier. Let us introduce a functional

$$\tilde{\mathcal{E}}_0(\mathbf{w},\lambda) = \mathcal{E}_0(\mathbf{w}) - \lambda\left(\frac{1}{2}\int_V \rho\mathbf{w}\cdot\mathbf{w}\,dV - 1\right).$$

Now $\mathbf{w}$ is not subjected to the constraint. We will show that $\mathbf{w}$ from a stationary point $(\mathbf{w}, \lambda)$ of $\tilde{\mathcal{E}}_0$ is a stationary point of $\mathcal{E}_0(\mathbf{w})$ subject to (4.52). Indeed, $\delta\tilde{\mathcal{E}}_0 = 0$ is

$$\delta\tilde{\mathcal{E}}_0 = \delta\mathcal{E}_0 - \lambda \int_V \rho\mathbf{w} \cdot \delta\mathbf{w}\, dV - (\delta\lambda)\left(\frac{1}{2}\int_V \rho\mathbf{w} \cdot \mathbf{w}\, dV - 1\right) = 0,$$

where $\delta\mathcal{E}_0$ is given by (4.53). Because $\mathbf{w}$ and λ are independent, $\delta\tilde{\mathcal{E}}_0 = 0$ implies two simultaneous equations:

$$\delta\mathcal{E}_0 - \lambda \int_V \rho\mathbf{w} \cdot \delta\mathbf{w}\, dV = 0, \qquad \frac{1}{2}\int_V \rho\mathbf{w} \cdot \mathbf{w}\, dV - 1 = 0. \tag{4.54}$$

This shows the required property possessed by stationary points of $\tilde{\mathcal{E}}_0$.

So we consider a stationary point $\tilde{\mathcal{E}}_0(\mathbf{w}, \lambda)$ without constraints on $\mathbf{w}$. From the first equation in (4.54) we get

$$-\int_V [(\nabla \cdot \boldsymbol{\sigma}) \cdot \delta\mathbf{w} + \lambda\rho\mathbf{w} \cdot \delta\mathbf{w}]\, dV - \int_{\Sigma_2} \mathbf{n} \cdot \boldsymbol{\sigma} \cdot \delta\mathbf{w}\, d\Sigma = 0.$$

As in the derivation of the minimum total energy principle, it follows from this integral equality for arbitrary $\delta\mathbf{w}$ that

$$\nabla \cdot \boldsymbol{\sigma} = -\rho\lambda\mathbf{w}, \qquad \mathbf{n} \cdot \boldsymbol{\sigma}\big|_{\Sigma_2} = \mathbf{0}.$$

If we change λ to ω^2, this equation coincides with (4.47). This explains the meaning of the Lagrange multiplier: it is equal to the squared eigenvalue ω whose non-negativeness was proved above. Thus the stationarity condition for $\mathcal{E}_0(\mathbf{w})$ holds on eigensolutions of (4.47).

Conversely, let $\mathbf{w}$ satisfy

$$\nabla \cdot \boldsymbol{\sigma} = -\rho\omega^2\mathbf{w}$$

in V for some ω. We dot-multiply it by $\mathbf{v}$, integrate over V, and apply the Gauss–Ostrogradsky theorem to get

$$\begin{aligned} 0 &= \int_V \left[(\nabla \cdot \boldsymbol{\sigma}) \cdot \mathbf{v} + \omega^2\rho\mathbf{w} \cdot \mathbf{v}\right] dV \\ &= \int_V \left[-\boldsymbol{\sigma} \cdot\cdot\, \boldsymbol{\varepsilon}(\mathbf{v}) + \omega^2\rho\mathbf{w} \cdot \mathbf{v}\right] dV + \int_{\Sigma_2} \mathbf{n} \cdot \boldsymbol{\sigma} \cdot \mathbf{v}\, d\Sigma, \end{aligned} \tag{4.55}$$

where

$$\boldsymbol{\varepsilon}(\mathbf{v}) = \tfrac{1}{2}(\nabla\mathbf{v} + (\nabla\mathbf{v})^T).$$

For each $\mathbf{v}$ such that

$$\int_V \rho\mathbf{w} \cdot \mathbf{v}\, dV = 0,$$

denoting $\mathbf{v} = \delta\mathbf{w}$, we transform the second line of (4.55) to $-\delta\mathcal{E}_0$ from (4.53). Thus $\delta\mathcal{E}_0$, which is calculated at the eigenmode, for any $\delta\mathbf{w}$ such that $\int_V \rho\mathbf{w} \cdot \delta\mathbf{w}\, dV = 0$, vanishes and so does the first variation of $\mathcal{E}_0$ at the solution $\mathbf{w}$. □

The next formulation of Rayleigh's principle does not require us to stipulate separate integral restrictions on the set of $\mathbf{w}$.

Theorem 4.11. *On the set of admissible vector-functions satisfying* $\mathbf{w}|_{\Sigma_1} = \mathbf{0}$, *the oscillation eigenmodes are stationary points of the functional*

$$R(\mathbf{w}) = \frac{\mathcal{E}_0}{K}, \quad \textit{where} \quad K = \frac{1}{2}\int_V \rho\mathbf{w} \cdot \mathbf{w}\, dV.$$

*$R(\mathbf{w})$ is called **Rayleigh's quotient**. Conversely, a normalized stationary point of $R(\mathbf{w})$ is an eigenmode corresponding to some eigenfrequency; the value of $R(\mathbf{w})$ on an eigenmode is the squared eigenfrequency.*

Proof. The first statement follows from the previous proof. Indeed, R does not change when we multiply $\mathbf{w}$ by a constant factor. Selecting such a factor that

$$\frac{1}{2}\int_V \rho\mathbf{w} \cdot \mathbf{w}\, dV = 1,$$

we see that $R = \mathcal{E}_0$.

Now we show that for the eigenmode $\mathbf{w}$ corresponding to the eigenfrequency ω, we have

$$R(\mathbf{w}) = \omega^2. \tag{4.56}$$

We dot-multiply (4.47) by $\mathbf{w}$, integrate over V, and apply the Gauss–Ostrogradsky theorem. We get

$$\int_V \boldsymbol{\sigma} \cdot\cdot\, \boldsymbol{\varepsilon}\, dV = \omega^2 \int_V \rho\mathbf{w} \cdot \mathbf{w}\, dV,$$

from which (4.56) follows. □

Equation (4.56) is widely used in mechanics to approximate eigenfrequencies. After using eigenmode orthogonality to approximate an eigenmode by $\tilde{\mathbf{w}}$, we can approximate the eigenfrequency through the equality

$$\tilde{\omega} = \sqrt{R(\tilde{\mathbf{w}})}.$$

A finite-dimensional analogue of Rayleigh's quotient spawned a class of numerical methods for large sparse systems of linear algebraic equations. The interested reader should consult specialized literature.

4.13 Plane Waves

One significant dynamics problem that can be solved analytically is the problem of plane waves traveling through an unbounded homogeneous anisotropic medium in the absence of body forces. The solutions describe acoustic propagation far from the source. Fortunately, the investigation reduces to an algebraic eigenvalue problem.

A *plane wave* is a solution of the form

$$\mathbf{u} = U(\mathbf{k} \cdot \mathbf{r} - \omega t)\mathbf{a},$$

where $\mathbf{k}$ is a constant *wave vector*, ω is the angular frequency, $\mathbf{a}$ is a constant vector, and $U = U(x)$ is an unknown function in the single variable $x = \mathbf{k} \cdot \mathbf{r} - \omega t$. The equation

$$\mathbf{k} \cdot \mathbf{r} - \omega t = \text{const}$$

describes a plane propagating through $\mathbb{R}^3$. At any instant of time, the normal to this plane lies along $\mathbf{k}$. The propagation velocity $c = \omega/|\mathbf{k}|$ is called the *wave velocity*. A solution of this form is constant on each propagating plane. The vector $\mathbf{k}$ specifies the direction of wave propagation, while $\mathbf{a}$ is the direction of the displacement. We have

$$\nabla\mathbf{u} = U'(\mathbf{k}\cdot\mathbf{r}-\omega t)\mathbf{ka}, \quad \boldsymbol{\varepsilon} = \tfrac{1}{2}U'(\mathbf{k}\cdot\mathbf{r}-\omega t)(\mathbf{ka}+\mathbf{ak}), \quad \ddot{\mathbf{u}} = \omega^2 U''(\mathbf{k}\cdot\mathbf{r}-\omega t)\mathbf{a},$$

where the prime denotes differentiation with respect to the argument x and the overdot differentiation with respect to t.

Using the symmetry properties of $\mathbf{C}$, we write out Hooke's law for this displacement field:

$$\boldsymbol{\sigma} = \mathbf{C} \cdot\cdot (\mathbf{ka})U'.$$

Substituting $\ddot{\mathbf{u}}$ and $\boldsymbol{\sigma}$ into the equations of motion, we get

$$\mathbf{k} \cdot \mathbf{C} \cdot\cdot (\mathbf{ka})U'' = \rho\omega^2 U''\mathbf{a}.$$

We set the trivial solution $U'' = 0$ aside and seek nontrivial solutions via the following algebraic problem:

Find a nontrivial solution of the equation

$$\mathbf{A}(\mathbf{k}) \cdot \mathbf{a} = \rho\omega^2\mathbf{a}, \quad \textit{where } \mathbf{A}(\mathbf{k}) = \mathbf{k} \cdot \mathbf{C} \cdot \mathbf{k}. \tag{4.57}$$

We call $\mathbf{A}$ *the* ***acoustic tensor****.*

We recall that $\mathbf{C}$ has some symmetry properties, and that $\boldsymbol{\varepsilon} \cdot\cdot \mathbf{C} \cdot\cdot \boldsymbol{\varepsilon}$ is positive for any $\boldsymbol{\varepsilon} \neq 0$. It is easy to verify that $\mathbf{A}(\mathbf{k})$ is a positive definite symmetric tensor. Considering $\rho\omega^2$ as an eigenvalue of $\mathbf{A}(\mathbf{k})$, we have an eigenvalue problem (4.57) with three positive eigenvalues. In other words, (4.57) implies that for any $\mathbf{k}$ there are three values of ω. A plane wave corresponds to each of these.

Let us consider the plane wave problem for an isotropic medium in more detail. Now

$$\mathbf{C} = \lambda \mathbf{E}\mathbf{E} + \mu(\mathbf{e}_k \mathbf{E} \mathbf{e}^k + \mathbf{I}),$$

where $\mathbf{E} = \mathbf{e}_k \mathbf{e}^k$ is the unit tensor and $\mathbf{I} = \mathbf{e}_k \mathbf{e}_m \mathbf{e}^k \mathbf{e}^m$. The acoustic tensor becomes

$$\mathbf{A} = \mathbf{k} \cdot \mathbf{C} \cdot \mathbf{k} = (\lambda + \mu)\mathbf{k}\mathbf{k} + \mu(\mathbf{k} \cdot \mathbf{k})\mathbf{E}. \tag{4.58}$$

We take a Cartesian frame with basis vectors $\mathbf{i}_1$, $\mathbf{i}_2$, $\tilde{\mathbf{k}} = \mathbf{k}/|\mathbf{k}|$, so $|\mathbf{i}_1| = |\mathbf{i}_2| = |\tilde{\mathbf{k}}| = 1$ and $\mathbf{i}_1 \cdot \mathbf{i}_2 = \mathbf{i}_1 \cdot \tilde{\mathbf{k}} = \mathbf{i}_2 \cdot \tilde{\mathbf{k}} = 0$. Then the matrix $\mathbf{A}(\mathbf{k})$ is diagonal with

$$\mathbf{A}(\mathbf{k}) = \begin{pmatrix} \mu|\mathbf{k}|^2 & 0 & 0 \\ 0 & \mu|\mathbf{k}|^2 & 0 \\ 0 & 0 & (2\mu + \lambda)|\mathbf{k}|^2 \end{pmatrix}.$$

The eigenvalues and eigenvectors from (4.57) are

$$\omega_1 = \sqrt{\frac{\mu}{\rho}}|\mathbf{k}| = \omega_2 \text{ for } \mathbf{a} = \mathbf{i}_1 \text{ and } \mathbf{i}_2, \qquad \omega_3 = \sqrt{\frac{\lambda + 2\mu}{\rho}}|\mathbf{k}| \text{ for } \mathbf{a} = \tilde{\mathbf{k}}.$$

The first two solutions describe *transverse waves*; their directions of displacement $\mathbf{a}$ are normal to the direction of propagation $\mathbf{k}$. This type of wave is also called a *shear* or *S wave*. The third solution describes a *longitudinal wave* with displacement along the direction of propagation (Fig. 4.7). This type of wave is also called a *dilatational*, *pressure*, or *P wave*.

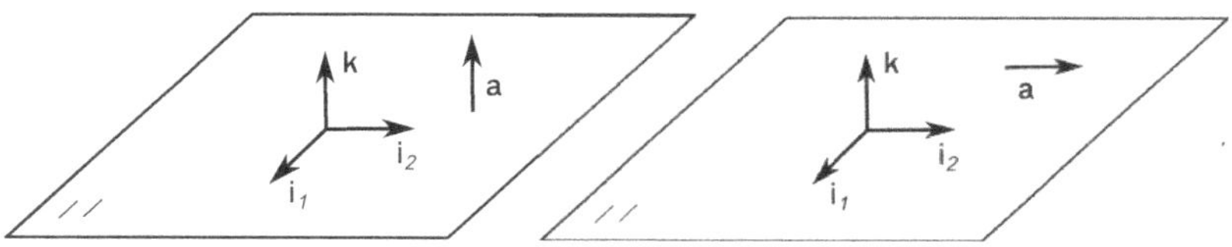

Fig. 4.7 Plane waves in an isotropic medium. Longitudinal wave is on the left. Transverse wave is on the right.

For an anisotropic medium, the spectral problem may yield waves that are neither transverse nor longitudinal.

Exercise 4.11. Verify (4.58).

Exercise 4.12. Prove that $\mathbf{A}(\mathbf{k})$ is a symmetric tensor.

Exercise 4.13. Show that $\mathbf{A}(\mathbf{k})$ is positive definite.

4.14 Plane Problems of Elasticity

Two important classes of deformation for which elastic problems strongly simplify are the *plane strain* and *plane stress* deformations. The corresponding problems are called the *plane problems of elasticity.*

First we consider the plane deformation problem, in which the displacements of all body points run parallel to a plane and depend only on the coordinates in this plane. Here the displacement vector takes the form

$$\mathbf{u} = u_1(x_1, x_2)\mathbf{i}_1 + u_2(x_1, x_2)\mathbf{i}_2,$$

and the body and contact forces should be similar:

$$\mathbf{f} = f_1(x_1, x_2)\mathbf{i}_1 + f_2(x_1, x_2)\mathbf{i}_2, \qquad \mathbf{t}^0 = t_1(x_1, x_2)\mathbf{i}_1 + t_2(x_1, x_2)\mathbf{i}_2.$$

In a Cartesian frame, the strain tensor components with indices 3, corresponding to $\mathbf{u}$, are zero: $\varepsilon_{13} = \varepsilon_{23} = \varepsilon_{33} = 0$, so

$$\boldsymbol{\varepsilon} = \varepsilon_{11}\mathbf{i}_1\mathbf{i}_1 + \varepsilon_{12}(\mathbf{i}_1\mathbf{i}_2 + \mathbf{i}_2\mathbf{i}_1) + \varepsilon_{22}\mathbf{i}_2\mathbf{i}_2.$$

Hooke's law yields the components of the stress tensor:

$$\begin{aligned} &\sigma_{11} = \lambda \operatorname{tr}\boldsymbol{\varepsilon} + 2\mu\varepsilon_{11}, &&\sigma_{22} = \lambda \operatorname{tr}\boldsymbol{\varepsilon} + 2\mu\varepsilon_{22}, &&\sigma_{33} = \lambda \operatorname{tr}\boldsymbol{\varepsilon},\\ &\sigma_{12} = \mu\varepsilon_{12}, &&\sigma_{23} = 0, &&\sigma_{13} = 0, \end{aligned}$$

where $\operatorname{tr}\boldsymbol{\varepsilon} = \varepsilon_{11} + \varepsilon_{22}$. Because $\varepsilon_{33} = 0$, we can express σ_{33} in terms of σ_{11} and σ_{22}:

$$\sigma_{33} = \frac{\lambda}{2(\lambda + \mu)}(\sigma_{11} + \sigma_{22}) = \nu(\sigma_{11} + \sigma_{22}).$$

As no component of $\boldsymbol{\sigma}$ depends on x_3, we can simplify the equilibrium equations:

$$\frac{\partial \sigma_{11}}{\partial x_1} + \frac{\partial \sigma_{21}}{\partial x_2} + \rho f_1 = 0, \qquad \frac{\partial \sigma_{12}}{\partial x_1} + \frac{\partial \sigma_{22}}{\partial x_2} + \rho f_2 = 0.$$

Consider the case when body forces are absent. Introducing the *Airy stress function* Φ via the relations

$$\sigma_{11} = \frac{\partial^2 \Phi}{\partial x_2^2}, \qquad \sigma_{22} = \frac{\partial^2 \Phi}{\partial x_1^2}, \qquad \sigma_{12} = -\frac{\partial^2 \Phi}{\partial x_1 \partial x_2}, \tag{4.59}$$

we identically satisfy the equilibrium equations. From the set of Beltrami–Michell equations, in the plane problem there remain only three nontrivial equations:

$$\begin{aligned}
&\Delta\sigma_{11} + \frac{1}{1+\nu}\frac{\partial^2\sigma}{\partial x_1^2} = 0,\\
&\Delta\sigma_{22} + \frac{1}{1+\nu}\frac{\partial^2\sigma}{\partial x_2^2} = 0,\\
&\Delta\sigma_{12} + \frac{1}{1+\nu}\frac{\partial^2\sigma}{\partial x_1 \partial x_2} = 0,
\end{aligned} \tag{4.60}$$

where

$$\sigma = \operatorname{tr}\boldsymbol{\sigma} = (1+\nu)(\sigma_{11} + \sigma_{22}).$$

Substitute (4.59) into (4.60). Since

$$\sigma = (1+\nu)\Delta\Phi,$$

we reduce the first equation of (4.60) to the biharmonic equation for Φ:

$$0 = \Delta\sigma_{11} + \frac{1}{1+\nu}\frac{\partial^2\sigma}{\partial x_1^2} = \Delta\frac{\partial^2\Phi}{\partial x_2^2} + \frac{\partial^2}{\partial x_1^2}\Delta\Phi = \Delta^2\Phi.$$

We can reduce the second equation from (4.60) to the same biharmonic equation, whereas the third equation holds identically. Hence

$$\Delta^2\Phi = 0. \tag{4.61}$$

In the plane problem, this plays the role of the compatibility equation; it is the equation we have to solve.

Similar simplifications can be done for the plane stress problem. This type of deformation occurs in a thin plate with unloaded planar faces. We choose the x_3 direction along the normal to the faces. Since the stress components $\sigma_{13}, \sigma_{23}, \sigma_{33}$ are small, we set them to zero. The equality $\sigma_{33} = 0$ distinguishes the plane strain deformation from the above plane deformation where $\sigma_{33} \neq 0$. For the plane stress deformation, the analysis of the equilibrium equations can be performed as above.

Exercise 4.14. Find the components of $\boldsymbol{\varepsilon}$ for the plane stress deformation.

Exercise 4.15. Find the components of the strain tensor for a plane stress state.

The results of Exercises 4.14 and 4.15 show that the equations of the plane strain and plain stress problems differ only in the values of the corresponding elastic moduli λ.

Special tools exist for use in plane elasticity theory. Some rely on complex variable theory [Muskhelishvili (1966)]. The complex potential method yields exact solutions to rather complex problems. Alternatively, numerical techniques based on the finite element method may be applied.

4.15 Problems

4.1 Write out the boundary conditions for the problem of plane elasticity on the square $ABCD$ shown in Fig. 4.8.

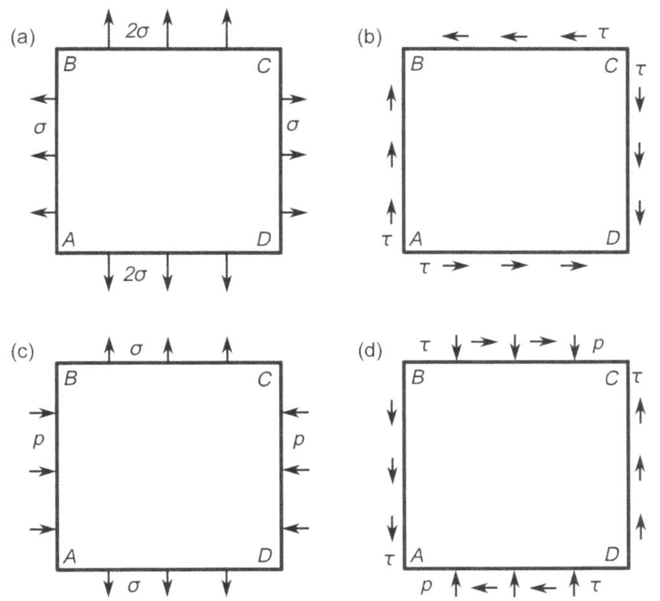

Fig. 4.8 Problem 4.1.

4.2 Write out the boundary conditions for the plane elasticity boundary value problem on the triangle ABC shown in Fig. 4.9.

4.3 Write out the boundary conditions for the plane elasticity problem on the triangle shown in Fig. 4.10. Assume the forces are normal to the sides and equal to p_1, p_2, p_3, respectively.

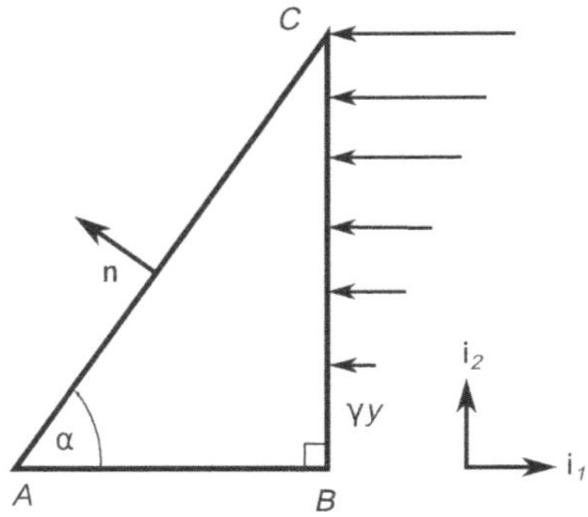

Fig. 4.9 Problem 4.2.

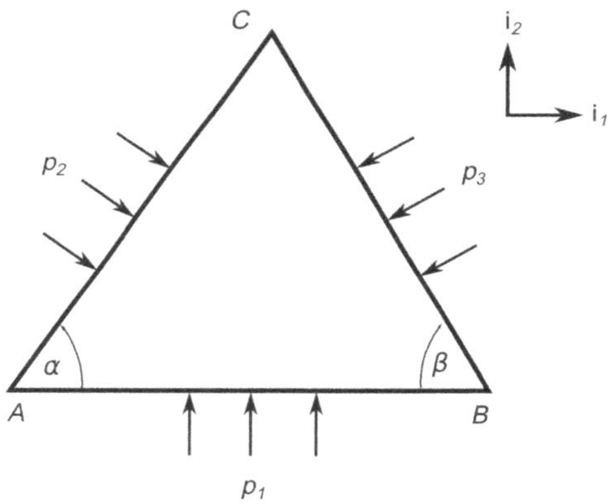

Fig. 4.10 Problem 4.3.

4.4 Write out the boundary conditions for the plane elasticity problem on the portion of a ring depicted in Fig. 4.11.

4.5 Write out the boundary conditions for the elasticity problem on the visible portions of the cubes shown in Fig. 4.12.

4.6 At a point of an elastic body the principal stresses are $\sigma_1 = 50$ MPa, $\sigma_2 = -50$ MPa, and $\sigma_3 = 75$ MPa. Referring to Fig. 4.13, find the stress vector on ABC that is equi-inclined to the principal axes.

4.7 Let γ, λ, k be given parameters. For the following displacement vectors, find the expressions for the strain tensors.

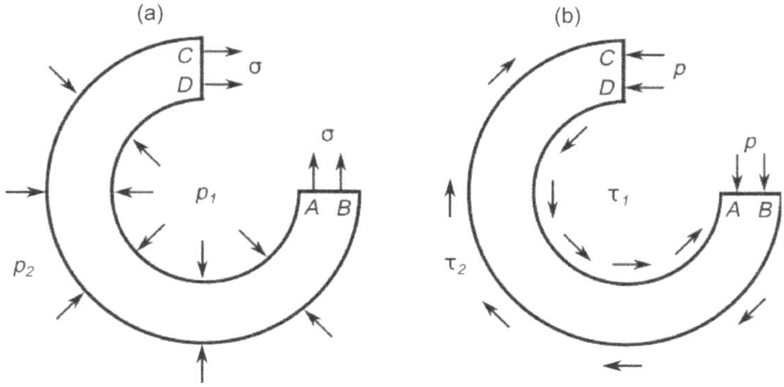

Fig. 4.11 Problem 4.4.

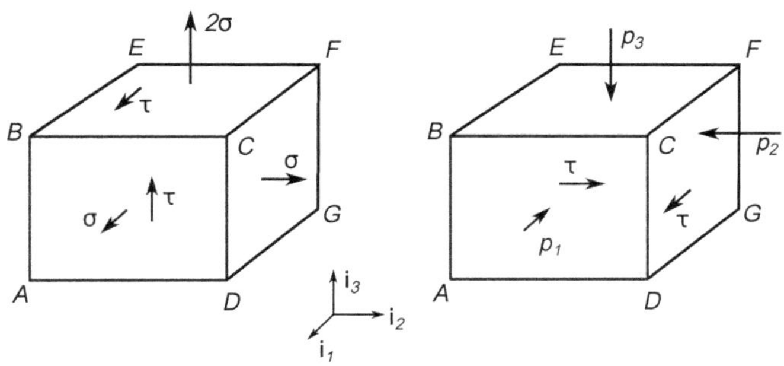

Fig. 4.12 Problem 4.5.

(a) $\mathbf{u} = \gamma x_2 \mathbf{i}_1$;
(b) $\mathbf{u} = \lambda x_1 \mathbf{i}_1$;
(c) $\mathbf{u} = \lambda \mathbf{r}$;
(d) $\mathbf{u} = u(r)\mathbf{e}_r + kz\mathbf{e}_z$ (use cylindrical coordinates);
(e) $\mathbf{u} = u(r)\mathbf{e}_\phi + kz\mathbf{e}_z$ (use cylindrical coordinates);
(f) $\mathbf{u} = (u(r) + kz)\mathbf{e}_z$ (use cylindrical coordinates);
(g) $\mathbf{u} = u(r)\mathbf{e}_r$ (use spherical coordinates).

4.8 Let $\boldsymbol{\varphi}$ be an arbitrary biharmonic function. Verify that the expression

$$2\mu\mathbf{u} = 2(1-\nu)\nabla^2\boldsymbol{\varphi} - \nabla\nabla\cdot\boldsymbol{\varphi}$$

is a solution of Lamé's equations (4.23) at $\mathbf{f} = \mathbf{0}$. It is the *Boussinesq–Galerkin* solution of the equilibrium problem.

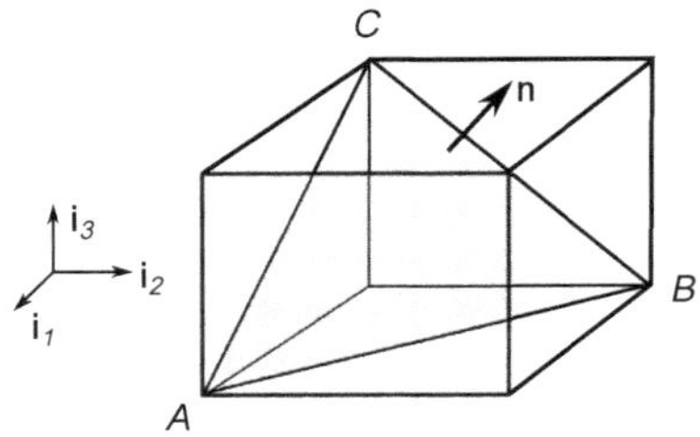

Fig. 4.13 Problem 4.6.

4.9 Let $\boldsymbol{\psi}$ be an arbitrary harmonic vector function and ψ_0 an arbitrary harmonic function. Verify that the expression

$$2\mu\mathbf{u} = 4(1-\nu)\boldsymbol{\psi} - \nabla(\mathbf{r}\cdot\boldsymbol{\psi} + \psi_0)$$

is a solution of the equilibrium equations (4.23) at $\mathbf{f} = \mathbf{0}$. It is called the *Papkovich–Neuber* solution.

4.10 Let $\boldsymbol{\psi}$ be an arbitrary harmonic vector function, and suppose the function η satisfies the equation

$$\nabla^2\eta = 2\nabla\cdot\boldsymbol{\psi}.$$

Verify that the expression

$$2\mu\mathbf{u} = 4(1-\nu)\boldsymbol{\psi} - \nabla\eta$$

is a solution to (4.23) at $\mathbf{f} = \mathbf{0}$.

4.11 Let $\boldsymbol{\psi}$ be an arbitrary harmonic vector function, and suppose the function ξ satisfies the equation

$$\frac{\partial\xi}{\partial z} = \frac{1}{4\nu-3}\nabla\cdot\boldsymbol{\psi}.$$

Verify that the expression

$$2\mu\mathbf{u} = \nabla\boldsymbol{\psi} + z\nabla\xi$$

represents a solution to equations (4.23) at $\mathbf{f} = \mathbf{0}$.

Chapter 5

Linear Elastic Shells

A *shell* is a three-dimensional solid with one thin dimension, typically capable of supporting large loads. The form of a shell is that of a surface "having" small thickness. Typical applications of shells are for hulls of ships, airplanes, cars, etc. There are very few problems for shells that admit analytical solutions. Moreover, treating a shell as a three dimensional structure presents difficulties for numerical solution, so engineers prefer to use two-dimensional models approximating three dimensional behavior of shell-like structures. They are constructed as surfaces that resist bending and stretching. There is no shell model which is applicable in all the circumstances; all the shell models are more or less approximate. However they serve well in many typical situations of engineering practice. A few popular shell models exist. We will present some linear models that are used in applications.

Historically, the first shell model was based on the Kirchhoff–Love assumptions. Models are also attributed to Timoshenko, Reissner, Cosserat, Naghdi, and others. In shell theory, we encounter the names of Euler, Lagrange, Poisson, Kelvin, etc. A particular case of shell theory is *plate theory*, where the object under consideration is akin to the top of a desk. In this chapter, some of the features of shell theory will be demonstrated only for plate models: our goal is an introduction to the theory and a demonstration of how tensor analysis is applied in its construction.

A mathematical model of a shell takes the form of a boundary value problem for simultaneous partial differential equations given on a base surface, usually the midsurface of the shell. Since the quantities describing shell behavior are functions of the surface coordinates, shell theory is closely related to the theory of surfaces from differential geometry (Chapter 5). The decisive point in any version of shell theory is the method of eliminating the third spatial coordinate — along the normal to the shell surface — from the three-dimensional model.

Before the invention of the computer, the numerical solution of three-dimensional elasticity problems was nearly impossible. Much interest was directed toward shell theories, where problems could be solved manually to required accuracy. Modern computer solution has not changed the situation for thin-walled problems. This is because numerical calculations of strains in shell-like bodies harbor difficulties related to huge mesh sizes in discrete models of a shell as a three-dimensional body and the ill-posedness of the corresponding three-dimensional problems. It is interesting to note that the application of powerful computers has increased interest in more accurate shell models which allow engineers to apply two-dimensional finite elements that incorporate a more realistic strain distribution along the shell thickness coordinate.

There are two approaches to the development of two-dimensional shell models. The first is based on direct approximation of a given three-dimensional boundary value problem for a shell-like body in terms of a boundary value problem given on the base surface of the body. This is done in a few ways. One is the *hypothesis method*, in which we assume the form of the dependence of the shell displacement and stress fields on the thickness coordinate. We will present an example of this type of model, based on the classical Kirchhoff–Love hypotheses. Another method involves asymptotic expansion of the solution of a three-dimensional problem with respect to a natural small parameter ε, the ratio of the shell thickness to a characteristic shell size such as the diameter of a circular plate. There are other principles for expanding a solution in a series with respect to ε. These models commonly use as a starting point the equilibrium equations for the shell as a three-dimensional body. Afterwards, the expansions reduce the equations to equations given on the shell surface. The models typically result in a system of equations on the base surface whose order exceeds the order of the initial system for a three-dimensional body.

Another approach to deriving shell models can be called the *direct approach*. In this case a shell is considered as a material surface having additional mechanical properties. So it is a surface that resists deformation, possesses strain energy, can have distributed mass, etc. The dynamical or equilibrium equations for such a surface are formulated directly through fundamental laws of mechanics such as the law of impulse conservation. In the direct approach, the equations of three-dimensional elasticity are not used; hence constitutive laws similar to Hooke's law should be formulated independently.

In this chapter, we will demonstrate how tensor calculus can be used to derive a few linear shell and plate models for the case of small strains. We will also discuss some properties of the boundary value problems of shell theory. We start with the Kirchhoff–Love model.

5.1 Some Equations From Surface Theory

First of all, we rewrite some classic formulas of surface theory from Chapter 3 in tensor form. We denote by $\boldsymbol{\rho}$ the position of a point P on a sufficiently smooth surface Σ in $\mathbb{R}^3$ (Fig. 5.1). We can then describe Σ by the equation $\boldsymbol{\rho} = \boldsymbol{\rho}(q^1, q^2)$ where q^1 and q^2 are the coordinates over Σ.

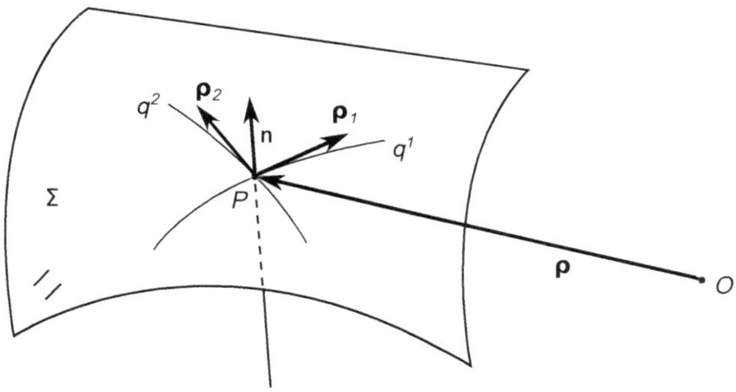

Fig. 5.1 Surface Σ near location $\boldsymbol{\rho}$.

The basis vectors $\boldsymbol{\rho}_1$ and $\boldsymbol{\rho}_2$ on Σ are tangent to the coordinate lines at P. We recall that

$$\boldsymbol{\rho}_1 = \frac{\partial \boldsymbol{\rho}}{\partial q^1}, \qquad \boldsymbol{\rho}_2 = \frac{\partial \boldsymbol{\rho}}{\partial q^2}, \qquad \boldsymbol{\rho}_\alpha \cdot \boldsymbol{\rho}^\beta = \delta_\alpha^\beta \quad (\alpha, \beta = 1, 2)$$

where $\boldsymbol{\rho}^1$ and $\boldsymbol{\rho}^2$ are the reciprocal basis vectors. In this chapter Greek indices take the values $1, 2$ while Roman indices take the values $1, 2, 3$. The unit normal at P is given by

$$\mathbf{n} = \frac{\boldsymbol{\rho}_1 \times \boldsymbol{\rho}_2}{|\boldsymbol{\rho}_1 \times \boldsymbol{\rho}_2|}.$$

Hence $\boldsymbol{\rho}_1, \boldsymbol{\rho}_2, \mathbf{n}$ constitute a right-hand basis at P in space. Mathematicians prefer to say that it is a basis of tangential space at point P of Σ, but in mechanics we consider it as a local basis at P that may change from point to point.

The metric tensor

$$\mathbf{A} = \boldsymbol{\rho}^\alpha \boldsymbol{\rho}_\alpha = \mathbf{E} - \mathbf{nn},$$

where $\mathbf{E}$ is the metric tensor of $\mathbb{R}^3$, plays the role of the unit tensor on Σ at P.

Exercise 5.1. Let a vector $\mathbf{v}$ lie in the tangent plane at P so that $\mathbf{v} \cdot \mathbf{n} = 0$. Show that $\mathbf{A} \cdot \mathbf{v} = \mathbf{v}$.

Exercise 5.2. Show that $\mathbf{A} = a_{\alpha\beta} \boldsymbol{\rho}^\alpha \boldsymbol{\rho}^\beta = a^{\alpha\beta} \boldsymbol{\rho}_\alpha \boldsymbol{\rho}_\beta$ where $a_{\alpha\beta} = \boldsymbol{\rho}_\alpha \cdot \boldsymbol{\rho}_\beta$ and $a^{\alpha\beta} = \boldsymbol{\rho}^\alpha \cdot \boldsymbol{\rho}^\beta$.

By Exercise 5.2, the components of $\mathbf{A}$ are the coefficients of the first fundamental form of Σ. So $\mathbf{A}$ is called the *first fundamental tensor* of Σ.

At point P, the vectors $(\boldsymbol{\rho}_1, \boldsymbol{\rho}_2, \mathbf{n})$ constitute a basis of $\mathbb{R}^3$ and the reciprocal basis is $(\boldsymbol{\rho}^1, \boldsymbol{\rho}^2, \mathbf{n})$. A vector field $\mathbf{v}$ given on Σ can be represented as

$$\begin{aligned}\mathbf{v} = \mathbf{v}(q^1, q^2) &= v_1(q^1, q^2)\boldsymbol{\rho}^1 + v_2(q^1, q^2)\boldsymbol{\rho}^2 + v_3(q^1, q^2)\mathbf{n} \\ &= v^1(q^1, q^2)\boldsymbol{\rho}_1 + v^2(q^1, q^2)\boldsymbol{\rho}_2 + v^3(q^1, q^2)\mathbf{n}\end{aligned}$$

where $v_3 = v^3$.

The *surface gradient* operator on Σ is

$$\widetilde{\nabla} = \boldsymbol{\rho}^\alpha \frac{\partial}{\partial q^\alpha}. \tag{5.1}$$

To apply the differential operations, we need the derivatives of the basis vectors $\boldsymbol{\rho}_1$, $\boldsymbol{\rho}_2$, $\mathbf{n}$, $\boldsymbol{\rho}^1$, $\boldsymbol{\rho}^2$. By (3.21) we have

$$\frac{\partial \mathbf{n}}{\partial q^\alpha} = -b_{\alpha\beta} \boldsymbol{\rho}^\beta.$$

The symmetric tensor

$$\mathbf{B} = b_{\alpha\beta} \boldsymbol{\rho}^\alpha \boldsymbol{\rho}^\beta.$$

is called the *curvature tensor* or *second fundamental tensor* of Σ; its components are the coefficients of the second fundamental form of Σ. By (5.1), we have

$$\mathbf{B} = -\widetilde{\nabla} \mathbf{n}.$$

Exercise 5.3. Show that $\mathbf{B} = b_{\alpha\beta} \boldsymbol{\rho}^\alpha \boldsymbol{\rho}^\beta = b^{\alpha\beta} \boldsymbol{\rho}_\alpha \boldsymbol{\rho}_\beta$ and that the expressions for $b_{\alpha\beta}$ coincide with the coefficients of the second fundamental form.

The derivatives of $\boldsymbol{\rho}_\alpha$ and $\boldsymbol{\rho}^\alpha$ are given by

$$\frac{\partial \boldsymbol{\rho}_\alpha}{\partial q^\beta} = \Gamma^\gamma_{\alpha\beta}\boldsymbol{\rho}_\gamma + b_{\alpha\beta}\mathbf{n}, \qquad \frac{\partial \boldsymbol{\rho}^\alpha}{\partial q^\beta} = -\Gamma^\alpha_{\beta\gamma}\boldsymbol{\rho}^\gamma + b^\alpha_\beta\mathbf{n}.$$

To express the gradient $\widetilde{\nabla}\mathbf{v}$ of a vector field $\mathbf{v}$ given on Σ, we split $\mathbf{v}$ into components tangent and normal to Σ:

$$\mathbf{v} = \tilde{\mathbf{v}} + w\mathbf{n}, \quad \tilde{\mathbf{v}} = v_1(q^1,q^2)\boldsymbol{\rho}^1 + v_2(q^1,q^2)\boldsymbol{\rho}^2, \quad w = v_3 = v^3 = \mathbf{v}\cdot\mathbf{n},$$

where $\tilde{\mathbf{v}}\cdot\mathbf{n} = 0$. Then

$$\widetilde{\nabla}\mathbf{v} = (\widetilde{\nabla}\tilde{\mathbf{v}})\cdot\mathbf{A} - w\mathbf{B} + (\widetilde{\nabla}w + \mathbf{B}\cdot\tilde{\mathbf{v}})\mathbf{n}. \tag{5.2}$$

Exercise 5.4. Establish (5.2) and use it to show that $\widetilde{\nabla}\cdot\mathbf{v} = \widetilde{\nabla}\cdot\tilde{\mathbf{v}} - w\,\mathrm{tr}\,\mathbf{B}$.

The "surface" divergence of tensor $\mathbf{T}$ is given by

$$\begin{aligned}
\widetilde{\nabla}\cdot\mathbf{T} &= \boldsymbol{\rho}^\gamma\frac{\partial}{\partial q^\gamma}\cdot\left(T^{\alpha\beta}\boldsymbol{\rho}_\alpha\boldsymbol{\rho}_\beta + T^{3\beta}\mathbf{n}\boldsymbol{\rho}_\beta + T^{\alpha 3}\boldsymbol{\rho}_\alpha\mathbf{n} + T^{33}\mathbf{n}\mathbf{n}\right)\\
&= \frac{\partial T^{\alpha\beta}}{\partial q^\alpha}\boldsymbol{\rho}_\beta + T^{\alpha\beta}\boldsymbol{\rho}^\gamma\cdot\frac{\partial\boldsymbol{\rho}_\alpha}{\partial q^\gamma}\boldsymbol{\rho}_\beta + T^{\alpha\beta}\frac{\partial\boldsymbol{\rho}_\beta}{\partial q^\alpha} + T^{3\beta}\boldsymbol{\rho}^\gamma\cdot\frac{\partial\mathbf{n}}{\partial q^\gamma}\boldsymbol{\rho}_\beta\\
&\quad + T^{33}\boldsymbol{\rho}^\gamma\cdot\frac{\partial\mathbf{n}}{\partial q^\gamma}\mathbf{n} + \frac{\partial T^{\alpha 3}}{\partial q^\alpha}\mathbf{n} + T^{\alpha 3}\boldsymbol{\rho}^\gamma\cdot\frac{\partial\boldsymbol{\rho}_\alpha}{\partial q^\gamma}\mathbf{n} + T^{\alpha 3}\frac{\partial\mathbf{n}}{\partial q^\alpha}\\
&= \frac{\partial T^{\alpha\beta}}{\partial q^\alpha}\boldsymbol{\rho}_\beta + T^{\alpha\beta}\Gamma^\gamma_{\alpha\gamma}\boldsymbol{\rho}_\beta + T^{\alpha\beta}\Gamma^\gamma_{\beta\alpha}\boldsymbol{\rho}_\gamma + T^{\alpha\beta}b_{\alpha\beta}\mathbf{n} - T^{3\beta}b^\gamma_\gamma\boldsymbol{\rho}_\beta\\
&\quad - T^{33}b^\gamma_\gamma\mathbf{n} + \frac{\partial T^{\alpha 3}}{\partial q^\alpha}\mathbf{n} + T^{\alpha 3}\Gamma^\gamma_{\alpha\gamma}\mathbf{n} - T^{\alpha 3}b^\beta_\alpha\boldsymbol{\rho}_\beta.
\end{aligned}$$

Exercise 5.5. Let $\mathbf{T}$, a second-order tensor given on Σ, satisfy $\mathbf{T}\cdot\mathbf{n} = \mathbf{n}\cdot\mathbf{T} = \mathbf{0}$. Show that $\widetilde{\nabla}\cdot\mathbf{T} = \mathbf{0}$ implies $T^{\alpha\beta}b_{\alpha\beta} = 0$.

5.2 Kinematics in a Neighborhood of Σ

A shell is a three-dimensional body occupying some neighborhood of a base surface Σ. A more precise description is given below. The quantities of § 5.2 serve to describe a spatial domain near Σ.

The coordinates of a point Q in a neighborhood of Σ are introduced in Fig. 5.2. Let $\mathbf{n}$ be normal to Σ and directed toward Q. The initial point P of $\mathbf{n}$ has surface coordinates q^1, q^2 and position vector $\boldsymbol{\rho}(q^1,q^2)$. The coordinates of Q are then defined as (q^1,q^2,z) where z is the distance from P to Q. The sign convention for z is that $z > 0$ when the vector $\overrightarrow{PQ}$ is codirected with $\mathbf{n}$.

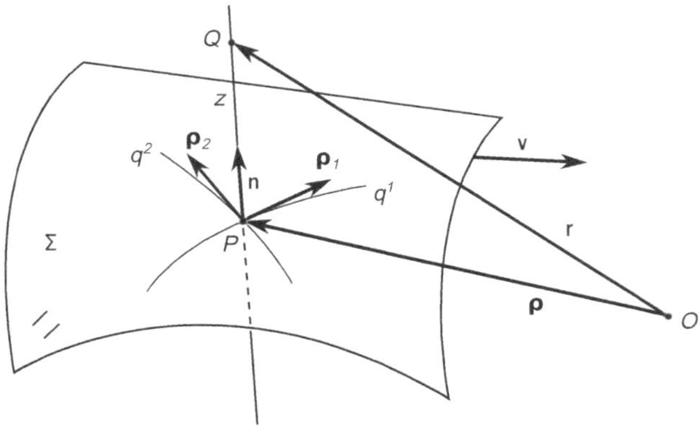

Fig. 5.2 Kinematics in a neighborhood of Σ.

The position vector of Q is

$$\mathbf{r} = \mathbf{r}(q^1, q^2, z) = \boldsymbol{\rho}(q^1, q^2) + z\mathbf{n}.$$

A shell of thickness h occupies the volume described $-h/2 \le z \le h/2$ where $h = h(q^1, q^2)$. At a point Q near Σ, the basis vectors are

$$\mathbf{r}_\alpha = \frac{\partial \mathbf{r}}{\partial q^\alpha} = \boldsymbol{\rho}_\alpha + z\frac{\partial \mathbf{n}}{\partial q^\alpha} = (\mathbf{A} - z\mathbf{B}) \cdot \boldsymbol{\rho}_\alpha, \qquad \mathbf{r}_3 = \mathbf{n}, \tag{5.3}$$

and the reciprocal basis vectors are

$$\mathbf{r}^\alpha = (\mathbf{A} - z\mathbf{B})^{-1} \cdot \boldsymbol{\rho}^\alpha, \qquad \mathbf{r}^3 = \mathbf{n}. \tag{5.4}$$

The notation $(\mathbf{A}-z\mathbf{B})^{-1}$ requires some explanation. Because $(\mathbf{A}-z\mathbf{B})\cdot\mathbf{n} = \mathbf{0}$, the tensor $\mathbf{A}-z\mathbf{B}$ is singular when regarded as an operator acting in three dimensions. Here we restrict the action of $\mathbf{A} - z\mathbf{B}$ to the two-dimensional subspace tangent to Σ. In this subspace, which has basis $\boldsymbol{\rho}^1, \boldsymbol{\rho}^2$, the tensor $\mathbf{A}$ serves as the unit operator. For $h\,\|\mathbf{B}\|$ sufficiently small, the contraction mapping theorem yields the existence of a unique inverse to $\mathbf{A} - z\mathbf{B}$ for small $|z| \le h/2$ on the tangent space. Therefore

$$(\mathbf{A} - z\mathbf{B})^{-1} \cdot (\mathbf{A} - z\mathbf{B}) = (\mathbf{A} - z\mathbf{B}) \cdot (\mathbf{A} - z\mathbf{B})^{-1} = \mathbf{A}. \tag{5.5}$$

Two examples will illustrate how $(\mathbf{A} - z\mathbf{B})^{-1}$ is obtained.

(1) For Σ a cylinder of radius R we have

$$\begin{aligned}
\mathbf{n} &= \mathbf{e}_r,\\
\mathbf{A} &= \mathbf{E} - \mathbf{n}\mathbf{n} = \mathbf{e}_\phi\mathbf{e}_\phi + \mathbf{e}_z\mathbf{e}_z,\\
\mathbf{B} &= -\widetilde{\nabla}\mathbf{e}_r = -\mathbf{e}_\phi\mathbf{e}_\phi/R,
\end{aligned}$$

so that

$$\mathbf{A} - z\mathbf{B} = \left(1 + \frac{z}{R}\right)\mathbf{e}_\phi\mathbf{e}_\phi + \mathbf{e}_z\mathbf{e}_z.$$

Equation (5.5) is satisfied if

$$(\mathbf{A} - z\mathbf{B})^{-1} = \left(1 + \frac{z}{R}\right)^{-1}\mathbf{e}_\phi\mathbf{e}_\phi + \mathbf{e}_z\mathbf{e}_z.$$

(2) For Σ a sphere of radius R we have

$$\begin{aligned}
&\mathbf{n} = \mathbf{e}_r,\\
&\mathbf{A} = \mathbf{E} - \mathbf{nn} = \mathbf{e}_\phi\mathbf{e}_\phi + \mathbf{e}_\theta\mathbf{e}_\theta,\\
&\mathbf{B} = -\widetilde{\nabla}\mathbf{e}_r = -(\mathbf{e}_\phi\mathbf{e}_\phi + \mathbf{e}_\theta\mathbf{e}_\theta)/R,
\end{aligned}$$

so that

$$\mathbf{A} - z\mathbf{B} = \left(1 + \frac{z}{R}\right)(\mathbf{e}_\phi\mathbf{e}_\phi + \mathbf{e}_\theta\mathbf{e}_\theta).$$

Equation (5.5) is satisfied if

$$(\mathbf{A} - z\mathbf{B})^{-1} = \left(1 + \frac{z}{R}\right)^{-1}(\mathbf{e}_\phi\mathbf{e}_\phi + \mathbf{e}_\theta\mathbf{e}_\theta).$$

In each example $(\mathbf{A} - z\mathbf{B})^{-1}$ exists for $|z| \le h/2$ provided that $h/2R < 1$. This is the domain in which each point is uniquely defined by the coordinate system introduced above. We shall interpret $(\mathbf{A} - z\mathbf{B})^{-1}$ in this restrictive sense.

Finally, using the reciprocal basis and (5.4), we find that the spatial nabla operator ∇ is given by

$$\begin{aligned}
\nabla &= \mathbf{r}^\alpha\frac{\partial}{\partial q^\alpha} + \mathbf{n}\frac{\partial}{\partial z} = (\mathbf{A} - z\mathbf{B})^{-1}\cdot\boldsymbol{\rho}^\alpha\frac{\partial}{\partial q^\alpha} + \mathbf{n}\frac{\partial}{\partial z}\\
&= (\mathbf{A} - z\mathbf{B})^{-1}\cdot\widetilde{\nabla} + \mathbf{n}\frac{\partial}{\partial z}.
\end{aligned}$$

5.3 Equations of Equilibrium for the Shell

The two-dimensional equilibrium equations follow directly from the three-dimensional equilibrium equations for the shell as a spatial body occupying a domain $V \subset \mathbb{R}^3$. In Fig. 5.3, V is bounded by three surfaces S_-, S_+, and S_ν. The two faces $S_\pm$ lie at distances $h/2$ from the midsurface Σ. The lateral ruled surface S_ν is formed by sweeping the normal vector $\mathbf{n}$ (to Σ) along the boundary of Σ. So the points of V are described by the position vector

$$\mathbf{r} = \boldsymbol{\rho}(q^1, q^2) + z\mathbf{n},$$

where $\boldsymbol{\rho}$ locates points on Σ and $-h/2 \le z \le h/2$. We will take h constant for simplicity. The surface Σ is often called the *midsurface* or *base surface* of the shell.

Denoting the respective normals to $S_\pm$ by $\mathbf{n}_\pm$, we have the boundary conditions

$$\mathbf{n}_+ \cdot \boldsymbol{\sigma}\Big|_{z=h/2} = \mathbf{t}^0_+, \qquad \mathbf{n}_- \cdot \boldsymbol{\sigma}\Big|_{z=-h/2} = \mathbf{t}^0_-, \tag{5.6}$$

where $\mathbf{t}^0_\pm$ are the respective surface loads on $S_\pm$. In the general case, on S_ν we will assign the boundary conditions induced by a mixed problem of elasticity.

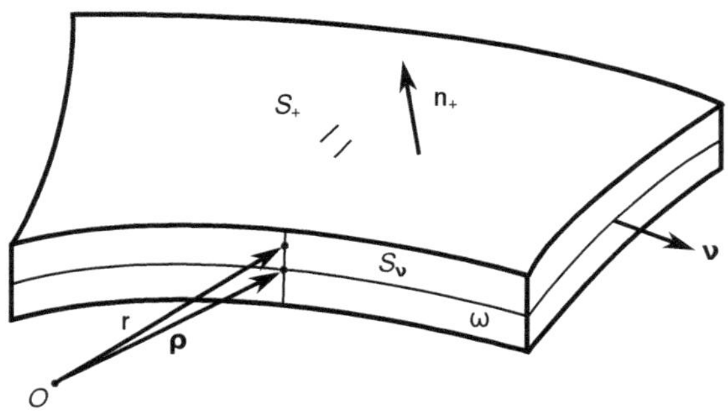

Fig. 5.3 A shell-like body.

The equilibrium of a three-dimensional elastic body occupying volume V is governed by (4.16) with $\partial^2\mathbf{u}/\partial t^2 = \mathbf{0}$:

$$\frac{1}{\sqrt{g}} \frac{\partial}{\partial q^i} \left(\sqrt{g} \sigma^{ij} \mathbf{r}_j \right) + \rho \mathbf{f} = \mathbf{0}. \tag{5.7}$$

With q^1, q^2, and $q^3 = z$ as the coordinates, we can isolate the derivative with respect to z in (5.7). Denoting

$$\boldsymbol{\sigma}^\alpha = \mathbf{r}^\alpha \cdot \boldsymbol{\sigma} = \sigma^{\alpha j} \mathbf{r}_j, \qquad \boldsymbol{\sigma}^3 = \mathbf{n} \cdot \boldsymbol{\sigma} = \sigma^{3j} \mathbf{r}_j,$$

we obtain

$$\frac{1}{\sqrt{g}} \left(\frac{\partial(\sqrt{g}\boldsymbol{\sigma}^\alpha)}{\partial q^\alpha} + \frac{\partial(\sqrt{g}\boldsymbol{\sigma}^3)}{\partial z} \right) + \rho \mathbf{f} = \mathbf{0}. \tag{5.8}$$

The quantity g, which is the determinant of the matrix $g_{ij} = \mathbf{r}_i \cdot \mathbf{r}_j$, can be represented in terms of the parameters of Σ. Since

$$g_{\alpha\beta} = \boldsymbol{\rho}_\alpha \cdot (\mathbf{A} - z\mathbf{B})^2 \cdot \boldsymbol{\rho}_\beta, \qquad g_{13} = 0, \qquad g_{23} = 0, \qquad g_{33} = 1,$$

we have

$$g = aG^2, \qquad a = a_{11}a_{22} - a_{12}^2, \qquad G = \det(\mathbf{A} - z\mathbf{B}), \tag{5.9}$$

where the $a_{\alpha\beta}$ are the values of $g_{\alpha\beta}$ taken on Σ. Here $\det(\mathbf{A} - z\mathbf{B})$ is calculated as the determinant of a two-dimensional tensor:

$$\det \mathbf{X} = X_1^1 X_2^2 - (X_1^2)^2,$$

where the X_α^β are the mixed components of $\mathbf{X} = \mathbf{A} - z\mathbf{B}$. In the new notation, (5.6) becomes

$$\boldsymbol{\sigma}^3\Big|_{z=\pm h/2} = \mathbf{t}_\pm^0.$$

Starting with (5.8) in the form

$$\frac{1}{\sqrt{a}} \frac{\partial(\sqrt{a} G \boldsymbol{\sigma}^\alpha)}{\partial q^\alpha} + \frac{\partial (G\boldsymbol{\sigma}^3)}{\partial z} + \rho G \mathbf{f} = \mathbf{0}, \tag{5.10}$$

we can obtain the two-dimensional equilibrium equations for the shell. Integrating (5.10) with respect to z from $-h/2$ to $h/2$ and using the boundary conditions, we get

$$\frac{1}{\sqrt{a}} \frac{\partial}{\partial q^\alpha} \left(\sqrt{a}\, [\![\boldsymbol{\sigma}^\alpha]\!]\right) + G_+ \mathbf{t}_+^0 - G_- \mathbf{t}_-^0 + [\![\rho \mathbf{f}]\!] = \mathbf{0} \tag{5.11}$$

where

$$G_\pm = G\Big|_{z=\pm h/2} \quad \text{and} \quad [\![f]\!] = \int_{-h/2}^{h/2} Gf\, dz.$$

Let us recast (5.11) in component-free form. The quantity

$$\mathbf{T} = \boldsymbol{\rho}_\alpha [\![\boldsymbol{\sigma}^\alpha]\!],$$

known as the *stress resultant tensor*, can be represented as

$$\begin{aligned} \mathbf{T} &= \boldsymbol{\rho}_\alpha [\![\mathbf{r}^\alpha \cdot \boldsymbol{\sigma}]\!] = \boldsymbol{\rho}_\alpha [\![\boldsymbol{\rho}^\alpha \cdot (\mathbf{A} - z\mathbf{B})^{-1} \cdot \boldsymbol{\sigma}]\!] = \boldsymbol{\rho}_\alpha \boldsymbol{\rho}^\alpha \cdot [\![(\mathbf{A} - z\mathbf{B})^{-1} \cdot \boldsymbol{\sigma}]\!] \\ &= \mathbf{A} \cdot [\![(\mathbf{A} - z\mathbf{B})^{-1} \cdot \boldsymbol{\sigma}]\!] = [\![(\mathbf{A} - z\mathbf{B})^{-1} \cdot \boldsymbol{\sigma}]\!]. \end{aligned}$$

Clearly $\mathbf{n} \cdot \mathbf{T} = \mathbf{0}$. The components of $\mathbf{T} \cdot \mathbf{A}$ are the *stress resultants* acting in the tangent plane of the shell. $\mathbf{T} \cdot \mathbf{n}$ is the vector of *transverse shear stress resultants*. Hence (5.11) is

$$\widetilde{\nabla} \cdot \mathbf{T} + \mathbf{q} = \mathbf{0}, \tag{5.12}$$

where the distributed load on the shell is

$$\mathbf{q} = G_+ \mathbf{t}_+^0 - G_- \mathbf{t}_-^0 + [\![\rho \mathbf{f}]\!].$$

The component $\mathbf{q} \cdot \mathbf{A}$ acts in the tangent plane to Σ, while $\mathbf{q} \cdot \mathbf{n}$ is the transverse load.

To derive the second equilibrium equation, we cross $z\mathbf{n}$ into (5.8) from the left:

$$\frac{1}{\sqrt{a}} z\mathbf{n} \times \frac{\partial(\sqrt{a}G\boldsymbol{\sigma}^\alpha)}{\partial q^\alpha} + z\mathbf{n} \times \frac{\partial(G\boldsymbol{\sigma}^3)}{\partial z} + \rho G z\mathbf{n} \times \mathbf{f} = \mathbf{0}. \tag{5.13}$$

Passing $\mathbf{n}$ through the derivative operators, we get

$$\frac{1}{\sqrt{a}} \frac{\partial}{\partial q^\alpha}(\mathbf{n} \times \sqrt{a}Gz\boldsymbol{\sigma}^\alpha) - \frac{\partial \mathbf{n}}{\partial q^\alpha} \times Gz\boldsymbol{\sigma}^\alpha + z\frac{\partial}{\partial z}(G\mathbf{n} \times \boldsymbol{\sigma}^3) + \rho G z\mathbf{n} \times \mathbf{f} = \mathbf{0}.$$

Since

$$\mathbf{r}_\alpha = \boldsymbol{\rho}_\alpha + z\frac{\partial \mathbf{n}}{\partial q^\alpha}$$

we have

$$z\frac{\partial \mathbf{n}}{\partial q^\alpha} = \mathbf{r}_\alpha - \boldsymbol{\rho}_\alpha$$

and hence

$$\frac{\partial \mathbf{n}}{\partial q^\alpha} \times z\boldsymbol{\sigma}^\alpha = (\mathbf{r}_\alpha - \boldsymbol{\rho}_\alpha) \times \boldsymbol{\sigma}^\alpha = -\mathbf{n} \times \boldsymbol{\sigma}^3 - \boldsymbol{\rho}_\alpha \times \boldsymbol{\sigma}^\alpha.$$

From

$$\begin{aligned}\mathbf{r}_i \times \boldsymbol{\sigma}^i &= \mathbf{r}_i \times (\mathbf{r}^i \cdot \boldsymbol{\sigma}) = \mathbf{r}_i \times (\mathbf{r}^i \cdot \sigma^{kj}\mathbf{r}_k\mathbf{r}_j) = \mathbf{r}_i \times \mathbf{r}_j\sigma^{ij} = -\mathbf{r}_j \times \mathbf{r}_i\sigma^{ij} \\ &= -\mathbf{r}_j \times \mathbf{r}_i\sigma^{ji} = -\mathbf{r}_i \times \mathbf{r}_j\sigma^{ij} = -\mathbf{r}_i \times \boldsymbol{\sigma}^i\end{aligned}$$

it follows that

$$\mathbf{r}_i \times \boldsymbol{\sigma}^i = \mathbf{0}$$

and from

$$\mathbf{r}_i \times \boldsymbol{\sigma}^i = \mathbf{r}_\alpha \times \boldsymbol{\sigma}^\alpha + \mathbf{n} \times \boldsymbol{\sigma}^3 = \mathbf{0}$$

it follows that

$$\mathbf{r}_\alpha \times \boldsymbol{\sigma}^\alpha = -\mathbf{n} \times \boldsymbol{\sigma}^3.$$

Equation (5.13) becomes

$$\frac{1}{\sqrt{a}} \frac{\partial}{\partial q^\alpha}(\mathbf{n} \times \sqrt{a}Gz\boldsymbol{\sigma}^\alpha) + \boldsymbol{\rho}_\alpha \times G\boldsymbol{\sigma}^\alpha + \frac{\partial}{\partial z}(G\mathbf{n} \times z\boldsymbol{\sigma}^3) + \rho G z\mathbf{n} \times \mathbf{f} = \mathbf{0}.$$

Integrating with respect to z from $-h/2$ to $h/2$ and using (5.6), we get the second vector equilibrium equation for the shell:

$$\frac{1}{\sqrt{a}}\frac{\partial}{\partial q^\alpha}\left(\sqrt{a}[\![\mathbf{n}\times z\boldsymbol{\sigma}^\alpha]\!]\right)+\boldsymbol{\rho}_\alpha\times[\![\boldsymbol{\sigma}^\alpha]\!]$$
$$+\frac{hG_+}{2}\mathbf{n}\times\mathbf{t}^0_+ +\frac{hG_-}{2}\mathbf{n}\times\mathbf{t}^0_- +[\![\rho z\mathbf{n}\times\mathbf{f}]\!]=\mathbf{0}. \tag{5.14}$$

Now we convert it to component-free form. We introduce the *stress couple tensor*

$$\mathbf{M}=\boldsymbol{\rho}_\alpha[\![\mathbf{n}\times z\boldsymbol{\sigma}^\alpha]\!]$$

and represent it as

$$\begin{aligned}\mathbf{M}&=-\boldsymbol{\rho}_\alpha[\![\mathbf{r}^\alpha\cdot z\boldsymbol{\sigma}\times\mathbf{n}]\!]=-\boldsymbol{\rho}_\alpha[\![\boldsymbol{\rho}^\alpha\cdot(\mathbf{A}-z\mathbf{B})^{-1}\cdot z\boldsymbol{\sigma}\times\mathbf{n}]\!]\\&=-\boldsymbol{\rho}_\alpha\boldsymbol{\rho}^\alpha\cdot[\![(\mathbf{A}-z\mathbf{B})^{-1}\cdot z\boldsymbol{\sigma}\times\mathbf{n}]\!]=-\mathbf{A}\cdot[\![(\mathbf{A}-z\mathbf{B})^{-1}\cdot z\boldsymbol{\sigma}\times\mathbf{n}]\!]\\&=-[\![(\mathbf{A}-z\mathbf{B})^{-1}\cdot z\boldsymbol{\sigma}\times\mathbf{n}]\!].\end{aligned}$$

By definition of $\mathbf{M}$, we see that $\mathbf{n}\cdot\mathbf{M}=\mathbf{M}\cdot\mathbf{n}=\mathbf{0}$. Let us denote

$$\mathbf{m}=\tfrac{1}{2}hG_+\mathbf{n}\times\mathbf{t}^0_+ +\tfrac{1}{2}hG_-\,\mathbf{n}\times\mathbf{t}^0_- +[\![\rho z\mathbf{n}\times\mathbf{f}]\!],$$

which represents the external bending couples acting on the shell surface. By its definition, $\mathbf{m}$ lies in the tangent plane of Σ so that $\mathbf{m}\cdot\mathbf{n}=0$.

In terms of the vectorial invariant of $\mathbf{T}$, which is

$$\mathbf{T}_\times=\boldsymbol{\rho}_\alpha\times[\![\boldsymbol{\sigma}^\alpha]\!],$$

we can rewrite (5.14) in component-free form:

$$\widetilde{\nabla}\cdot\mathbf{M}+\mathbf{T}_\times+\mathbf{m}=\mathbf{0}. \tag{5.15}$$

Equations (5.12) and (5.15), requiring that the resulting force and moment acting on a shell element must vanish, are the *exact* consequences of the equilibrium equations for the three-dimensional continuum of the shell. The unknowns $\mathbf{T}$ and $\mathbf{M}$ are defined on Σ.

5.4 Shell Deformation and Strains; Kirchhoff's Hypotheses

Kirchhoff's hypotheses on the laws of plate deformation extended Bernoulli's assumptions about beam bending. After their application to shell theory by A.E.H. Love, they became known as the *Kirchhoff–Love hypotheses*.

(1) (*Kirchhoff's kinematic hypothesis*) A shell cross section that is perpendicular to the midsurface before deformation remains perpendicular to the deformed midsurface. The transverse shear strains are negligibly small: $\mathbf{n} \cdot \boldsymbol{\varepsilon} \cdot \boldsymbol{\tau} = 0$ for all $\boldsymbol{\tau}$ such that $\boldsymbol{\tau} \cdot \mathbf{n} = 0$.
(2) (*Kirchhoff's static hypothesis*) The normal stress is negligible compared to the other stresses: $\sigma_{33} = \sigma^{33} = \mathbf{n} \cdot \boldsymbol{\sigma} \cdot \mathbf{n} = 0$.

These hypotheses are commonly confirmed by experiment. Shear strains and transverse compression of the shell are ignored. See, e.g., [Ciarlet (1997); Ciarlet (2000); Chróścielewski, Makowski, and Pietraszkiewicz (2004); Donnell (1976); Goldenveizer (1976); Libai and Simmonds (1998); Novozhilov, Chernykh, and Mikhailovskiy (1991); Timoshenko and Woinowsky-Krieger (1985); Vorovich (1999); Zhilin (2006); Zubov (1982)] for more thorough discussion.

Hypothesis (2) lets us eliminate $\varepsilon_{33} = \varepsilon^{33} \equiv \mathbf{n} \cdot \boldsymbol{\varepsilon} \cdot \mathbf{n}$ from Hooke's law (cf., Exercises 4.14 and 4.15 for the analogous situation in plane elasticity). By the hypothesis,

$$\sigma_{33} \equiv \mathbf{n} \cdot \boldsymbol{\sigma} \cdot \mathbf{n} = 0.$$

Let the shell material be isotropic. From

$$\sigma_{33} = \lambda \operatorname{tr} \boldsymbol{\varepsilon} + 2\mu\varepsilon_{33} = 0$$

it follows that

$$\varepsilon_{33} = -\frac{\lambda}{\lambda + 2\mu}(\varepsilon_{11} + \varepsilon_{22}).$$

The stress and strain tensors are

$$\begin{aligned} \boldsymbol{\sigma} &= \tilde{\boldsymbol{\sigma}} + \sigma_{\alpha 3}(\mathbf{r}^\alpha \mathbf{n} + \mathbf{n}\mathbf{r}^\alpha), \\ \boldsymbol{\varepsilon} &= \tilde{\boldsymbol{\varepsilon}} + \varepsilon_{\alpha 3}(\mathbf{r}^\alpha \mathbf{n} + \mathbf{n}\mathbf{r}^\alpha) - \frac{\lambda}{\lambda + 2\mu}(\varepsilon_{11} + \varepsilon_{22})\mathbf{n}\mathbf{n}, \end{aligned}$$

where $\tilde{\boldsymbol{\sigma}}$ and $\tilde{\boldsymbol{\varepsilon}}$ satisfy

$$\mathbf{n} \cdot \tilde{\boldsymbol{\sigma}} = \mathbf{0}, \qquad \tilde{\boldsymbol{\sigma}} \cdot \mathbf{n} = \mathbf{0}, \qquad \mathbf{n} \cdot \tilde{\boldsymbol{\varepsilon}} = \mathbf{0}, \qquad \tilde{\boldsymbol{\varepsilon}} \cdot \mathbf{n} = \mathbf{0}.$$

Hooke's law becomes

$$\boldsymbol{\sigma} = 2\mu \left[\boldsymbol{\varepsilon} + \frac{\lambda}{\lambda + 2\mu} \mathbf{E} \operatorname{tr} \tilde{\boldsymbol{\varepsilon}} \right] = \frac{E}{1+\nu} \left[\boldsymbol{\varepsilon} + \frac{\nu}{1-\nu} \mathbf{E} \operatorname{tr} \tilde{\boldsymbol{\varepsilon}} \right] \tag{5.16}$$

or in components

$$\sigma_1^1 = \frac{E}{1-\nu^2}(\varepsilon_1^1 + \nu\varepsilon_2^2), \qquad \sigma_2^2 = \frac{E}{1-\nu^2}(\varepsilon_2^2 + \nu\varepsilon_1^1),$$
$$\sigma_1^2 = \frac{E}{1+\nu}\varepsilon_1^2, \qquad \sigma_1^3 = \frac{E}{1+\nu}\varepsilon_1^3, \tag{5.17}$$
$$\sigma_2^3 = \frac{E}{1+\nu}\varepsilon_2^3,$$

where

$$\sigma_\alpha^\beta = \mathbf{r}_\alpha \cdot \boldsymbol{\sigma} \cdot \mathbf{r}^\beta, \qquad \sigma_\alpha^3 = \mathbf{r}_\alpha \cdot \boldsymbol{\sigma} \cdot \mathbf{n}.$$

Note that here $\sigma_\alpha^3 = \sigma_{\alpha 3}$. The components of $\boldsymbol{\varepsilon}$ are denoted similarly.

Exercise 5.6. Derive (5.16) and (5.17).

Now we consider hypothesis (1), to which can be added that the deformed cross-section remains planar. Fixing a point q^1, q^2 on Σ, we note that a cross-section through the point is planar after deformation if and only if the dependence of the displacement of the points on the normal at q^1, q^2 is linear in z. We conclude that the displacement vector has the general form

$$\mathbf{u}(q^1, q^2, z) = \tilde{\mathbf{v}}(q^1, q^2) + w(q^1, q^2)\mathbf{n} - z\boldsymbol{\vartheta}(q^1, q^2) \tag{5.18}$$

where

$$\boldsymbol{\vartheta}(q^1, q^2) = \tilde{\boldsymbol{\vartheta}} + \vartheta_n \mathbf{n}, \qquad \mathbf{n} \cdot \tilde{\boldsymbol{\vartheta}} = 0.$$

We now obtain the strain tensor $\boldsymbol{\varepsilon}$ corresponding to (5.18). We have

$$\begin{aligned}\nabla\mathbf{u} &= \left(\widetilde{\nabla} + \mathbf{n}\frac{\partial}{\partial z}\right)(\tilde{\mathbf{v}} + w\mathbf{n} - z\boldsymbol{\vartheta}) \\ &= \widetilde{\nabla}\tilde{\mathbf{v}} + (\widetilde{\nabla} w)\mathbf{n} - w\mathbf{B} - z\widetilde{\nabla}\tilde{\boldsymbol{\vartheta}} + z\vartheta_n\mathbf{B} - \mathbf{n}\boldsymbol{\vartheta}\end{aligned}$$

and

$$(\nabla\mathbf{u})^T = (\widetilde{\nabla}\tilde{\mathbf{v}})^T + \mathbf{n}\widetilde{\nabla} w - w\mathbf{B} - z(\widetilde{\nabla}\tilde{\boldsymbol{\vartheta}})^T + z\vartheta_n\mathbf{B} - \boldsymbol{\vartheta}\mathbf{n}.$$

Therefore

$$\begin{aligned}2\boldsymbol{\varepsilon} &= \nabla\mathbf{u} + \nabla\mathbf{u}^T \\ &= \widetilde{\nabla}\tilde{\mathbf{v}} + (\widetilde{\nabla}\tilde{\mathbf{v}})^T - 2w\mathbf{B} - z\left(\widetilde{\nabla}\tilde{\boldsymbol{\vartheta}} + (\widetilde{\nabla}\tilde{\boldsymbol{\vartheta}})^T\right) + 2z\vartheta_n\mathbf{B} \\ &\quad + (\widetilde{\nabla} w)\mathbf{n} - \mathbf{n}\boldsymbol{\vartheta} + \mathbf{n}\widetilde{\nabla} w - \boldsymbol{\vartheta}\mathbf{n}.\end{aligned} \tag{5.19}$$

But by (5.2),

$$\widetilde{\nabla}\tilde{\mathbf{v}} = (\widetilde{\nabla}\tilde{\mathbf{v}}) \cdot \mathbf{A} + (\mathbf{B} \cdot \tilde{\mathbf{v}})\mathbf{n}, \qquad \widetilde{\nabla}\tilde{\boldsymbol{\vartheta}} = (\widetilde{\nabla}\tilde{\boldsymbol{\vartheta}}) \cdot \mathbf{A} + (\mathbf{B} \cdot \tilde{\boldsymbol{\vartheta}})\mathbf{n}.$$

Substitution into (5.19) yields

$$\begin{aligned} 2\boldsymbol{\varepsilon} &= (\widetilde{\nabla}\tilde{\mathbf{v}}) \cdot \mathbf{A} + \mathbf{A} \cdot (\widetilde{\nabla}\tilde{\mathbf{v}})^T - 2w\mathbf{B} - z\left((\widetilde{\nabla}\tilde{\boldsymbol{\vartheta}}) \cdot \mathbf{A} + \mathbf{A} \cdot (\widetilde{\nabla}\tilde{\boldsymbol{\vartheta}})^T\right) \\ &\quad + (\mathbf{B} \cdot \tilde{\mathbf{v}})\mathbf{n} + \mathbf{n}(\mathbf{B} \cdot \tilde{\mathbf{v}}) - z(\mathbf{B} \cdot \tilde{\boldsymbol{\vartheta}})\mathbf{n} - z\mathbf{n}(\mathbf{B} \cdot \tilde{\boldsymbol{\vartheta}}) + 2z\vartheta_n\mathbf{B} \\ &\quad + (\widetilde{\nabla}w)\mathbf{n} - \mathbf{n}\boldsymbol{\vartheta} + \mathbf{n}\widetilde{\nabla}w - \boldsymbol{\vartheta}\mathbf{n}. \end{aligned} \tag{5.20}$$

Let us return to hypothesis (1):

$$\mathbf{n} \cdot \boldsymbol{\varepsilon} \cdot \boldsymbol{\tau} = 0, \tag{5.21}$$

where $\boldsymbol{\tau}$ is any vector orthogonal to $\mathbf{n}$; in components this is $\varepsilon_{31} = \varepsilon_{32} = 0$. Using (5.20) we find that

$$2\mathbf{n} \cdot \boldsymbol{\varepsilon} = \mathbf{B} \cdot \tilde{\mathbf{v}} - \tilde{\boldsymbol{\vartheta}} + \widetilde{\nabla}w - z\mathbf{B} \cdot \tilde{\boldsymbol{\vartheta}} - 2\vartheta_n\mathbf{n}.$$

Each term on the right-hand side except the last one is orthogonal to $\mathbf{n}$. It follows from (5.21) that

$$\mathbf{B} \cdot \tilde{\mathbf{v}} - \tilde{\boldsymbol{\vartheta}} + \widetilde{\nabla}w - z\mathbf{B} \cdot \tilde{\boldsymbol{\vartheta}} = \mathbf{0}.$$

Hence

$$\tilde{\boldsymbol{\vartheta}} = (\mathbf{A} + z\mathbf{B})^{-1} \cdot (\widetilde{\nabla}w + \mathbf{B} \cdot \tilde{\mathbf{v}}).$$

For a sufficiently thin and smooth shell we pose the following additional assumption.

(3) (*Assumption S.*) We have $h\,\|\mathbf{B}\| \ll 1$.

This means that the terms $z\mathbf{B}$ are negligible compared to unity. Then (5.9) implies that $G = \det(\mathbf{A} - z\mathbf{B}) \approx 1$. Moreover we find that

$$\tilde{\boldsymbol{\vartheta}} = \widetilde{\nabla}w + \mathbf{B} \cdot \tilde{\mathbf{v}} \tag{5.22}$$

and therefore

$$2\mathbf{n} \cdot \boldsymbol{\varepsilon} \cdot \boldsymbol{\tau} = -z\boldsymbol{\tau} \cdot \mathbf{B} \cdot \tilde{\boldsymbol{\vartheta}}.$$

By hypothesis (1), the left-hand side should vanish. By Assumption S the right-hand side is of a higher order of smallness than $\tilde{\boldsymbol{\vartheta}}$. So the choice of $\tilde{\boldsymbol{\vartheta}}$ given in (5.22) assures the accuracy of hypothesis (1). Note that ϑ_n is still undefined.

Together, (5.22) and (5.20) yield

$$2\varepsilon = (\widetilde{\nabla}\tilde{\mathbf{v}})\cdot\mathbf{A} + \mathbf{A}\cdot(\widetilde{\nabla}\tilde{\mathbf{v}})^T - 2w\mathbf{B} - z\left((\widetilde{\nabla}\tilde{\boldsymbol{\vartheta}})\cdot\mathbf{A} + \mathbf{A}\cdot(\widetilde{\nabla}\tilde{\boldsymbol{\vartheta}})^T\right)$$
$$- 2\vartheta_n\mathbf{nn} + \underline{2z\vartheta_n\mathbf{B} - z(\mathbf{B}\cdot\tilde{\boldsymbol{\vartheta}})\mathbf{n} - z\mathbf{n}(\mathbf{B}\cdot\tilde{\boldsymbol{\vartheta}})}.$$

The underlined terms are discarded by Assumption S, giving

$$2\varepsilon = (\widetilde{\nabla}\tilde{\mathbf{v}})\cdot\mathbf{A} + \mathbf{A}\cdot(\widetilde{\nabla}\tilde{\mathbf{v}})^T - 2w\mathbf{B} - z\left((\widetilde{\nabla}\tilde{\boldsymbol{\vartheta}})\cdot\mathbf{A} + \mathbf{A}\cdot(\widetilde{\nabla}\tilde{\boldsymbol{\vartheta}})^T\right) - 2\vartheta_n\mathbf{nn}$$
$$= 2\tilde{\varepsilon} - 2\vartheta_n\mathbf{nn}$$

where

$$2\tilde{\varepsilon} = (\widetilde{\nabla}\tilde{\mathbf{v}})\cdot\mathbf{A} + \mathbf{A}\cdot(\widetilde{\nabla}\tilde{\mathbf{v}})^T - 2w\mathbf{B} - z\left((\widetilde{\nabla}\tilde{\boldsymbol{\vartheta}})\cdot\mathbf{A} + \mathbf{A}\cdot(\widetilde{\nabla}\tilde{\boldsymbol{\vartheta}})^T\right).$$

This can be written as

$$\tilde{\varepsilon} = \boldsymbol{\epsilon} + z\text{æ}, \tag{5.23}$$

where

$$2\boldsymbol{\epsilon} = (\widetilde{\nabla}\tilde{\mathbf{v}})\cdot\mathbf{A} + \mathbf{A}\cdot(\widetilde{\nabla}\tilde{\mathbf{v}})^T - 2w\mathbf{B},$$
$$\text{æ} = -\tfrac{1}{2}\left((\widetilde{\nabla}\tilde{\boldsymbol{\vartheta}})\cdot\mathbf{A} + \mathbf{A}\cdot(\widetilde{\nabla}\tilde{\boldsymbol{\vartheta}})^T\right).$$

The tensor $\boldsymbol{\epsilon}$, called the *tangential strain measure* or *midsurface strain tensor*, describes deformation in the tangent plane. The tensor **æ**, the *bending strain measure* or *change-of-curvature tensor*, describes deformation due to bending.

In the Kirchhoff–Love theory, ϑ_n appears in the expression for ε_{33} but not in $\boldsymbol{\epsilon}$ or **æ**. Hence it does not affect the shell deformation or stress characteristics: indeed, if we set $\vartheta_n = 0$ we get the above expressions for $\boldsymbol{\epsilon}$ and **æ**. The quantity ϑ_n describes the rotation of the shell cross section, which is tangent to the midsurface, about the normal. Physically, at least for a homogeneous shell, these rotations are small compared to the other rotations. This is related to the fact that the shell twisting stiffness is much bigger than its bending stiffness. Thus, in the Kirchhoff–Love theory, a consideration of ϑ_n and the rotations of the cross sections about the normal related to ϑ_n is impossible. In much of the literature, it is assumed that ϑ_n is zero; alternatively some analysis of its smallness, similar to the above, is proposed. See [Novozhilov, Chernykh, and Mikhailovskiy (1991); Vorovich (1999); Zhilin (2006)]. The reader can find a rigorous asymptotic proof of the smallness of ϑ_n, under certain assumptions on the loads, the type of boundary conditions, and shell homogeneity, in [Goldenveizer (1976)].

Let us turn to Hooke's law, substituting $\boldsymbol{\varepsilon}$ into (5.17). The expression for $\tilde{\boldsymbol{\sigma}}$ becomes

$$\tilde{\boldsymbol{\sigma}} = \frac{E}{1+\nu}\left[\tilde{\boldsymbol{\varepsilon}} + \frac{\nu}{1-\nu}\mathbf{A}\,\mathrm{tr}\,\tilde{\boldsymbol{\varepsilon}}\right]. \tag{5.24}$$

In component form this is

$$\sigma_\alpha^\beta = \frac{E}{1+\nu}\left[\varepsilon_\alpha^\beta + \frac{\nu}{1-\nu}\delta_\alpha^\beta\,\mathrm{tr}\,\tilde{\boldsymbol{\varepsilon}}\right] \tag{5.25}$$

so that

$$\sigma_1^1 = \frac{E}{1-\nu^2}(\epsilon_1^1 + \nu\epsilon_2^2) + \frac{Ez}{1-\nu^2}(\text{æ}_1^1 + \nu\text{æ}_2^2),$$
$$\sigma_2^2 = \frac{E}{1-\nu^2}(\epsilon_2^2 + \nu\epsilon_1^1) + \frac{Ez}{1-\nu^2}(\text{æ}_2^2 + \nu\text{æ}_1^1),$$
$$\sigma_1^2 = \frac{E}{1+\nu}\epsilon_1^2 + \frac{Ez}{1+\nu}\text{æ}_1^2.$$

These yield the resultants $\mathbf{T}\cdot\mathbf{A}$ and $\mathbf{M}$ via the formulas

$$\mathbf{T} = [\![(\mathbf{A} - z\mathbf{B})^{-1}\cdot\boldsymbol{\sigma}]\!], \qquad \mathbf{M} = -[\![(\mathbf{A} - z\mathbf{B})^{-1}\cdot z\boldsymbol{\sigma}\times\mathbf{n}]\!].$$

In the context of Kirchhoff's theory, knowing the deformation, we cannot recover the transverse shear stress resultants $\mathbf{T}\cdot\mathbf{n}$; indeed by Kirchhoff's hypothesis, the shear strains ${\varepsilon_1}^3$ and ${\varepsilon_2}^3$ vanish. But the vector $\mathbf{T}\cdot\mathbf{n}$, which represents the shear stress components, is not zero. Until now it has been left indefinite, but $\mathbf{T}\cdot\mathbf{n}$ can be found via (5.12) and (5.15).

Assumption S allows us to overcome the difficulties of integrating $(\mathbf{A} - z\mathbf{B})^{-1}$ over the thickness. This simplifies the formulas for the stress resultants and stress couples to

$$\mathbf{T}\cdot\mathbf{A} = [\![\mathbf{A}\cdot\tilde{\boldsymbol{\sigma}}]\!], \qquad \mathbf{M} = -[\![\mathbf{A}\cdot z\tilde{\boldsymbol{\sigma}}\times\mathbf{n}]\!]. \tag{5.26}$$

The simplification (5.26) implies that the two-dimensional equilibrium equations (5.12) and (5.15) with these versions of $\mathbf{T}$ and $\mathbf{M}$ are not exact consequences of the three-dimensional equilibrium equations as earlier. Rather, they are only first approximations to the equilibrium equations that differ in terms of the second order of smallness. Integrating with respect to z over the thickness, we get

$$\mathbf{T}\cdot\mathbf{A} = \frac{Eh}{1+\nu}\left[\boldsymbol{\epsilon} + \frac{\nu}{1-\nu}\mathbf{A}\,\mathrm{tr}\,\boldsymbol{\epsilon}\right] \tag{5.27}$$

and

$$\mathbf{M} = -\frac{Eh^3}{12(1+\nu)}\left[\boldsymbol{\text{æ}} + \frac{\nu}{1-\nu}\mathbf{A}\,\mathrm{tr}\,\boldsymbol{\text{æ}}\right]\times\mathbf{n}. \tag{5.28}$$

In the basis $\boldsymbol{\rho}^\alpha$, we have

$$\mathbf{T} \cdot \mathbf{A} = T_{\alpha\beta}\boldsymbol{\rho}^\alpha\boldsymbol{\rho}^\beta \qquad \text{and} \qquad \boldsymbol{\epsilon} = \epsilon_{\alpha\beta}\boldsymbol{\rho}^\alpha\boldsymbol{\rho}^\beta.$$

Then the component representation of the stress resultant tensor is

$$T_{\alpha\beta} = \frac{Eh}{1+\nu}\left[\epsilon_{\alpha\beta} + \frac{\nu}{1-\nu}a_{\alpha\beta}\epsilon^\gamma_\gamma\right]. \tag{5.29}$$

Comparing this with (5.25), we see that

$$T_{\alpha\beta} = [\![\sigma_{\alpha\beta}]\!] \equiv \int_{-h/2}^{h/2} \sigma_{\alpha\beta}\, dz. \tag{5.30}$$

Here we have used the fact that, under Assumption S, we can take $G = 1$; hence we redefined

$$[\![f]\!] \equiv \int_{-h/2}^{h/2} f(z)\, dz.$$

Similarly, we find the components of the stress couple tensor:

$$M_{\alpha\beta} = [\![z\sigma_{\alpha\beta}]\!] \equiv \int_{-h/2}^{h/2} z\sigma_{\alpha\beta}\, dz. \tag{5.31}$$

These are related to the components of æ through the formulas

$$M_{\alpha\beta} = \frac{Eh}{1+\nu}\left[æ_{\alpha\beta} + \frac{\nu}{1-\nu}a_{\alpha\beta}æ^\gamma_\gamma\right]. \tag{5.32}$$

Comparing (5.30) and (5.31) with (5.26), we see that

$$\mathbf{T} \cdot \mathbf{A} = T_{\alpha\beta}\boldsymbol{\rho}^\alpha\boldsymbol{\rho}^\beta, \qquad \mathbf{M} = -M_{\alpha\beta}\boldsymbol{\rho}^\alpha\boldsymbol{\rho}^\beta \times \mathbf{n}.$$

These expressions for the components of stress resultants and stress couple tensors were used in [Koiter (1970)], [Donnell (1976)], [Libai and Simmonds (1998)], and many others.

Donnell analyzed the conditions under which a particular class of shallow shells that obey Assumption S can be used. The *theory of shallow shells* is usable for moderate deformations where the shell shape remains close to planar. For example, this is the case when the midsurface is a spherical cap which is a relatively small part of the complete sphere. This theory is attributed to Donnell, Marguerre, and Vlasov. In writing out æ, they neglect $\mathbf{B} \cdot \tilde{\mathbf{v}}$ in the expression for $\boldsymbol{\vartheta} = \widetilde{\nabla} w$ as a small quantity of higher order. Thus, the strains in shallow shell theory are given by

$$2\boldsymbol{\epsilon} = (\widetilde{\nabla}\tilde{\mathbf{v}}) \cdot \mathbf{A} + \mathbf{A} \cdot (\widetilde{\nabla}\tilde{\mathbf{v}})^T - 2w\mathbf{B}, \qquad æ = -\widetilde{\nabla}\widetilde{\nabla} w.$$

This simplification for æ was first introduced by the Russian engineer and mechanicist Vasilij Z. Vlasov [Vlasov (1949)].

In shallow shell theory, in the expressions for the stress resultants and stress couples, we use a "strange" plane geometry; we formally set the curvature components **B** to zero in æ but retain them in $\boldsymbol{\epsilon}$. The differential operations in this theory are produced in the curvilinear coordinate system. Note that Vlasov's initial equations were derived when the geometry of a shell was taken to be plane, but some nonzero curvatures were artificially included in $\boldsymbol{\epsilon}$. A particular case of shallow shell theory is the plate theory considered in § 5.8. It may be appropriate to note that the shallow shell theory has a physical defect: because of simplification of the formulas for the change of curvatures, the equations are not invariant with respect to small rotations of the shell as an entire body. So applying the theory to the problem for a shell free from geometrical constraints is not recommended.

It is worth mentioning some aspects of the application of tensor analysis to shell theory. In the literature, one can find many versions of linear shell theory using various assumptions on the smallness of certain quantities. Some of these were derived in special coordinate systems like the ones constituted by the principal curvature lines. In a certain sense, for an isotropic material, such equations may lose their isotropic properties when we express them in arbitrary curvilinear coordinates. Such a change of the isotropic property was noted by [Zhilin (2006)] for an early version of the shell equations by Novozhilov. Tensor analysis allows us to avoid similar accidents.

5.5 Shell Energy

In the theory of elasticity, the strain and total energy functionals are used to prove uniqueness of solution for boundary value problems and to formulate variational principles. These functions also play important roles in shell theory. As Kirchhoff's hypotheses affect the form of the stresses and strains in a shell as a three-dimensional body, we should derive the form of the strain energy density for the shell.

The strain energy density W of the shell as a three-dimensional body takes the form

$$W = \tfrac{1}{2}\boldsymbol{\sigma} \cdot\cdot\, \boldsymbol{\varepsilon} = \tfrac{1}{2}\sigma^i_j \varepsilon^j_i = \tfrac{1}{2}\sigma^\alpha_\beta \varepsilon^\beta_\alpha.$$

This follows from Kirchhoff's hypotheses that $\varepsilon^1_3 = 0$, $\varepsilon^2_3 = 0$, and $\sigma^3_3 = 0$.

Substituting (5.25) into W we get

$$W = \frac{E}{2(1+\nu)}\left[\tilde{\varepsilon}\cdot\cdot\,\tilde{\varepsilon} + \frac{\nu}{1-\nu}\operatorname{tr}^2\tilde{\varepsilon}\right] = \frac{E}{2(1+\nu)}\left[\varepsilon_\alpha^\beta\varepsilon_\beta^\alpha + \frac{\nu}{1-\nu}(\varepsilon_\alpha^\alpha)^2\right]$$

or, by (5.23),

$$\begin{aligned} W &= \frac{E}{2(1+\nu)}\left[\boldsymbol{\epsilon}\cdot\cdot\,\boldsymbol{\epsilon} + \frac{\nu}{1-\nu}\operatorname{tr}^2\boldsymbol{\epsilon}\right] + \frac{Ez^2}{2(1+\nu)}\left[\text{æ}\cdot\cdot\,\text{æ} + \frac{\nu}{1-\nu}\operatorname{tr}^2\text{æ}\right] \\ &\quad + \frac{Ez}{1+\nu}\left[\boldsymbol{\epsilon}\cdot\cdot\,\text{æ} + \frac{\nu}{1-\nu}\operatorname{tr}\boldsymbol{\epsilon}\operatorname{tr}\text{æ}\right]. \end{aligned} \tag{5.33}$$

So Kirchhoff's hypotheses yield an expression for W in terms of $\boldsymbol{\epsilon}$, æ, and z.

Consider the strain energy functional

$$\mathcal{E} = \int_V W\,dV.$$

We recall that in curvilinear coordinates q^1, q^2, z the volume element is

$$dV = \sqrt{g}\,dq^1\,dq^2\,dz = \sqrt{a}G\,dq^1\,dq^2\,dz, \qquad G = \det(\mathbf{A} - z\mathbf{B}).$$

Under our assumptions on smallness of the displacements, W is a quantity of the second order of smallness because it is a quadratic form in the strains. By Assumption S we can neglect $z\mathbf{B}$ compared to $\mathbf{A}$ and use $G = 1$ since $\det\mathbf{A} = 1$. Integrating (5.33) over z explicitly, we get

$$\mathcal{E} = \int_V W\sqrt{a}\,dq^1\,dq^2\,dz = \int_\Sigma U\sqrt{a}\,dq^1\,dq^2$$

where $U = [\![W]\!]$ denotes the strain energy density per unit area:

$$U = \frac{Eh}{2(1+\nu)}\left[\boldsymbol{\epsilon}\cdot\cdot\,\boldsymbol{\epsilon} + \frac{\nu}{1-\nu}\operatorname{tr}^2\boldsymbol{\epsilon}\right] + \frac{Eh^3}{24(1+\nu)}\left[\text{æ}\cdot\cdot\,\text{æ} + \frac{\nu}{1-\nu}\operatorname{tr}^2\text{æ}\right]. \tag{5.34}$$

U is a quadratic form in two tensors, $\boldsymbol{\epsilon}$ and æ, defined on Σ. It splits into two terms, the first depending only on $\boldsymbol{\epsilon}$ and second only on æ. Physically, this decomposition splits the strain energy density into two parts: one due to tangential deformation (shell deformations in the tangent plane) and one due to bending.

From (5.34) it follows that $U \geq 0$, with $U = 0$ if and only if $\varepsilon = \mathbf{0} =$ æ. Indeed, we recall that $-1 < \nu \leq 1/2$ for an elastic material. Then

$$\boldsymbol{\epsilon}\cdot\cdot\,\boldsymbol{\epsilon} + \frac{\nu}{1-\nu}\operatorname{tr}^2\boldsymbol{\epsilon} > 0 \quad \text{if} \quad \boldsymbol{\epsilon} \neq \mathbf{0}. \tag{5.35}$$

Similarly

$$æ \cdot\cdot æ + \frac{\nu}{1-\nu} \operatorname{tr}^2 æ > 0 \quad \text{if} \quad æ \neq \mathbf{0}. \tag{5.36}$$

Exercise 5.7. Prove (5.35).

Expressing $\mathbf{T}$ and $\mathbf{M}$ in components, we can write

$$U = \tfrac{1}{2} T_{\alpha\beta} \epsilon^{\alpha\beta} + \tfrac{1}{2} M_{\alpha\beta} æ^{\alpha\beta}. \tag{5.37}$$

From (5.37), with use of (5.29) and (5.32), it follows that

$$T_{\alpha\beta} = \frac{\partial U}{\partial \epsilon^{\alpha\beta}}, \qquad M_{\alpha\beta} = \frac{\partial U}{\partial æ^{\alpha\beta}}. \tag{5.38}$$

In component-free form these are

$$U = \tfrac{1}{2}\mathbf{T} \cdot\cdot \boldsymbol{\epsilon} + \tfrac{1}{2}\mathbf{M} \cdot\cdot (\mathbf{n} \times æ) = \tfrac{1}{2}\mathbf{T} \cdot\cdot \boldsymbol{\epsilon} + \tfrac{1}{2}(\mathbf{M} \times \mathbf{n}) \cdot\cdot æ \tag{5.39}$$

and

$$\mathbf{T} = \frac{\partial U}{\partial \boldsymbol{\epsilon}}, \qquad \mathbf{M} \times \mathbf{n} = \frac{\partial U}{\partial æ}. \tag{5.40}$$

U inherits the positive definiteness of the energy density in three-dimensional linear elasticity, where the energy density $W = 0$ if and only if $\boldsymbol{\varepsilon} = \mathbf{0}$. We recall that $\boldsymbol{\varepsilon} = \mathbf{0}$ if and only if the displacement field is a small rigid-body displacement, i.e., a parallel translation plus a rotation of the body in space. The situation in shell theory is similar. As the expressions for $\boldsymbol{\epsilon}$ and æ in Kirchhoff's theory are cumbersome, however, we will merely illustrate this fact for the case of plate theory later on.

Exercise 5.8. Prove (5.39).

5.6 Boundary Conditions

To uniquely define a solution of the equilibrium problem for an elastic shell, we should supplement the equilibrium equations with boundary conditions. This is no elementary task: in the beginnings of shell theory, the question of boundary conditions was debated for a long time. Even the great Poisson proposed an erroneous number of boundary conditions.

The equilibrium equations are defined on the midsurface Σ, so the boundary conditions are given on its boundary contour ω.

First we will introduce the *kinematic boundary conditions*: on the boundary we assign the displacement $\mathbf{v}$ that defines the position of the boundary contour after deformation. In the Kirchhoff–Love theory we

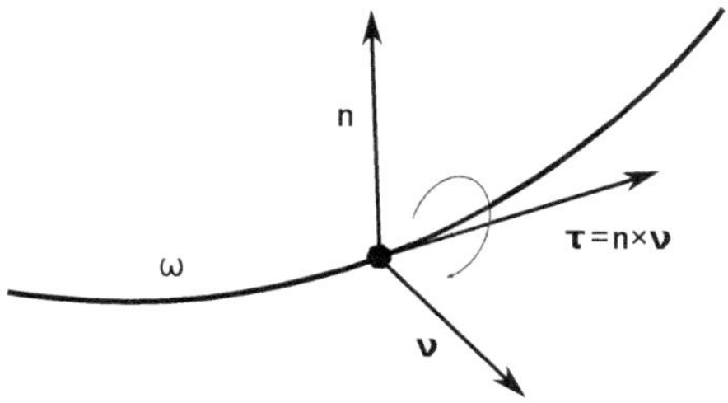

Fig. 5.4 The trihedron $\mathbf{n}, \boldsymbol{\nu}, \boldsymbol{\tau} = \mathbf{n} \times \boldsymbol{\nu}$ used to formulate the boundary conditions.

should also assign the rotation of the normal $\mathbf{n}$ about the tangential to the boundary shell contour vector — that is, about the vector $\mathbf{n} \times \boldsymbol{\nu}$ (Fig. 5.4).

The rotation is defined by the formula $\boldsymbol{\vartheta} = \widetilde{\nabla} w + \mathbf{B} \cdot \tilde{\mathbf{v}}$, so the rotation about $\boldsymbol{\tau} = \mathbf{n} \times \boldsymbol{\nu}$ is

$$\vartheta_\nu \equiv \boldsymbol{\nu} \cdot \boldsymbol{\vartheta} = \boldsymbol{\nu} \cdot \widetilde{\nabla} w + \boldsymbol{\nu} \cdot \mathbf{B} \cdot \tilde{\mathbf{v}} = \frac{\partial w}{\partial \nu} + \boldsymbol{\nu} \cdot \mathbf{B} \cdot \tilde{\mathbf{v}},$$

where $\boldsymbol{\nu} \cdot \widetilde{\nabla} w = \partial w / \partial \nu$, the derivative with respect to the normal. Thus, in the Kirchhoff–Love theory, the *kinematic boundary conditions* consist of four scalar equations:

$$\mathbf{v}\big|_\omega = \mathbf{v}^0(s), \qquad \frac{\partial w}{\partial \nu} + \boldsymbol{\nu} \cdot \mathbf{B} \cdot \tilde{\mathbf{v}}\big|_\omega = \vartheta^0(s), \tag{5.41}$$

where $\mathbf{v}^0(s)$ and $\vartheta^0(s)$ are given on ω.

Some difficulties are involved in formulating static boundary conditions in the Kirchhoff–Love theory. At first glance, it seems we could assign the quantities $\boldsymbol{\nu} \cdot \mathbf{T}$ and $\boldsymbol{\nu} \cdot (\mathbf{M} \times \mathbf{n})$ on the boundary. But this provides five, not four, scalar conditions which should arise according to the number of kinematic conditions (their number should be equal). Earlier we mentioned Poisson's error in constructing plate theory: he proposed too many static conditions on the boundary.

The derivation of the static boundary conditions can be based on the Lagrange variational principle for the shell, which states that in equilibrium the total energy functional takes its minimum value on the set of kinematically admissible displacements. The static boundary conditions become the natural boundary conditions for the problem of minimizing the total energy; they are dual to the kinematic conditions. For the shell, the proof of the Lagrange principle and the derivation of the static boundary conditions are cumbersome, so we refer to more specialized literature (e.g., [Green and Zerna (1954); Novozhilov, Chernykh, and Mikhailovskiy (1991)]) and merely sketch the procedure for deriving the boundary conditions. Later, this will be given in more detail for the technically simpler case of a plate.

A consequence of the Lagrange variational principle is the statement that the variation of the shell strain energy is equal to the work of external forces on the kinematically admissible displacements:

$$\delta\mathcal{E} = \delta A.$$

It turns out that on the set of admissible displacements of the shell edge, the quantities $\mathbf{v} = \tilde{\mathbf{v}} + w\mathbf{n}$ and ϑ_ν, which constitute four scalar quantities, can be assigned independently. Hence the work of external forces on the shell edge takes the form

$$\delta A = \int_\omega \left(\boldsymbol{\varphi}^0(s) \cdot \delta\mathbf{v} - \ell^0(s)\delta\vartheta_\nu\right) ds,$$

where $\boldsymbol{\varphi}^0(s)$ and $\ell^0(s)$ are given functions on ω. It is seen that δA depends on $\delta\mathbf{v}$ and $\delta\vartheta_\nu$ linearly. In view of (5.38) we get

$$\delta\mathcal{E} = \int_\Sigma \delta U \sqrt{a}\, dq^1\, dq^2 = \int_\Sigma (T_{\alpha\beta}\, \delta\epsilon^{\alpha\beta} + M_{\alpha\beta}\, \delta æ^{\alpha\beta})\sqrt{a}\, dq^1\, dq^2.$$

The further transformations of $\delta\mathcal{E}$ are done using integration by parts and the Gauss–Ostrogradsky theorem. We should represent $\delta\epsilon^{\alpha\beta}$ and $\delta æ^{\alpha\beta}$ through $\delta\mathbf{v}$ and $\delta\boldsymbol{\vartheta}$ and eliminate their derivatives from the surface integrals. Unlike the case with three-dimensional elasticity, $\delta æ^{\alpha\beta}$ contains the second derivatives of w and hence integration by parts should be done twice. For the boundary integrals we should also apply further transformations in such a way that they contain only the factors $\delta\mathbf{v}$ and $\delta\vartheta_\nu$; the techniques will be presented in the section for the plate. The equation $\delta\mathcal{E} = \delta A$, which also expresses the virtual work principle for the shell, holds for any admissible $\delta\mathbf{v}$ and $\delta\boldsymbol{\vartheta}$. Selecting only $\delta\mathbf{v}$ and $\delta\boldsymbol{\vartheta}$ satisfying the homogeneous kinematic conditions, we find that the minimum point $\mathbf{v}$ and $\boldsymbol{\vartheta}$ satisfies the equilibrium equations on Σ. Next, extending the set of $\delta\mathbf{v}$ and $\delta\boldsymbol{\vartheta}$ to all the admissible $\delta\mathbf{v}$ and $\delta\boldsymbol{\vartheta}$, we derive the following set of static boundary conditions for $\mathbf{v}$ and $\boldsymbol{\vartheta}$. They are defined on ω:

$$\begin{aligned} \boldsymbol{\nu} \cdot \mathbf{T} \cdot \mathbf{A}\big|_\omega &= \boldsymbol{\varphi}^0(s) \cdot \mathbf{A}, \\ \boldsymbol{\nu} \cdot \mathbf{T} \cdot \mathbf{n}\big|_\omega - \frac{\partial}{\partial s}\left(\boldsymbol{\nu} \cdot (\mathbf{M} \times \mathbf{n}) \cdot \boldsymbol{\tau}\right)\big|_\omega &= \boldsymbol{\varphi}^0(s) \cdot \mathbf{n}, \\ \boldsymbol{\nu} \cdot (\mathbf{M} \times \mathbf{n}) \cdot \boldsymbol{\nu}\big|_\omega &= \ell^0(s). \end{aligned} \tag{5.42}$$

As with the kinematic boundary conditions, these are also four scalar equations. Here $\boldsymbol{\varphi}^0(s)$ and $\ell^0(s)$ are given functions on ω that represent the stress resultants and the bending couples on the shell edge.

Some common principal boundary conditions are as follows.

(1) Clamped (fixed) edge:

$$\mathbf{v}\big|_{\omega} = \mathbf{0}, \qquad \frac{\partial w}{\partial \nu}\bigg|_{\omega} = 0.$$

(2) Simply supported edge:

$$\mathbf{v}\big|_{\omega} = \mathbf{0}, \qquad \boldsymbol{\nu} \cdot (\mathbf{M} \times \mathbf{n}) \cdot \boldsymbol{\nu}\big|_{\omega} = 0.$$

(3) Free edge:

$$\boldsymbol{\nu} \cdot \mathbf{T} \cdot \mathbf{A}\big|_{\omega} = \mathbf{0}, \qquad \boldsymbol{\nu} \cdot \mathbf{T} \cdot \mathbf{n}\big|_{\omega} - \frac{\partial}{\partial s}\left(\boldsymbol{\nu} \cdot (\mathbf{M} \times \mathbf{n}) \cdot \boldsymbol{\tau}\right)\big|_{\omega} = 0,$$

$$\boldsymbol{\nu} \cdot (\mathbf{M} \times \mathbf{n}) \cdot \boldsymbol{\nu}\big|_{\omega} = 0.$$

Mixed boundary conditions are also possible; we might, for instance specify (5.41) on part of ω and (5.42) on the rest. Other combinations of boundary conditions, four at each point of ω, define various boundary value problems. Admissible combinations of boundary conditions are defined by the variational setups of these problems.

5.7 A Few Remarks on the Kirchhoff–Love Theory

Various approaches within the Kirchhoff–Love theory use different simplifying assumptions that depend on the choice of constitutive relations for U, $\mathbf{T}$, and $\mathbf{M}$, as well as on the choice of deformation measures $\boldsymbol{\epsilon}$ and æ. In a linear theory, equation (5.34) for U is a common form of dependence of U on the deformation measures (see, e.g., [Donnell (1976); Koiter (1970); Novozhilov, Chernykh, and Mikhailovskiy (1991); Timoshenko and Woinowsky-Krieger (1985)]).

The versions may differ in their definitions of $\mathbf{T}$. We introduced $\mathbf{T}$ as

$$\mathbf{T} = [\![(\mathbf{A} - z\mathbf{B})^{-1} \cdot \boldsymbol{\sigma}]\!]$$

and used Assumption S to recast it in the form

$$\mathbf{T} = [\![\mathbf{A} \cdot \boldsymbol{\sigma}]\!].$$

In the Kirchhoff–Love literature, one can find both the initial and simplified definitions. Let us consider this difference in more detail. We introduce a special coordinate system on the shell surface, defined by the principal curvature curves on Σ. Now $\mathbf{B}$ takes the diagonal form

$$\mathbf{B} = \mathbf{e}_1\mathbf{e}_1/R_1 + \mathbf{e}_2\mathbf{e}_2/R_2,$$

where R_1 and R_2 are principal radii of curvature of Σ. Then

$$\mathbf{A} = \mathbf{e}_1\mathbf{e}_1 + \mathbf{e}_2\mathbf{e}_2$$

and

$$G = (1 - z/R_1)(1 - z/R_2).$$

The tensor $G(\mathbf{A} - z\mathbf{B})^{-1}$ takes the form

$$G(\mathbf{A} - z\mathbf{B})^{-1} = \left(1 - \frac{z}{R_2}\right)\mathbf{e}_1\mathbf{e}_1 + \left(1 - \frac{z}{R_1}\right)\mathbf{e}_2\mathbf{e}_2.$$

Taking

$$\mathbf{T} = [\![(\mathbf{A} - z\mathbf{B})^{-1} \cdot \boldsymbol{\sigma}]\!],$$

we get the following formulas for the components of $\mathbf{T}$:

$$T_{11} = \int_{-h/2}^{h/2} \left(1 - \frac{z}{R_2}\right)\sigma_{11}\,dz, \quad T_{22} = \int_{-h/2}^{h/2} \left(1 - \frac{z}{R_1}\right)\sigma_{22}\,dz,$$

$$T_{12} = T_{21} = \int_{-h/2}^{h/2} \sigma_{12}\,dz.$$

The alternative definition $\mathbf{T} = [\![\mathbf{A} \cdot \boldsymbol{\sigma}]\!]$ yields

$$T_{11} = \int_{-h/2}^{h/2} \sigma_{11}\,dz, \quad T_{22} = \int_{-h/2}^{h/2} \sigma_{22}\,dz, \quad T_{12} = T_{21} = \int_{-h/2}^{h/2} \sigma_{12}\,dz.$$

So the two definitions of $\mathbf{T}$ give equilibrium equations that differ by quantities that are second-order small. The simplified form for $\mathbf{T}$ was used by a number of authors, e.g., [Goldenveizer (1976); Novozhilov, Chernykh, and Mikhailovskiy (1991); Timoshenko and Woinowsky-Krieger (1985)]. Koiter [Koiter (1970)] used (5.34) and (5.38) to relate the stress resultants and the bending couples with U, but he presented more complex expressions to calculate them through the stresses in a three-dimensional body.

The components of $M_{\alpha\beta}$ are also commonly calculated using (5.31), and are related to U through (5.38). As is the case for $\mathbf{T}$, however, in the literature one can find other ways to introduce $\mathbf{M}$. In [Goldenveizer (1976)], for instance, the expressions for $M_{\alpha\beta}$ include terms that are quadratic in z.

Some works on shell theory feature a "mixed" formulation of the boundary value problems, introducing the stress resultants via the simplified formula $\mathbf{T} = [\![\mathbf{A} \cdot \boldsymbol{\sigma}]\!]$. The constitutive equations take the form (5.34), but the equilibrium equations include additional terms arising from the complete definition $\mathbf{T} = [\![(\mathbf{A} - z\mathbf{B})^{-1} \cdot \boldsymbol{\sigma}]\!]$. In this way, the couple stresses appear in all the equilibrium equations.

In [Koiter (1970); Novozhilov, Chernykh, and Mikhailovskiy (1991)] and certain other works, the above tensor $\boldsymbol{\epsilon}$ is also used. The second deformation measure æ, which was derived by the simplifying replacement of $\boldsymbol{\vartheta}$ by $\widetilde{\nabla} w$, also has a few different forms (see the discussion in [Koiter (1970)]).

In the linear Kirchhoff–Love shell theory, the reader can encounter various forms of the equilibrium equations in displacements, as well as differences in the stress resultants and bending couples. Clearly, in different versions of shell theory the forms of the static boundary conditions can differ, and this is also reflected in the literature.

It is worth noting that in different references, the same equilibrium equations in stresses can lead to different equilibrium equations in displacements. For a thin shell with a sufficiently smooth midsurface and smooth loads, this difference, as a rule, is of the second order of smallness in comparison with the main terms in the equations.

5.8 Plate Theory

Resultant forces and moments in a plate; the equilibrium equations

A thin plate is a special case of a shell. Since the base surface Σ of the plate is planar, $\mathbf{B} = \mathbf{0}$. The plate relations follow from shell theory. The equilibrium equations are

$$\widetilde{\nabla} \cdot \mathbf{T} + \mathbf{q} = \mathbf{0}, \qquad \widetilde{\nabla} \cdot \mathbf{M} + \mathbf{T}_{\times} + \mathbf{m} = \mathbf{0}, \tag{5.43}$$

where

$$\mathbf{T} = [\![\mathbf{A} \cdot \boldsymbol{\sigma}]\!], \qquad \mathbf{M} = -[\![\mathbf{A} \cdot z\boldsymbol{\sigma} \times \mathbf{n}]\!]. \tag{5.44}$$

In a Cartesian basis $\mathbf{i}_1$, $\mathbf{i}_2$, $\mathbf{i}_3 = \mathbf{n}$, we have

$$\boldsymbol{\sigma} = \sigma_{\alpha\beta}\mathbf{i}_\alpha\mathbf{i}_\beta + \sigma_{\alpha 3}(\mathbf{i}_\alpha\mathbf{n} + \mathbf{n}\mathbf{i}_\alpha) + \sigma_{33}\mathbf{n}\mathbf{n}, \quad \mathbf{A} = \mathbf{i}_\alpha\mathbf{i}_\alpha, \quad \widetilde{\nabla} = \mathbf{i}_\alpha \frac{\partial}{\partial x_\alpha}, \quad G = 1,$$

where x_1, x_2 are Cartesian coordinates in the plane. The stress resultants are

$$\mathbf{T} = T_{\alpha\beta}\mathbf{i}_\alpha\mathbf{i}_\beta + T_{\alpha n}\mathbf{i}_\alpha\mathbf{n}, \tag{5.45}$$

where

$$T_{\alpha\beta} = [\![\sigma_{\alpha\beta}]\!] \equiv \int_{-h/2}^{h/2} \sigma_{\alpha\beta}\, dz, \qquad T_{\alpha n} = [\![\sigma_{\alpha 3}]\!] \equiv \int_{-h/2}^{h/2} \sigma_{\alpha 3}\, dz.$$

Note that σ_{33} does not appear in the definition of the stress resultants or the transverse shear stress resultants. It follows that $T_{\alpha\beta} = T_{\beta\alpha}$. We have

$$\mathbf{T}_{\times} = T_{\alpha\beta}\mathbf{i}_\alpha \times \mathbf{i}_\beta + T_{\alpha n}\mathbf{i}_\alpha \times \mathbf{n} = T_{\alpha n}\mathbf{i}_\alpha \times \mathbf{n} = T_{2n}\mathbf{i}_1 - T_{1n}\mathbf{i}_2.$$

The couple matrix

$$M_{\alpha\beta} = [\![z\sigma_{\alpha\beta}]\!] \equiv \int_{-h/2}^{h/2} z\sigma_{\alpha\beta}\,dz. \tag{5.46}$$

is also symmetric. In view of the fact that

$$\boldsymbol{\sigma} \times \mathbf{n} = \sigma_{\alpha\beta}\mathbf{i}_\alpha\mathbf{i}_\beta \times \mathbf{n} = -\sigma_{11}\mathbf{i}_1\mathbf{i}_2 + \sigma_{12}\mathbf{i}_1\mathbf{i}_1 - \sigma_{21}\mathbf{i}_2\mathbf{i}_2 + \sigma_{22}\mathbf{i}_2\mathbf{i}_1,$$

relation (5.44) yields

$$\mathbf{M} = [\![z\sigma_{11}]\!]\mathbf{i}_1\mathbf{i}_2 - [\![z\sigma_{12}]\!]\mathbf{i}_1\mathbf{i}_1 + [\![z\sigma_{21}]\!]\mathbf{i}_2\mathbf{i}_2 - [\![z\sigma_{22}]\!]\mathbf{i}_2\mathbf{i}_1.$$

By (5.46),

$$\mathbf{M} = M_{11}\mathbf{i}_1\mathbf{i}_2 - M_{12}\mathbf{i}_1\mathbf{i}_1 + M_{21}\mathbf{i}_2\mathbf{i}_2 - M_{22}\mathbf{i}_2\mathbf{i}_1 = -M_{\alpha\beta}\mathbf{i}_\alpha\mathbf{i}_\beta \times \mathbf{n}.$$

So the equilibrium equations in Cartesian coordinates are

$$\frac{\partial T_{\alpha\beta}}{\partial x_\alpha}\mathbf{i}_\beta + \frac{\partial T_{\alpha n}}{\partial x_\alpha}\mathbf{n} + \mathbf{q} = \mathbf{0}, \qquad -\frac{\partial M_{\alpha\beta}}{\partial x_\alpha}\mathbf{i}_\beta \times \mathbf{n} + \mathbf{T}_\times + \mathbf{m} = \mathbf{0}. \tag{5.47}$$

In component form, these are

$$\begin{aligned}
&\frac{\partial T_{11}}{\partial x_1} + \frac{\partial T_{21}}{\partial x_2} + q_1 = 0,\\
&\frac{\partial T_{12}}{\partial x_1} + \frac{\partial T_{22}}{\partial x_2} + q_2 = 0,\\
&\frac{\partial T_{1n}}{\partial x_1} + \frac{\partial T_{2n}}{\partial x_2} + q_n = 0,
\end{aligned} \tag{5.48}$$

and

$$\begin{aligned}
&\frac{\partial M_{11}}{\partial x_1} + \frac{\partial M_{21}}{\partial x_2} - T_{1n} + m_1 = 0,\\
&\frac{\partial M_{12}}{\partial x_1} + \frac{\partial M_{22}}{\partial x_2} - T_{2n} + m_2 = 0,
\end{aligned} \tag{5.49}$$

where

$$q_\alpha = \mathbf{i}_\alpha \cdot \mathbf{q}, \qquad q_n = \mathbf{n} \cdot \mathbf{q}, \qquad m_\alpha = \mathbf{i}_\alpha \cdot (\mathbf{n} \times \mathbf{m}).$$

Note that $T_{12} = T_{21}$ and $M_{12} = M_{21}$; we keep the different notation for symmetry.

In applications, the influence of $\mathbf{m}$ on the strains in the shell is often much smaller than that of the external forces. Especially for thin shells, the values of surface moments are small compared with surface external forces. So we set $\mathbf{m} = \mathbf{0}$. Let us eliminate T_{1n} and T_{2n} from (5.49). First we differentiate the second equation of (5.49) with respect to x_2, the first equation from (5.49) with respect to x_1, and add the results. Then, using the third equation from (5.48), we obtain the familiar relation

$$\frac{\partial^2 M_{11}}{\partial {x_1}^2} + 2\frac{\partial^2 M_{12}}{\partial x_1 \partial x_2} + \frac{\partial^2 M_{22}}{\partial {x_2}^2} + q_n = 0 \tag{5.50}$$

which, in terms of deflection w, equation (5.55), is considered in many textbooks on mathematical physics as well. In index notation this is

$$\frac{\partial^2 M_{\alpha\beta}}{\partial x_\alpha \partial x_\beta} + q_n = 0,$$

whereas in component-free notation it is

$$\widetilde{\nabla} \cdot [\widetilde{\nabla} \cdot (\mathbf{M} \times \mathbf{n})] + \mathbf{q} \cdot \mathbf{n} = 0. \tag{5.51}$$

Indeed,

$$\mathbf{M} = -M_{\alpha\beta}\mathbf{i}_\alpha \mathbf{i}_\beta \times \mathbf{n}.$$

Using the identity

$$(\mathbf{i}_\beta \times \mathbf{n}) \times \mathbf{n} = -\mathbf{i}_\beta,$$

the proof of which is left to the reader, we get

$$\mathbf{M} \times \mathbf{n} = -M_{\alpha\beta}\mathbf{i}_\alpha(\mathbf{i}_\beta \times \mathbf{n}) \times \mathbf{n} = M_{\alpha\beta}\mathbf{i}_\alpha\mathbf{i}_\beta.$$

Next we obtain

$$\widetilde{\nabla} \cdot (\mathbf{M} \times \mathbf{n}) = \frac{\partial M_{\alpha\beta}}{\partial x_\alpha}\mathbf{i}_\beta, \qquad \widetilde{\nabla} \cdot [\widetilde{\nabla} \cdot (\mathbf{M} \times \mathbf{n})] = \frac{\partial^2 M_{\alpha\beta}}{\partial x_\alpha \partial x_\beta},$$

from which (5.51) follows.

In plate theory, Kirchhoff's kinematic hypotheses reduce to an assumption on the form of the displacement field:

$$\begin{aligned} \mathbf{u}(x_1, x_2, z) &= \mathbf{v}(x_1, x_2) - z\widetilde{\nabla} w(x_1, x_2) \\ &= \tilde{\mathbf{v}} + w\mathbf{n} - z\widetilde{\nabla} w, \qquad \tilde{\mathbf{v}} \cdot \mathbf{n} = 0. \end{aligned} \tag{5.52}$$

Let us find the expression for the strain tensor that corresponds to $\mathbf{u}$. We have

$$\nabla \mathbf{u} = \left(\widetilde{\nabla} + \mathbf{n}\frac{\partial}{\partial z}\right)(\tilde{\mathbf{v}} + w\mathbf{n} - z\widetilde{\nabla} w) = \widetilde{\nabla}\tilde{\mathbf{v}} - z\widetilde{\nabla}\widetilde{\nabla} w + (\widetilde{\nabla} w)\mathbf{n} - \mathbf{n}\widetilde{\nabla} w,$$

and

$$(\nabla \mathbf{u})^T = (\widetilde{\nabla}\tilde{\mathbf{v}})^T - z\widetilde{\nabla}\widetilde{\nabla}w + \mathbf{n}\widetilde{\nabla}w - (\widetilde{\nabla}w)\mathbf{n}.$$

Thus,

$$2\boldsymbol{\varepsilon} = \widetilde{\nabla}\tilde{\mathbf{v}} + (\widetilde{\nabla}\tilde{\mathbf{v}})^T - 2z\widetilde{\nabla}\widetilde{\nabla}w.$$

It follows that $\boldsymbol{\varepsilon} \cdot \mathbf{n} = \mathbf{n} \cdot \boldsymbol{\varepsilon} = \mathbf{0}$.

Using the notation

$$2\boldsymbol{\epsilon} = \widetilde{\nabla}\tilde{\mathbf{v}} + (\widetilde{\nabla}\tilde{\mathbf{v}})^T, \qquad \boldsymbol{æ} = -\widetilde{\nabla}\widetilde{\nabla}w,$$

we find that the plate strains split into two parts:

$$\boldsymbol{\varepsilon} = \boldsymbol{\epsilon} + z\boldsymbol{æ}. \tag{5.53}$$

Here $\boldsymbol{\epsilon}$ represents the deformation in the plate plane, and $\boldsymbol{æ}$ is related to plate bending. Geometrically, $\boldsymbol{æ}$ describes an infinitesimal change of the plate curvatures due to bending.

Now let us use Kirchhoff's static hypothesis $\sigma_{33} = 0$. From Hooke's law it follows that

$$\begin{aligned}
\sigma_{11} &= \frac{E}{1-\nu^2}(\varepsilon_{11} + \nu\varepsilon_{22}), \\
\sigma_{22} &= \frac{E}{1-\nu^2}(\varepsilon_{22} + \nu\varepsilon_{11}), \\
\sigma_{12} &= \frac{E}{1+\nu}\varepsilon_{12}.
\end{aligned} \tag{5.54}$$

Substituting (5.53) into these, we get

$$\begin{aligned}
\sigma_{11} &= \frac{E}{1-\nu^2}(\epsilon_{11} + \nu\epsilon_{22}) + \frac{Ez}{1-\nu^2}(æ_{11} + \nu æ_{22}), \\
\sigma_{22} &= \frac{E}{1-\nu^2}(\epsilon_{22} + \nu\epsilon_{11}) + \frac{Ez}{1-\nu^2}(æ_{22} + \nu æ_{11}), \\
\sigma_{12} &= \frac{E}{1+\nu}\epsilon_{12} + \frac{Ez}{1+\nu}æ_{12}.
\end{aligned}$$

Now by (5.45) and (5.46), integrating the expressions for $\sigma_{\alpha\beta}$ with respect to z over thickness, we derive the expressions for the stress resultants and stress couples:

$$T_{11} = \frac{Eh}{1-\nu^2}(\epsilon_{11} + \nu\epsilon_{22}), \qquad M_{11} = \frac{Eh^3}{12(1-\nu^2)}(\text{æ}_{11} + \nu\text{æ}_{22}),$$

$$T_{22} = \frac{Eh}{1-\nu^2}(\epsilon_{22} + \nu\epsilon_{11}), \qquad M_{22} = \frac{Eh^3}{12(1-\nu^2)}(\text{æ}_{22} + \nu\text{æ}_{11}),$$

$$T_{12} = \frac{Eh}{1+\nu}\epsilon_{12}, \qquad M_{12} = \frac{Eh^3}{12(1+\nu)}\text{æ}_{12}.$$

The constant

$$D = Eh^3/12(1-\nu^2)$$

is the *bending stiffness.* Substituting the expressions for $M_{\alpha\beta}$ into (5.50) and taking into account that

$$\text{æ}_{11} = -\frac{\partial^2 w}{\partial x_1^2}, \qquad \text{æ}_{22} = -\frac{\partial^2 w}{\partial x_2^2}, \qquad \text{æ}_{12} = -\frac{\partial^2 w}{\partial x_1 \partial x_2},$$

we obtain the equilibrium equation

$$D\left(\frac{\partial^4 w}{\partial x_1^4} + 2\frac{\partial^4 w}{\partial x_1^2 \partial x_2^2} + \frac{\partial^4 w}{\partial x_2^4}\right) = q_n. \tag{5.55}$$

In component-free form this is

$$D\widetilde{\nabla}^4 w = q_n. \tag{5.56}$$

Because

$$\widetilde{\nabla}^4 = (\widetilde{\nabla}^2)^2 = \Delta^2,$$

we call (5.55) (or (5.56)) the *biharmonic equation.*

This equation has a long history in plate theory. The bending equation was published by Marie-Sophie Germain, who received a prize from the Paris Academy of Sciences in 1816. Later some of Germain's errors were corrected by Lagrange, hence the equation is now called *Germain's equation* or the *Germain–Lagrange equation.* The plate theory we have considered is attributed to Kirchhoff, who formulated correct boundary conditions for a plate. Moreover, Kirchhoff clearly introduced physically meaningful hypotheses regarding the distribution of displacements and stresses in the plate.

Note that the $T_{\alpha\beta}$ do not depend on w. Thus, the plate equilibrium equations split into two groups. The first consists of the two first equations of (5.48) with respect to $\tilde{\mathbf{v}}$, whereas the second consists of the equation (5.56) for variable w.

Let us write the equations for $\mathbf{T} \cdot \mathbf{A}$ in terms of the tangential displacement vector $\tilde{\mathbf{v}}$. In tensor form, equations (5.48) are

$$\widetilde{\nabla} \cdot (\mathbf{T} \cdot \mathbf{A}) + \mathbf{q} \cdot \mathbf{A} = \mathbf{0}. \tag{5.57}$$

The constitutive equation for $\mathbf{T} \cdot \mathbf{A}$ is

$$\mathbf{T} \cdot \mathbf{A} = \frac{Eh}{1+\nu} \left[\boldsymbol{\epsilon} + \frac{\nu}{1-\nu} \mathbf{A} \operatorname{tr} \boldsymbol{\epsilon} \right].$$

So we obtain

$$\begin{aligned}
\widetilde{\nabla} \cdot (\mathbf{T} \cdot \mathbf{A}) &= \frac{Eh}{1+\nu} \left[\widetilde{\nabla} \cdot \boldsymbol{\epsilon} + \frac{\nu}{1-\nu} \widetilde{\nabla} \operatorname{tr} \boldsymbol{\epsilon} \right] \\
&= \frac{Eh}{2(1+\nu)} \left[\widetilde{\nabla} \cdot \widetilde{\nabla} \tilde{\mathbf{v}} + \widetilde{\nabla} \widetilde{\nabla} \cdot \tilde{\mathbf{v}} + \frac{2\nu}{1-\nu} \widetilde{\nabla} \widetilde{\nabla} \cdot \tilde{\mathbf{v}} \right] \\
&= \frac{Eh}{2(1+\nu)} \left[\widetilde{\nabla} \cdot \widetilde{\nabla} \tilde{\mathbf{v}} + \frac{1+\nu}{1-\nu} \widetilde{\nabla} \widetilde{\nabla} \cdot \tilde{\mathbf{v}} \right].
\end{aligned}$$

In terms of $\tilde{\mathbf{v}}$, equation (5.57) takes the form

$$\frac{Eh}{2(1+\nu)} \left[\widetilde{\nabla} \cdot \widetilde{\nabla} \tilde{\mathbf{v}} + \frac{1+\nu}{1-\nu} \widetilde{\nabla} \widetilde{\nabla} \cdot \tilde{\mathbf{v}} \right] + \mathbf{q} \cdot \mathbf{A} = \mathbf{0}. \tag{5.58}$$

There may be some concern over equations (5.49), which contain the transverse shear stress resultants T_{1n} and T_{2n}. It may seem that by the kinematic hypothesis $\mathbf{n} \cdot \boldsymbol{\varepsilon} = \mathbf{0}$ and the last equation from (5.45), both T_{1n} and T_{2n} should vanish. But Kirchhoff's kinematic hypothesis restricts only the form of deformation. So in Kirchhoff's plate theory, T_{1n} and T_{2n} are not defined by the strains; they are the reactions due to the strain constraints and the equations (5.49) are used to determine T_{1n} and T_{2n}.

Boundary conditions

In plate theory as a particular case of shell theory, on the boundary contour ω we can assign four scalar kinematic conditions

$$\tilde{\mathbf{v}} = \mathbf{v}^0(s), \qquad w = w^0(s), \qquad \frac{\partial w}{\partial \nu} = \vartheta^0(s), \tag{5.59}$$

or the static conditions

$$\begin{gathered}
\boldsymbol{\nu} \cdot \mathbf{T} \cdot \mathbf{A} = \boldsymbol{\varphi}^0(s) \cdot \mathbf{A}, \qquad \boldsymbol{\nu} \cdot \mathbf{T} \cdot \mathbf{n} - \frac{\partial}{\partial s} \left(\boldsymbol{\nu} \cdot (\mathbf{M} \times \mathbf{n}) \cdot \boldsymbol{\tau} \right) = \varphi_n^0(s), \\
\boldsymbol{\nu} \cdot (\mathbf{M} \times \mathbf{n}) \cdot \boldsymbol{\nu} = \ell^0(s).
\end{gathered} \tag{5.60}$$

For a mixed boundary value problem, these sets of conditions can be given on different portions of ω. Alternatively, at a point we can include conditions from (5.59) and the dual conditions from (5.60). At each point, four scalar conditions must be appointed.

We split the boundary load into tangential and normal parts:

$$\boldsymbol{\varphi}^0 = \boldsymbol{\varphi}^0 \cdot \mathbf{A} + \varphi_n^0 \mathbf{n}, \qquad \varphi_n^0 = \boldsymbol{\varphi}^0 \cdot \mathbf{n}.$$

From equilibrium equations (5.49), we find the transverse shear stress resultants in terms of the stress couples. As in our earlier treatment of shell theory, we set $m_1 = m_2 = 0$. We have

$$T_{1n} = \frac{\partial M_{11}}{\partial x_1} + \frac{\partial M_{21}}{\partial x_2}, \qquad T_{2n} = \frac{\partial M_{12}}{\partial x_1} + \frac{\partial M_{22}}{\partial x_2}.$$

These can be written as the single vector equation

$$\mathbf{T} \cdot \mathbf{n} = \widetilde{\nabla} \cdot (\mathbf{M} \times \mathbf{n}).$$

There are various types of mixed boundary conditions. We restrict ourselves to the two defined by (5.59) and (5.60). Note that $\mathbf{T} \cdot \mathbf{A}$ depends only on $\tilde{\mathbf{v}}$, whereas $\mathbf{M}$ is defined through w. Thus, like the equilibrium equations, the boundary conditions in plate theory split into two sets. One is for normal displacement w:

$$\begin{gathered} w\Big|_{\omega_1} = w^0(s), \qquad \frac{\partial w}{\partial \nu}\Big|_{\omega_1} = \vartheta^0(s), \\ \boldsymbol{\nu} \cdot (\widetilde{\nabla} \cdot (\mathbf{M} \times \mathbf{n}))\Big|_{\omega_2} - \frac{\partial}{\partial s} (\boldsymbol{\nu} \cdot (\mathbf{M} \times \mathbf{n}) \cdot \boldsymbol{\tau})\Big|_{\omega_2} = \varphi_n^0(s), \\ \boldsymbol{\nu} \cdot (\mathbf{M} \times \mathbf{n}) \cdot \boldsymbol{\nu}\Big|_{\omega_2} = \ell^0(s). \end{gathered} \tag{5.61}$$

The other is for tangential displacements $\tilde{\mathbf{v}}$:

$$\tilde{\mathbf{v}}\Big|_{\omega_3} = \mathbf{v}^0(s), \qquad \boldsymbol{\nu} \cdot \mathbf{T} \cdot \mathbf{A}\Big|_{\omega_4} = \boldsymbol{\varphi}^0(s) \cdot \mathbf{A}. \tag{5.62}$$

We have supposed the plate contour to be partitioned in such a way that $\omega_2 = \omega \setminus \omega_1$ and $\omega_4 = \omega \setminus \omega_3$. Let us note that the variational approach is not only useful for deriving natural, i.e., static, boundary conditions. In some sense it gives a straightforward tool for deriving these conditions for complex models such as the plate model. Indeed, the first and possibly naive idea of how to formulate static boundary conditions yields conditions expressing given forces and moments, that leads to more boundary conditions than necessary. This constitutes the famous Poisson's error in the theory of plates. In fact, it is not so simple to forecast the combination of transverse force and moments in one static condition $(5.60)_2$. Recently, the formulation of the static boundary conditions in the classic theory of plates without using variational principles is one of the difficult questions for students.

In Appendix A we list some common homogeneous boundary conditions in terms of w for the plate.

Strain energy for the plate

Let us consider the strain energy density for a plate in more detail. It is

$$U = \frac{Eh}{2(1+\nu)}\left[\boldsymbol{\epsilon}\cdot\cdot\,\boldsymbol{\epsilon} + \frac{\nu}{1-\nu}\operatorname{tr}^2\boldsymbol{\epsilon}\right] + \frac{Eh^3}{24(1+\nu)}\left[\text{æ}\cdot\cdot\,\text{æ} + \frac{\nu}{1-\nu}\operatorname{tr}^2\text{æ}\right]. \tag{5.63}$$

A plate is a particular case of a shell considered above, but with $\mathbf{B} = \mathbf{0}$. Hence we can use the expression for shell strain density

$$U = \tfrac{1}{2}T_{\alpha\beta}\epsilon^{\alpha\beta} + \tfrac{1}{2}M_{\alpha\beta}\text{æ}^{\alpha\beta} = \tfrac{1}{2}\mathbf{T}\cdot\cdot\,\boldsymbol{\varepsilon} + \tfrac{1}{2}\mathbf{M}\cdot\cdot\,(\mathbf{n}\times\text{æ}).$$

Let us rewrite U in terms of the displacements:

$$U = \frac{Eh}{2(1+\nu)}\left[\frac{\partial u_\alpha}{\partial x_\beta}\frac{\partial u_\alpha}{\partial x_\beta} + \frac{\nu}{1-\nu}\left(\frac{\partial u_\alpha}{\partial x_\alpha}\right)^2\right]$$
$$+ \frac{Eh^3}{24(1+\nu)}\left[\frac{\partial^2 w}{\partial x_\alpha \partial x_\beta}\frac{\partial^2 w}{\partial x_\alpha \partial x_\beta} + \frac{\nu}{1-\nu}\left(\frac{\partial^2 w}{\partial x_\alpha \partial x_\alpha}\right)^2\right].$$

Rigid motions

Now we consider the consequences of the equation $U = 0$. Because U is positive definite, it is zero if and only if both $\boldsymbol{\epsilon}$ and æ are zero. But $\boldsymbol{\epsilon} = \mathbf{0}$ implies that the plate can move in its plane as a rigid body only. In other words, it can translate and rotate about a normal to its midplane. The corresponding displacement field takes the form

$$\tilde{\mathbf{v}} = \mathbf{v}_0 + \omega_0\mathbf{n}\times(\boldsymbol{\rho} - \boldsymbol{\rho}_0),$$

where $\mathbf{v}_0$ is an arbitrary fixed vector in the midplane, i.e., $\mathbf{v}_0\cdot\mathbf{n} = 0$, the angle ω_0 is an arbitrary but fixed rotation angle about $\mathbf{n}$, the vector $\boldsymbol{\rho}$ is the position vector of a point on the midplane, and $\boldsymbol{\rho}_0$ is an arbitrary but fixed vector in the same plane. This formula is a particular case of (4.32) for the plane deformation.

The equation

$$\text{æ} \equiv -\widetilde{\nabla}\widetilde{\nabla}w = \mathbf{0}$$

implies that w is a linear function in x_1 and x_2:

$$w = w_0 + w_1 x_1 + w_2 x_2,$$

where w_0, w_1, and w_2 are arbitrary fixed scalars. Physically, w corresponds to a translation of the whole plate in the normal direction and a rotation with respect to the axes $\mathbf{i}_1, \mathbf{i}_2$. Thus, as for three-dimensional elasticity, $U = 0$ only for those displacements that represent small displacements of the plate as a rigid body.

Lagrange variational principle in plate theory

In shell theory, the Lagrange variational principle of elasticity changes somewhat because of Kirchhoff's hypotheses. We will formulate Lagrange's variational principle for the plate. Recall that in three-dimensional elasticity, the total energy functional takes the form

$$\mathcal{E}(\mathbf{u}) = \int_V W(\boldsymbol{\varepsilon})\, dV - \int_V \rho \mathbf{f} \cdot \mathbf{u}\, dV - \int_S \mathbf{t}^0 \cdot \mathbf{u}\, dS.$$

We found that for a thin shell, the integral over V is represented through the integral over the midplane Σ. The same is true for the plate:

$$\int_V W(\boldsymbol{\varepsilon})\, dV = \int_\Sigma U\, d\Sigma,$$

where U is defined by (5.63). Let us consider the remaining terms in $\mathcal{E}$, which represent the work of external forces. In terms of the plate theory, they take the form

$$\int_\Sigma \mathbf{q} \cdot \mathbf{v}\, d\Sigma + \int_{\omega_4} \boldsymbol{\varphi}^0(s) \cdot \tilde{\mathbf{v}}\, dS + \int_{\omega_2} \left(\varphi_n^0 w - \ell^0 \frac{\partial w}{\partial \nu} \right) dS.$$

So the total energy functional for the plate splits into two parts:

$$\mathcal{E} = \mathcal{E}_{\tilde{\mathbf{v}}}(\tilde{\mathbf{v}}) + \mathcal{E}_w(w),$$

where

$$\begin{aligned} \mathcal{E}_{\tilde{\mathbf{v}}}(\tilde{\mathbf{v}}) = {} & \frac{Eh}{2(1+\nu)} \int_\Sigma \left[\boldsymbol{\epsilon} \cdot\cdot\, \boldsymbol{\epsilon} + \frac{\nu}{1-\nu} \operatorname{tr}^2 \boldsymbol{\epsilon} \right] d\Sigma \\ & - \int_\Sigma \mathbf{q} \cdot \tilde{\mathbf{v}}\, d\Sigma - \int_{\omega_4} \boldsymbol{\varphi}^0(s) \cdot \tilde{\mathbf{v}}\, dS \end{aligned}$$

is due to tangential displacements, and

$$\mathcal{E}_w(w) = \frac{Eh^3}{24(1+\nu)} \int_\Sigma \left[æ \cdot\cdot æ + \frac{\nu}{1-\nu} \operatorname{tr}^2 æ \right] d\Sigma$$
$$- \int_\Sigma w\mathbf{q} \cdot \mathbf{n} \, d\Sigma - \int_{\omega_2} \varphi_n^0 w \, dS + \int_{\omega_2} \ell^0 \frac{\partial w}{\partial \nu} \, dS$$

accounts for bending.

The arguments $\tilde{\mathbf{v}}$ and w of $\mathcal{E}_{\tilde{\mathbf{v}}}$ and $\mathcal{E}_w$ are independent, so the problem of minimizing the total energy naturally splits into two minimization problems. Moreover, we can formulate two variational principles: one for tangential deformations, and one for plate bending.

Lagrange's variational principle for tangential deformation

Theorem 5.1. *On the set of admissible tangential displacement fields, a stationary point* $\tilde{\mathbf{v}}$ *of* $\mathcal{E}_{\tilde{\mathbf{v}}}$ *satisfies the plate equilibrium equations* (5.57) *on* Σ, *or, equivalently* (5.58) *and the boundary condition*

$$\boldsymbol{\nu} \cdot \mathbf{T} \cdot \mathbf{A}|_{\omega_4} = \boldsymbol{\varphi}^0(s) \cdot \mathbf{A},$$

and conversely. The stationary point, if it exists, minimizes $\mathcal{E}_{\tilde{\mathbf{v}}}$.

The set of admissible tangential displacements $\tilde{\mathbf{v}}$ consists of vector functions twice differentiable in Σ and satisfying the kinematic boundary condition on ω_3 from (5.62). The proof mimics that of Lagrange's principle in three-dimensional elasticity (Exercise 5.13).

Note that we represented the work of external forces using the general definition of work and common sense. The fact that the equilibrium equations and the kinematic boundary conditions really define a stationary point of the energy functional shows that the work functional was written correctly.

Lagrange's variational principle for plate bending

Theorem 5.2. *On the set of admissible deflection fields, a stationary point* w *of* $\mathcal{E}_w$ *satisfies the plate equilibrium equations* (5.51) *or* (5.56) *in* Σ *and the static boundary conditions from* (5.61) *on* ω_2, *and conversely. The stationary point, if it exists, minimizes* $\mathcal{E}_w$.

Note that the set of admissible deflections w consists of the functions that possess continuous derivatives in Σ up to order four and satisfy two kinematic boundary conditions on ω_1 from (5.61).

Proof. Let us define the variation of $\mathcal{E}_w$. We have

$$\delta\left[\frac{Eh^3}{24(1+\nu)}\left[æ \cdot\cdot\, æ + \frac{\nu}{1-\nu}\operatorname{tr}^2 æ\right]\right] = \frac{Eh^3}{12(1+\nu)}\left[æ + \frac{\nu}{1-\nu}\operatorname{tr} æ\right] \cdot\cdot\, \delta æ$$

$$= M_{\alpha\beta}\,\delta æ^{\alpha\beta} = -M_{\alpha\beta}\frac{\partial^2\,\delta w}{\partial x_\alpha x_\beta}.$$

Using this and the Gauss–Ostrogradsky theorem, we get

$$\begin{aligned}
-\int_\Sigma M_{\alpha\beta}\frac{\partial^2\,\delta w}{\partial x_\alpha x_\beta}\,d\Sigma &= \int_\Sigma \frac{\partial M_{\alpha\beta}}{\partial x_\alpha}\frac{\partial\,\delta w}{\partial x_\beta}\,d\Sigma - \int_\omega \nu_\alpha M_{\alpha\beta}\frac{\partial\,\delta w}{\partial x_\beta}\,ds \\
&= -\int_\Sigma \frac{\partial^2 M_{\alpha\beta}}{\partial x_\alpha \partial x_\beta}\delta w\,d\Sigma + \int_\omega \nu_\beta\frac{\partial M_{\alpha\beta}}{\partial x_\alpha}\delta w\,ds \\
&\quad - \int_\omega \nu_\alpha M_{\alpha\beta}\frac{\partial\,\delta w}{\partial x_\beta}\,ds \\
&= -\int_\Sigma \widetilde{\nabla}\cdot(\widetilde{\nabla}\cdot(\mathbf{M}\times\mathbf{n}))\,\delta w\,d\Sigma \\
&\quad + \int_\omega \boldsymbol{\nu}\cdot(\widetilde{\nabla}\cdot(\mathbf{M}\times\mathbf{n}))\,\delta w\,ds \\
&\quad - \int_\omega \boldsymbol{\nu}\cdot(\mathbf{M}\times\mathbf{n})\cdot\widetilde{\nabla}\delta w\,ds,
\end{aligned} \tag{5.64}$$

where $\nu_\alpha = \boldsymbol{\nu}\cdot\mathbf{i}_\alpha$ and s is the length parameter over ω.

Let us consider the last integral in (5.64). On ω the expression $\widetilde{\nabla}\delta w$ can be represented as

$$\widetilde{\nabla}\delta w = \boldsymbol{\nu}\frac{\partial\,\delta w}{\partial \nu} + \boldsymbol{\tau}\frac{\partial\,\delta w}{\partial \tau},$$

where $\boldsymbol{\tau} = \mathbf{n}\times\boldsymbol{\nu}$ is the vector tangent to ω. Using the contour length parameter s over ω, we get

$$\frac{\partial\,\delta w}{\partial\tau} = \frac{\partial\,\delta w}{\partial s}.$$

Note that on ω the expressions $\partial\,\delta w/\partial\nu$ and δw are independent. The expressions $\partial\,\delta w/\partial s$ and δw are dependent because the value $\partial\,\delta w/\partial s$ is uniquely defined when δw is known on ω, whereas $\partial\,\delta w/\partial\nu$ is not defined by the values of δw on ω. To exclude the derivative $\partial\,\delta w/\partial s$, we integrate by parts in the integral over the closed contour ω:

$$\int_\omega f\frac{\partial g}{\partial s}\,ds = -\int_\omega g\frac{\partial f}{\partial s}\,ds,$$

where f and g are arbitrary functions of s. Applying this to the last integral in (5.64), we get

$$-\int_\omega \boldsymbol{\nu}\cdot(\mathbf{M}\times\mathbf{n})\cdot\widetilde{\nabla}\delta w\,ds = -\int_\omega\left[\boldsymbol{\nu}\cdot(\mathbf{M}\times\mathbf{n})\cdot\boldsymbol{\nu}\frac{\partial\,\delta w}{\partial\nu} - \frac{\partial}{\partial s}\left(\boldsymbol{\nu}\cdot(\mathbf{M}\times\mathbf{n})\cdot\boldsymbol{\tau}\right)\delta w\right]ds.$$

On ω_1 we have

$$\delta w = 0 = \frac{\partial\,\delta w}{\partial\nu},$$

hence

$$\begin{aligned}\delta\mathcal{E}_w = &-\int_\Sigma\left[\widetilde{\nabla}\cdot(\widetilde{\nabla}\cdot(\mathbf{M}\times\mathbf{n})) + \mathbf{q}\cdot\mathbf{n}\right]\delta w\,d\Sigma\\ &+\int_{\omega_2}\left[\boldsymbol{\nu}\cdot\left(\widetilde{\nabla}\cdot(\mathbf{M}\times\mathbf{n})\right)\right.\\ &\left.-\frac{\partial}{\partial s}(\boldsymbol{\nu}\cdot(\mathbf{M}\times\mathbf{n})\cdot\boldsymbol{\tau}) - \varphi_n^0\right]\delta w\,ds\\ &+\int_{\omega_2}\left[\boldsymbol{\nu}\cdot(\mathbf{M}\times\mathbf{n})\cdot\boldsymbol{\nu} - \ell^0\right]\frac{\partial\,\delta w}{\partial\nu}\,ds.\end{aligned}$$

Using a standard procedure in the calculus of variations (see, e.g., [Lebedev and Cloud (2003)]), we can show that the equation $\delta\mathcal{E}_w = 0$, which holds for all admissible δw, implies both the equilibrium equation for the plate

$$\widetilde{\nabla}\cdot\left(\widetilde{\nabla}\cdot(\mathbf{M}\times\mathbf{n})\right) + \mathbf{q}\cdot\mathbf{n} = 0 \quad \text{in } \Sigma$$

and the static boundary conditions

$$\boldsymbol{\nu}\cdot\left(\widetilde{\nabla}\cdot(\mathbf{M}\times\mathbf{n})\right) - \frac{\partial}{\partial s}\left(\boldsymbol{\nu}\cdot(\mathbf{M}\times\mathbf{n})\cdot\boldsymbol{\tau}\right) - \varphi_n^0 = 0$$

and

$$\boldsymbol{\nu}\cdot(\mathbf{M}\times\mathbf{n})\cdot\boldsymbol{\nu} - \ell^0 = 0 \quad \text{on } \omega_2.$$

This derivation also indicates that we have introduced the work for the transverse load correctly: a stationary point of $\mathcal{E}_w$ satisfies (5.51) on Σ and the boundary conditions (5.61) on ω_2.

Now we can prove the converse statement: a solution w of the boundary value problem (5.51), (5.61) is a stationary point of $\mathcal{E}_w$. Multiplying (5.51) by an admissible δw, integrating appropriately by parts and therefore doing the above transformations in the reverse order, we arrive at the equation $\delta\mathcal{E}_w = 0$. This completes the proof.

In a manner similar to the proof of minimality in the Lagrange principle in elasticity, we can show minimality of a stationary point in the bending problem as well. The reader is encouraged to produce a complete proof. Note that in the proof for three-dimensional elasticity, the only thing that mattered was the structure of the total energy functional: it is a sum of quadratic and linear functionals, and its quadratic portion is positive definite. □

Note that by considering a plate as a three-dimensional body, we obtain a system of simultaneous equations in displacements of the second order, accompanied by three boundary conditions at each point of the boundary. Transformation of the problem to a two-dimensional plate problem brings us again to three differential equations in displacements, but the equation for w is of fourth order; it is supplemented with two conditions at each point of the plate edge, and the other two equations are of second order. The increase in order of the system is the cost of reducing a three-dimensional problem to a two-dimensional one.

In shell theory there are other variational principles. Their importance in this theory is even more than in elasticity. They are the basis for numerical methods and investigations of qualitative questions of the theory. Moreover, variational principles are used to construct various versions of shell theory. For example, we can introduce the type of distribution of the stresses and displacements along the thickness, and derive the equilibrium equations, find admissible sets of boundary conditions, and justify well-posedness of the corresponding boundary value problems.

Uniqueness of solution

For an equilibrium problem in plate theory, a solution is unique in the same meaning as for a problem in three-dimensional elasticity. That is, if some part of the boundary contour is clamped, then the solution is truly unique. When the plate can move as a rigid body, the solution is uniquely defined by the equations and boundary conditions up to a small rigid motion. We will briefly outline a proof.

Suppose to the contrary the existence of two solutions $\mathbf{v}_1$ and $\mathbf{v}_2$ to the boundary value problem (5.57), (5.62), (5.51), (5.61). The difference $\mathbf{v} = \mathbf{v}_2 - \mathbf{v}_1$ is a solution of this homogeneous boundary value problem.

It is easy to understand that it should be a stationary point of the strain energy functional

$$\int_{\Sigma} U \, d\Sigma.$$

This implies that

$$\delta \int_{\Sigma} U \, d\Sigma = 0.$$

But U is a homogeneous quadratic form, so putting $\delta \mathbf{v} = \mathbf{v}$ in δU we get $\delta U = 2U(\mathbf{v})$ and thus

$$2 \int_{\Sigma} U \, d\Sigma = 0.$$

But earlier we considered the question of when the strain energy functional takes zero value: the equation

$$\int_{\Sigma} U \, d\Sigma = 0$$

implies that $\mathbf{v}$ is a small rigid displacement of the plate. So the uniqueness of a solution for an equilibrium problem is demonstrated: a solution is uniquely defined up to a rigid displacement if there are no kinematic boundary constraints.

Exercise 5.9. Derive (5.48) from (5.47).

Exercise 5.10. Show that $\operatorname{tr} \boldsymbol{\varepsilon} = \widetilde{\nabla} \cdot \tilde{\mathbf{v}} - z \widetilde{\nabla} \cdot \widetilde{\nabla} w$.

Exercise 5.11. Let w and its first and second derivatives be small. Demonstrate that æ is the curvature tensor of the bent surface of the plate. Use the results of Exercise 3.16.

Exercise 5.12. Derive (5.55).

Exercise 5.13. Demonstrate the variational principle for tangential displacements in the theory of plates.

Exercise 5.14. Show that at a point of plate equilibrium, $\mathcal{E}_w$ takes its minimum value.

Remarks

Kirchhoff's plate theory was extended to shell theory. In shell theory, Kirchhoff's hypotheses are called the Kirchhoff–Love hypotheses. In the Kirchhoff–Love shell theory, the equations are written on the midsurface. The strain energy density for the shell also splits into two parts, for tangential and bending deformations. However, unlike the splitting of the plate energy, these terms are dependent. Moreover, all three equilibrium equations in displacements contain all components of the displacement vector. Technically, the Kirchhoff–Love shell theory is more complicated than plate theory, but qualitatively they are quite similar. In shell theory, we have Lagrange's variational principle and uniqueness of solution to a boundary value problem. In both cases, we can also prove existence of a solution, but in shell theory this requires more advanced mathematical tools.

5.9 On Non-Classical Theories of Plates and Shells

Reissner's approach to plate and shell theory

We recall that Kirchhoff's hypotheses yield only an approximation to the real deformations of thin-walled bodies. In the mechanics of plates and shells, there are other approaches to the representation of three-dimensional deformation, where shear deformation, normal extension, and other factors are taken into account. In particular, the Reissner and Mindlin approaches allow us to construct more precise two-dimensional models for the solution of a three-dimensional elastic problem for a shell. The main features of the theory will be demonstrated using Reissner's plate equations. We call the Reissner or Mindlin model non-classical theory, but it is worth noting that these models are implemented in commercial software and widely used in engineering. From this point of view these models — also known as shear deformable plate or shell theory — can be treated as classical ones, reserving the term "non-classical theory" for more recent theories.

Unlike (5.52), in Reissner's approach we take the displacement field to be of the form

$$\begin{aligned}\mathbf{u}(x_1, x_2, z) &= \mathbf{v}(x_1, x_2) - z\boldsymbol{\vartheta}(x_1, x_2)\\ &= \tilde{\mathbf{v}} + w\mathbf{n} - z\boldsymbol{\vartheta}, \qquad \boldsymbol{\vartheta}\cdot\mathbf{n} = 0.\end{aligned} \tag{5.65}$$

The components of $\boldsymbol{\vartheta}$ can be interpreted as the rotation angles of the shell cross section. In the general case, unlike in the Kirchhoff theory, $\boldsymbol{\vartheta} \neq \widetilde{\nabla} w$ and $\boldsymbol{\vartheta}$ is considered to be independent of $\mathbf{v}$. Thus, in Reissner's theory there are five unknown variables: the components v_1, v_2, w of $\mathbf{v}$ and the components ϑ_1, ϑ_2 of $\boldsymbol{\vartheta}$. In Reissner's theory, the transverse shear stress resultants are on equal footing with the stress resultants and the bending and twisting stress couples. Unlike what happens in Kirchhoff's model, they are independent and so it is necessary to formulate additional constitutive equations along the lines of (5.45). The equilibrium equations for Reissner's plate take the same form as in Kirchhoff's theory, i.e., (5.43):

$$\widetilde{\nabla} \cdot \mathbf{T} + \mathbf{q} = \mathbf{0}, \qquad \widetilde{\nabla} \cdot \mathbf{M} + \mathbf{T}_\times + \mathbf{m} = \mathbf{0}, \tag{5.66}$$

where $\mathbf{T}$ and $\mathbf{M}$ are the stress resultant tensor and the stress couple tensor, respectively, and $\mathbf{q}$ and $\mathbf{m}$ represent the forces and couples distributed over Σ. Note that $\mathbf{m} \cdot \mathbf{n} = 0$. In the component representation, they coincide with (5.48) and (5.49).

The dynamic equations of the theory are

$$\begin{aligned} \widetilde{\nabla} \cdot \mathbf{T} + \mathbf{q} \quad &= \rho \ddot{\mathbf{v}} + \rho \boldsymbol{\Theta}_1 \cdot \ddot{\boldsymbol{\vartheta}}, \\ \widetilde{\nabla} \cdot \mathbf{M} + \mathbf{T}_\times + \mathbf{m} &= \rho \boldsymbol{\Theta}_1^{\mathrm{T}} \cdot \ddot{\mathbf{v}} + \rho \boldsymbol{\Theta}_2 \cdot \ddot{\boldsymbol{\vartheta}}, \end{aligned} \tag{5.67}$$

where ρ is the surface shell density, $\boldsymbol{\Theta}_1$ and $\boldsymbol{\Theta}_2$ are the inertia tensors, and

$$\boldsymbol{\Theta}_2^T = \boldsymbol{\Theta}_2.$$

The boundary conditions for the plate are

$$\mathbf{v}\big|_{\omega_1} = \mathbf{v}^0(s), \qquad \boldsymbol{\vartheta}\big|_{\omega_3} = \boldsymbol{\vartheta}^0(s), \tag{5.68}$$

$$\boldsymbol{\nu} \cdot \mathbf{T}\big|_{\omega_2} = \boldsymbol{\varphi}(s), \qquad \boldsymbol{\nu} \cdot \mathbf{M}\big|_{\omega_4} = \boldsymbol{\ell}(s), \tag{5.69}$$

where $\mathbf{v}^0(s)$ and $\boldsymbol{\vartheta}^0(s)$ are given vector functions of the length parameter s such that $\boldsymbol{\vartheta}^0 \cdot \mathbf{n} = 0$. They define the displacements and the rotation on some part of the boundary contour, respectively. A given $\boldsymbol{\varphi}(s)$ and $\boldsymbol{\ell}(s)$ would define the stress resultants and the stress couples acting on the rest of the plate edge, $\boldsymbol{\ell} \cdot \mathbf{n} = 0$.

In Reissner's plate theory, the constitutive equations are

$$\mathbf{T} \cdot \mathbf{A} = \mathbf{C} \cdot\cdot\, \boldsymbol{\mu}, \qquad \mathbf{T} \cdot \mathbf{n} = \boldsymbol{\Gamma} \cdot \boldsymbol{\gamma}, \qquad \mathbf{M} = \mathbf{D} \cdot\cdot\, \boldsymbol{\kappa}. \tag{5.70}$$

Here $\mathbf{T} \cdot \mathbf{A}$ is the in-plane stress resultant tensor, $\mathbf{T} \cdot \mathbf{n}$ represents the transverse shear stress resultants, and $\mathbf{M}$ is the stress couple tensor. The strain measures are denoted as follows: $\boldsymbol{\mu}$ is the in-plane strain tensor, $\boldsymbol{\gamma}$ is the vector of transverse shear strains, and $\boldsymbol{\kappa}$ is the tensor of out-of-plane strains. Their definitions are

$$\boldsymbol{\mu} = \tfrac{1}{2}\left(\widetilde{\nabla}\tilde{\mathbf{v}} + (\widetilde{\nabla}\tilde{\mathbf{v}})^T\right), \qquad \boldsymbol{\gamma} = \widetilde{\nabla} w - \mathbf{n} \times \boldsymbol{\vartheta}, \qquad \boldsymbol{\kappa} = \widetilde{\nabla}\boldsymbol{\vartheta}.$$

The remaining notation is for the fourth-order tensors $\mathbf{C}$ and $\mathbf{D}$ and the second-order tensor $\boldsymbol{\Gamma}$ that describe the effective stiffness properties of the plate; they depend on the material properties of the plate and on the cross-section geometry. In the case of an isotropic plate having properties symmetric about the midplane, the effective stiffness tensors take the form

$$\mathbf{C} = C_{11}\mathbf{A}\mathbf{A} + C_{22}(\mathbf{A}_2\mathbf{A}_2 + \mathbf{A}_4\mathbf{A}_4),$$
$$\mathbf{D} = D_{22}(\mathbf{A}_2\mathbf{A}_2 + \mathbf{A}_4\mathbf{A}_4) + D_{33}\mathbf{A}_3\mathbf{A}_3,$$
$$\boldsymbol{\Gamma} = \Gamma\mathbf{A},$$

where

$$\mathbf{A} = \mathbf{i}_1\mathbf{i}_1 + \mathbf{i}_2\mathbf{i}_2, \qquad \mathbf{A}_2 = \mathbf{i}_1\mathbf{i}_1 - \mathbf{i}_2\mathbf{i}_2,$$
$$\mathbf{A}_3 = \mathbf{i}_1\mathbf{i}_2 - \mathbf{i}_2\mathbf{i}_1, \qquad \mathbf{A}_4 = \mathbf{i}_1\mathbf{i}_2 + \mathbf{i}_2\mathbf{i}_1,$$

and $\mathbf{i}_1$ and $\mathbf{i}_2$ are the unit basis vectors with $\mathbf{i}_1 \cdot \mathbf{i}_2 = 0$ on the midplane. Denote $\mathbf{A}_1 = \mathbf{A}$. The following orthogonality relation can be obtained:

$$\tfrac{1}{2}\mathbf{A}_i \cdot\cdot\, \mathbf{A}_j = \delta_{ij} \qquad (i, j = 1, 2, 3, 4).$$

For an isotropic homogeneous plate, the components of the stiffness tensors are defined as follows:

$$C_{11} = \frac{Eh}{2(1-\nu)}, \qquad C_{22} = \frac{Eh}{2(1+\nu)} = \mu h,$$
$$D_{33} = \frac{Eh^3}{24(1-\nu)}, \qquad D_{22} = \frac{Eh^3}{24(1+\nu)} = \frac{\mu h^3}{12}.$$

The classical bending stiffness for a plate is

$$D = D_{33} + D_{22} = \frac{Eh^3}{12(1-\nu^2)}. \tag{5.71}$$

The surface density and the inertia tensors are

$$\rho = \rho_0 h, \qquad \boldsymbol{\Theta}_1 = \mathbf{0}, \qquad \boldsymbol{\Theta}_2 = \Theta\mathbf{A}, \qquad \Theta = \frac{\rho_0 h^3}{12}, \tag{5.72}$$

where ρ_0 is the plate material density.

The transverse shear stiffness is given by

$$\Gamma = k\mu h, \tag{5.73}$$

where k is a *shear correction factor* first introduced by Timoshenko in the theory of beams[1]. For the value of k, Mindlin[2] proposed $k = \pi^2/12$ while

[1]Timoshenko, S. P. On the correction for shear of the differential equation for transverse vibrations of prismatic bars. *Phil. Mag. Ser.* 6, **41**, 744–746, 1921.

[2]Mindlin, R.D. Influence of rotatory inertia and shear on flexural motions of isotropic, elastic plates. *Trans. ASME J. Appl. Mech.*, **18**, 31–38, 1951.

Reissner[3] proposed a similar value of $k = 5/6$. The literature mentions other values of k; for example, for plates that are strongly nonhomogeneous in thickness, k can significantly differ from the above values[4].

The strain energy density for Reissner's plate is given by

$$U = \tfrac{1}{2}\boldsymbol{\mu} \cdot\cdot\, \mathbf{C} \cdot\cdot\, \boldsymbol{\mu} + \tfrac{1}{2}\boldsymbol{\kappa} \cdot\cdot\, \mathbf{D} \cdot\cdot\, \boldsymbol{\kappa} + \tfrac{1}{2}\boldsymbol{\gamma} \cdot \boldsymbol{\Gamma} \cdot \boldsymbol{\gamma}.$$

It is seen that

$$\mathbf{T} \cdot \mathbf{A} = \frac{\partial U}{\partial \boldsymbol{\mu}}, \qquad \mathbf{T} \cdot \mathbf{n} = \frac{\partial U}{\partial \boldsymbol{\gamma}}, \qquad \mathbf{M} = \frac{\partial U}{\partial \boldsymbol{\kappa}}.$$

The five scalar equations of (5.48) and (5.49) contain five scalar unknowns v_1, v_2, w, ϑ_1, and ϑ_2. In Reissner's model, equation (5.56) for w changes to the following:

$$D\widetilde{\nabla}^4 w = q_n - \frac{D}{\Gamma}\Delta q_n. \tag{5.74}$$

Because of the additional term $(D/\Gamma)\Delta q_n$, for strongly nonhomogeneous loads the results obtained from Reissner's theory can differ significantly from those obtained from Kirchhoff's model.

For advanced readers, we propose

Exercise 5.15. Derive (5.74).

Reissner's theory is usually used for non-thin plates, for dynamic problems, and for cases of anisotropic materials having small shear stiffness. A more detailed presentation of the theory can be found in [Wang, Reddy, and Lee (2000); Zhilin (2006)]; see also Altenbach[5] and Grigolyuk and Selezov[6].

Plate and shell theories of higher order

The approximation of the displacement field (5.65) is linear in z. In the literature on shell theory, we encounter displacement approximations in z of higher order. The natural idea of expanding the displacement in a series of

[3] Reissner, E. On the theory of bending of elastic plates. *J. Math. Physics*, **23**, 184–194, 1944.

[4] Altenbach, H., and Eremeyev, V.A. Direct approach based analysis of plates composed of functionally graded materials. *Arch. Appl. Mech.*, **78**, 775–794, 2008.

[5] Altenbach, H. An alternative determination of transverse shear stiffnesses for sandwich and laminated plates. *Int. J. Solids Struct.*, **37**, 3503–3520, 2000. Altenbach, H. On the determination of transverse shear stiffnesses of orthotropic plates. *ZAMP*, **51**, 629–649, 2000.

[6] Grigolyuk, E.I., and Selezov, I.T. *Nonclassical theories of vibration of beams, plates and shells* (in Russian). In: Itogi nauki i tekhniki, Mekhanika tverdogo deformiruemogo tela, vol 5, VINITI, Moskva, 1973.

powers in z belongs to Cauchy and Poisson. We could say that Germain's equation for plate deflection represents the zero-order theory, as here the deflection field does not depend on z. Equation (5.65) can be considered as a basis of the first-order theory. Some versions of higher-order theory were developed in a number of works; certain difficulties arise, however, such as how one should define the boundary conditions and physically interpret moments of higher order.

As an example, we will consider the *third-order plate theory*. In [Wang, Reddy, and Lee (2000)] there is a review of the literature on this version of the theory. The displacement field is approximated by the expressions

$$\begin{aligned}\mathbf{u}(x_1, x_2, z) &= \mathbf{v}(x_1, x_2) + z\boldsymbol{\phi}(x_1, x_2) + z^3\boldsymbol{\psi}(x_1, x_2) \\ &= \tilde{\mathbf{v}} + w\mathbf{n} + z\boldsymbol{\phi} + z^3\boldsymbol{\psi},\end{aligned}$$

where $\tilde{\mathbf{v}} \cdot \mathbf{n} = 0$, and $\boldsymbol{\phi}$ and $\boldsymbol{\psi}$ are functions such that $\boldsymbol{\phi} \cdot \mathbf{n} = 0$ and $\boldsymbol{\psi} \cdot \mathbf{n} = 0$. In particular, Reddy proposed the following expression for $\boldsymbol{\psi}$:

$$\boldsymbol{\psi} = -\alpha\left(\boldsymbol{\phi} + \widetilde{\nabla} w\right),$$

where $\alpha = 4/(3h^2)$ and h is the plate thickness. The strain tensor becomes

$$\boldsymbol{\varepsilon} = \tilde{\boldsymbol{\varepsilon}} + \tfrac{1}{2}(\boldsymbol{\gamma}\mathbf{n} + \mathbf{n}\boldsymbol{\gamma}), \tag{5.75}$$

where

$$\begin{aligned}\tilde{\boldsymbol{\varepsilon}} &= \boldsymbol{\epsilon} + \frac{z}{2}\left(\widetilde{\nabla}\boldsymbol{\phi} + (\widetilde{\nabla}\boldsymbol{\phi})^T\right) + \frac{z^3}{2}\left(\widetilde{\nabla}\boldsymbol{\psi} + (\widetilde{\nabla}\boldsymbol{\psi})^T\right) \\ &= \boldsymbol{\epsilon} + \frac{z}{2}\left(\widetilde{\nabla}\boldsymbol{\phi} + (\widetilde{\nabla}\boldsymbol{\phi})^T\right) - \alpha\frac{z^3}{2}\left(\widetilde{\nabla}\boldsymbol{\phi} + (\widetilde{\nabla}\boldsymbol{\phi})^T + 2\widetilde{\nabla}\widetilde{\nabla} w\right), \\ \boldsymbol{\gamma} &= \boldsymbol{\phi} + 3z^2\boldsymbol{\psi} = \boldsymbol{\phi} - 3z^2\alpha\boldsymbol{\phi} - 3z^2\alpha\widetilde{\nabla} w, \\ 2\boldsymbol{\epsilon} &= \widetilde{\nabla}\tilde{\mathbf{v}} + (\widetilde{\nabla}\tilde{\mathbf{v}})^T.\end{aligned}$$

In the third-order theory, we introduce the stress resultants $T_{\alpha\beta}$, the transverse shear stress resultants $T_{\alpha n}$, and the stress couples $M_{\alpha\beta}$. But we introduce some additional quantities such as the higher-order stress resultants $P_{\alpha\beta}$ and R_α:

$$\begin{gathered}T_{\alpha\beta} = [\![\sigma_{\alpha\beta}]\!] \equiv \int_{-h/2}^{h/2} \sigma_{\alpha\beta}\, dz, \qquad T_{\alpha n} = [\![\sigma_{\alpha n}]\!] \equiv \int_{-h/2}^{h/2} \sigma_{\alpha 3}\, dz, \\ M_{\alpha\beta} = [\![z\sigma_{\alpha\beta}]\!] \equiv \int_{-h/2}^{h/2} z\sigma_{\alpha\beta}\, dz, \\ P_{\alpha\beta} = [\![z^3\sigma_{\alpha\beta}]\!] \equiv \int_{-h/2}^{h/2} z^3\sigma_{\alpha\beta}\, dz, \quad R_\alpha = [\![z^2\sigma_{\alpha n}]\!] \equiv \int_{-h/2}^{h/2} z^2\sigma_{\alpha 3}\, dz.\end{gathered} \tag{5.76}$$

Exercise 5.16. Using (5.75) and (5.54), express (5.76) in terms of the components of $\boldsymbol{\phi}$ and w.

The equilibrium equations split into two systems. The first is for *the plane state*; it is of the form we had in Kirchhoff's theory:

$$\frac{\partial T_{11}}{\partial x_1} + \frac{\partial T_{21}}{\partial x_2} + q_1 = 0, \qquad \frac{\partial T_{12}}{\partial x_1} + \frac{\partial T_{22}}{\partial x_2} + q_2 = 0. \tag{5.77}$$

Equations (5.77) are supplemented with the boundary conditions

$$\tilde{\mathbf{v}}\big|_{\omega_1} = \tilde{\mathbf{v}}^0(s), \qquad \boldsymbol{\nu}\cdot\mathbf{T}\cdot\mathbf{A}\big|_{\omega_2} = \boldsymbol{\varphi}^0(s), \tag{5.78}$$

where

$$\mathbf{n}\cdot\tilde{\mathbf{v}}^0 = 0, \qquad \mathbf{n}\cdot\boldsymbol{\varphi}^0 = 0, \qquad \omega_1\cup\omega_2 = \omega, \qquad \omega_1\cap\omega_2 = \emptyset.$$

The constitutive equations are

$$T_{11} = \frac{Eh}{1-\nu^2}(\epsilon_{11} + \nu\epsilon_{22}),$$
$$T_{22} = \frac{Eh}{1-\nu^2}(\epsilon_{22} + \nu\epsilon_{11}),$$
$$T_{12} = \frac{Eh}{1+\nu}\epsilon_{12}.$$

The *boundary value problem for plate bending* is presented by the equilibrium equations

$$\frac{\partial T_{1n}}{\partial x_1} + \frac{\partial T_{2n}}{\partial x_2} + \alpha\left(\frac{\partial^2 P_{11}}{\partial x_1^2} + 2\frac{\partial^2 P_{12}}{\partial x_1\partial x_2} + \frac{\partial^2 P_{22}}{\partial x_2^2}\right) + q_n = 0,$$
$$\frac{\partial \widehat{M}_{11}}{\partial x_1} + \frac{\partial \widehat{M}_{21}}{\partial x_2} - \widehat{T}_{1n} = 0,$$
$$\frac{\partial \widehat{M}_{12}}{\partial x_1} + \frac{\partial \widehat{M}_{22}}{\partial x_2} - \widehat{T}_{2n} = 0,$$

where

$$\widehat{M}_{\alpha\beta} = M_{\alpha\beta} - \alpha P_{\alpha\beta}, \qquad \widehat{T}_{\alpha n} = T_{\alpha n} - 3\alpha R_\alpha.$$

The kinematic boundary conditions for the bending problem consist of four scalar equations:

$$w\big|_\omega = w^0(s), \qquad \left.\frac{\partial w}{\partial n}\right|_\omega = w_n^0(s), \qquad \boldsymbol{\phi}\big|_\omega = \boldsymbol{\phi}^0(s),$$

where $\mathbf{n}\cdot\boldsymbol{\phi}^0 = 0$. The static boundary conditions are expressed in terms of stress resultants, stress couples, and moments of higher order; the interested reader should consult the original sources [Wang, Reddy, and Lee (2000)] and Kienzler[7], Levinson[8], and Reddy[9].

Note that the third- and higher-order shell and plate theories show why variational formulations are needed. These are the only way to introduce correct static boundary conditions.

Micropolar shells or six-parameter shell theory

As an example of a non-classical shell theory, we will sketch the linear theory of micropolar shells. Its roots originate in the work of the Cosserat brothers, Eugèn and François [Cosserat and Cosserat (1909)]. The micropolar theory of shells is presented in the works of Eremeyev and Zubov [Eremeyev and Zubov (2008)], Zhilin [Zhilin (2006)], Eremeyev et al. [Eremeyev et al. (2013)] and others.

In this theory, the dynamical equations and shell kinematics coincide with those of a six-parameter shell theory, a nonlinear version of which is presented in the books by Libai and Simmonds [Libai and Simmonds (1998)] and Chróścielewski et al [Chróścielewski, Makowski, and Pietraszkiewicz (2004)]. There exists a micropolar plate theory by Eringen [Eringen (1999)]. It is based on integration over the thickness of the plate when its material is considered as a Cosserat continuum. This differs from six-parameter shell-plate theory, as it contains more than 6 unknown scalar functions. For six-parameter shells and other Cosserat-type theories of plates and shells, we refer the reader to Altenbach et al.[10], Pietraszkiewicz[11].

The kinematics of the shell surface are described by six scalar quantities, three of which are the components of the displacement vector $\mathbf{v}$ and the rest of which are the components of the microrotation vector $\boldsymbol{\vartheta}$. So a shell particle has six degrees of freedom described by the components of $\mathbf{v}$ and $\boldsymbol{\vartheta}$. The vector $\boldsymbol{\vartheta}$ satisfies $\boldsymbol{\vartheta} \cdot \mathbf{n} \neq 0$; the vectors $\mathbf{v}$ and $\boldsymbol{\vartheta}$ are mutually

[7] Kienzler, R. On consistent plate theories. *Arch. Appl. Mech.*, **72**, 229–247, 2002.

[8] Levinson, M. An accurate, simple theory of the statics and dynamics of elastic plates. *Mech. Res. Comm.*, **7**, 343–350, 1980.

[9] Reddy, J.N. A simple higher-order theory for laminated composite plates. *Trans. ASME J. Appl. Mech.*, **51**, 745–752, 1984.

[10] Altenbach, J., Altenbach, H., and Eremeyev, V. A. On generalized Cosserat-type theories of plates and shells: a short review and bibliography. *Arch. Appl. Mech.*, **80**, 73-92, 2010; Eremeyev, V., Altenbach, H. (2017). *Basics of Mechanics of Micropolar Shells.* In: Shell-like Structures. Advanced Theories and Applications. Springer, Cham, pp. 63-111.

[11] Pietraszkiewicz, W. The resultant linear six-field theory of elastic shells: What it brings to the classical linear shell models? *ZAMM*, **96**(8), 899-915, 2016.

independent. This is analogous to a two-dimensional version of the Cosserat medium with couple stresses and the rotation interaction of the particles.

For a micropolar shell, we can assign the couple load acting on the shell surface. Here the order of the equilibrium equations is 12, so we should supplement them with 6 conditions on the shell edge. On the part of the edge that is free of geometrical constraints, we can assign the distribution of forces and couples. This extends the range of applicability of the theory in comparison with the theories considered above: we can, say, assign boundary conditions that describe a shell clamped to a rigid body.

The micropolar theory is used, in particular, to describe branching shells and thin-walled bodies with complex intrinsic structure that include multilayered or composite plates and shells, shells with inner partition, with stringers and those similar to honeycomb or made of highly porous materials.

We present the micropolar shell equations for small deformation. The dynamical equations are

$$\begin{aligned} \widetilde{\nabla} \cdot \mathbf{T} + \mathbf{q} \quad &= \rho \ddot{\mathbf{v}} + \rho \boldsymbol{\Theta}_1 \cdot \ddot{\boldsymbol{\vartheta}}, \\ \widetilde{\nabla} \cdot \mathbf{M} + \mathbf{T}_\times + \mathbf{m} &= \rho \boldsymbol{\Theta}_1^{\mathrm{T}} \cdot \ddot{\mathbf{v}} + \rho \boldsymbol{\Theta}_2 \cdot \ddot{\boldsymbol{\vartheta}}, \end{aligned} \tag{5.79}$$

where $\mathbf{T}$ and $\mathbf{M}$ are surface stress and couple stress tensors analogous to those in Kirchhoff's theory, ρ is the shell surface density, $\boldsymbol{\Theta}_1$ and $\boldsymbol{\Theta}_2$ are the inertia tensors, $\boldsymbol{\Theta}_2$ is symmetric, $\boldsymbol{\Theta}_2^T = \boldsymbol{\Theta}_2$, and $\mathbf{q}$ and $\mathbf{m}$ are distributed surface forces and couples, respectively. Besides, in this theory, we can assign a twisting couple (or drilling moment) on the shell surface. The equilibrium equations take the form

$$\widetilde{\nabla} \cdot \mathbf{T} + \mathbf{q} = \mathbf{0}, \qquad \widetilde{\nabla} \cdot \mathbf{M} + \mathbf{T}_\times + \mathbf{m} = \mathbf{0}. \tag{5.80}$$

The conditions at a boundary point can be kinematic

$$\mathbf{v}\big|_{\omega_1} = \mathbf{v}^0(s), \qquad \boldsymbol{\vartheta}\big|_{\omega_3} = \boldsymbol{\vartheta}^0(s), \tag{5.81}$$

or static

$$\boldsymbol{\nu} \cdot \mathbf{T}\big|_{\omega_2} = \boldsymbol{\varphi}(s), \qquad \boldsymbol{\nu} \cdot \mathbf{M}\big|_{\omega_4} = \boldsymbol{\ell}(s), \tag{5.82}$$

where $\mathbf{v}^0(s)$ and $\boldsymbol{\vartheta}^0(s)$ are given vector functions of the length parameter s; they define the displacements and microrotation of the shell edge. The functions $\boldsymbol{\varphi}(s)$ and $\boldsymbol{\ell}(s)$ determine the surface stresses and stress couples on the edge. Here

$$\omega = \omega_1 \cup \omega_2 = \omega_3 \cup \omega_4, \qquad \omega_2 = \omega \setminus \omega_1, \qquad \omega_4 = \omega \setminus \omega_3.$$

The form of the equilibrium and dynamical equations for a micropolar shell does not differ from (5.66) and (5.67). In general, in this theory $\mathbf{m} \cdot \mathbf{n} \neq 0$ and $\boldsymbol{\ell} \cdot \mathbf{n} \neq 0$. In particular, $\mathbf{M} \cdot \mathbf{n} \neq \mathbf{0}$.

The strain measures take the form

$$\boldsymbol{\epsilon} = \widetilde{\nabla}\mathbf{v} + \mathbf{A} \times \boldsymbol{\vartheta}, \qquad \boldsymbol{\varkappa} = \widetilde{\nabla}\boldsymbol{\vartheta}, \tag{5.83}$$

where $\boldsymbol{\epsilon}$ and $\boldsymbol{\varkappa}$ are non-symmetric strain and bending strain tensors respectively.

The constitutive equations for an elastic shell are represented though the strain energy density $U = U(\boldsymbol{\epsilon}, \boldsymbol{\varkappa})$:

$$\mathbf{T} = \frac{\partial U}{\partial \boldsymbol{\epsilon}}, \qquad \mathbf{M} = \frac{\partial U}{\partial \boldsymbol{\varkappa}}. \tag{5.84}$$

For an isotropic shell, U is a quadratic form in its variables:

$$\begin{aligned} 2U &= \alpha_1 \operatorname{tr}^2 \boldsymbol{\epsilon} + \alpha_2 \operatorname{tr} \tilde{\boldsymbol{\epsilon}}^2 + \alpha_3 \operatorname{tr}\left(\tilde{\boldsymbol{\epsilon}} \cdot \tilde{\boldsymbol{\epsilon}}^T\right) + \alpha_4 \mathbf{n} \cdot \boldsymbol{\epsilon}^T \cdot \boldsymbol{\epsilon} \cdot \mathbf{n} \\ &\quad + \beta_1 \operatorname{tr}^2 \boldsymbol{\varkappa} + \beta_2 \operatorname{tr} \tilde{\boldsymbol{\varkappa}}^2 + \beta_3 \operatorname{tr}\left(\tilde{\boldsymbol{\varkappa}} \cdot \tilde{\boldsymbol{\varkappa}}^T\right) + \beta_4 \mathbf{n} \cdot \boldsymbol{\varkappa}^T \cdot \boldsymbol{\varkappa} \cdot \mathbf{n}, \\ \tilde{\boldsymbol{\epsilon}} &= \boldsymbol{\epsilon} \cdot \mathbf{A}, \quad \tilde{\boldsymbol{\varkappa}} = \boldsymbol{\varkappa} \cdot \mathbf{A}, \end{aligned} \tag{5.85}$$

where α_k and β_k $(k = 1, 2, 3, 4)$ are elastic moduli. Substituting (5.85) into (5.84), we get

$$\begin{aligned} \mathbf{T} &= \alpha_1 \mathbf{A} \operatorname{tr} \boldsymbol{\epsilon} + \alpha_2 \tilde{\boldsymbol{\epsilon}}^T + \alpha_3 \tilde{\boldsymbol{\epsilon}} + \alpha_4 (\boldsymbol{\epsilon} \cdot \mathbf{n})\mathbf{n}, \\ \mathbf{M} &= \beta_1 \mathbf{A} \operatorname{tr} \boldsymbol{\varkappa} + \beta_2 \tilde{\boldsymbol{\varkappa}}^T + \beta_3 \tilde{\boldsymbol{\varkappa}} + \beta_4 (\boldsymbol{\varkappa} \cdot \mathbf{n})\mathbf{n}. \end{aligned} \tag{5.86}$$

Equations (5.80)–(5.82) and (5.86) constitute a linear boundary value problem with respect to the fields of displacement and microrotation; they describe the equilibrium of a micropolar shell in the case of small deformation. For dynamic problems, equations (5.80) change to (5.79). Under some additional assumptions, Reissner's shell theory is a consequence of the micropolar theory.

As in the other shell theories, we can formulate some variational principles. *Lagrange's variational principle* for an elastic micropolar shell starts with formulation of the total energy functional

$$\mathcal{E}(\mathbf{v}, \boldsymbol{\vartheta}) = \int_\Sigma U(\boldsymbol{\epsilon}, \boldsymbol{\varkappa})\, d\Sigma - A(\mathbf{v}, \boldsymbol{\vartheta}), \tag{5.87}$$

where the potential of external loads $A(\mathbf{v}, \boldsymbol{\vartheta})$ is

$$A(\mathbf{v}, \boldsymbol{\vartheta}) = \int_\Sigma (\mathbf{q} \cdot \mathbf{v} + \mathbf{m} \cdot \boldsymbol{\vartheta})\, d\Sigma + \int_{\omega_2} \boldsymbol{\varphi} \cdot \mathbf{v}\, ds + \int_{\omega_4} \boldsymbol{\ell} \cdot \boldsymbol{\vartheta}\, ds.$$

The functional $\mathcal{E}(\mathbf{v}, \boldsymbol{\vartheta})$ is considered on the set of twice continuously differentiable fields of displacements and microrotations that satisfy (5.81). The pair $(\mathbf{v}, \boldsymbol{\vartheta})$ that satisfies (5.80) and (5.82) is a stationary point of $\mathcal{E}(\mathbf{v}, \boldsymbol{\vartheta})$. Lagrange's stationary principle is minimal: on the equilibrium solution, the functional (5.87) attains its minimum.

Rayleigh's variational principle now takes the following form.

On the set of functions with boundary conditions $\mathbf{v}\big|_{\omega_1} = \mathbf{0}$, $\boldsymbol{\vartheta}\big|_{\omega_3} = \mathbf{0}$ *that obey the constraint*

$$K(\mathbf{v}^\circ, \boldsymbol{\vartheta}^\circ) \equiv \int_\Sigma \rho \left(\tfrac{1}{2}\mathbf{v}^\circ \cdot \mathbf{v}^\circ + \mathbf{v}^\circ \cdot \boldsymbol{\Theta}_1 \cdot \boldsymbol{\vartheta}^\circ + \tfrac{1}{2}\boldsymbol{\vartheta}^\circ \cdot \boldsymbol{\Theta}_2 \cdot \boldsymbol{\vartheta}^\circ \right) d\Sigma = 1,$$

the eigenmodes of the shell are stationary points of the strain energy functional

$$\mathcal{E}(\mathbf{v}^\circ, \boldsymbol{\vartheta}^\circ) = \int_\Sigma U(\boldsymbol{\epsilon}^\circ, \boldsymbol{\varkappa}^\circ)\, d\Sigma,$$

where $\boldsymbol{\epsilon}^\circ = \widetilde{\nabla}\mathbf{v}^\circ + \mathbf{A} \times \boldsymbol{\vartheta}^\circ$ *and* $\boldsymbol{\varkappa}^\circ = \widetilde{\nabla}\boldsymbol{\vartheta}^\circ$.

Rayleigh's principle also includes the reverse statement: on the set of functions satisfying the above restrictions, the stationary points of $\mathcal{E}$ are the eigenmodes. Here $\mathbf{v}^\circ$ and $\boldsymbol{\vartheta}^\circ$ are the amplitudes of the oscillations of the displacements and microrotations as the solutions of the eigenproblem are sought in the form $\mathbf{v} = \mathbf{v}^\circ e^{i\omega t}$, $\boldsymbol{\vartheta} = \boldsymbol{\vartheta}^\circ e^{i\omega t}$. Rayleigh's quotient takes the form

$$R(\mathbf{v}^\circ, \boldsymbol{\vartheta}^\circ) = \frac{\mathcal{E}(\mathbf{v}^\circ, \boldsymbol{\vartheta}^\circ)}{K(\mathbf{v}^\circ, \boldsymbol{\vartheta}^\circ)}.$$

Now the lowest eigenfrequency of the shell, which is the minimal frequency, is equal to the minimum value of the functional R. For applications of Rayleigh's quotient in the micropolar shell theory we refer the reader to [Eremeyev, Lebedev, and Cloud (2015)].

Finally, we would like to repeat that, contrary to the situation with linear elasticity, shell theory is still under development; investigators continue to seek better models and numerical methods for the solution of practical problems. This is a consequence of the evident fact that it is impossible to find an accurate approximation to the results of a three-dimensional elasticity problem under all the circumstances when one solves the same problem using one of the two-dimensional models of shell theory.

Chapter 6

Mechanics of Generalized Media

This chapter is devoted to some applications of tensor analysis to the mechanics of so-called generalized continua. The notion of "generalized medium" has had a relatively long history over which the meaning of this term changed significantly a few times. Hence it can refer to very different objects in the literature of various time periods.

So we should first clarify what we mean by a *generalized continuum*. The term would seem to imply a generalization of the classical continuum model first developed by Cauchy, who elaborated the ideas of the stress and small strain tensors and their relationships. Comparing the elasticity for isotropic solids under small deformations with anisotropic elasticity, we may call the latter a generalization of the former. Indeed, in the latter we use the "generalized" Hooke's law. Another, wider generalization of isotropic linear elasticity is the elasticity under finite deformations. This leads to nonlinear equations and more complex boundary-value problems. But now the anisotropic and nonlinear elasticities are regarded as classical models. So as a generalized continuum model we treat a model beyond classical elasticity, whether it is linear or nonlinear, isotropic or anisotropic.

Historically, the development of the models of continuum mechanics first brought us so-called *weak* generalizations of elasticity. These were the theories of thermoelasticity, linear piezoelectricity, and the model of a continuum with diffusion. These models additionally employed non-mechanical scalar variables — the temperature θ, the electrical potential φ, the concentration c — but they still used the deformation apparatus of classical elasticity.

An early generalization of classical continuum theory that significantly changed how deformations of a body were described was the model of the medium with *couple stresses* and rigidly rotated microstructure. It was developed by brothers E. and F. Cosserat [Cosserat and Cosserat (1909)] over a century ago. In the Cosserats' theory, the microstructure of the material is determined by three orthogonal vectors called the *directors*. Their rotations are described by the *vector of rotations* or by an *orthogonal tensor*. So in addition to the classical translations of a point, as new kinematic variables the model introduces rotations attached to the point. Physically, each material point of the Cosserat medium has six degrees of freedom, the same six degrees of freedom as a rigid body of classic dynamics [Bauchau (2010)] has. At present the Cosserat continuum is also known as the *micropolar* or *polar* or *oriented* medium [Kafadar and Eringen (1971); Eringen (1999); Eringen (2001); Eremeyev et al. (2013); Maugin (2017)]. The Cosserat continuum can be regarded as an early *true* or *strong* generalization of classical elasticity. Unlike the above-mentioned weak generalizations, here we see the introduction of additional internal variables and additional degrees of freedom of a mechanical nature: the rotations.

The idea of introducing additional degrees of freedom was used in other generalized models of a similar nature. We could mention the model of the *micromorphic continuum* introduced by Mindlin and Eringen [Eringen (1999)], that of the *liquid crystals* [Chandrasekhar (1977); de Gennes and Prost (1993)], and those of the microstretched medium, dilatational elasticity, mixtures, and some others; see [Eringen (1999); Eringen (2001); Maugin (2017)] and the references therein.

Some other types of generalized models of continuum were developed in *gradient theories*, which take into account the higher-order gradients of strains or stresses, and in *nonlocal theories* [Eringen (2002); Maugin (2017)].

Developing the foundations of continuum mechanics, scientists have proposed various approaches to the construction of models. Many models are now known as models of generalized media; see for example the analysis of Piola's contribution[1]. For some history of development in the field of generalized continua, we refer to Maugin's recent works[2].

---

[1]dell'Isola, F., Andreaus, U., and Placidi, L. (2015). At the origins and in the vanguard of peridynamics, non-local and higher-gradient continuum mechanics: An underestimated and still topical contribution of Gabrio Piola. *Math. Mech. Solids*, 20(8), 887–928.

[2]Maugin, G. A. (2010). Generalized continuum mechanics: what do we mean by that?. In: Maugin, G.A., Metrikine, A.V. (eds.) *Mechanics of Generalized Continua: One Hundred Years After the Cosserats*, pp. 3–13. (Springer, New York); Maugin, G. A. (2013). *Continuum mechanics through the twentieth century.* (Berlin, Springer); Maugin, G. A. (2015). Some remarks on generalized continuum mechanics. *Math. Mech. Solids*, 20(3), 280–291.

This chapter considers some applications of tensor analysis to models for materials and structural behaviors more complex than those of linear elasticity or the theory of linear plates and shells.

6.1 Some Basic Principles for Modeling Complex Media

Now we will briefly discuss certain broad principles useful in describing generalized media. These will be illustrated through their applications to nonlinear elasticity, micropolar and micromorphic media, strain gradient models, and liquid crystals.

Let us consider some simplified mechanical versions of the models for a complex medium. Here we will neglect temperature effects, chemical reactions, and other non-mechanical processes. Consideration will be restricted to elastic materials only.

We start by *assuming* the existence of a strain energy density W which is a *function* of properly chosen kinematic variables. W is the deformation energy stored in a unit volume of the elastic body. Its dependence on the kinematic variables is captured in a *constitutive relation*. The so-called kinematical variables will include all the parameters necessary to describe the body's motion. Constitutive equations describe the behavior of materials, and the form of such an equation will differ for such materials as steel, rubber, and oil. It is often formulated empirically; tension test determine Young's modulus, torsion tests determine shear modulus, and so on. For purely elastic materials, a single constitutive equation should suffice for the strain energy density. But in general more than one such equation may be required, interrelating the stress tensor, free energy, internal energy, entropy, heat flux, or other material characteristics.

Constitutive equations and experimentally obtained material parameters are always *approximate*. Nevertheless, regardless of medium, all constitutive equations must obey certain general principles that serve to restrict their form [Truesdell and Noll (2004)]. These are of a different nature compared to such principles as energy conservation or momentum balance. We will formulate them in terms of the strain energy function W, but they apply *mutatis mutandis* to other functions.

The first restrictive principle for W is

The Principle of Determinism. *At each point of a body and any instant of time, the strain energy density is uniquely determined by the history of the body's motion.*

In other words a constitutive equation cannot depend on the future; rather, the present state of the body is uniquely determined by the past motion. The history of the body's motion will in general include all past states of the body up to the current state.

The Principle of Local Action. *At each point of a body and any instant of time, the strain energy density is uniquely determined by the motion of an arbitrarily small neighborhood of the point.*

This expresses a natural property of a constitutive equation: the stored energy at a point of the body depends on the deformations of a (very) small neighborhood of the point and is independent of the deformations at remote points. In other words, the energy depends on the interaction between the nearest neighbors of the point under consideration. However, while the Principle of Determinism is absolute (holding under all circumstances), the Principle of Local Action fails for certain materials. Indeed, for some models the value of the energy function at a point may depend on the deformation of the whole body. Such non-local continuum models are presented, for example, in [Eringen (2002)].

Next we turn attention to the *Principle of Material Frame-Indifference*, starting with the notion of frame-indifference for an objective quantity. Recall how equivalent motions are defined. In classical mechanics, a motion is described by the position vector $\mathbf{r}(t)$. We call two motions $\mathbf{r}(t)$ and $\mathbf{r}^*(t)$ *equivalent* if they are related by the formula

$$\mathbf{r}^*(t) = \mathbf{a}(t) + \mathbf{Q}(t) \cdot (\mathbf{r}(t) - \mathbf{r}_0) \tag{6.1}$$

where $\mathbf{Q}(t)$ is an arbitrary orthogonal tensor, $\mathbf{a}(t)$ is an arbitrary vector function, and the constant vector $\mathbf{r}_0$ represents a fixed point position (a pole). We can treat all the equivalent motions as the motions of a body considered in different reference frames which can translate and rotate relative to one another. Introducing a so-called observer with respect to a related reference frame, we can say that we change the position and orientation of the observer who considers the motions of the body.

The mapping $\mathbf{r} \to \mathbf{r}^*$ given by (6.1) is called the *Euclidean transformation* because it preserves lengths in the sense that $|\mathbf{r}_1 - \mathbf{r}_2| = |\mathbf{r}_1^* - \mathbf{r}_2^*|$. Another interpretation is that (6.1) describes a superposition of two motions: the original motion described by $\mathbf{r}(t)$, and a rigid body motion. The latter is determined by the translation vector $\mathbf{a}(t)$ and the rotation tensor $\mathbf{Q}(t)$ about the "pole" at position $\mathbf{r}_0$. So (6.1) can be regarded as a *superimposed rigid-body motion.*

We call a scalar x, a vector $\mathbf{x}$, and a second-order tensor $\mathbf{X}$ *indifferent* or *objective* if for any two equivalent motions

$$\begin{aligned} x^* &= x, \\ \mathbf{x}^* &= \mathbf{Q} \cdot \mathbf{x}, \\ \mathbf{X}^* &= \mathbf{Q} \cdot \mathbf{X} \cdot \mathbf{Q}^T, \end{aligned} \tag{6.2}$$

where $*$ tags quantities in the equivalent motion defined by (6.1).

As an example, let us consider the *mirror reflection* described by $\mathbf{Q} = -\mathbf{E}$. From (6.2) it follows that the objective quantities satisfy

$$x^* = x, \quad \mathbf{x}^* = -\mathbf{x}, \quad \mathbf{X}^* = \mathbf{X}.$$

To consider more general media, we must extend the material frame-indifference principle to axial (or pseudo) vectors, tensors and scalars. As mentioned in § 2.10, axial quantities change sign under mirror reflections. So according to our previous definition axial scalars, vectors, and tensors are not "real" frame-indifferent quantities. Of course, we could require $\mathbf{Q}$ in (6.1) to be a proper orthogonal tensor for which there is no difference between axial and polar quantities. But a non-proper orthogonal tensor corresponds to a coordinate rotation and a change of orientation. Such transformations are admissible for some models. So we modify the definition of indifferent or objective quantities as follows.

We call an axial scalar (pseudoscalar) ω, an axial vector $\boldsymbol{\omega}$, and an axial tensor $\boldsymbol{\Omega}$ indifferent or objective if for two equivalent motions

$$\begin{aligned} \omega^* &= (\det \mathbf{Q})\omega, \\ \boldsymbol{\omega}^* &= (\det \mathbf{Q})\mathbf{Q} \cdot \boldsymbol{\omega}, \\ \boldsymbol{\Omega}^* &= (\det \mathbf{Q})\mathbf{Q} \cdot \boldsymbol{\Omega} \cdot \mathbf{Q}^T. \end{aligned} \tag{6.3}$$

In particular, for the mirror reflection $\mathbf{Q} = -\mathbf{E}$ we get

$$\omega^* = -\omega, \quad \boldsymbol{\omega}^* = \boldsymbol{\omega}, \quad \boldsymbol{\Omega}^* = -\boldsymbol{\Omega}.$$

At first glance the transformation rules (6.3) may contradict a naive understanding of the independence of an objective quantity from the change of coordinates. For example, a pseudoscalar changes sign under a mirror reflection. However, the dependence on the mirror reflections is a consequence of our agreement on the choice of right-hand or left-hand rules when we define such quantities as cross product or curvature. It is similar to the convention by which the two poles of a magnet are designated as north and south. Being described by indifferent axial quantities, physical processes do not depend on the choice of coordinates; in particular, they do not depend on change of frame orientation. An interesting discussion of the physical nature of axial vectors and the mirror reflection can be found in [Feynman et al. (1977)]. In what follows, we will encounter mechanical quantities which represent pseudoscalars, pseudovectors and pseudotensors.

Now we are able to formulate

The Principle of Material Frame-Indifference. *The strain energy density is material frame-indifferent.*

Also known as the *Principle of Objectivity*, it was proposed by Walter Noll [Truesdell and Noll (2004)]. The concise formulation implies serious consequences; even recently the literature has featured extensive discussion on how the principle should be interpreted, as its different formulations seem to reflect different physical contents (see, e.g., Murdoch[3], Rivlin[4], [Bertram (2005)], and others[5]). Svenden and Bertram stated that the principle of material frame-indifference is related to three different principles:

(1) the Principle of Invariance under Euclidean transformations,
(2) the Principle of Invariance under superimposed rigid-body motions, and
(3) the Principle of Form-Invariance of the constitutive equations under the change of observer.

They proved that any two of these principles implies the third one. Each places certain constraints on the possible form of a constitutive equation.

Anisotropy

Other forms of constitutive restriction follows from the consideration of the anisotropy properties of a material. In short, they follow from the invariance of the constitutive equations under certain changes of reference

[3]Murdoch, A.I. (2000). On objectivity and material symmetry for simple elastic solids. *J. Elast.* **60**(3), 233–242; (2003). Objectivity in classical continuum physics: a rationale for discarding the 'principle of invariance under superposed rigid body motions' in favour of purely objective considerations. *Continuum Mech. Thermodynamics* **15**(3), 309–320; (2005). On criticism of the nature of objectivity in classical continuum physics. *Continuum Mech. Thermodynamics* **17**(2), 135–148

[4]Rivlin, R.S. (2005). Frame indifference and relative frame indifference. *Math. Mech. Solids* **10**(2), 145–154; (2006). Some thoughts on frame indifference. *Math. Mech. Solids* **11**(2), 113–122

[5]Muschik, W., Restuccia, L. (2002) Changing the observer and moving materials in continuum physics: objectivity and frame-indifference. *Technische Mechanik* **22**(2), 152–160; Svendsen, B., Bertram, A. (1999) On frame-indifference and form-invariance in constitutive theory. *Acta Mech.***132**(1–4), 195–207.

configuration (or reference placement). Recall that in cases of large (finite) deformation we should use two states of a body. The *actual* or *deformed configuration* (or placement) describes the current state of the body at time t. To measure the strains and other changes in motion, we introduce another state called the *reference* or *initial configuration* (or placement). A deformation can be considered as a mapping from the reference configuration to the actual configuration. As the initial configuration, the stress-free reference configuration (or natural configuration) is typically used. But important exceptions do exist. Obviously, the constitutive equations depend on the choice of reference configuration. Assuming certain symmetries for the material, we find that some transformations of the reference configuration should preserve certain forms of the constitutive equations. So we can speak of the invariance of the constitutive equations under certain transformations of the reference configuration. Such transformations constitute the *local material symmetry group*. For example, the notion of an isotropic material is related to the invariance of its constitutive equations under all orthogonal transformations of the reference configuration.

The literature also describes other restrictions that can simplify or reduce the form of constitutive equations. Among the restrictive principles we should mention Hadamard's inequality and strong ellipticity conditions, which are based on mathematical considerations. Unlike the principles of determinism, locality, and material frame-indifference, these restrictions are not absolute, in general. However, in some cases a loss of ellipticity of the equilibrium equations of nonlinear elasticity appears to be unacceptable.

6.2 Introducing a Generalized Medium Model

Now we sketch the steps for introducing a model of a generalized medium and studying its properties. The model commences with a set of kinematic variables describing the deformations.

Step 1. Kinematic variables. This important step includes the introduction of all variables, and their gradients up to the desired order, used to describe motions of the medium. We introduce the kinematic variables $\mathbf{r}, \nabla\mathbf{r}, \nabla\nabla\mathbf{r}, \ldots$, and $\mathbf{p}, \nabla\mathbf{p}, \nabla\nabla\mathbf{p}, \ldots$, where $\mathbf{r}$ is the position vector of a medium point and $\mathbf{p}$ stands for additional parameters called the *internal variables*. This $\mathbf{p}$ can be a vectorial parameter, a tensor of any order, or even a set of vectors and tensors. The choice is related to the additional degrees of freedom we wish to assign to particles in the model. Note that the gradient operator ∇ is unspecified in the sense that we can use ∇ defined in the initial state, in the deformed state, or even in an intermediate state of the medium. *All constitutive equations should be presented in these variables.*

So here we assume that the form of the strain energy density is

$$W = W(\mathbf{r}, \nabla\mathbf{r}, \nabla\nabla\mathbf{r}, \ldots, \mathbf{p}, \nabla\mathbf{p}, \nabla\nabla\mathbf{p}, \ldots). \tag{6.4}$$

In this step we can also impose additional *a priori* constraints like the incompressibility condition.

Step 2. Constitutive equations. Here we apply to W the principles of determinism, locality, and material frame-indifference. As we have restricted ourselves to elastic media, applying the determinism and locality principles is simple. The principle of determinism holds if we use the constitutive equation in the form (6.4): *W is a function of kinematic variables taken at current instant t; there is no dependence of W on the past or future.* With finitely many gradients in the set of variables, the principle of locality also holds. More restrictive is the principle of material frame-indifference. Assuming W to be a scalar (not a pseudoscalar!) valued function, we get the following constraint:

$$W^* = W, \quad W^* = W^*(\mathbf{r}^*, \nabla\mathbf{r}^*, \ldots, \mathbf{p}^*, \nabla\mathbf{p}^*, \ldots). \tag{6.5}$$

This implies a reduced form of the constitutive equations in general. In particular, we may obtain some proper strain measures. At this stage the material symmetry considerations may be useful as well.

Step 3. Derivation of static equations. This is usually done as follows. Find the first variation δW of W expressed through the kinematic variables and their variations. Introduce the first variation of the energy functional $\delta\mathcal{E} = \int \delta W \, dV$. Then apply the Gauss–Ostrogradsky (divergence) theorem to $\delta\mathcal{E}$ to obtain $\delta\mathcal{E}$ in the form containing only kinematically independent variables. For the materials with internal constraints, one should use the Lagrange multiplier method. Application of the divergence theorem results at least in two consequences. First, we establish the form of the equilibrium equations and the natural (static) boundary conditions. Second, the form of $\delta\mathcal{E}$ dictates the possible form of the work of external loads which is consistent with the choice of kinematical variables. In this way we come to the *Virtual Work Principle.* For a demonstration of how to apply the virtual work principle to a model of generalized media, we refer to original papers by Sedov[6], Germain[7], and Maugin[8].

Steps 1–3 have been formulated for static problems. But they are also used for the dynamic approach.

Step 4. Dynamics. In extending a static approach to dynamical models, we should also introduce velocities and generalized velocities as variables, relating to the time rates of the additional degrees of freedom. As shown in Chapter 4, a direct way to extend static problems to dynamics is to apply d'Alembert's principle, by which we formally add the inertia forces to the forces in the equilibrium equations. In this way we come to the equations of motion. A more general and more physically reasonable way is to use the Least Action Principle for the generalized medium. Its formulation requires the introduction of the kinetic energy density K as a function of generalized velocities. Commonly K is taken to be be a positive quadratic form in the velocities. Then we can apply the least action principle (or *Hamilton's principle*) which is

$$\delta\mathcal{H} = 0, \quad \mathcal{H} = \int_{t_1}^{t_2} \int_V (K - W)\, dV\, dt$$

for the parameter functions which, at time instants t_1 and t_2, coincide with the values of the variables in the real motion. From this, using standard tools from the calculus of variations, we can derive the equations of motion and again, as in statics, the natural boundary conditions. Obviously, for dynamics, the assumed form of K plays a crucial role.

Step 5. Inelastic media. There are numerous types of inelastic behavior, with various mechanisms responsible. Here we can mention plasticity, creep, fatigue, and viscoelasticity among others. Unlike linear elasticity, the models for these theories are still under development. For these media, the principle of virtual work is also a fruitful tool despite the fact that the strain energy density may not exist here.

Now we consider the introduction of some simple mechanical models.

[6]Sedov, L.I. (1968). Models of continuous media with internal degrees of freedom. *J. Appl. Math. Mech.* 32(5), 803–819

[7]Germain, P. (1973). La méthode des puissances virtuelles en mécanique des milieux continus, Première partie: théorie du second gradient. *J. de Mécanique (Paris)* 12, 235–274; The method of virtual power in continuum mechanics-II: microstructure. *SIAM J. Appl. Math.* 25, 556–575 (1973)

[8]Maugin, G.A. (2013). The principle of virtual power: from eliminating metaphysical forces to providing an efficient modelling tool. *Continuum Mech. Thermodyn.* 25(2–4), 127–146.

6.3 Nonlinear Elasticity

Although nonlinear elasticity cannot be treated as a generalized continuum model, we review its basic elements in order to demonstrate the application of the aforementioned principles and to provide background for subsequent sections. Nonlinear elasticity is also known as the elasticity of large deformations, or finite elasticity. It is a well developed branch of continuum mechanics. The nonlinear theory is free from the assumption of small deformations; it can describe the deformations of rubber-like materials and biological tissues, which may exhibit thousand percent deformations. Many textbooks and monographs treat nonlinear elasticity [Antman (2005); Bertram (2005); Ciarlet (1988); Green and Adkins (1960); Green and Zerna (1954); Lurie (1990); Ogden (1997); Ogden and Holzapfel (2006)]. For a novice, the short introductions in [Gurtin (1983)] and [Ogden (2001)] may be quite useful.

Kinematics

An elastic body occupies a closed connected subset of the three-dimensional (3D) space.

Each material point of the body is labeled by its position vector. As indicated above, a deformation is defined as a one-to-one mapping between two states of the body. The first is called the *reference* or *initial configuration* (or *placement*); the second, deformed, state is the *current* or *actual configuration.* It is usual to choose the undeformed state as the initial configuration, while the actual configuration is the state of the body under the action of external loads. We will denote the reference and actual configurations by κ and χ, respectively, and the volumes occupied by the body in these configurations by $V \in \mathbb{R}^3$ and $v \in \mathbb{R}^3$. The deformation of an elastic body has a simple geometric meaning as a mapping from V into v:

$$\mathbf{r} = \mathbf{r}(\mathbf{x}) \qquad (\mathbf{x} \in V). \tag{6.6}$$

Here $\mathbf{x}$ and $\mathbf{r}$ are the position vectors of the body in the reference and actual configurations, respectively (Fig. 6.1). In what follows, we assume that $\mathbf{r}(\mathbf{x})$ is a twice continuously differentiable vector-valued function.

Following [Gurtin (1983); Ogden (1997); Ogden (2001)], we slightly change the definition of the *deformation gradient tensor*

$$\mathbf{F} = (\nabla \mathbf{r})^T. \tag{6.7}$$

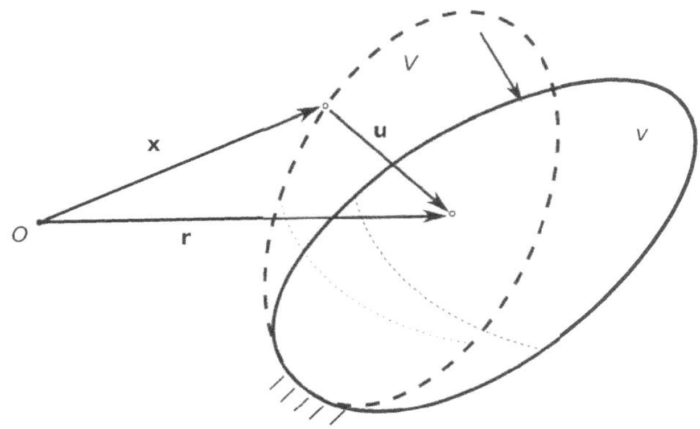

Fig. 6.1 A large deformation of a three-dimensional body. The dashed contour indicates the shape of the body in the reference configuration.

Here ∇ is the gradient operator defined in the initial configuration, i.e., defined via $d\mathbf{r} = d\mathbf{x} \cdot \nabla\mathbf{r} = \mathbf{F} \cdot d\mathbf{x}$ (see the discussion below). As $\mathbf{r}(\mathbf{x})$ is assumed to be invertible, $\mathbf{F}$ is nonsingular except possibly at certain singular points, so $\det \mathbf{F} \neq 0$ in general and we adopt the convention that $\det \mathbf{F} > 0$ everywhere in the body.

Note that for non-small deformations we should distinguish the reference and actual configurations, κ and χ, respectively. As they occupy different regions in $\mathbb{R}^3$, in both regions we can introduce different coordinate systems and hence different gradient operators. Let ∇_κ be the referential gradient operator and ∇_χ the gradient operator in the actual configuration. Using the identities which follow from the definition of the gradient operator

$$df = d\mathbf{r} \cdot \nabla_\chi f = d\mathbf{x} \cdot \nabla_\kappa f$$

for any scalar function f which can be considered as a function of argument $\mathbf{r}$ or of $\mathbf{x}$, we find the relation between ∇_χ and ∇_κ:

$$\nabla_\chi = (\nabla_\kappa \mathbf{r}) \cdot \nabla_\kappa \tag{6.8}$$

and in consequence the identities

$$\nabla_\kappa \mathbf{r} = \mathbf{F}^T, \qquad \nabla_\kappa \mathbf{x} = \mathbf{E}, \qquad \nabla_\chi \mathbf{r} = \mathbf{E}, \qquad \nabla_\chi \mathbf{x} = (\mathbf{F}^{-1})^T.$$

But for brevity, in this section we will simply write $\nabla_\kappa = \nabla$.

Using the polar decompositions (2.39)–(2.40), we can uniquely represent the deformation gradient, a nonsingular second-order tensor, in two forms:

$$\mathbf{F} = \mathbf{R} \cdot \mathbf{U} = \mathbf{V} \cdot \mathbf{R}, \tag{6.9}$$

where $\mathbf{R}$ is a proper orthogonal tensor (we have supposed that $\det \mathbf{F} > 0$!) and $\mathbf{U}, \mathbf{V}$ are positive definite symmetric tensors. Each decomposition in (6.9) is unique. $\mathbf{U}$ and $\mathbf{V}$ are the *right* and *left stretch tensors*, respectively. They are obtained from the equations

$$\mathbf{U}^2 = \mathbf{F}^T \cdot \mathbf{F}, \qquad \mathbf{V}^2 = \mathbf{F} \cdot \mathbf{F}^T. \tag{6.10}$$

Tensors $\mathbf{C} \equiv \mathbf{F}^T \cdot \mathbf{F}$ and $\mathbf{B} \equiv \mathbf{F} \cdot \mathbf{F}^T$ are the *right* and *left Cauchy–Green deformation tensors*; $\mathbf{B}$ is also called the *Finger strain measure*.

For the body deformation we can obtain useful formulas that describe the changes of certain geometric quantities. In Chapter 4 we discussed the change in length of a segment during deformation. Similarly, considering the change in the elementary volume dV of an infinitesimal part of a body in the reference placement, we get

$$J\, dV = dv$$

where $J = \det \mathbf{F}$ and dv denotes the corresponding elementary volume after deformation (i.e., in the actual configuration).

For oriented elementary areas in the reference and actual configurations, we have *Nanson's formula*

$$\hat{\mathbf{n}}\, \widehat{dS} = J \mathbf{n} \cdot \mathbf{F}^{-1}\, dS, \tag{6.11}$$

where dS, $\widehat{dS}$ are corresponding elementary areas and $\mathbf{n}, \hat{\mathbf{n}}$ are their unit normals in the reference and actual configurations. Equation (6.11) implies the following relation between the elementary areas:

$$\widehat{dS} = J\sqrt{\mathbf{n} \cdot \mathbf{F}^{-1} \cdot \mathbf{F}^{-T} \cdot \mathbf{n}}\, dS = J\sqrt{\mathbf{n} \cdot \mathbf{C}^{-1} \cdot \mathbf{n}}\, dS. \tag{6.12}$$

An important tool for transforming equations written in the reference configuration into equations associated with the actual configuration is the pair of *Piola identities*

$$\nabla \cdot \left[J(\nabla \mathbf{r})^{-T}\right] = \mathbf{0}, \qquad \nabla_{\chi} \cdot \left[J^{-1}(\nabla \mathbf{r})^{T}\right] = \mathbf{0}. \tag{6.13}$$

Exercise 6.1. Prove Nanson's formula.

Exercise 6.2. Prove (6.13) using Nanson's formula.

The strain energy density and the strains

As in linear elasticity, we introduce the strain energy function W. Let us consider a *nonlinear hyperelastic material.* The term *hyperelasticity* simply indicates the existence of a strain energy function. Hence the difference between *Cauchy elasticity*, in which the existence of the strain energy function is not assumed, and the *Green elasticity* of hyperelasticity in which there exists a strain energy function.

Now we demonstrate the use of the general principles in hyperelasticity. First, according to the determinism principle, we assume W depends on $\mathbf{r}$ as we have only one kinematic variable. This means that W is a functional dependent on $\mathbf{r}(\mathbf{x})$. Now the principle of local action obviously holds if W depends on $\mathbf{r}$ and a finite number of its gradients. So W takes the form

$$W = W(\mathbf{r}, \nabla\mathbf{r}, \nabla^2\mathbf{r} \ldots, \nabla^N\mathbf{r}). \tag{6.14}$$

If W depends only on the first gradient,

$$W = W(\mathbf{r}, \nabla\mathbf{r}), \tag{6.15}$$

the material is said to be *simple*. Let us proceed to the principle of material frame-indifference. By this principle, W does not change form if we replace $\mathbf{r}$ by $\mathbf{r}^* = \mathbf{a} + \mathbf{Q} \cdot (\mathbf{r} - \mathbf{r}_0)$ for any constant $\mathbf{a}$, $\mathbf{Q}$, and $\mathbf{r}_0$. Namely,

$$W(\mathbf{r}, \nabla\mathbf{r}) = W(\mathbf{r}^*, \nabla\mathbf{r}^*). \tag{6.16}$$

It follows that W cannot depend on $\mathbf{r}$. Indeed, taking $\mathbf{Q} = \mathbf{E}$ and $\mathbf{r}_0 = \mathbf{0}$, we transform (6.16) into

$$W(\mathbf{r}, \nabla\mathbf{r}) = W(\mathbf{r} + \mathbf{a}, \nabla\mathbf{r}) \qquad \forall\, \mathbf{a}, \mathbf{r}, \nabla\mathbf{r}.$$

Replacing $\mathbf{a}$ by $\tau\mathbf{a}$ for some scalar τ, and then differentiating the latter equation with respect to τ, we find that

$$0 = \frac{d}{d\tau} W(\mathbf{r} + \tau\mathbf{a}, \nabla\mathbf{r})\bigg|_{\tau=0} \equiv W_{,\mathbf{r}} \cdot \mathbf{a}.$$

Because this holds for all $\mathbf{a}$, we get $W_{,\mathbf{r}} = \mathbf{0}$ and W depends on the deformation gradient but not on $\mathbf{r}$:

$$W = W(\nabla\mathbf{r}) = W(\mathbf{F}^T).$$

Consequently

$$W(\mathbf{F}^T) = W(\mathbf{F}^{*T}), \qquad \mathbf{F}^* = (\nabla\mathbf{r}^*)^T.$$

As $\nabla\mathbf{r}^* = \nabla\mathbf{r} \cdot \mathbf{Q}^T$, this transforms to

$$W(\mathbf{F}^T) = W(\mathbf{F}^T \cdot \mathbf{Q}^T) \tag{6.17}$$

for all $\mathbf{Q}$. Equation (6.16) must hold for any orthogonal tensor $\mathbf{Q}$, in particular for $\mathbf{Q} = \mathbf{R}^T$ where $\mathbf{R}$ is the rotation tensor from the polar decomposition $\mathbf{F} = \mathbf{R} \cdot \mathbf{U}$. Therefore

$$W = W(\mathbf{F}^T \cdot \mathbf{R}) = W(\mathbf{U} \cdot \mathbf{R}^T \cdot \mathbf{R}) = W(\mathbf{U} \cdot \mathbf{E}) = W(\mathbf{U}). \tag{6.18}$$

Hence the material frame-indifference principle implies that the strain energy density is a function of the right stretch tensor. Obviously any scalar function of just the argument $\mathbf{U}$ accords with this principle. Thus, applying the principle of material frame-indifference, we have significantly reduced the admissible arguments in W to just the argument $\mathbf{U}$: starting with (6.14) we came to the dependence (6.18).

By (6.10), the tensor $\mathbf{U}$ is a positive square root of $\mathbf{C}$; it is uniquely determined through $\mathbf{C}$. On the other hand, the calculation of $\mathbf{U}$ entails some difficulties; see for example (2.47) or (2.52). So as an argument of W, it is more convenient to use $\mathbf{C}$ instead of $\mathbf{U}$:

$$W = W(\mathbf{C}). \tag{6.19}$$

Here we have kept the same notation for the strain energy.

Instead of $\mathbf{C}$ or $\mathbf{U}$, one can use any invertible tensor-valued function of $\mathbf{C}$, e.g., any member of the family of the strain tensors

$$\mathbf{E}^{(m)} = \begin{cases} \frac{1}{m}(\mathbf{U}^m - \mathbf{E}), & m > 0, \\ \ln \mathbf{U}, & m = 0, \end{cases} \tag{6.20}$$

where m is a real number. All of these strain tensors vanish if $\mathbf{U} = \mathbf{E}$. Note that $\mathbf{E}^{(0)} = \ln \mathbf{U}$ is also called the *Hencky strain measure* or the *logarithmic strain measure*.

Exercise 6.3. Let $f(s)$ be an arbitrary scalar-valued function. Prove that the functional

$$W(\mathbf{r}(\mathbf{x})) = \int_V f(|\mathbf{r}(\mathbf{x}') - \mathbf{r}(\mathbf{x})|)\, dV',$$

where the integration is done with respect to the variable $\mathbf{x}'$, satisfies the principles of determinism and material frame-indifference.

Equilibrium equations

To derive the equilibrium conditions, we will use the variational approach. Let us find the first variation of the strain energy functional

$$\mathcal{E} = \int_V W\, dV,$$

where V is the volume occupied by a nonlinear elastic solid in the reference configuration. For simplicity we consider W of the form $W = W(\nabla \mathbf{r})$.

Integrating by parts, we get

$$\delta\mathcal{E} = \int_V \delta W\, dV = \int_V \frac{dW}{d\nabla\mathbf{r}} \cdot\cdot\, (\delta\nabla\mathbf{r})^T\, dV$$
$$= -\int_V \left(\nabla \cdot \frac{dW}{d\nabla\mathbf{r}}\right) \cdot \delta\mathbf{r}\, dV + \int_{\partial V} \mathbf{n} \cdot \frac{dW}{d\nabla\mathbf{r}} \cdot \delta\mathbf{r}\, dS.$$

Now we recall the definition (2.62) of the derivative of a scalar function with respect to a tensor argument. Here $\frac{dW}{d\nabla\mathbf{r}}$ is defined as a second-order tensor such that the equation

$$\frac{d}{d\varepsilon} W(\nabla\mathbf{r} + \varepsilon\mathbf{B})\bigg|_{\varepsilon=0} \equiv \lim_{\varepsilon\to 0} \frac{W(\nabla\mathbf{r} + \varepsilon\mathbf{B}) - W(\nabla\mathbf{r})}{\varepsilon} = \frac{dW}{d\nabla\mathbf{r}} \bullet \mathbf{B}$$

holds for all second-order tensors $\mathbf{B}$.

For brevity we denote $\delta\mathbf{r}$ by $\mathbf{w}$. The first variation $\delta\mathcal{E}$ of $\mathcal{E}$ is a sum of volume and surface integrals with integrands linearly dependent on $\mathbf{w}$. Its form dictates that for the virtual work of external forces:

$$\delta\mathcal{A} = \int_V \rho\mathbf{f} \cdot \mathbf{w}\, dV + \int_\Sigma \mathbf{t} \cdot \mathbf{w}\, dS. \tag{6.21}$$

Here $\mathbf{f}$ is the vector of external forces per unit mass, $\mathbf{t}$ is the vector of traction, and $\Sigma = \partial V$.

As a result the principle of virtual work, stating that

$$\delta\mathcal{E} - \delta\mathcal{A} = 0$$

for all admissible virtual displacements $\mathbf{w}$, transforms into the equation

$$\delta\mathcal{E} - \delta\mathcal{A} = -\int_V \left[\left(\nabla \cdot \frac{dW}{d\nabla\mathbf{r}}\right) + \rho\mathbf{f}\right] \cdot \mathbf{w}\, dV$$
$$+ \int_\Sigma \left[\mathbf{n} \cdot \frac{dW}{d\nabla\mathbf{r}} - \mathbf{t}\right] \cdot \mathbf{w}\, dS = 0. \tag{6.22}$$

A standard technique in the calculus of variations is the application of the so called main lemma. One version of the lemma states that if a continuous function f satisfies the equation

$$\int_V f w\, dV = 0$$

for all smooth w, then $f = 0$ in V (see, e.g., [Lebedev and Cloud (2003)]). Using the main lemma, we get two equations

$$\left(\nabla \cdot \frac{dW}{d\nabla\mathbf{r}}\right) + \rho\mathbf{f} = \mathbf{0} \quad (\mathbf{x} \in V); \qquad \mathbf{n} \cdot \frac{dW}{d\nabla\mathbf{r}} = \mathbf{t} \quad (\mathbf{x} \in \Sigma = \partial V).$$

The first equation expresses the equilibrium condition in the volume. The second presents the natural boundary condition for the boundary load $\mathbf{t}$.

Let us introduce the second-order *nominal stress tensor*

$$\mathbf{S} = \frac{dW}{d\nabla\mathbf{r}}. \tag{6.23}$$

Note that $\mathbf{S}$ is not in general a symmetric tensor. With this, the equilibrium equation and natural (static) boundary condition take the forms

$$\nabla \cdot \mathbf{S} + \rho\mathbf{f} = \mathbf{0} \qquad (\mathbf{x} \in V), \tag{6.24}$$

$$\mathbf{n} \cdot \mathbf{S} = \mathbf{t} \qquad (\mathbf{x} \in \Sigma = \partial V). \tag{6.25}$$

Together with the kinematic boundary conditions, these equations constitute the *Lagrangian boundary value problem* of statics for a nonlinearly elastic body occupying volume V.

From (6.25) it follows that $\mathbf{n}\cdot\mathbf{S}$ describes the traction per unit area in the reference configuration. As we consider finite deformations, note that here we should distinguish the reference and actual configurations. For example, when we first introduced stress as force divided by area (Fig. 4.1), we did not specify which of the areas is considered, before or after deformation. This issue was unimportant under small deformations. Now, considering uniaxial finite deformations in Fig. 4.1, we should say that the nominal stress is the force f divided by the initial area, while the real stress is f divided by the area of the cross-section in the actual configuration after deformation. These ratios can differ greatly under large finite deformations. For practical purposes nominal stresses can be more convenient as in experiments we can control the applied force but cannot control and measure simultaneously the area during deformation.

Using Piola's identity (6.13) and introducing the *Cauchy stress tensor* $\mathbf{T}$ by the formula

$$\mathbf{T} = J^{-1}\mathbf{F} \cdot \mathbf{S}^T, \tag{6.26}$$

we transform (6.24)–(6.25) into a boundary-value problem in the coordinates of the actual configuration:

$$\nabla_\chi \cdot \mathbf{T} + \hat{\rho}\mathbf{f} = \mathbf{0} \qquad (\mathbf{r} \in v), \tag{6.27}$$

$$\hat{\mathbf{n}} \cdot \mathbf{T} = \mathbf{t}\frac{dS}{\widehat{dS}} \qquad (\mathbf{r} \in \partial v). \tag{6.28}$$

These equations constitute the *Eulerian boundary value problem* of nonlinear elasticity. Of the two boundary value problem setups, Lagrangian and

Eulerian, the most popular at present is the Lagrangian. This seems natural because one knows the initial shape of the body and tries to determine the changes in geometry upon application of the external forces. With the Eulerian formulation we consider static boundary conditions on an *a priori* unknown surface. So here we can say that we know the deformed state and seek the initial state from which it was produced. Nevertheless, the Eulerian statement may be fruitful for contact problems; consider, for example, an indentation problem where the indentor is rigid with a known shape so that we know the shape of the body under the indenter (this can be useful for fluid-solid interaction problems among others). In particular, the Eulerian approach is applied for the incremental formulation in nonlinear elasticity used for computations.

Stress tensors

Nonlinear elasticity can employ stress tensors other than $\mathbf{S}$ and $\mathbf{T}$. Let us consider a few along with their interrelations. As the objective strain energy density depends on $\mathbf{C}$, let us find $\mathbf{S}$ within this dependence $W = W(\mathbf{C})$. Using the formulas

$$dW = \frac{dW}{d\mathbf{C}} \cdot\cdot d\mathbf{C} = \frac{dW}{d(\nabla\mathbf{r})} \cdot\cdot d(\nabla\mathbf{r})^T \equiv \mathbf{S} \cdot\cdot d(\nabla\mathbf{r})^T, \quad d\mathbf{C} = (d\mathbf{F})^T \cdot \mathbf{F} + \mathbf{F}^T \cdot d\mathbf{F}$$

we get

$$\mathbf{S} = 2\frac{dW}{d\mathbf{C}} \cdot \nabla\mathbf{r}.$$

As a derivative with respect to a symmetric tensor, $\frac{dW}{d\mathbf{C}}$ is itself a symmetric second-order tensor (Chapter 2). It follows that $\mathbf{S}$ satisfies

$$(\nabla\mathbf{r})^T \cdot \mathbf{S} = \mathbf{S}^T \cdot \nabla\mathbf{r}. \tag{6.29}$$

Indeed,

$$\mathbf{S} \cdot (\nabla\mathbf{r})^{-1} = 2\frac{dW}{d\mathbf{C}}.$$

Since $\frac{dW}{d\mathbf{C}}$ is symmetric, so is $\mathbf{S} \cdot (\nabla\mathbf{r})^{-1}$ and

$$\mathbf{S} \cdot (\nabla\mathbf{r})^{-1} = (\nabla\mathbf{r})^{-T} \cdot \mathbf{S}^T.$$

Multiplying by $\nabla\mathbf{r}$ from the right and by $(\nabla\mathbf{r})^T$ from the left, we get (6.29).

In the literature, $\mathbf{F} = (\nabla\mathbf{r})^T$ is often used as an argument of W. This leads to the stress tensor

$$\mathbf{P} = \frac{dW}{d\mathbf{F}},$$

called the *first Piola–Kirchhoff stress tensor* or simply the *Piola stress tensor*. Obviously $\mathbf{P} = \mathbf{S}^T$ and it follows that

$$\mathbf{P} \cdot \mathbf{F}^T = \mathbf{F} \cdot \mathbf{P}^T. \tag{6.30}$$

Note that for (6.29) or (6.30) it follows that $\mathbf{T}$ (see (6.26)) is a symmetric tensor: $\mathbf{T}^T = \mathbf{T}$. So the symmetry of the Cauchy stress tensor $\mathbf{T}$ follows from the objectivity requirement.

Still other stress tensors can be introduced. One frequently used tensor is the symmetric *second Piola–Kirchhoff stress tensor*

$$\mathbf{K} = 2\frac{dW}{d\mathbf{C}}.$$

There is also Biot's stress tensor, among others [Ogden (2001)].

On the stationary points of the total energy functional

While discussing the problems of linear elasticity, we proved Theorem 4.5 which states that a solution to the equilibrium equations is a stationary point of the total energy functional. Things are not as simple with finite deformations, as the external forces $\mathbf{f}$ and $\mathbf{t}$ may depend on placement $\mathbf{r}$. Hence the virtual work of external forces (6.21) is not the first variation of a functional $\mathcal{A}$. In other words, this work functional may not exist. Here we simply assume the existence of a functional $\mathcal{A}$ such that its first variation is given by (6.21). External loads for which the work functional $\mathcal{A}$ exists are said to be *conservative*. When $\mathcal{A}$ does not exist, we speak of *nonconservative problems* of elasticity. In particular, $\mathcal{A}$ exists if the external forces are *dead*, meaning that they have no dependence on $\mathbf{r}$. In this case the form of $\mathcal{A}$ coincides with its form in linear elasticity

$$\mathcal{A} = \int_V \rho \mathbf{f} \cdot \mathbf{w}\, dV + \int_{\Sigma_2} \mathbf{t} \cdot \mathbf{w}\, dS,$$

where Σ_2 is a part of the boundary where the traction is given; on the remainder $\Sigma_1 = \Sigma \backslash \Sigma_1$ the placement is given as $\mathbf{r}\big|_{\Sigma_2} = \mathbf{r}_0$. Another example of the existence of $\mathcal{A}$ occurs for the potential mass forces for which there exists a function π such that $\mathbf{f} = -\frac{d\pi}{d\mathbf{r}}$. A typical example of a potential force is the gravitational force with $\pi = gz$, where z is the Cartesian vertical coordinate and g is the free fall acceleration. A more complex example of conservative traction is the external hydrostatic pressure p. Physically this load corresponds to the action of a fluid on the body boundary. If the external loads consist of a uniform hydrostatic pressure p acting on the whole body surface, the work functional is given by

$$\mathcal{A} = \int_V p \det(\nabla \mathbf{r})\, dV = pv,$$

where v is the volume of the body in the actual configuration. Note that if a uniform hydrostatic pressure p acts only on part of the body surface, such a load is non-conservative in general [Lurie (1990)].

For conservative forces there is an analogue of Theorem 4.5:

Theorem 6.1. *A stationary point* $\mathbf{r}$ *of* $\mathcal{E}$*, where*

$$\mathcal{E} = \int_V W\, dV - \mathcal{A}, \qquad \delta\mathcal{A} = \int_V \rho \mathbf{f} \cdot \delta\mathbf{r}\, dV + \int_{\Sigma_2} \mathbf{t} \cdot \delta\mathbf{r}\, dS,$$

on the set of admissible placements satisfies the equilibrium equations of the elastic body undergoing finite deformations in volume V *and the boundary condition* $\mathbf{n} \cdot \mathbf{S}|_{\Sigma_2} = \mathbf{t}$*, and vice versa.*

The proof mimics that of Theorem 4.5.

Exercise 6.4. Prove that

$$\delta \int_V p \det(\nabla \mathbf{r})\, dV = \int_{\partial V} p(\det \nabla \mathbf{r})\mathbf{n} \cdot (\nabla \mathbf{r})^{-T} \cdot \delta\mathbf{r}\, dS,$$

and the natural boundary condition corresponding to the action of external hydrostatic pressure $p =$ constant is $\mathbf{n} \cdot \mathbf{S} = -p\mathbf{n} \cdot \left[(\det \nabla \mathbf{r})(\nabla \mathbf{r})^{-T}\right]$.

Material symmetry group

Considering finite deformation as a mapping from a chosen reference configuration κ into an actual configuration χ, we usually assume κ is given and χ is sought. Let us discuss how the choice of κ can change the governing equations. Let κ_1 and κ_2 be two different reference configurations with the position vectors $\mathbf{x}_1$ and $\mathbf{x}_2$, respectively. So we also have two referential nabla-operators ∇_1 and ∇_2. From the identities $df = d\mathbf{x}_1 \cdot \nabla_1 f = d\mathbf{x}_2 \cdot \nabla_2 f$, where f is an arbitrary scalar function considered in both reference configurations, we get the relations between ∇_1 and ∇_2:

$$\nabla_1 = \nabla_1 \mathbf{x}_2 \cdot \nabla_2, \qquad \nabla_2 = \nabla_2 \mathbf{x}_1 \cdot \nabla_1. \tag{6.31}$$

Indeed, (6.31) immediately follows from the formulas $d\mathbf{x}_2 = d\mathbf{x}_1 \cdot \nabla_1 \mathbf{x}_2$ and $d\mathbf{x}_1 = d\mathbf{x}_2 \cdot \nabla_2 \mathbf{x}_1$. This means we have two deformation gradients and the strain energy density can be written in terms of both. The objective form of W is given by (6.19) and can be written for the two reference configurations as $W = W_1(\mathbf{C}_1)$ and $W = W_2(\mathbf{C}_2)$ where $\mathbf{C}_1 = \nabla_1\mathbf{r}\cdot(\nabla_1\mathbf{r})^T$ and $\mathbf{C}_2 = \nabla_2\mathbf{r}\cdot(\nabla_2\mathbf{r})^T$. Using (6.31) we find the dependence between $\mathbf{C}_1$ and $\mathbf{C}_2$:

$$\mathbf{C}_1 = \nabla_1\mathbf{x}_2\cdot\mathbf{C}_2\cdot(\nabla_1\mathbf{x}_2)^T, \qquad \mathbf{C}_2 = \nabla_2\mathbf{x}_1\cdot\mathbf{C}_1\cdot(\nabla_2\mathbf{x}_1)^T. \tag{6.32}$$

For brevity let us denote $\mathbf{H} = \nabla_1\mathbf{x}_2$ which describes deformation from κ_1 into κ_2. So $\mathbf{C}_1 = \mathbf{H}\cdot\mathbf{C}_2\cdot\mathbf{H}^T$.

In general, the response of the material — the dependence W on $\mathbf{C}$ — may depend on the choice of reference configuration. In other words, $W_1(\mathbf{C}_1)$ relative to κ_1 differs from $W = W_2(\mathbf{C}_2)$ relative to κ_2. However, for a specific $\mathbf{H}$ we can have W invariant. Such transformations constitute a multiplicative group called the *local material symmetry group*.

The set of second-order nonsingular tensors $\mathbf{H}$ for which

$$W(\mathbf{C}) = W(\mathbf{H}\cdot\mathbf{C}\cdot\mathbf{H}^T) \tag{6.33}$$

holds for any admissible values of $\mathbf{C}$ forms the local material symmetry group $\mathcal{G}$ of the material.

The definition states that the material does not "feel" the transformations of the reference configuration belonging to its material symmetry group. This group characterizes the physical symmetries of the material at a point. That is why it is called local. From the definition it follows that $\mathcal{G}$ depends on the reference configuration. It is also clear that the transformations from $\mathcal{G}$ should be unimodular, that is $\det\mathbf{H} = \pm 1$. As $\det\mathbf{H}$ relates to the volumetric change during transformation from κ_1 into κ_2, it is hard to imagine that the strain energy does not "feel" such deformations. In particular, using this definition we may introduce solids, fluids, and various types of anisotropy.

A material is called *isotropic* at a point if there is a reference (said to be *undistorted*) such that the material symmetry group $\mathcal{G}$ includes the orthogonal group: $\mathrm{Orth} \subset \mathcal{G}$.

A material is *solid* at a given point if there is a reference (again said to be *undistorted*) such that the material symmetry group $\mathcal{G}$ is a subset of the orthogonal group: $\mathcal{G} \subset \mathrm{Orth}$.

It follows that $\mathcal{G} = \mathrm{Orth}$ for an isotropic solid material. Thus (6.33) means that W is an isotropic function of $\mathbf{C}$:

$$W(\mathbf{C}) = W(\mathbf{Q}\cdot\mathbf{C}\cdot\mathbf{Q}^T), \quad \forall\mathbf{Q}:\ \mathbf{Q}\cdot\mathbf{Q}^T = \mathbf{E}.$$

So W is a function of the invariants of $\mathbf{C}$,

$$W = W(I_1, I_2, I_3), \qquad I_1 = \operatorname{tr}\mathbf{C}, \quad I_2 = \tfrac{1}{2}[\operatorname{tr}^2\mathbf{C} - \operatorname{tr}\mathbf{C}^2], \quad I_3 = \det\mathbf{C},$$

or of its eigenvalues,

$$W = W(\lambda_1, \lambda_2, \lambda_3).$$

In the literature both formulas are used; see [Gurtin (1983); Lurie (1990); Ogden (1997); Ogden (2001); Truesdell and Noll (2004)].

A material is said to be *fluid* at a point if there is a reference (said to be *undistorted*) such that the material symmetry group $\mathcal{G}$ is the full unimodular group: $\mathcal{G} = \text{Unim}$. Clearly a fluid is also isotropic.

The last definition implies that the strain energy for a fluid depends only on $J = \det\mathbf{C}$:

$$W = W(J). \tag{6.34}$$

This constitutive relation corresponds to the compressible fluids and gases, called an *elastic fluid* or a *barotropic fluid.* This is used in Euler's fluid dynamics. To prove (6.34), let us take tensor $\mathbf{H} = (\det\mathbf{F})\mathbf{F}^{-T}$ which is unimodular and can be used in (6.33). Substituting $\mathbf{H}$ into (6.33) we get

$$W(\mathbf{C}) = W(\mathbf{H}\cdot\mathbf{C}\cdot\mathbf{H}^T) = W((\det\mathbf{F})^2\mathbf{F}^{-T}\cdot\mathbf{F}^T\cdot\mathbf{F}\cdot\mathbf{F}^{-1}) = W(J^2\mathbf{E}).$$

Finally, let us consider how the material symmetry group may depend on the choice of reference configuration. Again, let κ_1 and κ_2 be two reference configurations related by the transformation $\mathbf{H} : \kappa_1 \to \kappa_2$. The corresponding material symmetry groups are $\mathcal{G}_1$ and $\mathcal{G}_2$, respectively. Then the dependence between $\mathcal{G}_1$ and $\mathcal{G}_2$ is given by *Noll's rule*:

$$\mathcal{G}_1 = \mathbf{H}\cdot\mathcal{G}_2\cdot\mathbf{H}^{-1}, \qquad \mathcal{G}_2 = \mathbf{H}^{-1}\cdot\mathcal{G}_1\cdot\mathbf{H}. \tag{6.35}$$

Equation (6.35) means that if $\mathbf{H}_1 \in \mathcal{G}_1$ then $\mathbf{H}_2$ defined as $\mathbf{H}_2 = \mathbf{H}^{-1}\cdot\mathbf{H}_1\cdot\mathbf{H}$ belongs to $\mathcal{G}_2$.

As for the fluids $\mathcal{G}$ is the full unimodular group which is the maximal group, Noll's rule implies that for a fluid the material symmetry group does not depend on the choice of reference placement.

Besides solids and fluids, there is another possibility when the material symmetry group contains Orth but does not coincide with Unim. In other words, there are multiplicative groups which include all the orthogonal tensors and contain non-orthogonal unimodular tensors. The corresponding materials are called *liquid crystals* or *subfluids*; see [Truesdell and Noll (2004)] and Wang[9], where the classification of these groups is presented.

Exercise 6.5. Prove Noll's rule.

On the deformations with constraints

To derive the equilibrium equations we assumed arbitrary deformations. However, sometimes we can neglect certain types of deformations in comparison with the rest. This leads us to introduce some constraints for the deformations. The most important example is the incompressibility condition which states that all body deformations conserve the volume of any portion. So we have

$$J \equiv \det \mathbf{F} = 1. \tag{6.36}$$

Physically, this means that we can neglect the change in volume of any part of the body in comparison with other types of deformations. Typical materials that can be considered as practically incompressible are water, rubber and rubber-like materials, and biological tissues. Of course the incompressibility assumption is only an approximation, but it can help to find some exact solutions of nonlinear elasticity; see universal deformations in [Lurie (1990); Truesdell and Noll (2004)].

For an incompressible material, $I_3 = J^2 = 1$ and the strain energy density of an isotropic material depends only on two invariants of $\mathbf{C}$: $W = W(I_1, I_2)$. Here the derivation of the equilibrium equations should be modified as virtual placements are not arbitrary. For the variation we should use the *Lagrange multiplier* technique. Now we should find the variation of the functional

$$\mathcal{E} = \int_V [W - p(J-1)]\, dV,$$

where p denotes a Lagrange multiplier. As $J = 1$ the additional term does not change the energy functional. Then we calculate $\delta\mathcal{E}$ for independent variations of $\mathbf{r}$ and p:

$$\delta\mathcal{E} = -\int_V \left[\nabla \cdot \frac{dW}{d\nabla\mathbf{r}} - pJ(\nabla\mathbf{r})^{-T} - \delta p(J-1)\right] \cdot \delta\mathbf{r}\, dV$$
$$+ \int_{\partial V} \left[\mathbf{n} \cdot \frac{dW}{d\nabla\mathbf{r}} - pJ(\nabla\mathbf{r})^{-T}\right] \cdot \delta\mathbf{r}\, dS.$$

[9]Wang, C.-C. (1965). A general theory of subfluids. *Arch. Ration. Mech. Analysis* 20(1), 1–40

Taking first $\delta \mathbf{r} = \mathbf{0}$ and $\delta p \neq 0$, from $\delta \mathcal{E} = 0$ we get the constraint $J = 1$. So considering as earlier $\delta \mathcal{E} - \delta \mathcal{A} = 0$, we obtain (6.24) and (6.25) where $\mathbf{S}$ is given by

$$\mathbf{S} = \frac{dW}{d\nabla \mathbf{r}} - p(\nabla \mathbf{r})^{-T}. \tag{6.37}$$

The Lagrange multiplier p plays a role of hydrostatic pressure in fluids, it is a reaction to the constraint (6.36). Note that p does not depend on the deformation; it should be determined from the boundary-value problem together with $\mathbf{r}$.

Examples of incompressible material models

To a large extent, nonlinear elasticity was developed to describe almost incompressible rubber-like materials which exhibit large deformations. For an isotropic incompressible elastic material, the strain energy W depends on two principal invariants of $\mathbf{C}$: $W = W(I_1, I_2)$. One possible form of W is a polynomial approximation of W which uses the Taylor series expansion in a neighborhood of the undeformed state:

$$W = \sum_{m=0}^{M} \sum_{n=0}^{N} C_{mn}(I_1 - 3)^m (I_2 - 3)^n. \tag{6.38}$$

Here C_{mn} are elastic moduli with $C_{00} = 0$, and the integers M and N describe the order of approximation[10]. Some examples of (6.38) are given in Table 6.1.

Note that the literature may use the term "material" for the name of a model, not for the description of real materials such as steels or alloys. For example, the Treloar material describes various rubber-like materials for not very large (average) deformations; its name does not refer to any real material.

Non-polynomial approximations to $W(I_1, I_2)$ are also known. For example, Hart-Smith and Crisp[11] use an exponential–hyperbolic constitutive equation

$$W = C \left[\int e^{k_1 (I_1 - 3)^2} \, dI_1 + k_2 \ln \frac{I_2}{3} \right]. \tag{6.39}$$

[10] Oden, J. T. (1972). *Finite Elements of Nonlinear Continua.* (McGraw-Hill, New York).

[11] Hart-Smith, L.J. and Crisp, J.D.C. (1967). Large elastic deformations in thin rubber membranes. *Int. J. Engng Sci.* 5(1), 1–24.

Table 6.1 Constitutive equations for certain incompressible isotropic materials; I_1, I_2, I_3 are the invariants of $\mathbf{C}$.

Model	Strain energy	Elastic moduli
Treloar (neo-Hookean)	$W = \mu(I_1 - 3)$	μ
Yeoh	$W = C_1(I_1 - 3) + C_2(I_1 - 3)^2 + C_3(I_1 - 3)^3$	C_1, C_2, C_3
Mooney-Rivlin	$W = C_1(I_1 - 3) + C_2(I_2 - 3)$	C_1, C_2
Rivlin-Saunders	$W = C_1(I_1 - 3) + F(I_2 - 3)$	C_1, F is polynom in I_2
Klosner-Segal	$W = C_1(I_1 - 3) + C_2(I_2 - 3) + C_3(I_2 - 3)^2 + C_4(I_2 - 3)^3$	C_k $(k = 1, 2, 3, 4)$
Biderman	$W = C_1(I_1 - 3) + B_1(I_1 - 3)^2 + B_2(I_1 - 3)^3 + C_2(I_2 - 3)$	C_1, C_2, B_1, B_2

Alexander[12] introduced the equations

$$W = C_1(I_1 - 3) + C_2(I_2 - 3) + C_3 \ln \frac{I_2 - 3 + k}{k} \tag{6.40}$$

and

$$W = C_1 \int e^{k_1(I_1-3)^2}\, dI_1 + C_2(I_2 - 3) + C_3 \ln \frac{I_2 - 3 + k_2}{k_2}, \tag{6.41}$$

where C, C_1, C_2, C_3, k, k_1, and k_2 are elastic moduli.

Another type of of constitutive equation is based on the representation of the strain energy density as a function of the principal stretches $\lambda_1, \lambda_2, \lambda_3$:

$$W = W(\lambda_1, \lambda_2, \lambda_3)$$

with the constraint $\lambda_1\lambda_2\lambda_3 = 1$. One of the most popular models is the two-parameter *Ogden material*[13]

$$W = \frac{2\mu}{\alpha}(\lambda_1^\alpha + \lambda_2^\alpha + \lambda_3^\alpha - 3) = \frac{2\mu}{\alpha}\left[\operatorname{tr}\mathbf{U}^\alpha - 3\right] \tag{6.42}$$

and its generalization

$$W = \sum_{n=1}^{N} \frac{2\mu_n}{\alpha_n}(\lambda_1^{\alpha_n} + \lambda_2^{\alpha_n} + \lambda_3^{\alpha_n} - 3) = \sum_{n=1}^{N} \frac{2\mu_n}{\alpha_n}(\operatorname{tr}\mathbf{U}^{\alpha_n} - 3), \tag{6.43}$$

[12] Alexander, H. (1968). A constitutive relation for rubber-like materials. *Int. J. Engng Sci.* 6(9), 549–563.

[13] Ogden, R. W. (1972). Large deformation isotropic elasticity – on the correlation of theory and experiment for incompressible rubberlike solids. *Proc. Roy. Soc. London A.* 326(1567), 565–584.

which is also called an *Ogden material*. Here μ, μ_n, α, and α_n are material parameters. The model (6.43) is similar to the polynomial approximation (6.38); it contains a sufficient number of material parameters to fit the experimental data. Obviously for analytical calculations with use of Ogden's model, powers of $\mathbf{U}$ must be calculated.

As a specific case of Ogden's model, we should mention the two-parameter constitutive equation by *Chernykh and Shubina*[14]

$$\begin{aligned} W &= \mu\left[(1+\beta)(\lambda_1+\lambda_2+\lambda_3-3)+(1-\beta)(\lambda_1^{-1}+\lambda_2^{-1}+\lambda_3^{-1}-3)\right] \\ &= \mu\left[(1+\beta)(\operatorname{tr}\mathbf{U}-3)+(1-\beta)(\operatorname{tr}\mathbf{U}^{-1}-3)\right], \end{aligned} \tag{6.44}$$

where μ and β are elastic moduli.

Another case of Ogden's model is the *Varga material*[15] or *Bartenev–Khazanovich material*[16]

$$W = 2\mu(\lambda_1+\lambda_2+\lambda_3-3) = 2\mu(\operatorname{tr}\mathbf{U}-3). \tag{6.45}$$

As a *neo-Hookean material*, this model contains only one elastic modulus μ. In the case of small strains, μ plays the role of shear modulus.

Linear elasticity of isotropic materials employs only two elastic moduli which can be measured (appointed) via uniaxial tests. The behavior of nonlinear elastic materials is more complex. To fit experimental data, the nonlinear theory needs more than two parameters, in general. Moreover, nowadays it is known that to determine constitutive functions of nonlinear elastic materials, it is insufficient to produce only uniaxial experimental tests. The constitutive equations of nonlinear elasticity are deduced with use of biaxial or even more complex tests.

Examples of the models for compressible materials

The constitutive equation for a compressible nonlinear elastic solid takes the form (6.19). For an isotropic material, W is a function of the invariants I_k of $\mathbf{C}$ or of the principal stretches λ_k. So W can have two forms

$$W = W(I_1, I_2, I_3) \quad \text{or} \quad W = W(\lambda_1, \lambda_2, \lambda_3). \tag{6.46}$$

Here for brevity we use the same notation for different functions W.

[14] Chernykh, K.F. (1986). *Nonlinear Theory of Elasticity in Engineering Calculations* (in Russian). (Mashinostroenie, Moscow).
[15] Varga, O.H. (1966). *Stress-Strain Behaviour of Elastic Materials.* (Interscience, New York).
[16] Bartenev, G.M. and Khazanovich, T.N. (1960). On a constitutive law for hyperelastic strains of cross-linked polymer (in Russian). *Polymer Science*, 2(1), 20–28.

A straightforward generalization of Hooke's law to finite strains is the *semilinear material*, also called *harmonic* or *John's* material [Lurie (1990)]. Here the strain energy is

$$W = \tfrac{1}{2}\lambda(\lambda_1 + \lambda_2 + \lambda_3 - 3)^2 + \mu(\lambda_1^2 + \lambda_2^2 + \lambda_3^2 - 3)$$
$$= \tfrac{1}{2}\lambda(\operatorname{tr}\mathbf{U} - 3)^2 + \mu(\operatorname{tr}\mathbf{U}^2 - 3), \tag{6.47}$$

where λ and μ are elastic moduli playing the role of the Lamé moduli in the theory for small strains.

A similar two-parameter model generalizing Hooke's law is the *Saint Venant–Kirchhoff model*

$$W = \tfrac{1}{2}\lambda \operatorname{tr}^2\left[\tfrac{1}{2}(\mathbf{C} - \mathbf{E})\right] + \mu \operatorname{tr}\left[\tfrac{1}{2}(\mathbf{C} - \mathbf{E})\right]^2, \tag{6.48}$$

where elastic moduli λ and μ again play the role of the Lamé moduli in case of small strains.

Another model was introduced by *Blatz and Ko*[17] for certain rubber-like porous materials:

$$W = \tfrac{1}{2}\mu\beta\left[I_1 + \frac{1}{\alpha}(I_3^{-\alpha} - 1) - 3\right]$$
$$+ \tfrac{1}{2}\mu(1-\beta)\left[I_2 I_3^{-1} + \frac{1}{\alpha}(I_3^{\alpha} - 1) - 3\right], \quad \alpha = \frac{\lambda}{2\mu}. \tag{6.49}$$

This model contains three moduli μ, α, β. The simplified one-parameter form of (6.49) is also in use:

$$W = \tfrac{1}{2}\mu\left[\frac{I_2}{I_3} + 2\sqrt{I_3} - 5\right]. \tag{6.50}$$

Another way to construct constitutive equations uses the assumption that the material is weakly compressible. Obviously, the assumption of incompressibility is only an approximation. Moreover, for calculations it is more efficient to use the models of compressible materials, for example, it is to avoid locking related with incompressibility. In the case of small strains, the decomposition of a deformation as a sum of spherical and deviatoric parts is widely used. In the case of finite deformations, its analogue is the multiplicative decomposition of the deformation gradient $\mathbf{F}$ as a product of spherical and unimodular tensors:

$$\mathbf{F} = J^{1/3}\mathbf{E}\cdot\bar{\mathbf{F}},$$

[17] Blatz, P.J., and Ko, W.L. (1962). Application of finite elastic theory to the deformation of rubbery materials. *Trans. Soc. Rheology*, 6(1), 223–251.

where $J = \det \mathbf{F}$ and $\bar{\mathbf{F}} = J^{-1/3}\mathbf{F}$. It follows that $\det \bar{\mathbf{F}} = 1$. In addition, let us introduce the normalized Cauchy–Green strain measure $\bar{\mathbf{C}}$:

$$\bar{\mathbf{C}} = \bar{\mathbf{F}}^T \cdot \bar{\mathbf{F}}. \tag{6.51}$$

Here, instead of the invariants I_1, I_2, I_3, we should use the normalized invariants of $\bar{\mathbf{C}}$ and J:

$$\bar{I}_1 = J^{-1/3} I_1, \qquad \bar{I}_2 = J^{-2/3} I_2, \qquad J. \tag{6.52}$$

Thus, we have introduced the strain energy density as a function of the normalized invariants:

$$W = W(\bar{I}_1, \bar{I}_2, J).$$

The invariants (6.52) are used to construct constitutive equations for weakly compressible materials. They can be expressed through the eigenvalues $\bar{\lambda}_i^2$ of the strain measure of $\bar{\mathbf{C}}$. Instead of $\bar{I}_1$ and $\bar{I}_2$, we also can use eigenvalues $\bar{\lambda}_i^2$ of $\bar{\mathbf{C}}$:

$$\bar{I}_1 = \bar{\lambda}_1^2 + \bar{\lambda}_2^2 + \bar{\lambda}_3^2, \qquad \bar{I}_2 = \bar{\lambda}_1^2\bar{\lambda}_2^2 + \bar{\lambda}_2^2\bar{\lambda}_3^2 + \bar{\lambda}_1^2\bar{\lambda}_3^2. \tag{6.53}$$

One of the simplest examples of this kind is the *compressible neo-Hookean material* with strain energy

$$W = C_1(\bar{I}_1 - 3) + \frac{1}{d}(J-1)^2, \tag{6.54}$$

where d is a modulus characterizing the compressibility. In case of small strains, the bulk modulus k is related to d through the equation $k = 2/d$. Formally, for an incompressible material $d = 0$. So for small d, the last term in (6.54) can be treated as a penalty preserving non-isochoric deformations.

The next series of constitutive equations follows from generalizations of polynomial approximations (6.38):

$$W = \sum_{m=0}^{M}\sum_{n=0}^{N} C_{mn}(\bar{I}_1 - 3)^m(\bar{I}_2 - 3)^n + \sum_{n=1}^{N}\frac{1}{d_n}(J-1)^{2n} \tag{6.55}$$

with the additional terms relating to volumetric strains. Here C_{mn} and d_n are elastic moduli and $C_{00} = 0$.

An example of (6.55) is the *two-parameter Mooney–Rivlin material*

$$W = C_{10}(\bar{I}_1 - 3) + C_{01}(\bar{I}_2 - 3) + \frac{1}{d}(J-1)^2. \tag{6.56}$$

Such polynomial approximations are widely implemented in finite element codes, as they contain a number of material parameters for describing complex behaviors of elastomers. Among them are the *three-parameter Mooney–Rivlin material*

$$W = C_{10}(\bar{I}_1 - 3) + C_{01}(\bar{I}_2 - 3) + C_{11}(\bar{I}_1 - 3)(\bar{I}_2 - 3) + \frac{1}{d}(J-1)^2, \quad (6.57)$$

the *five-parameter Mooney–Rivlin material*

$$\begin{aligned} W &= C_{10}(\bar{I}_1 - 3) + C_{01}(\bar{I}_2 - 3) \\ &\quad + C_{20}(\bar{I}_1 - 3)^2 + C_{02}(\bar{I}_2 - 3)^2 + C_{11}(\bar{I}_1 - 3)(\bar{I}_2 - 3) + \frac{1}{d}(J-1)^2, \end{aligned} \quad (6.58)$$

and the *nine-parameter Mooney–Rivlin material*

$$\begin{aligned} W &= C_{10}(\bar{I}_1 - 3) + C_{01}(\bar{I}_2 - 3) + C_{20}(\bar{I}_1 - 3)^2 + C_{02}(\bar{I}_2 - 3)^2 \\ &\quad + C_{11}(\bar{I}_1 - 3)(\bar{I}_2 - 3) + C_{30}(\bar{I}_1 - 3)^3 + C_{21}(\bar{I}_1 - 3)^2(\bar{I}_2 - 3) \\ &\quad + C_{12}(\bar{I}_1 - 3)(\bar{I}_2 - 3)^2 + dsC_{03}(\bar{I}_2 - 3)^3 + \frac{1}{d}(J-1)^2. \end{aligned} \quad (6.59)$$

Based on uniaxial, biaxial, triaxial, and other tests, Arruda and Boyce[18] proposed a model for compressible and incompressible rubber and polymer materials. The form of the strain energy for the *Arruda–Boyce material* for compressible materials is

$$\begin{aligned} W &= \mu \left[\frac{1}{2}(\bar{I}_1 - 3) + \frac{1}{20\lambda_L^2}(\bar{I}_1^2 - 9) + \frac{11}{1050\lambda_L^4}(\bar{I}_1^3 - 279) \right. \\ &\quad \left. + \frac{19}{7000\lambda_L^6}(\bar{I}_1^4 - 81) + \frac{519}{673750\lambda_L^8}(\bar{I}_1^5 - 243) \right] + \frac{1}{d}\left(\frac{J^2 - 1}{2} - \ln J \right). \end{aligned} \quad (6.60)$$

Here μ, λ_L, and d are elastic moduli. For $\lambda_L \to \infty$ and $d \to 0$ equation (6.60) transforms to the one for the neo-Hookean material.

Gent[19] also introduced a model for compressible and incompressible materials. In the *Gent model*, the strain energy density for compressible material is

$$W = -\frac{\mu I_m}{2} \ln\left(1 - \frac{I_1 - 3}{I_m}\right) + \frac{k}{2}\left(\frac{J^2 - 1}{2} - \ln J\right)^4, \quad (6.61)$$

where μ, I_m, and k are elastic moduli.

[18]Arruda, E.M. and Boyce, M.C. (1993). A three-dimensional constitutive model for the large stretch behavior of rubber elastic materials. *J. Mech. Phys. Solids*, 41(2), 389–412.

[19]Gent, A.N. (1996). A new constitutive relation for rubber. *Rubber Chemistry and Technology*, 69(1), 59–61.

Ogden's model (6.43) also can be extended to compressible materials. The form of the energy for the *compressible Ogden material* is

$$W = \sum_{n=1}^{N} \frac{2\mu_n}{\alpha_n}(\bar{\lambda}_1^{\alpha_n} + \bar{\lambda}_2^{\alpha_n} + \bar{\lambda}_3^{\alpha_n} - 3) + \sum_{n=1}^{N} \frac{1}{d_n}(J-1)^{2n}, \tag{6.62}$$

with an additional set of parameters d_n ($n = 1, \ldots N$). It is interesting to note that the Ogden model (6.62) is useful for weakly compressible materials and also for highly compressible materials such as polymer foams[20].

As an example of an anisotropic nonlinear constitutive equation we present the *Fung material*[21]

$$W = \tfrac{1}{2}\left[\mathbf{G}\cdot\cdot\mathbf{A}\cdot\cdot\mathbf{G} + c\left(e^{\mathbf{G}\cdot\cdot\mathbf{B}\cdot\cdot\mathbf{G}} - 1\right)\right], \quad \mathbf{G} = \tfrac{1}{2}(\mathbf{C} - \mathbf{E}), \tag{6.63}$$

where $\mathbf{A}$ and $\mathbf{B}$ are fourth-order tensors of elastic moduli and c is a material parameter. This model is used to model soft tissues. For isotropic materials, (6.63) reduces to a three-parameter model

$$\begin{aligned} W &= \tfrac{1}{2}\left[a\mathbf{G}\cdot\cdot\mathbf{G} + c\left(e^{b\mathbf{G}\cdot\cdot\mathbf{G}} - 1\right)\right] \\ &= \tfrac{1}{2}\left[a(\lambda_1^2 + \lambda_2^2 + \lambda_3^2 - 3) + c\left(e^{b(\lambda_1^2+\lambda_2^2+\lambda_3^2-3)} - 1\right)\right] \end{aligned}$$

with elastic moduli a, b, and c.

In the literature there are other models of compressible nonlinear elastic materials related with the names of Seth, Signorini, Murnaghan, and others; see [Lurie (1990); Lurie (2005); Ogden (1997); Ogden (2001)]. The above selection of constitutive equations shows that the world of models for nonlinear elastic materials is rich and complex in comparison with the model of linear elasticity.

Dynamics

For dynamical problems, the deformation of an elastic solid depends on time t:

$$\mathbf{r} = \mathbf{r}(\mathbf{x}, t) \qquad (\mathbf{x} \in V). \tag{6.64}$$

The velocity and acceleration of a material particle with position $\mathbf{x}$ are defined by the formulas

$$\mathbf{v} = \dot{\mathbf{r}} = \frac{d\mathbf{r}}{dt}, \qquad \mathbf{a} = \dot{\mathbf{v}} = \frac{d\mathbf{v}}{dt} = \ddot{\mathbf{r}} = \frac{d^2\mathbf{v}}{d^2t}.$$

[20] Mills, N. (2007). *Polymer Foams Handbook. Engineering and Biomechanics Applications and Design Guide.* (Butterworth-Heinemann, Amsterdam).

[21] Fung, Y.-C. (1993). *Biomechanics: Mechanical Properties of Living Tissues.* (Springer, New York).

The simplest way to derive the equation of motion is to use d'Alembert's principle of mechanics. Introducing the inertia forces and replacing the body forces in the static equation as

$$\mathbf{f} \to \mathbf{f} - \rho\frac{\partial^2 \mathbf{r}}{\partial t^2},$$

we obtain

$$\nabla \cdot \mathbf{S} + \rho\mathbf{f} = \rho\frac{\partial^2 \mathbf{r}}{\partial t^2}. \tag{6.65}$$

To formulate an initial-boundary value problem, the equation of motion should be supplemented by the initial data

$$\mathbf{r}\Big|_{t=0} = \mathbf{r}^0, \qquad \mathbf{v}\Big|_{t=0} = \mathbf{v}^0,$$

where $\mathbf{r}^0$ and $\mathbf{v}^0$ are known functions, and the boundary conditions coinciding in form with the static boundary conditions.

Now we demonstrate how the equation of motion can be derived using the least action principle (Hamilton's principle). We introduce the kinetic energy density

$$K = \tfrac{1}{2}\rho\mathbf{v}\cdot\mathbf{v}$$

and consider the action functional

$$\mathcal{H} = \int_{t_1}^{t_2}\int_V (K - W)\,dV\,dt.$$

Here t_1 and t_2 are some time instants. Looking for stationary points of $\mathcal{H}$ over the set of deformations coinciding with the real deformations at time instants t_1, t_2, we assume the variation $\mathbf{w} = \delta\mathbf{r}$ vanishes at $t = t_1, t_2$:

$$\mathbf{w}\Big|_{t=t_1} = \mathbf{w}\Big|_{t=t_2} = \mathbf{0}.$$

To derive the first variation $\delta\mathcal{H}$, let us first find the variation of the kinetic energy. We get

$$\delta\int_{t_1}^{t_2}\int_V K\,dV\,dt = \int_{t_1}^{t_2}\int_V \delta K\,dV\,dt = \int_{t_1}^{t_2}\int_V \rho\mathbf{v}\cdot\delta\mathbf{v}\,dV\,dt.$$

As the variation and differentiation with respect to t are commutative we have $\delta\mathbf{v} = \delta\dot{\mathbf{r}} = \dot{\overline{\delta\mathbf{r}}} = \dot{\mathbf{w}}$. So

$$\delta\int_{t_1}^{t_2}\int_V K\,dV\,dt = \int_{t_1}^{t_2}\int_V \rho\mathbf{v}\cdot\dot{\mathbf{w}}\,dV\,dt = \int_V\int_{t_1}^{t_2} \rho\mathbf{v}\cdot\dot{\mathbf{w}}\,dt\,dV.$$

Integration by parts yields

$$\int_{t_1}^{t_2} \rho\mathbf{v}\cdot\dot{\mathbf{w}}\,dt = \rho\mathbf{v}\cdot\mathbf{w}\Big|_{t_1}^{t_2} - \int_{t_1}^{t_2} \rho\dot{\mathbf{v}}\cdot\mathbf{w}\,dt = -\int_{t_1}^{t_2} \rho\dot{\mathbf{v}}\cdot\mathbf{w}\,dt.$$

Using the earlier derived variation of $\delta\mathcal{E}$ we get

$$\delta\mathcal{H} = -\int_{t_1}^{t_2}\int_V (\rho\dot{\mathbf{v}} - \nabla\cdot\mathbf{S})\cdot\mathbf{w}\,dV\,dt - \int_{t_1}^{t_2}\int_\Sigma \mathbf{n}\cdot\mathbf{S}\cdot\mathbf{w}\,dS\,dt.$$

Finally, it is easy to see that the equations of motion (6.65) follow from the variational equation

$$\delta\mathcal{H} + \int_{t_1}^{t_2} \delta\mathcal{A}\,dt = 0,$$

where $\delta\mathcal{A}$ is defined by (6.21). The details are left to the reader.

Transition to linear elasticity

We have now introduced some strain and stress tensors along with various nonlinear constitutive equations. Let us consider what happens in the case of small deformations. In other words, we wish to learn how nonlinear elasticity relates to linear elasticity. Introducing displacements $\mathbf{u}(\mathbf{x}) = \mathbf{r}(\mathbf{x}) - \mathbf{x}$ we find that

$$\mathbf{F} = \mathbf{E} + (\nabla\mathbf{u})^T, \qquad \mathbf{C} = \mathbf{E} + (\nabla\mathbf{u})^T + \nabla\mathbf{u} + (\nabla\mathbf{u})\cdot(\nabla\mathbf{u})^T.$$

Assuming $\|\mathbf{u}\| \ll 1$ and $\|\nabla\mathbf{u}\| \ll 1$, we will linearize all the necessary quantities. In particular, up to linear terms we have

$$\mathbf{F}^{-1} = \mathbf{E} - (\nabla\mathbf{u})^T, \qquad \det\mathbf{F} = 1 + \nabla\cdot\mathbf{u}, \qquad \mathbf{C} = \mathbf{E} + 2\boldsymbol{\varepsilon},$$

where $\boldsymbol{\varepsilon}$ is the linear strain tensor

$$\boldsymbol{\varepsilon} = \tfrac{1}{2}[(\nabla\mathbf{u})^T + \nabla\mathbf{u}].$$

Let us assume the strain energy density and stresses vanish if $\mathbf{C} = \mathbf{E}$:

$$W(\mathbf{E}) = 0, \qquad \frac{dW}{d\mathbf{C}}\bigg|_{\mathbf{C}=\mathbf{E}} = \mathbf{0}.$$

Then W becomes a quadratic function of $\boldsymbol{\varepsilon}$,

$$W = \tfrac{1}{2}\boldsymbol{\varepsilon}\cdot\cdot\,\mathbf{C}\cdot\cdot\,\boldsymbol{\varepsilon}, \quad \text{where } \mathbf{C} = \frac{d^2W}{d\mathbf{C}^2}\bigg|_{\mathbf{C}=\mathbf{E}},$$

and all the stress tensors coincide: $\mathbf{S} = \mathbf{P} = \mathbf{T} = \mathbf{C}\cdot\cdot\,\boldsymbol{\varepsilon}$.

6.4 Cosserat Continuum

One of the first generalized media was the model proposed by the Cosserats brothers, Eugèn and François [Cosserat and Cosserat (1909)]. Unlike classical elasticity they assumed that interactions between parts of a deformed body are possible with both forces and moments. To describe these interactions they introduced not only a stress vector $\mathbf{t}$ as shown in Fig. 4.3 but also a couple stress vector. This leads to the appearance of the couple stresses in addition to classical stresses. Each material particle in the Cosserat continuum has six degrees of freedom as in rigid body dynamics [Bauchau (2010)]. Now called the *Cosserat continuum*, *micropolar medium*, or simply *polar medium*, the model is used to describe solid materials with complex microstructures like soils, composite and granular materials, masonries, etc. Typical examples of such microstructured solids are foams and beam lattices (Fig. 6.2). These materials, when treated as homogeneous elastic media, require us to account for the moments acting in foam struts or in beams. This leads to non-classical models of elastic media such as the Cosserat continuum.

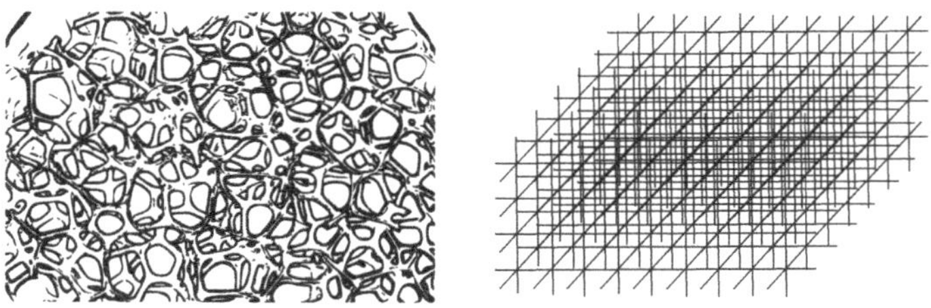

Fig. 6.2 Examples of microstructured solids: irregular microstructure (open-cell foam) on the left and regular structure (beam lattice) on the right.

Kinematics

In micropolar mechanics, the kinematics are described by the position vector $\mathbf{r}$ of a mass-point of a body in an actual configuration and by three unit orthogonal vectors $\mathbf{d}_k$ ($k = 1, 2, 3$; $\mathbf{d}_k \cdot \mathbf{d}_m = \delta_{km}$) called *directors*, which play roles of additional, rotational, degrees of freedom. Physically the Cosserat continuum can be viewed as a medium consisting of infinitesimal particles that undergo independent translations and rotations. So each such particle can be treated as a rigid body. Comparing the model with the classical "Cauchy" continuum, we see that in the classical continuum each material particle is a point for which rotation is negligible. Introducing the position vector $\mathbf{x}$ and directors $\mathbf{D}_k$ in a reference configuration, we now define the deformation as a mapping

$$\mathbf{r} = \mathbf{r}(\mathbf{x}), \qquad \mathbf{d}_k = \mathbf{d}_k(\mathbf{x}). \tag{6.66}$$

To describe the relative rotation we introduce an orthogonal tensor

$$\mathbf{H} = \mathbf{D}_k \mathbf{d}_k$$

called the *microrotation tensor.* So instead of (6.66), the two fields $\mathbf{r}$ and $\mathbf{H}$ are used to describe the deformations of a micropolar body:

$$\mathbf{r} = \mathbf{r}(\mathbf{x}), \qquad \mathbf{H} = \mathbf{H}(\mathbf{x}). \tag{6.67}$$

The tensor $\mathbf{H}$ serves as an additional kinematical descriptor or internal variable.

For small deformations (i.e., for small translations and small rotations) we introduce the vectors of displacement $\mathbf{u}$ and rotation $\boldsymbol{\varphi}$ defined by

$$\mathbf{u}(\mathbf{x}) = \mathbf{r}(\mathbf{x}) - \mathbf{x}, \qquad \mathbf{H}(\mathbf{x}) = \mathbf{E} - \mathbf{E} \times \boldsymbol{\varphi}(\mathbf{x}), \qquad \boldsymbol{\varphi}(\mathbf{x}) = \tfrac{1}{2}(\mathbf{H} - \mathbf{E})_{\times}.$$

So instead of a vector and an orthogonal tensor, the linear micropolar kinematics is based on two vectorial fields.

Strain energy density

For a hyperelastic micropolar solid we introduce the strain energy density W as a function of $\mathbf{r}$ and $\mathbf{H}$ along with their first gradients:

$$W = W(\mathbf{r}, \mathbf{H}, \nabla\mathbf{r}, \nabla\mathbf{H}).$$

As in nonlinear elasticity, application of the material-frame-indifference principle leads to the statement that W does not depend on $\mathbf{r}$: $W = W(\mathbf{H}, \nabla\mathbf{r}, \nabla\mathbf{H})$. To further reduce the number of arguments we assume that in the equivalent motion the directors $\mathbf{d}_k$ rotate similarly to $\mathbf{r}$ [Kafadar and Eringen (1971)]:

$$\mathbf{d}_k^* = \mathbf{Q} \cdot \mathbf{d}_k.$$

This results in the following transformation of the microrotation tensor:

$$\mathbf{H}^* = \mathbf{H} \cdot \mathbf{Q}^T. \tag{6.68}$$

Thus, from the material-frame-indifference principle the invariance property

$$W(\mathbf{H}, \nabla\mathbf{r}, \nabla\mathbf{H}) = W(\mathbf{H} \cdot \mathbf{Q}^T, \nabla\mathbf{r} \cdot \mathbf{Q}^T, \nabla\mathbf{H} \cdot \mathbf{Q}^T), \quad \forall \mathbf{Q} : \mathbf{Q}^T \cdot \mathbf{Q} = \mathbf{E}$$

follows. The substitution $\mathbf{Q} = \mathbf{H}$ yields

$$W(\mathbf{H}, \nabla\mathbf{r}, \nabla\mathbf{H}) = W(\mathbf{H}\cdot\mathbf{H}^T, \nabla\mathbf{r}\cdot\mathbf{H}^T, \nabla\mathbf{H}\cdot\mathbf{H}^T).$$

Taking into account that $\mathbf{H}$ is an orthogonal tensor, we find the representation for the third-order tensor

$$\nabla\mathbf{H}\cdot\mathbf{H}^T = -\text{æ}\times\mathbf{E}, \tag{6.69}$$

where **æ** is a second-order tensor. Indeed, using Cartesian coordinates we have the relation

$$\nabla\mathbf{H}\cdot\mathbf{H}^T = \mathbf{i}_k\frac{\partial\mathbf{H}}{\partial x_k}\cdot\mathbf{H}^T.$$

As $\mathbf{H}$ is orthogonal, $\frac{\partial\mathbf{H}}{\partial x_k}\cdot\mathbf{H}^T$ is an antisymmetric tensor and as such it can be represented as

$$\frac{\partial\mathbf{H}}{\partial x_k}\cdot\mathbf{H}^T = \text{æ}_k\times\mathbf{E}$$

for some vector æ_k. With $\text{æ} = -\mathbf{i}_k\text{æ}_k$ we get (6.69). Note that the negative sign in (6.69) was chosen by convention. The tensor **æ** is called the *wryness tensor* [Eringen (1999); Eremeyev et al. (2013)]. Introducing the *stretch tensor* $\boldsymbol{\epsilon}$ by the formula

$$\boldsymbol{\epsilon} = \nabla\mathbf{r}\cdot\mathbf{H}^T - \mathbf{E}, \tag{6.70}$$

we obtain the form of W satisfying the material frame-indifference principle:

$$W = W(\boldsymbol{\epsilon}, \text{æ}). \tag{6.71}$$

The tensors $\boldsymbol{\epsilon}$ and **æ**, in general non-symmetric, are called the natural Lagrangian strain measures; see Pietraszkiewicz and Eremeyev[22] where the introduction of strain measures in micropolar elasticity is discussed in detail. In particular, let us note that according to its definition where the cross product was involved, **æ** is a pseudotensor. So it changes sign under mirror reflection transformations applied to a coordinate system.

For the theory of constitutive equations of the micropolar medium under finite deformations, including various statements of the boundary value problems, we refer to [Kafadar and Eringen (1971); Eringen (1999); Eremeyev et al. (2013)]. In particular, the representations of W for micropolar elastic solids and fluids using the theory of invariants is given by

[22]Pietraszkiewicz, W., Eremeyev, V.A. (2009). On natural strain measures of the non-linear micropolar continuum. *Int. J. Solids Struct.*, 46(3), 774–787.

Eremeyev and Pietraszkiewicz[23]. For example, in the case of an isotropic solid, W depends on 15 invariants of $\boldsymbol{\epsilon}$ and æ:

$$\begin{aligned}
&J_1(\boldsymbol{\epsilon}), \quad J_2(\boldsymbol{\epsilon}), \quad J_3(\boldsymbol{\epsilon}), \quad J_4(\boldsymbol{\epsilon}), \quad J_5(\boldsymbol{\epsilon}), \quad J_6(\boldsymbol{\epsilon}),\\
&J_1(\text{æ}), \quad J_2(\text{æ}), \quad J_3(\text{æ}), \quad J_4(\text{æ}), \quad J_5(\text{æ}), \quad J_6(\text{æ}),\\
&\operatorname{tr}\boldsymbol{\epsilon}\cdot\text{æ}, \quad \operatorname{tr}\boldsymbol{\epsilon}^2\cdot\text{æ}, \quad \operatorname{tr}\boldsymbol{\epsilon}\cdot\text{æ}^2.
\end{aligned}$$

See (2.118) and (2.119). So in comparison with nonlinear elasticity, here we have more complicated constitutive relations even for isotropic materials.

For brevity, the discussion below is restricted to linear micropolar elasticity.

Kinetic energy

Considering motion of a micropolar body in addition to the *linear velocity* $\mathbf{v} = \dot{\mathbf{r}}$, we also introduce the *angular velocity* or *microgyration vector*. It follows from differentiation of the directors by the formula

$$\dot{\mathbf{d}}_k = \boldsymbol{\omega} \times \mathbf{d}_k.$$

Thus $\dot{\mathbf{H}} = -\mathbf{H} \times \boldsymbol{\omega}$. Using $(\mathbf{E} \times \boldsymbol{\omega})_\times = -2\boldsymbol{\omega}$, we express $\boldsymbol{\omega}$ via $\dot{\mathbf{H}}$ as

$$\boldsymbol{\omega} = \tfrac{1}{2}(\dot{\mathbf{H}} \cdot \mathbf{H}^T)_\times. \tag{6.72}$$

For small deformations we have the simpler formula $\boldsymbol{\omega} = \dot{\boldsymbol{\varphi}}$.

The kinetic energy density is assumed as a general nonnegative quadratic function of $\mathbf{v}$ and $\boldsymbol{\omega}$:

$$K = \tfrac{1}{2}(\mathbf{v} \cdot \mathbf{I}_1 \cdot \mathbf{v} + \mathbf{v} \cdot \mathbf{I}_2 \cdot \boldsymbol{\omega} + \boldsymbol{\omega} \cdot \mathbf{I}_3 \cdot \mathbf{v} + \boldsymbol{\omega} \cdot \mathbf{I}_4 \cdot \boldsymbol{\omega}), \tag{6.73}$$

where $\mathbf{I}_k$ $(k = 1, 2, 3, 4)$ are four *tensors of inertia*. The simplest and often used form of (6.73) is

$$K = \tfrac{1}{2}\rho\mathbf{v} \cdot \mathbf{v} + \tfrac{1}{2}\rho j\boldsymbol{\omega} \cdot \boldsymbol{\omega}, \tag{6.74}$$

where ρ is the mass density in an actual configuration and j is a scalar measure of rotatory inertia.

[23] Eremeyev, V. A., Pietraszkiewicz, W. (2012). Material symmetry group of the nonlinear polar-elastic continuum. *Int. J. Solids Struct.*, 49(14), 1993–2005; (2016). Material symmetry group and constitutive equations of micropolar anisotropic elastic solids. *Math. Mech. Solids*, 21(2), 210–221.

Linear theory

For small deformations of the Cosserat continuum we use two kinematical descriptors: the *infinitesimal translation vector* $\mathbf{u}$, and the *infinitesimal rotation vector* $\boldsymbol{\varphi}$. The strain measures transform into

$$\boldsymbol{\epsilon} = \nabla\mathbf{u} + \mathbf{E}\times\boldsymbol{\varphi}, \qquad \text{æ} = \nabla\boldsymbol{\varphi}. \tag{6.75}$$

We call $\boldsymbol{\epsilon}$ the *linear stretch tensor* and æ the *linear wryness tensor.*

As in classic linear elasticity, the various stress tensors coincide. We introduce the Cauchy stress and *couple stress tensors* by the formulas

$$\mathbf{T} = \frac{\partial W}{\partial \boldsymbol{\epsilon}}, \qquad \mathbf{M} = \frac{\partial W}{\partial \text{æ}}. \tag{6.76}$$

Physically, $\mathbf{T}$ is similar to the stress resultant tensor or surface stress tensor used in the theory of shells, while $\mathbf{M}$ is a 3D analog to the tensor of moments or surface couple stress tensor. Moreover, the linear theory of micropolar shells (Chapter 5) can be treated as a 2D Cosserat continuum. Note that $\mathbf{T}$ and $\mathbf{M}$ are not symmetric, in general. Detailed introduction of stress and couple stress tensors (similar to the presentation in Chapter 4) is given, for example, in [Eremeyev et al. (2013)].

The equations of motion take the form

$$\nabla\cdot\mathbf{T} + \rho\mathbf{f} = \rho\ddot{\mathbf{u}}, \tag{6.77}$$

$$\nabla\cdot\mathbf{M} + \mathbf{T}_\times + \rho\mathbf{m} = \rho j\ddot{\boldsymbol{\varphi}}. \tag{6.78}$$

Here $\mathbf{f}$ and $\mathbf{m}$ are the vectors of mass forces and couples. The kinematic boundary conditions are

$$\mathbf{u}\big|_{\Sigma_1} = \mathbf{u}_0, \qquad \boldsymbol{\varphi}\big|_{\Sigma_1} = \boldsymbol{\varphi}_0, \tag{6.79}$$

while the static boundary conditions are

$$\mathbf{n}\cdot\mathbf{T}\big|_{\Sigma_2} = \boldsymbol{\tau}, \qquad \mathbf{n}\cdot\mathbf{M}\big|_{\Sigma_2} = \boldsymbol{\mu}, \tag{6.80}$$

where $\boldsymbol{\tau}$ and $\boldsymbol{\mu}$ are given surface forces and couples at Σ_2, and $\mathbf{u}_0$ and $\boldsymbol{\varphi}_0$ are translations and rotations given on the complementary part Σ_1 of the boundary $\Sigma = \partial V = \Sigma_1 \cup \Sigma_2$. Obviously, various mixed boundary conditions are also possible. For non-stationary dynamic problems we also introduce the initial conditions

$$\mathbf{u}(\mathbf{x},t)\big|_{t=0} = \mathbf{u}^0(\mathbf{x}), \quad \dot{\mathbf{u}}(\mathbf{x},t)\big|_{t=0} = \mathbf{v}^0(\mathbf{x}),$$
$$\boldsymbol{\varphi}(\mathbf{x},t)\big|_{t=0} = \boldsymbol{\varphi}^0(\mathbf{x}), \quad \dot{\boldsymbol{\varphi}}(\mathbf{x},t)\big|_{t=0} = \boldsymbol{\omega}^0(\mathbf{x}),$$

where the $\mathbf{u}^0$ and $\boldsymbol{\varphi}^0$ are initial translation and rotation, whereas $\mathbf{v}^0$ and $\boldsymbol{\omega}^0$ are the initial linear and angular velocities, respectively. As a result, within micropolar elasticity we have six scalar equations of equilibrium or motion complemented with six boundary conditions. In the theory the action of external forces and couples is possible.

For an isotropic material the strain energy density is an quadratic form of $\boldsymbol{\epsilon}$ and æ:

$$W = \frac{\lambda}{2}\operatorname{tr}^2\boldsymbol{\epsilon} + \frac{\mu+\kappa}{2}\operatorname{tr}(\boldsymbol{\epsilon}\cdot\boldsymbol{\epsilon}^T) + \frac{\mu}{2}\operatorname{tr}(\boldsymbol{\epsilon}\cdot\boldsymbol{\epsilon}) + \frac{\beta_1}{2}\operatorname{tr}^2\text{æ} + \frac{\beta_2}{2}\operatorname{tr}(\text{æ}\cdot\text{æ}) + \frac{\beta_3}{2}\operatorname{tr}(\text{æ}\cdot\text{æ}^T), \qquad (6.81)$$

where λ, μ, κ, β_i $(i = 1, 2, 3)$ are the elastic moduli. Positive definiteness of W requires the following inequalities:

$$3\lambda + 2\mu + \kappa > 0, \quad 2\mu + \kappa > 0, \quad \kappa > 0, \qquad (6.82)$$

$$3\beta_1 + \beta_2 + \beta_3 > 0, \quad \beta_2 + \beta_3 > 0, \quad \beta_3 - \beta_2 > 0. \qquad (6.83)$$

The form (6.81) can be also used for finite deformations if we apply the nonlinear strain measures defined in (6.69) and (6.70). In this case the model is called the *physically linear micropolar material.*

The stress and couple stress tensors are given by

$$\mathbf{T} = \lambda\mathbf{E}\operatorname{tr}\boldsymbol{\epsilon} + (\mu+\kappa)\boldsymbol{\epsilon} + \mu\boldsymbol{\epsilon}^T,$$
$$\mathbf{M} = \beta_1\mathbf{E}\operatorname{tr}\text{æ} + \beta_2\text{æ}^T + \beta_3\text{æ}.$$

Substituting $\mathbf{T}$ and $\mathbf{M}$ into (6.77) and (6.78), we get the equations of motion in terms of translations and rotations:

$$(\lambda+\mu)\nabla\nabla\cdot\mathbf{u} + (\mu+\kappa)\nabla\cdot\nabla\mathbf{u} + \kappa\nabla\times\boldsymbol{\varphi} + \rho\mathbf{f} = \rho\ddot{\mathbf{u}}, \qquad (6.84)$$

$$(\beta_1+\beta_2)\nabla\nabla\cdot\boldsymbol{\varphi} + \beta_3\nabla\cdot\nabla\boldsymbol{\varphi} + \kappa(\nabla\times\mathbf{u} - 2\boldsymbol{\varphi}) + \rho\mathbf{m} = \rho j\ddot{\boldsymbol{\varphi}}. \qquad (6.85)$$

In the linear Cosserat elasticity we have six elastic moduli which should be determined from experiments, such as provided by Lakes[24] or from theoretical analysis such as homogenization techniques[25]. Note that for $\kappa = 0$

[24] Lakes, R.S. (1986). Experimental microelasticity of two porous solids. *Int. J. Solids Struct.*, 22(1), 55–63; Rueger, Z., Lakes, R.S. (2016). Experimental Cosserat elasticity in open-cell polymer foam. *Phil. Magazine*, 96(2), 93–111.

[25] Bigoni, D., Drugan, W.J. (2007). Analytical derivation of Cosserat moduli via homogenization of heterogeneous elastic materials. *J. Appl. Mech.*, 74(4), 741–753; Dos Reis, F., Ganghoffer, J.F. (2012). Construction of micropolar continua from the asymptotic homogenization of beam lattices. *Computers & Structures*, 112, 354–363; Reda, H., Rahali, Y., Ganghoffer, J.F., Lakiss, H. (2016). Wave propagation in 3D viscoelastic auxetic and textile materials by homogenized continuum micropolar models. *Composite Struct.*, 141, 328–345.

the stress tensor is symmetric with $\mathbf{T} = \mathbf{T}^T$ and the equilibrium equations decouple. So κ is often called the *Cosserat couple modulus.* The linear micropolar elasticity is discussed in many works, see e.g. [Eringen (1999)], Nowacki[26], and Dyszlewicz[27] where many classic solutions were extended to the micropolar solids.

The Cosserat medium with constrained rotations

Another model related to the Cosserat medium is *Cosserat's pseudocontinuum*, also referred to under the names *medium with constrained rotations* or the *theory of couple stresses.* Within this model the microrotations are not independent of translations but coincide with macrorotations. Let us recall that the macrorotation of a body particle that is determined by displacement $\mathbf{u}$ is given by the formula $-1/2\nabla \times \mathbf{u}$; see, e.g., [Lurie (2005)]. So equating $\boldsymbol{\varphi}$ with the vector of rotations of $\mathbf{u}$, we get the relation

$$\boldsymbol{\varphi} = -\tfrac{1}{2}\nabla \times \mathbf{u}. \tag{6.86}$$

This equation is an example of a constraint similar to the incompressibility condition, but it is vectorial. From (6.86) it follows that $\boldsymbol{\varepsilon}$ and $\mathbf{T}$ are both symmetric, whereas $\operatorname{tr} \boldsymbol{æ} = 0$ and $\operatorname{tr} \mathbf{M} = 0$.

Relation (6.86) also implies that $\boldsymbol{æ} = -\frac{1}{2}\nabla\nabla \times \mathbf{u}$ is expressed through the second spatial derivatives of $\mathbf{u}$. Thus the theory of couple stresses provides another example of generalized media called the strain gradient elasticity. The reader is referred to the original paper[28].

6.5 Micromorphic Continuum

A further generalization of the Cosserat continuum is the *micromorphic continuum* or the *medium with microstrains.* The model was proposed by Mindlin[29] for small deformations and by Eringen and Suhubi[30] for finite deformations; see also [Eringen (1999); Maugin (2017)]. Within the model, the medium kinematics is described by two fields: that of the placement vector $\mathbf{r}$, and that of the *micro-distortion* second-order tensor $\mathbf{P}$. The

[26] Nowacki, W. (1986). *Theory of Asymmetric Elasticity* (Pergamon Press, Oxford).
[27] Dyszlewicz, J. (2004). *Micropolar Theory of Elasticity* (Springer, Berlin).
[28] Mindlin, R.D., Tiersten, H.F. (1962). Effects of couple-stresses in linear elasticity. *Arch. Ration. Mech. Analysis*, 11(1), 415–448.
[29] Mindlin, R.D. (1964). Micro-structure in linear elasticity, *Arch. Ration. Mech. Analysis* 16(1) 51–78.
[30] Eringen, A.C., Suhubi, E.S. (1964). Nonlinear theory of simple micro-elastic solids–I, *Int. J. Engng. Sci.* 2 (2) 189–203.

latter corresponds to the modelling of microdeformations such as, for example, microstrains of foam cells, of grains in granular media, etc. Thus the medium kinematics is given by two mappings

$$\mathbf{r} = \mathbf{r}(\mathbf{x}, t) \quad \text{and} \quad \mathbf{P} = \mathbf{P}(\mathbf{x}, t). \tag{6.87}$$

For a hyperelastic medium we introduce the form

$$W = W(\nabla\mathbf{r}, \mathbf{P}, \nabla\mathbf{P}), \tag{6.88}$$

where ∇ is the Lagrangian nabla–operator.

In what follows we assume that $\mathbf{P}$ behaves in equivalent motion similarly to $\nabla\mathbf{r}$, that is

$$\mathbf{P}^* = \mathbf{P} \cdot \mathbf{Q}$$

for any orthogonal tensor $\mathbf{Q} = \mathbf{Q}^{-T}$. As a result, the principle of material frame indifference yields the invariance property

$$W(\mathbf{F}, \mathbf{P}, \nabla\mathbf{P}) = W(\mathbf{F} \cdot \mathbf{Q}, \mathbf{P} \cdot \mathbf{Q}, \nabla\mathbf{P} \cdot \mathbf{Q}). \tag{6.89}$$

Here for brevity we write $\mathbf{F} = \nabla\mathbf{r}$. Considering the polar decomposition of $\mathbf{F}$ in the form $\mathbf{F} = \mathbf{U} \cdot \mathbf{R}$, where $\mathbf{U} = (\mathbf{F} \cdot \mathbf{F}^T)^{1/2}$ is a positive definite symmetric second-order tensor and $\mathbf{R} = \mathbf{U}^{-1} \cdot \mathbf{F}$ is an orthogonal tensor, in (6.89) we can set $\mathbf{Q} = \mathbf{R}^T$. Thus

$$\begin{aligned} W(\mathbf{F}, \mathbf{P}, \nabla\mathbf{P}) &= W(\mathbf{F} \cdot \mathbf{R}^T, \mathbf{P} \cdot \mathbf{R}^T, \nabla\mathbf{P} \cdot \mathbf{R}^T) \\ &= W(\mathbf{U}, \mathbf{P} \cdot \mathbf{R}^{-1}, \nabla\mathbf{P} \cdot \mathbf{R}^{-1}) \\ &= W(\mathbf{U}, \mathbf{P} \cdot \mathbf{F}^{-1} \cdot \mathbf{U}, \nabla\mathbf{P} \cdot \mathbf{F}^{-1} \cdot \mathbf{U}) \\ &= W(\mathbf{C}, \mathbf{P} \cdot \mathbf{F}^{-1}, \nabla\mathbf{P} \cdot \mathbf{F}^{-1}). \end{aligned} \tag{6.90}$$

Here $\mathbf{C} = \mathbf{U}^2 = \mathbf{F} \cdot \mathbf{F}^T$ is the Cauchy–Green strain measure. Note that retain the notation W for different functions having different arguments, in general. So we introduce the following Lagrangian strain measures:

$$\mathbf{C}, \qquad \mathbf{P} \cdot \mathbf{F}^{-1}, \qquad \nabla\mathbf{P} \cdot \mathbf{F}^{-1}. \tag{6.91}$$

Instead of $\mathbf{R}$ which follows from the polar decomposition of $\mathbf{F}$, we can use another orthogonal tensor coming from the polar decomposition of $\mathbf{P}$. Indeed, let us consider the polar decomposition of $\mathbf{P}$:

$$\mathbf{P} = \tilde{\mathbf{U}} \cdot \tilde{\mathbf{R}}, \qquad \tilde{\mathbf{U}} = (\mathbf{P} \cdot \mathbf{P}^T)^{1/2}, \qquad \tilde{\mathbf{R}} = \tilde{\mathbf{U}}^{-1} \cdot \mathbf{P}.$$

Then, substituting $\mathbf{Q} = \tilde{\mathbf{R}}^T$ into (6.89), we obtain the indifferent form

$$W = W(\mathbf{F} \cdot \mathbf{P}^{-1}, \mathbf{P} \cdot \mathbf{P}^T, \nabla\mathbf{P} \cdot \mathbf{P}^{-1}), \tag{6.92}$$

with another set of strain measures

$$\boldsymbol{\varepsilon} = \mathbf{F} \cdot \mathbf{P}^{-1}, \qquad \mathbf{e} = \mathbf{P} \cdot \mathbf{P}^T, \qquad \boldsymbol{\kappa} = \nabla \mathbf{P} \cdot \mathbf{P}^{-1}. \tag{6.93}$$

This set may be more useful than (6.91), as it contains two strain measures depending only on $\mathbf{P}$. In particular, if $\mathbf{P}$ is an orthogonal tensor like the one in micropolar media, (6.93) can be transformed to the Cosserat strain measures (6.69)–(6.70).

For dynamics of micromorphic continuum we define the kinetic energy density

$$K = \tfrac{1}{2}\rho\, \mathbf{v} \cdot \mathbf{v} + \tfrac{1}{2} j\, \dot{\mathbf{P}} \bullet \dot{\mathbf{P}} \tag{6.94}$$

where ρ is the mass density in the reference configuration, $\mathbf{v} = \dot{\mathbf{r}}$, and $j \geq 0$ is a scalar measure of microinertia.

The least action principle yields the Lagrangian equations of motion in the form

$$\nabla \cdot \mathbf{T} + \rho \mathbf{f} = \rho \dot{\mathbf{v}}, \qquad \nabla \cdot \mathbf{M} - \frac{\partial W}{\partial \mathbf{P}} + \rho \mathbf{m} = j\, \ddot{\mathbf{P}}, \tag{6.95}$$

where $\mathbf{T}$ is the first Piola–Kirchhoff second-order stress tensor, $\mathbf{M}$ is the first Piola–Kirchhoff third-order *hyper-stress tensor*

$$\mathbf{T} = \frac{\partial W}{\partial \mathbf{F}}, \qquad \mathbf{M} = \frac{\partial W}{\partial \nabla \mathbf{P}}, \tag{6.96}$$

and $\mathbf{f}$ and $\mathbf{m}$ are given external loads. Note that by "hyper-stresses" we usually denote internal forces and related tensors which are dual to additional kinematical variables.

The corresponding kinematic and static boundary conditions are

$$\mathbf{r}\big|_{\Sigma_k} = \mathbf{r}^0, \qquad \mathbf{P}\big|_{\Sigma_k} = \mathbf{P}^0, \tag{6.97}$$

$$\mathbf{n} \cdot \mathbf{T}\big|_{\Sigma_s} = \mathbf{t}^0, \qquad \mathbf{n} \cdot \mathbf{M}\big|_{\Sigma_s} = \mathbf{m}^0, \tag{6.98}$$

with a given field of placement vector $\mathbf{r}^0$ on Σ_k and a given field of microdistortion tensor $\mathbf{P}^0$, whereas on the complementary part of the boundary, Σ_s, external given loads $\mathbf{t}^0$ and $\mathbf{m}^0$ are prescribed. Note that $\mathbf{m}$ and $\mathbf{m}^0$ are second-order tensors.

The boundary-value problem (6.95), (6.97), (6.98) presents a common difficulty in the mechanics of generalized media, viz., the appearance and mechanical interpretation of external loads connected to the additional degrees of freedom. Unlike the Cosserat continuum where the external loadings are forces and moments already known in structural mechanics, here we need to attract more "exotic" external actions. This poses a serious challenge in the numerical and experimental implementation of such extended models. But higher-order moments, which are similar to the earlier introduced hyperstresses, are also known in the structural mechanics; see, for example, *bimoments* in the theory of thin-walled beams of open cross-section[31] and *double-forces*[32]. In particular, a bimoment relates with the stress distribution at a cross-section with zero values of both total force and total moment. So we refer to the literature for detailed discussion of hyperstresses in the mechanics of generalized media; see, e.g., [Eringen (1999); Eringen (2001); Eringen (2002); Maugin (2017)].

Let us calculate $\mathbf{T}$ and $\mathbf{M}$ for the strain energy (6.92). According to (6.96) we have

$$\begin{aligned} dW &= \frac{\partial W}{\partial \mathbf{F}} \bullet d\mathbf{F} + \frac{\partial W}{\partial \nabla \mathbf{P}} \bullet d(\nabla \mathbf{P}) + \frac{\partial W}{\partial \mathbf{P}} \bullet d\mathbf{P} \\ &= \mathbf{T} \bullet d\mathbf{F} + \mathbf{M} \bullet d(\nabla \mathbf{P}) + \frac{\partial W}{\partial \mathbf{P}} \bullet d\mathbf{P}. \end{aligned} \tag{6.99}$$

On the other hand, for (6.92) dW takes the form

$$dW = \frac{\partial W}{\partial \boldsymbol{\varepsilon}} \bullet d\boldsymbol{\varepsilon} + \frac{\partial W}{\partial \boldsymbol{\kappa}} \bullet d\boldsymbol{\kappa} + \frac{\partial W}{\partial \mathbf{e}} \bullet d\mathbf{e}.$$

Substituting into the latter equation the expressions

$$\begin{aligned} d\boldsymbol{\varepsilon} &= d\mathbf{F} \cdot \mathbf{P}^{-1} - \mathbf{F} \cdot \mathbf{P}^{-1} \cdot d\mathbf{P} \cdot \mathbf{P}^{-1}, \\ d\boldsymbol{\kappa} &= d(\nabla \mathbf{P}) \cdot \mathbf{P}^{-1} - (\nabla \mathbf{P}) \cdot \mathbf{P}^{-1} \cdot d\mathbf{P} \cdot \mathbf{P}^{-1}, \\ d\mathbf{e} &= d\mathbf{P} \cdot \mathbf{P}^{T} + \mathbf{P} \cdot d\mathbf{P}^{T}, \end{aligned}$$

we obtain

$$\begin{aligned} dW = {} & \frac{\partial W}{\partial \boldsymbol{\varepsilon}} \bullet (d\mathbf{F} \cdot \mathbf{P}^{-1} - \mathbf{F} \cdot \mathbf{P}^{-1} \cdot d\mathbf{P} \cdot \mathbf{P}^{-1}) \\ & + \frac{\partial W}{\partial \boldsymbol{\kappa}} \bullet \left[d(\nabla \mathbf{P}) \cdot \mathbf{P}^{-1} - (\nabla \mathbf{P}) \cdot \mathbf{P}^{-1} \cdot d\mathbf{P} \cdot \mathbf{P}^{-1} \right] \\ & + \frac{\partial W}{\partial \mathbf{e}} \bullet (d\mathbf{P} \cdot \mathbf{P}^{T} + \mathbf{P} \cdot d\mathbf{P}^{T}) \\ = {} & \left(\frac{\partial W}{\partial \boldsymbol{\varepsilon}} \cdot \mathbf{P}^{-T} \right) \bullet d\mathbf{F} + \left(\frac{\partial W}{\partial \boldsymbol{\kappa}} \cdot \mathbf{P}^{-T} \right) \bullet d(\nabla \mathbf{P}) \\ & + \left[2 \frac{\partial W}{\partial \mathbf{e}} \cdot \mathbf{P} - \mathbf{P}^{-T} \cdot \mathbf{F}^{T} \cdot \frac{\partial W}{\partial \boldsymbol{\varepsilon}} \cdot \mathbf{P}^{-T} - \boldsymbol{\kappa} : \frac{\partial W}{\partial \boldsymbol{\kappa}} \cdot \mathbf{P}^{-T} \right] \bullet d\mathbf{P}. \end{aligned} \tag{6.100}$$

---

[31]Vlasov, V.Z. (1940). *Thin-walled Elastic Rods* (in Russian) (Gosstroyizdat, Moscow); Chen, W.F., Atsuta, T. (2008). *Theory of Beam-Columns: Space Behavior and Design* (Vol. 2) (J. Ross Publishing, Fort Lauderdale).
[32]Ziegler, F. (1991). *Mechanics of Solids and Fluids* (Springer, New York).

Here we apply the operation ":" to any two third-order tensors $\mathbf{A}$ and $\mathbf{B}$ as

$$\mathbf{A} : \mathbf{B} = (A_{mnk}\mathbf{i}_m\mathbf{i}_n\mathbf{i}_k) : (B_{pqr}\mathbf{i}_p\mathbf{i}_q\mathbf{i}_r) = A_{mnk}B_{mnr}\mathbf{i}_k\mathbf{i}_r,$$

and use the identity

$$\mathbf{A} \bullet (\mathbf{B} \cdot \mathbf{X} \cdot \mathbf{Y}) = (\mathbf{B} : \mathbf{A} \cdot \mathbf{Y}^T) \bullet \mathbf{X} \tag{6.101}$$

where $\mathbf{A}, \mathbf{B}$ are third-order tensors and $\mathbf{X}, \mathbf{Y}$ are second-order tensors.

Comparison of (6.99) and (6.100) yields

$$\mathbf{T} = \frac{\partial W}{\partial \boldsymbol{\varepsilon}} \cdot \mathbf{P}^{-T}, \qquad \mathbf{M} = \frac{\partial W}{\partial \boldsymbol{\kappa}} \cdot \mathbf{P}^{-T}, \tag{6.102}$$

whereas

$$\frac{\partial W}{\partial \mathbf{P}} = 2\frac{\partial W}{\partial \mathbf{e}} \cdot \mathbf{P} - \boldsymbol{\varepsilon}^T \cdot \frac{\partial W}{\partial \boldsymbol{\varepsilon}} \cdot \mathbf{P}^{-T} - \boldsymbol{\kappa} : \frac{\partial W}{\partial \boldsymbol{\kappa}} \cdot \mathbf{P}^{-T}.$$

Exercise 6.6. Using the least action variational principle, derive (6.95) and (6.98); take the virtual work of external loads in the form

$$\delta \mathcal{A} = \int_V \rho\,(\mathbf{f} \cdot \delta\mathbf{r} + \mathbf{m} \bullet \delta\mathbf{P})\,dv + \int_{\Sigma_s} (\mathbf{t}^0 \cdot \delta\mathbf{r} + \mathbf{m}^0 \bullet \delta\mathbf{P})\,dS.$$

Exercise 6.7. Prove (6.101).

Exercise 6.8. Find $\mathbf{T}$ and $\mathbf{M}$ for W in the form (6.90).

Small deformations of a micromorphic solid

Within small deformations, we introduce the usual displacement vector $\mathbf{u} = \mathbf{r} - \mathbf{R}$ and small second-order tensor $\mathbf{p}$ such that

$$\mathbf{P} = \mathbf{E} + \mathbf{p}.$$

In this case the two sets of strain measures used in (6.91) and (6.93) transform into a single set

$$\boldsymbol{\varepsilon} = \nabla\mathbf{u} - \mathbf{p}, \qquad \mathbf{e} = \tfrac{1}{2}(\mathbf{p} + \mathbf{p}^T), \qquad \boldsymbol{\kappa} = \nabla\mathbf{p}. \tag{6.103}$$

Indeed, up to terms linear in $\nabla\mathbf{u}$, $\mathbf{p}$, and $\nabla\mathbf{p}$ we have

$$\mathbf{F} = \mathbf{E} + \nabla\mathbf{u}, \quad \mathbf{F}^{-1} = \mathbf{E} - \nabla\mathbf{u}, \quad \mathbf{P}^{-1} = \mathbf{E} - \mathbf{p},$$
$$\mathbf{C} = \mathbf{E} + \nabla\mathbf{u} + (\nabla\mathbf{u})^T, \quad \mathbf{P} \cdot \mathbf{F}^{-1} = \mathbf{E} + \mathbf{p} - \nabla\mathbf{u}, \quad \nabla\mathbf{P} \cdot \mathbf{F}^{-1} = \nabla\mathbf{p},$$
$$\mathbf{F} \cdot \mathbf{P}^{-1} = \mathbf{E} + \nabla\mathbf{u} - \mathbf{p}, \quad \mathbf{P} \cdot \mathbf{P}^T = \mathbf{E} + \mathbf{p} + \mathbf{p}^T, \quad \nabla\mathbf{P} \cdot \mathbf{P}^{-1} = \nabla\mathbf{p}.$$

Following Mindlin, $\boldsymbol{\varepsilon}$ is called a *relative deformation* and $\mathbf{e}$ a *micro-strain tensor*, whereas $\frac{1}{2}[\nabla\mathbf{u} + (\nabla\mathbf{u})^T]$ is now called a *macro-strain tensor* and $\boldsymbol{\kappa}$ a *micro-deformation gradient*.

For infinitesimal deformations of an isotropic micromorphic solid, the following form of strain energy can be used:

$$W = \frac{\lambda}{2}\operatorname{tr}^2 \boldsymbol{\varepsilon} + \frac{\mu+\kappa}{2}\boldsymbol{\varepsilon} \bullet \boldsymbol{\varepsilon} + \frac{\mu-\kappa}{2}\boldsymbol{\varepsilon} \bullet \boldsymbol{\varepsilon}^T + \frac{\alpha}{2}\operatorname{tr}^2 \mathbf{e} + \beta \mathbf{e} \bullet \mathbf{e} + \frac{\gamma}{2}\boldsymbol{\kappa} \bullet \boldsymbol{\kappa}, \tag{6.104}$$

where λ, μ, κ, α, β, and γ are elastic moduli. The moduli λ, μ, α, β are similar to Lamé's moduli in classic elasticity. We call $2\kappa \geq 0$ the *Cosserat couple modulus*. If $\kappa = 0$, the strain energy density depends only on the symmetric part of $\boldsymbol{\varepsilon}$ and there is no coupling between $\mathbf{u}$ and $\mathbf{p}$.

Let us find the restrictions on λ, μ, κ, α, β, and γ which follow from positivity of W for all possible values of its arguments:

$$W(\boldsymbol{\varepsilon}, \mathbf{e}, \boldsymbol{\kappa}) > 0, \quad \forall\, \boldsymbol{\varepsilon},\, \mathbf{e},\, \boldsymbol{\kappa} \neq \mathbf{0}.$$

First, considering the inequality $W(\mathbf{0}, \mathbf{0}, \boldsymbol{\kappa}) > 0$ for $\boldsymbol{\kappa} \neq \mathbf{0}$, we find that $\gamma > 0$. Second, let us set $\boldsymbol{\varepsilon} = \mathbf{0}$ and $\boldsymbol{\kappa} = \mathbf{0}$. Splitting the symmetric tensor $\mathbf{e}$ into spherical and deviatoric parts as

$$\mathbf{e} = \tfrac{1}{3}\mathbf{E}\operatorname{tr}\mathbf{e} + \operatorname{dev}\mathbf{e},$$

we get

$$\begin{aligned} W(\mathbf{0}, \mathbf{e}, \mathbf{0}) &= \frac{\alpha}{2}\operatorname{tr}^2 \mathbf{e} + \beta \mathbf{e} \bullet \mathbf{e} \\ &= \frac{\alpha}{2}\operatorname{tr}^2 \mathbf{e} + \beta \left[\tfrac{1}{3}\mathbf{E}\operatorname{tr}\mathbf{e} + \operatorname{dev}\mathbf{e}\right] \bullet \left[\tfrac{1}{3}\mathbf{E}\operatorname{tr}\mathbf{e} + \operatorname{dev}\mathbf{e}\right] \\ &= \frac{\alpha}{2}\operatorname{tr}^2 \mathbf{e} + \frac{\beta}{3}\operatorname{tr}^2 \mathbf{e} + \beta(\operatorname{dev}\mathbf{e}) \bullet (\operatorname{dev}\mathbf{e}) > 0. \end{aligned}$$

As the spherical and deviatoric parts of a second-order tensor are independent and constitute orthogonal subspaces in the space of second-order tensors, it follows that $\beta > 0$ and $3\alpha + 2\beta > 0$. Finally, taking $\mathbf{e} = \mathbf{0}$ and $\boldsymbol{\kappa} = \mathbf{0}$, we get

$$W = \frac{\lambda}{2}\operatorname{tr}^2 \boldsymbol{\varepsilon} + \frac{\mu+\kappa}{2}\boldsymbol{\varepsilon} \bullet \boldsymbol{\varepsilon} + \frac{\mu-\kappa}{2}\boldsymbol{\varepsilon} \bullet \boldsymbol{\varepsilon}^T. \tag{6.105}$$

Representing $\boldsymbol{\varepsilon}$ as the sum of its symmetric and antisymmetric parts as

$$\boldsymbol{\varepsilon} = \boldsymbol{\varepsilon}_A + \boldsymbol{\varepsilon}_S, \qquad \boldsymbol{\varepsilon}_A = \tfrac{1}{2}(\boldsymbol{\varepsilon} - \boldsymbol{\varepsilon}^T), \qquad \boldsymbol{\varepsilon}_S = \tfrac{1}{2}(\boldsymbol{\varepsilon} + \boldsymbol{\varepsilon}^T)$$

and again using the deviator and ball tensor representation (2.53) for $\boldsymbol{\varepsilon}_S$,

$$\boldsymbol{\varepsilon}_S = \tfrac{1}{3}\mathbf{E}\operatorname{tr}\boldsymbol{\varepsilon}_S + \operatorname{dev}\boldsymbol{\varepsilon}_S,$$

we split $\boldsymbol{\varepsilon}$ into three parts:

$$\boldsymbol{\varepsilon} = \tfrac{1}{3}\mathbf{E}\,\mathrm{tr}\,\boldsymbol{\varepsilon}_S + \mathrm{dev}\,\boldsymbol{\varepsilon}_S + \boldsymbol{\varepsilon}_A\,.$$

Substitution into (6.105) gives

$$\begin{aligned}
W &= \frac{\lambda}{2}\,\mathrm{tr}^2\,\boldsymbol{\varepsilon} \\
&\quad + \frac{\mu+\kappa}{2}\left[\tfrac{1}{3}\mathbf{E}\,\mathrm{tr}\,\boldsymbol{\varepsilon}_S + \mathrm{dev}\,\boldsymbol{\varepsilon}_S + \boldsymbol{\varepsilon}_A\right] \bullet \left[\tfrac{1}{3}\mathbf{E}\,\mathrm{tr}\,\boldsymbol{\varepsilon}_S + \mathrm{dev}\,\boldsymbol{\varepsilon}_S + \boldsymbol{\varepsilon}_A\right] \\
&\quad + \frac{\mu-\kappa}{2}\left[\tfrac{1}{3}\mathbf{E}\,\mathrm{tr}\,\boldsymbol{\varepsilon}_S + \mathrm{dev}\,\boldsymbol{\varepsilon}_S + \boldsymbol{\varepsilon}_A\right] \bullet \left[\tfrac{1}{3}\mathbf{E}\,\mathrm{tr}\,\boldsymbol{\varepsilon}_S + \mathrm{dev}\,\boldsymbol{\varepsilon}_S - \boldsymbol{\varepsilon}_A\right] \\
&= \frac{\lambda}{2}\,\mathrm{tr}^2\,\boldsymbol{\varepsilon} + \frac{\mu}{3}\,\mathrm{tr}^2\,\boldsymbol{\varepsilon} + \mu(\mathrm{dev}\,\boldsymbol{\varepsilon}_S) \bullet (\mathrm{dev}\,\boldsymbol{\varepsilon}_S) + \kappa\boldsymbol{\varepsilon}_A \bullet \boldsymbol{\varepsilon}_A > 0,
\end{aligned}$$

which results in $\mu > 0$, $\kappa > 0$, and $3\lambda + 2\mu > 0$.

In summary, positive definiteness of W implies the inequalities

$$\mu > 0, \qquad 3\lambda + 2\mu > 0, \qquad \kappa > 0, \tag{6.106}$$

$$\beta > 0, \qquad 3\alpha + 2\beta > 0, \qquad \gamma > 0. \tag{6.107}$$

Under these conditions, $W = 0$ if and only if

$$\mathbf{u} = \mathbf{a} + \mathbf{x} \times \boldsymbol{\omega}, \qquad \mathbf{p} = \mathbf{E} \times \boldsymbol{\omega} \tag{6.108}$$

with constant vectors $\mathbf{a}$ and $\boldsymbol{\omega}$. Indeed, since $\gamma > 0$, from $W = 0$ it follows that $\boldsymbol{\kappa} = \nabla\mathbf{p} = \mathbf{0}$. Thus $\mathbf{p}$ is a constant second-order tensor. As $\beta > 0$ and $3\alpha + 2\beta > 0$, we have $\mathbf{e} = \frac{1}{2}(\mathbf{p} + \mathbf{p}^T) = \mathbf{0}$. So $\mathbf{p}$ is a constant antisymmetric (skew) tensor. As with any skew tensor, using (2.18) $\mathbf{p}$ can be represented as $\mathbf{p} = \mathbf{E} \times \boldsymbol{\omega}$ with a constant vector $\boldsymbol{\omega}$. Finally, as $\mu > 0$, $\kappa > 0$, and $3\lambda + 2\mu > 0$, from $W = 0$ it follows that $\boldsymbol{\varepsilon} = \nabla\mathbf{u} - \mathbf{p} = \mathbf{0}$. Thus $\nabla\mathbf{u} = \mathbf{p} = \mathbf{E} \times \boldsymbol{\omega}$. Integration gives $\mathbf{u} = \mathbf{a} + \mathbf{x} \times \boldsymbol{\omega}$ for some constant $\mathbf{a}$.

Equation (6.108) describes the *infinitesimal rigid body motions* for a micromorphic continuum. Thus, as in linear elasticity, W vanishes only for rigid body motions.

For small deformations, the equations of motion become

$$\nabla \cdot \mathbf{T} + \rho\mathbf{f} = \rho\dot{\mathbf{v}}, \qquad \nabla \cdot \mathbf{M} - \frac{\partial W}{\partial \mathbf{p}} + \rho\mathbf{m} = j\,\ddot{\mathbf{p}}, \tag{6.109}$$

where $\mathbf{T}$ and $\mathbf{M}$ are

$$\mathbf{T} = \frac{\partial W}{\partial \boldsymbol{\varepsilon}} = \lambda\mathbf{E}\,\mathrm{tr}\,\boldsymbol{\varepsilon} + (\mu+\kappa)\boldsymbol{\varepsilon} + (\mu-\kappa)\boldsymbol{\varepsilon}^T, \qquad \mathbf{M} = \frac{\partial W}{\partial \boldsymbol{\kappa}} = \gamma\boldsymbol{\kappa}, \tag{6.110}$$

and

$$\frac{\partial W}{\partial \mathbf{p}} = -\frac{\partial W}{\partial \boldsymbol{\varepsilon}}.$$

Physically linear micromorphic materials

The constitutive equation (6.104) can be easily extended to the case of large deformations considering, for example, the following strain tensors from (6.93):

$$\boldsymbol{\varepsilon} = \mathbf{F}\cdot\mathbf{P}^{-1} - \mathbf{E}, \qquad \mathbf{e} = \tfrac{1}{2}(\mathbf{P}\cdot\mathbf{P}^T - \mathbf{E}), \qquad \boldsymbol{\kappa} = \nabla\mathbf{P}\cdot\mathbf{P}^{-1}.$$

As a result, we have

$$W = \tfrac{\lambda}{2}\operatorname{tr}^2\boldsymbol{\varepsilon} + \tfrac{\mu+\kappa}{2}\boldsymbol{\varepsilon}\bullet\boldsymbol{\varepsilon} + \tfrac{\mu-\kappa}{2}\boldsymbol{\varepsilon}\bullet\boldsymbol{\varepsilon}^T + \tfrac{\alpha}{2}\operatorname{tr}^2\mathbf{e} + \beta\mathbf{e}\bullet\mathbf{e} + \tfrac{\gamma}{2}\boldsymbol{\kappa}\bullet\boldsymbol{\kappa}. \quad (6.111)$$

The corresponding micromorphic model can be called a *physically linear micromorphic material*. Using (6.102) we obtain the expressions for $\mathbf{T}$ and $\mathbf{M}$ for this model:

$$\mathbf{T} = \left[\lambda\mathbf{E}\operatorname{tr}\boldsymbol{\varepsilon} + (\mu+\kappa)\boldsymbol{\varepsilon} + (\mu-\kappa)\boldsymbol{\varepsilon}^T\right]\cdot\mathbf{P}^{-T}, \quad \mathbf{M} = \gamma\boldsymbol{\kappa}\cdot\mathbf{P}^{-T}. \quad (6.112)$$

Obviously the strain measures from (6.91) would yield another micromorphic model.

Relaxed micromorphic medium

A modification of the general micromorphic continuum called the *relaxed micromorphic medium* model was introduced by Neff *et al.*[33]. For small deformations, the strain energy density is

$$\begin{aligned} W = {} & \frac{\lambda}{2}\operatorname{tr}^2\boldsymbol{\varepsilon} + \frac{\mu+\kappa}{2}\boldsymbol{\varepsilon}\bullet\boldsymbol{\varepsilon} + \frac{\mu-\kappa}{2}\boldsymbol{\varepsilon}\bullet\boldsymbol{\varepsilon}^T \\ & + \frac{\alpha}{2}\operatorname{tr}^2\mathbf{e} + \beta\mathbf{e}\bullet\mathbf{e} + \frac{\gamma}{2}(\nabla\times\mathbf{p})\bullet(\nabla\times\mathbf{p}). \end{aligned} \quad (6.113)$$

The nonlinear counterpart of (6.113) is

$$W = W(\mathbf{F}\cdot\mathbf{P}^{-1}, \mathbf{P}\cdot\mathbf{P}^T, (\nabla\times\mathbf{P})\cdot\mathbf{P}^{-1}) \quad (6.114)$$

and the Lagrangian equations of motion transform into

$$\nabla\cdot\mathbf{T} + \rho\mathbf{f} = \rho\dot{\mathbf{v}}, \qquad \nabla\times\mathbf{M} - \frac{\partial W}{\partial\mathbf{P}} + \rho\mathbf{m} = j\,\ddot{\mathbf{P}}, \quad (6.115)$$

where

$$\mathbf{T} = \frac{\partial W}{\partial\mathbf{F}}, \qquad \mathbf{M} = \frac{\partial W}{\partial\nabla\times\mathbf{P}}. \quad (6.116)$$

[33] Neff, P., Ghiba, I. D., Madeo, A., Placidi, L., Rosi, G. (2014). A unifying perspective: the relaxed linear micromorphic continuum, *Continuum Mech. Thermodynamics* 26(5) 639–681.

Note that unlike the general micromorphic model, here $\mathbf{M}$ is a second-order tensor, whereas $\mathbf{m}$ is a vector. As $\nabla \times \mathbf{P}$ is a pseudotensor, the hyper-stress second-order tensor $\mathbf{M}$ is also a pseudotensor as is the couple-stress tensor in the Cosserat continuum.

For various applications of the micromorphic model to modelling of complex materials, including the study of wave propagation, we also refer to [34].

6.6 Strain-Gradient Elasticity

Introducing nonlinearly elastic materials, we have mentioned a possible form of constitutive equation (6.14) in which the strain energy density can depend on the deformation gradients to any finite order. This theory is called *strain gradient elasticity*. It can be characterized as a *weak nonlocal theory* because the strain energy at a point depends only on the deformations in an infinitesimal neighborhood of the point. The initial ideas arose around the same time as the basics of continuum mechanics; see, for example, the analysis of Piola's contribution[35], Maugin's historical remarks, and [Maugin (2017)]. Further developments in this field were contributed by Toupin[36], Mindlin[37] and many others. Recently, for example, strain gradient elasticity found applications in modelling composite materials[38].

[34]Forest, S. (2013). Micromorphic media. In: *Generalized Continua from the Theory to Engineering Applications* pp. 249-300 (Springer, Vienna); Madeo, A., Neff, P., Ghiba, I. D., Rosi, G. (2016). Reflection and transmission of elastic waves in non-local band-gap metamaterials: a comprehensive study via the relaxed micromorphic model. *J. Mech. Phys. Solids*, 95, 441-479. Eremeyev, V. A., Lebedev, L. P., Cloud, M. J. (2017). Acceleration waves in the nonlinear micromorphic continuum. *Mech. Res. Communications.*

[35]dell'Isola, F., Corte, A.D., Giorgio, I. (2017). Higher-gradient continua: The legacy of Piola, Mindlin, Sedov and Toupin and some future research perspectives. *Mathematics and Mechanics of Solids*, 22(4), 852–872.

[36]Toupin, R.A. (1962). Elastic materials with couple-stresses. *Arch. Ration. Mech. Analysis*, 11(1), 385–414; (1964). Theories of elasticity with couple-stress. *Arch. Ration. Mech. Analysis*, 17(2), 85–112.

[37]Mindlin, R.D. (1965). Second gradient of strain and surface-tension in linear elasticity. *Int. J. Solids Struct.*, 1(4), 417–438; Mindlin, R.D., Eshel, N.N. (1968). On first strain-gradient theories in linear elasticity. *Int. J. Solids Struct.*, 4(1), 109–124.

[38]dell'Isola, F., Giorgio, I., Pawlikowski, M., Rizzi, N. (2016). Large deformations of planar extensible beams and pantographic lattices: Heuristic homogenisation, experimental and numerical examples of equilibrium. *Proc. Roy. Soc. London A.* 472(2185), 20150790; Rahali, Y., Giorgio, I., Ganghoffer, J.F., dell'Isola, F. (2015). Homogenization à la Piola produces second gradient continuum models for linear pantographic lattices. *Int. J. Engng. Sci.*, 97, 148–172.

Now we will consider what happens if W depends on the second deformation gradient:

$$W = W(\mathbf{r}, \nabla\mathbf{r}, \nabla^2\mathbf{r}). \tag{6.117}$$

Applying to (6.117) the material frame-indifference principle, we find that W does not depend on $\mathbf{r}$ and has the invariance property

$$W(\nabla\mathbf{r}, \nabla^2\mathbf{r}) = W(\nabla\mathbf{r}\cdot\mathbf{Q}^T, \nabla^2\mathbf{r}\cdot\mathbf{Q}^T) \tag{6.118}$$

for any orthogonal tensor $\mathbf{Q}$. Using $\mathbf{Q} = \mathbf{R}$ with $\mathbf{R}$ from the polar decomposition $\mathbf{F} \equiv \nabla\mathbf{r} = \mathbf{U}\cdot\mathbf{R}$, we come to the following form for the objective strain energy:

$$W = W(\mathbf{C}, \nabla\mathbf{F}\cdot\mathbf{F}^T), \tag{6.119}$$

where $\mathbf{C} = \mathbf{F}\cdot\mathbf{F}^T$. Instead of (6.119), the objective form

$$W = W(\mathbf{C}, \nabla\mathbf{C}). \tag{6.120}$$

was used by dell'Isola *et al*[39].

To deduce the static equilibrium equations and natural boundary conditions, we derive the first variation of

$$\mathcal{E} = \int_V W(\nabla\mathbf{r}, \nabla\nabla\mathbf{r})\, dV.$$

First we obtain

$$\begin{aligned}
\delta\mathcal{E} &= \int_V \delta W(\nabla\mathbf{r}, \nabla\nabla\mathbf{r})\, dV \\
&= \int_V \left[\frac{\partial W}{\partial\nabla\mathbf{r}} \bullet \nabla\mathbf{w} + \frac{\partial W}{\partial\nabla\nabla\mathbf{r}} \bullet \nabla\nabla\mathbf{w}\right] dV \\
&= \int_V \left[-\nabla\cdot\frac{\partial W}{\partial\nabla\mathbf{r}}\cdot\mathbf{w} - \nabla\cdot\frac{\partial W}{\partial\nabla\nabla\mathbf{r}} \bullet \nabla\mathbf{w}\right] dV \\
&\quad + \int_\Sigma \left(\mathbf{n}\cdot\frac{\partial W}{\partial\nabla\mathbf{r}}\cdot\mathbf{w} + \mathbf{n}\cdot\frac{\partial W}{\partial\nabla\nabla\mathbf{r}} \bullet \nabla\mathbf{w}\right) dS \\
&= \int_V \left[-\nabla\cdot\frac{\partial W}{\partial\nabla\mathbf{r}} - \nabla\cdot\left(\nabla\cdot\frac{\partial W}{\partial\nabla\nabla\mathbf{r}}\right)\right]\cdot\mathbf{w}\, dV \\
&\quad + \int_\Sigma \left[\mathbf{n}\cdot\frac{\partial W}{\partial\nabla\mathbf{r}}\cdot\mathbf{w} + \underline{\mathbf{n}\cdot\frac{\partial W}{\partial\nabla\nabla\mathbf{r}} \bullet \nabla\mathbf{w}} - \mathbf{n}\cdot\left(\nabla\cdot\frac{\partial W}{\partial\nabla\nabla\mathbf{r}}\right)\cdot\mathbf{w}\right] dS.
\end{aligned}$$

Here $\mathbf{w} = \delta\mathbf{r}$ and $\Sigma = \partial V$; the underlined term will be referenced below.

[39]dell'Isola, F., Sciarra, G., Vidoli, S. (2009) Generalized Hooke's law for isotropic second gradient materials. *Proc. Roy. Soc. London A*, 465 (2107) 2177–2196.

Note that the surface integrals contain the derivatives of $\mathbf{w}$. Remembering that when we change $\mathbf{w}$ and $\nabla\mathbf{w}$ only the normal derivatives can be considered as independent of $\mathbf{w}$, we should exclude the tangential derivatives of $\mathbf{w}$ over Σ from the surface integrals. So first we introduce the surface nabla-operator

$$\nabla_S = \nabla - \mathbf{n}\frac{\partial}{\partial n}, \tag{6.121}$$

where $\mathbf{n}$ is the unit (external) normal vector to Σ. Let us transform the underlined term:

$$\int_\Sigma \mathbf{n} \cdot \frac{\partial W}{\partial \nabla\nabla\mathbf{r}} \bullet \nabla\mathbf{w}\, dS = \int_\Sigma \left[\mathbf{n} \cdot \frac{\partial W}{\partial \nabla\nabla\mathbf{r}} \bullet \nabla_S\mathbf{w} + \mathbf{n} \cdot \frac{\partial W}{\partial \nabla\nabla\mathbf{r}} \bullet \left(\mathbf{n}\frac{\partial \mathbf{w}}{\partial n}\right)\right] dS.$$

For integration by parts we use the surface divergence theorem (3.29)

$$\int_\Sigma (\nabla_S \cdot \mathbf{T} + 2H\mathbf{n} \cdot \mathbf{T})\, dS = \oint_{\partial\Sigma} \boldsymbol{\nu} \cdot \mathbf{T}\, ds, \tag{6.122}$$

where $\mathbf{T}$ is a tensor-valued field, Σ is a smooth surface with contour $\partial\Sigma$, H is the mean curvature of Σ, and $\boldsymbol{\nu}$ is the unit normal to $\partial\Sigma$ such that $\boldsymbol{\nu} \cdot \mathbf{n} = 0$.

First, let us assume that $\Sigma = \partial V$ is smooth, i.e., $\mathbf{n}$ is a continuous function of surface coordinates on Σ. As Σ is closed, $\partial\Sigma$ is empty and the surface divergence theorem takes the form

$$\int_\Sigma (\nabla_S \cdot \mathbf{T} + 2H\mathbf{n} \cdot \mathbf{T})\, dS = 0. \tag{6.123}$$

Then

$$\int_\Sigma \mathbf{n} \cdot \frac{\partial W}{\partial \nabla\nabla\mathbf{r}} \bullet \nabla_S\mathbf{w} = -\int_\Sigma \left[\nabla_S \cdot \left(\mathbf{n} \cdot \frac{\partial W}{\partial \nabla\nabla\mathbf{r}}\right) + 2H\mathbf{n} \cdot \left(\mathbf{n} \cdot \frac{\partial W}{\partial \nabla\nabla\mathbf{r}}\right)\right] \cdot \mathbf{w}\, dS.$$

Thus

$$
\begin{aligned}
\delta\mathcal{E} = &\int_V \left[-\nabla\cdot\frac{\partial W}{\partial\nabla\mathbf{r}} - \nabla\cdot\left(\nabla\cdot\frac{\partial W}{\partial\nabla\nabla\mathbf{r}}\right)\right]\cdot\mathbf{w}\,dV \\
&+\int_\Sigma \left[\mathbf{n}\cdot\frac{\partial W}{\partial\nabla\mathbf{r}} - \nabla_S\cdot\left(\mathbf{n}\cdot\frac{\partial W}{\partial\nabla\nabla\mathbf{r}}\right) - 2H\mathbf{n}\cdot\left(\mathbf{n}\cdot\frac{\partial W}{\partial\nabla\nabla\mathbf{r}}\right)\right. \\
&\left.-\mathbf{n}\cdot\left(\nabla\cdot\frac{\partial W}{\partial\nabla\nabla\mathbf{r}}\right)\right]\cdot\mathbf{w}\,dS \\
&+\int_\Sigma \mathbf{n}\cdot\left(\mathbf{n}\cdot\frac{\partial W}{\partial\nabla\nabla\mathbf{r}}\right)\cdot\frac{\partial\mathbf{w}}{\partial n}\,dS.
\end{aligned}
$$

Second, let us consider a volume V with boundary $\Sigma = \bigcup_{k=1}^m \Sigma_k$ consisting of a finite number of smooth surfaces separated by edges L_j, where $\mathbf{n}$ undergoes a jump discontinuity, or where $\frac{\partial\mathbf{w}}{\partial n}$ is discontinuous. Applying (6.122) to each Σ_k, we have

$$
\int_{\Sigma_k} (\nabla_S\cdot\mathbf{T} + 2H\mathbf{n}\cdot\mathbf{T})\,dS = \oint_{\partial\Sigma_k} \boldsymbol{\nu}_k\cdot\mathbf{T}\,ds \tag{6.124}
$$

where $\boldsymbol{\nu}_k$ is a unit normal to $\partial\Sigma_k$ lying in the tangent plane to Σ_k (Fig. 6.3).

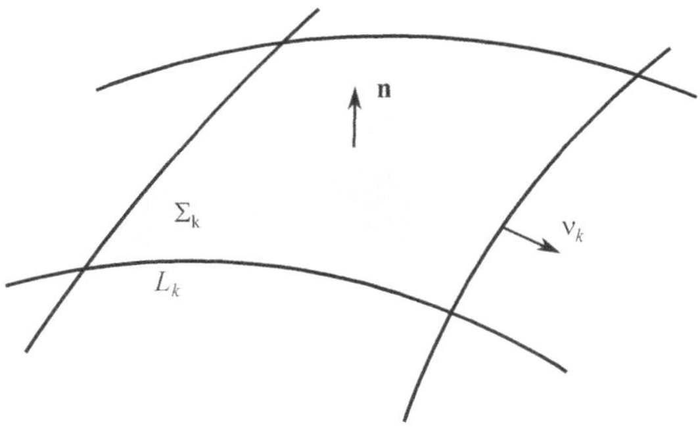

Fig. 6.3 Part Σ_k of the boundary with contour L_k.

Thus

$$
\begin{aligned}
\int_{\Sigma_k} \mathbf{n}\cdot\frac{\partial W}{\partial\nabla\nabla\mathbf{r}}\bullet\nabla_S\mathbf{w} = &-\int_{\Sigma_k}\left[\nabla_S\cdot\left(\mathbf{n}\cdot\frac{\partial W}{\partial\nabla\nabla\mathbf{r}}\right)\right. \\
&\left.+2H\mathbf{n}\cdot\left(\mathbf{n}\cdot\frac{\partial W}{\partial\nabla\nabla\mathbf{r}}\right)\right]\cdot\mathbf{w}\,dS \\
&-\oint_{\partial\Sigma_k}\boldsymbol{\nu}_k\cdot\left(\mathbf{n}\cdot\frac{\partial W}{\partial\nabla\nabla\mathbf{r}}\right)\cdot\mathbf{w}\,ds.
\end{aligned}
$$

Finally, $\delta\mathcal{E}$ is a sum of volume, surface, and contour integrals:

$$\begin{aligned}\delta\mathcal{E} = &\int_V \left[-\nabla\cdot\frac{\partial W}{\partial\nabla\mathbf{r}} - \nabla\cdot\left(\nabla\cdot\frac{\partial W}{\partial\nabla\nabla\mathbf{r}}\right)\right]\cdot\mathbf{w}\,dV\\ &+\sum_{k=1}^m\int_{\Sigma_k}\left[\mathbf{n}\cdot\frac{\partial W}{\partial\nabla\mathbf{r}} - \nabla_S\cdot\left(\mathbf{n}\cdot\frac{\partial W}{\partial\nabla\nabla\mathbf{r}}\right) - 2H\mathbf{n}\cdot\left(\mathbf{n}\cdot\frac{\partial W}{\partial\nabla\nabla\mathbf{r}}\right)\right.\\ &\left.-\mathbf{n}\cdot\left(\nabla\cdot\frac{\partial W}{\partial\nabla\nabla\mathbf{r}}\right)\right]\cdot\mathbf{w}\,dS\\ &+\sum_{k=1}^m\int_{\Sigma_k}\mathbf{n}\cdot\left(\mathbf{n}\cdot\frac{\partial W}{\partial\nabla\nabla\mathbf{r}}\right)\cdot\frac{\partial\mathbf{w}}{\partial n}\,dS\\ &-\sum_{k=1}^m\oint_{\partial\Sigma_k}\boldsymbol{\nu}_k\cdot\left(\mathbf{n}\cdot\frac{\partial W}{\partial\nabla\nabla\mathbf{r}}\right)\cdot\mathbf{w}\,ds.\end{aligned}$$

Let us assume that, in the following, the form of the virtual work $\delta\mathcal{A}$ is consistent with the form of $\delta\mathcal{E}$, that is

$$\delta\mathcal{A} = \int_V \rho\,\mathbf{f}\cdot\mathbf{w}\,dV + \sum_{k=1}^m\int_{\Sigma_k}\left[\mathbf{t}\cdot\mathbf{w} + \mathbf{c}\cdot\frac{\partial\mathbf{w}}{\partial n}\right]dS + \sum_{k=1}^m\oint_{\partial\Sigma_k}\mathbf{t}_k\cdot\mathbf{w}\,ds.$$

Here $\mathbf{f}$ is the vector of external forces per unit mass, ρ is the mass density in the reference configuration, $\mathbf{t}$ is the vector of external surface traction, $\mathbf{c}$ is the vector external moment, and $\mathbf{t}_k$ can be interpreted as the edge forces acting along the contours $L_k = \partial\Sigma_k$.

Thus the equation $\delta\mathcal{E} - \delta\mathcal{A} = 0$ implies the equilibrium equation

$$\nabla\cdot\mathbf{P} - \nabla\cdot(\nabla\cdot\mathbf{M}) + \rho\mathbf{f} = \mathbf{0}, \tag{6.125}$$

where we introduce the *intrinsic stress tensor* $\mathbf{P}$ and the *hyper-stress tensor* $\mathbf{M}$ of the first Piola–Kirchhoff type

$$\mathbf{P} = \frac{\partial W}{\partial\nabla\mathbf{r}}, \qquad \mathbf{M} = \frac{\partial W}{\partial\nabla\nabla\mathbf{r}}. \tag{6.126}$$

Then we deduce the natural boundary conditions on the surface $\Sigma = \cup_{k=1}^m\Sigma_k$,

$$\mathbf{n}\cdot\mathbf{P} - \nabla_S\cdot(\mathbf{n}\cdot\mathbf{M}) - 2H\mathbf{n}\cdot(\mathbf{n}\cdot\mathbf{M}) - \mathbf{n}\cdot(\nabla\cdot\mathbf{M}) = \mathbf{t}, \tag{6.127}$$

$$\mathbf{n}\cdot(\mathbf{n}\cdot\mathbf{M}) = \mathbf{c}, \tag{6.128}$$

and the compatibility conditions at contours L_j:

$$\boldsymbol{\nu}_+\cdot(\mathbf{n}_+\cdot\mathbf{M}_+) + \boldsymbol{\nu}_-\cdot(\mathbf{n}_-\cdot\mathbf{M}_-) + \mathbf{t}_j = \mathbf{0}. \tag{6.129}$$

In (6.129), indices $\pm$ relate with the values defined on the neighboring parts of the surface denoted by $\Sigma_\pm$ separated by L_j; see Fig. 6.4.

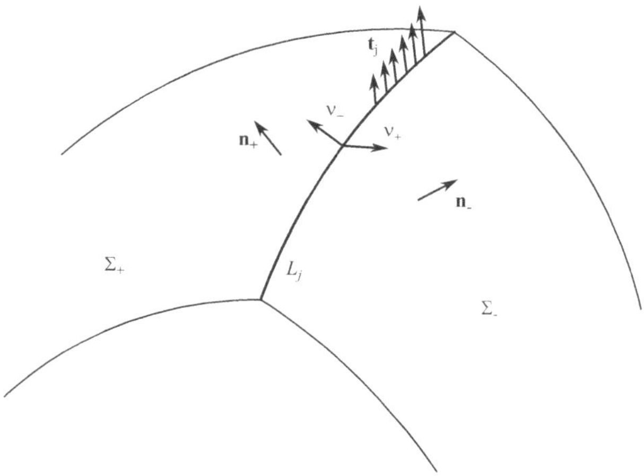

Fig. 6.4 Edge L_j with the normals $\boldsymbol{\nu}_-$ and $\boldsymbol{\nu}_+$.

Introducing the *effective first Piola–Kirchhoff tensor* $\mathbf{T} = \mathbf{P} - \nabla \cdot \mathbf{M}$ we transform (6.125) and (6.127) to a more compact form

$$\nabla \cdot \mathbf{T} + \rho \mathbf{f} = \mathbf{0}, \tag{6.130}$$

$$\mathbf{n} \cdot \mathbf{T} - \nabla_S \cdot (\mathbf{n} \cdot \mathbf{M}) - 2H\mathbf{n} \cdot (\mathbf{n} \cdot \mathbf{M}) = \mathbf{t}. \tag{6.131}$$

Note that within the strain gradient elasticity, the boundary conditions depend on the boundary curvature.

For small deformations, the strain energy density (6.120) transforms into

$$W = W(\boldsymbol{\varepsilon}, \nabla\boldsymbol{\varepsilon}), \qquad \boldsymbol{\varepsilon} = \tfrac{1}{2}(\nabla \mathbf{u} + (\nabla \mathbf{u})^T), \tag{6.132}$$

where $\mathbf{u}$ is the vector of displacements. Now W is a quadratic function of its arguments:

$$W = \tfrac{1}{2}\boldsymbol{\varepsilon} \cdot\cdot \mathbf{C} \cdot\cdot \boldsymbol{\varepsilon} + \boldsymbol{\varepsilon} \cdot\cdot \mathbf{G} \bullet \nabla\boldsymbol{\varepsilon} + \tfrac{1}{2}\nabla\boldsymbol{\varepsilon} \bullet \mathbf{D} \bullet \nabla\epsilon$$

where $\mathbf{C}$, $\mathbf{G}$, and $\mathbf{D}$ are fourth-order, fifth-order and sixth-order tensors of elastic moduli, respectively. Even in the case of isotropic materials, the form of W is rather complex. In particular, in addition to the two classical Lamé moduli λ and μ, there are five elastic moduli[39]. This presents serious challenges for numerical and experimental evaluation of the general strain-gradient model.

Surprisingly, we have already encountered a strain gradient continuum among the models of classical structural mechanics. This is the the classical theory of plates and shells based on the Kirchhoff hypotheses. In particular comparing (6.131) with the boundary conditions for the Kirchhoff–Love shell (5.42), one can easily see the similarity. Indeed, unlike classical or micropolar elasticity, here we see a combination of stresses and couple stresses in the same boundary condition. So Kirchhoff–Love shells and plates present the first example of a two-dimensional strain gradient continuum. Indeed for bending, the shell strain energy (5.34) depends on the second derivatives of deflection w, that is, $W = W(\nabla\nabla w)$.

Another example of a strain-gradient medium is the theory of couple stresses (Cosserat continuum with constraint rotations). Indeed, for this theory in the case of small deformation, the strain energy density takes the form

$$W = W(\boldsymbol{\varepsilon}, \nabla(\nabla \times \mathbf{u}))\,;$$

see § 6.4 for details. As $\nabla(\nabla\times\mathbf{u})$ is a second-order tensor, the corresponding hyper-stress tensor — the couple stress tensor — is also of second order.

Finally, let us note that there is a straightforward correspondence between the micromorphic and strain-gradient continua. Indeed, considering the micromorphic continuum with the constraint when the micro-distortion tensor $\mathbf{P}$ coincides with the deformation gradient $\mathbf{F}$, we come to the strain gradient continuum.

Exercise 6.9. Prove the formula of integration by parts

$$\int_\Sigma \mathbf{T} \bullet \nabla_S \mathbf{w}\, dS = -\int_\Sigma [(\nabla_S \cdot \mathbf{T}) \cdot \mathbf{w} + 2H\mathbf{n}\cdot\mathbf{T}\cdot\mathbf{w}]\, dS + \oint_{\partial\Sigma} \boldsymbol{\nu}\cdot\mathbf{T}\cdot\mathbf{w}\, ds,$$

where $\mathbf{T}$ is a second-order tensor, $\mathbf{w}$ is a vector, and $\mathbf{n}\cdot\boldsymbol{\nu} = 0$.

Exercise 6.10. Find the equilibrium equations for an Nth-order strain gradient solid with energy density $W = W(\nabla\mathbf{r}, \nabla^2\mathbf{r}, \ldots, \nabla^N\mathbf{r})$.

6.7 Linear Stress- and Strain-Gradient Elasticity

The general form of constitutive equations for strain-gradient elasticity is rather complex even in the case of small deformations and isotropic solids. For small deformations Aifantis[40] proposed a simpler model of strain-gradient elasticity with just one additional parameter. Within Aifantis's model, the strain energy density is assumed to be

$$W = \tfrac{1}{2}\boldsymbol{\sigma}\cdot\cdot\,\boldsymbol{\varepsilon} + \tfrac{1}{2}\ell^2\nabla\boldsymbol{\sigma}\bullet\nabla\boldsymbol{\varepsilon}, \quad \boldsymbol{\sigma} = \mathbf{C}\cdot\cdot\,\boldsymbol{\varepsilon}, \quad \boldsymbol{\varepsilon} = \tfrac{1}{2}(\nabla\mathbf{u} + (\nabla\mathbf{u})^T), \quad (6.133)$$

[40]Aifantis, E. C. (1992). On the role of gradients in the localization of deformation and fracture. *Int. J. Engng. Sci.*, 30(10), 1279-1299.

where $\boldsymbol{\sigma}$ is the *intrinsic stress tensor*, $\mathbf{C}$ is the classical stiffness tensor (4.18), $\boldsymbol{\varepsilon}$ is the strain tensor, and ℓ is an additional parameter having the sense of a characteristic length.

In particular, for isotropic solids

$$W = \tfrac{1}{2}\lambda(\operatorname{tr}\boldsymbol{\varepsilon})^2 + \mu\boldsymbol{\varepsilon}\cdot\cdot\,\boldsymbol{\varepsilon} + \tfrac{1}{2}\lambda\ell^2\nabla\operatorname{tr}\boldsymbol{\varepsilon}\cdot\nabla\operatorname{tr}\boldsymbol{\varepsilon} + \mu\ell^2\boldsymbol{\varepsilon}\bullet\boldsymbol{\varepsilon},$$
$$\boldsymbol{\sigma} = \lambda\mathbf{E}\operatorname{tr}\boldsymbol{\varepsilon} + 2\mu\boldsymbol{\varepsilon},$$

where λ, μ are Lamé's moduli. So the model has three material parameters.

The equilibrium equations now take the form

$$\nabla\cdot(\boldsymbol{\sigma} - \ell^2\nabla^2\boldsymbol{\sigma}) + \rho\mathbf{f} = \mathbf{0}, \tag{6.134}$$

where $\nabla^2 = \nabla\cdot\nabla$, ρ is the mass density, and $\mathbf{f}$ is the external force vector. Introducing the *effective stress tensor* $\mathbf{T}$ by

$$\mathbf{T} = \boldsymbol{\sigma} - \ell^2\nabla^2\boldsymbol{\sigma} \tag{6.135}$$

we transform (6.134) into the classical form

$$\nabla\cdot\mathbf{T} + \rho\mathbf{f} = \mathbf{0}.$$

The form of (6.134) provides an approach to the corresponding boundary value problem. Following Ru and Aifantis[41] we decompose it into two vectorial equations

$$\nabla\cdot\boldsymbol{\sigma} + \rho\mathbf{f} = \mathbf{0}, \quad \boldsymbol{\sigma} = \mathbf{C}\cdot\cdot\,\tilde{\boldsymbol{\varepsilon}}, \quad \tilde{\boldsymbol{\varepsilon}} = \tfrac{1}{2}(\nabla\tilde{\mathbf{u}} + (\nabla\tilde{\mathbf{u}})^T), \tag{6.136}$$
$$\mathbf{u} - \ell^2\nabla^2\mathbf{u} = \tilde{\mathbf{u}}. \tag{6.137}$$

This decomposition is often called the *Ru–Aifantis theorem.* According (6.136) we first solve the classical problem of linear elasticity with proper boundary conditions. The result is $\tilde{\mathbf{u}}$. Then (6.137) yields the real displacement field $\mathbf{u}$. Thus we can apply many known solutions from linear elasticity to the strain-gradient elasticity.

For generalization of the linear strain-gradient elasticity for dynamics we refer to the literature[42].

We emphasize that the linear or nonlinear versions of strain-gradient elasticity satisfy the principle of locality. Obviously, the strain energy $W =$

[41] Ru, C.Q., Aifantis, E.C. (1993). A simple approach to solve boundary-value problems in gradient elasticity. *Acta Mech.*, 101(1-4), 59–68.

[42] Askes, H., Aifantis, E.C. (2011). Gradient elasticity in statics and dynamics: an overview of formulations, length scale identification procedures, finite element implementations and new results. *Int. J. Solids. Struct.*, 48(13), 1962–1990; Askes, H., Aifantis, E.C. (2006). Gradient elasticity theories in statics and dynamics — a unification of approaches. *Int. J. Fracture*, 139(2), 297–304.

$W(\nabla\mathbf{r}, \nabla^2\mathbf{r}, \ldots)$ defined at a certain point is "less" local than the energy of a simple material $W = W(\nabla\mathbf{r})$. But formally it is still determined by an infinitesimal neighborhood of the chosen point. So we can call the model a *weak nonlocal model* of a continuum. Eringen introduced a nonlocal elastic material [Eringen (2002)] which we call a *strong nonlocal model*. The general form of Eringen's energy functional for finite deformations is

$$\mathcal{E} = \int_V \int_V W(\mathbf{C}, \mathbf{C}')\, dV'\, dV, \tag{6.138}$$

where $\mathbf{C} = \mathbf{C}(\mathbf{x}) = \nabla\mathbf{r}(\mathbf{x}) \cdot (\nabla\mathbf{r}(\mathbf{x}))^T$ is the Cauchy–Green strain measure at the point with vector $\mathbf{x}$ whereas $\mathbf{C}' = \mathbf{C}(\mathbf{x}')$ is taken at $\mathbf{x}'$. In (6.138) we apply double integration over volume V with respect to $\mathbf{x}$ and $\mathbf{x}'$. Here V is a volume of the considered body in the reference configuration. Since the order of integration does not affect the value of $\mathcal{E}$, we assume W is a symmetric function of its arguments with $W(\mathbf{C}, \mathbf{C}') = W(\mathbf{C}', \mathbf{C})$. Note that unlike classic nonlinear elasticity we have a more complex constitutive dependence as W depends on two tensorial arguments.

For small deformations, (6.138) transforms into

$$\mathcal{E} = \frac{1}{2} \int_V \int_V \boldsymbol{\varepsilon}(\mathbf{x}') \cdot\cdot\, \mathbf{C}(\mathbf{x}', \mathbf{x}) \cdot\cdot\, \boldsymbol{\varepsilon}(\mathbf{x})\, dV'\, dV, \tag{6.139}$$

where $\mathbf{C} = \mathbf{C}(\mathbf{x}', \mathbf{x}) = \mathbf{C}(\mathbf{x}, \mathbf{x}')$ plays the role of the stiffness moduli. For Eringen's model, instead of the set of material parameters, we have a set of material functions that brings more difficulties in both the theoretical analysis and the experimental validation of the model. The corresponding stress tensor is given by

$$\mathbf{T} = \int_V \mathbf{C}(\mathbf{x}', \mathbf{x}) \cdot\cdot\, \boldsymbol{\epsilon}'\, dV'. \tag{6.140}$$

Note that the nonlocal stress-strain relation (6.140) transforms formally into the classical relation $\boldsymbol{\sigma}(\mathbf{x}) = \mathbf{C}(\mathbf{x}) \cdot\cdot\, \boldsymbol{\varepsilon}(\mathbf{x})$ if we replace the kernel $\mathbf{C}(\mathbf{x}', \mathbf{x})$ by

$$\mathbf{C}(\mathbf{x}', \mathbf{x}) = \mathbf{C}(\mathbf{x})\delta(\mathbf{x}' - \mathbf{x})$$

where δ is the Dirac delta. Recall that

$$\int_V \delta(\mathbf{x})\, dV = 1, \qquad \int_V \delta(\mathbf{x}' - \mathbf{x}) f(\mathbf{x}')\, dV' = f(\mathbf{x}),$$

for any continuous function $f(\mathbf{x})$ if $\mathbf{x} \in V$ [Korn and Korn (1968)].

To simplify the model, we restrict ourselves to homogeneous isotropic materials. Homogeneity results in the restriction

$$\mathbf{C}(\mathbf{x}', \mathbf{x}) = \mathbf{C}(\mathbf{x}' - \mathbf{x}),$$

whereas isotropy leads to

$$\mathbf{C} = \lambda(|\mathbf{x}' - \mathbf{x}|)\mathbf{E}\mathbf{E} + \mu(|\mathbf{x}' - \mathbf{x}|)(\mathbf{e}_k\mathbf{E}\mathbf{e}^k + \mathbf{I}).$$

Here λ and μ depend only on the distance $\varkappa = |\mathbf{x}' - \mathbf{x}|$ from $\mathbf{x}'$ to $\mathbf{x}$.

Moreover, motivated by the considerations of interatomic interactions, these dependencies can be replaced by

$$\lambda(\varkappa) = \alpha(\varkappa)\lambda, \qquad \mu(\varkappa) = \alpha(\varkappa)\mu,$$

where $\alpha(\varkappa)$ is a kernel function and λ and μ are the classical Lamé moduli [Eringen (2002)]. Thus we need only one additional function describing the nonlocality of the material behavior. The kernel function $\alpha(\varkappa)$ should be normalized over the volume of the body similarly to the Dirac-delta function,

$$\int_V \alpha(|\mathbf{x}|)\, dV = 1, \tag{6.141}$$

hence it carries units m^{-3}. In [Eringen (2002)] various kernel functions are discussed, including

$$\alpha(|\mathbf{x}|) = \alpha_0 \exp^{-\alpha_1|\mathbf{x}|^2}$$

where α_0 and α_1 are material parameters subjected to constraint (6.141).

Another way of constructing of the kernel function is based on consideration of α as a Green's function corresponding to some linear differential equation. Recall that a *Green's function* is a solution of a differential equation $L\alpha(\mathbf{x}) = \delta(\mathbf{x})$, where L is a linear differential operator [Korn and Korn (1968)]. Considering lattice dynamics and dispersion curves, Eringen proposed to represent L as

$$L = 1 - \alpha_1^2\nabla^2 + \alpha_2^4\nabla^4, \qquad \nabla^2 = \nabla \cdot \nabla, \qquad \nabla^4 = (\nabla \cdot \nabla)^2 \tag{6.142}$$

or in the simplified form

$$L = 1 - \alpha^2\nabla^2. \tag{6.143}$$

Here α, α_1, and α_2 are material parameters having units m^{-1}. For the proper choice of kernel function, Eringen suggested that one minimize the number of additional parameters used and take into account the dispersion behavior from experiments or from lattice dynamics.

With the introduced kernel function, we write the nonlocal stress-strain relation in the form

$$\mathbf{T} = \int_V \alpha(|\mathbf{x}' - \mathbf{x}|)\boldsymbol{\sigma}(\mathbf{x}')\, dV', \quad \boldsymbol{\sigma}(\mathbf{x}) = \lambda \mathbf{E} \operatorname{tr} \boldsymbol{\varepsilon}(\mathbf{x}) + 2\mu\boldsymbol{\varepsilon}(\mathbf{x}). \tag{6.144}$$

Now the equilibrium equations and static boundary conditions take the classical form:

$$\nabla \cdot \mathbf{T} + \rho\mathbf{f} = \mathbf{0} \qquad (\mathbf{x} \in V), \tag{6.145}$$

$$\mathbf{n} \cdot \mathbf{T} = \mathbf{t} \qquad (\mathbf{x} \in \partial V), \tag{6.146}$$

where $\mathbf{f}$ and $\mathbf{t}$ are the external mass forces and surface traction, respectively. So for determining the displacements and stresses, we have a system of linear integro-differential equations.

If the kernel function was introduced as a Green function of the differential operator L, we apply L to Eq. (6.144) and (6.145). First we get

$$\begin{aligned} L\mathbf{T} &= L \int_V \alpha(|\mathbf{x}' - \mathbf{x}|)\boldsymbol{\sigma}(\mathbf{x}')\, dV' = \int_V L\alpha(|\mathbf{x}' - \mathbf{x}|)\boldsymbol{\sigma}(\mathbf{x}')\, dV' \\ &= \int_V \delta(\mathbf{x}' - \mathbf{x})\boldsymbol{\sigma}(\mathbf{x}')\, dV' = \boldsymbol{\sigma}(\mathbf{x}). \end{aligned}$$

Applying L to (6.145) we obtain the equilibrium equation in the form

$$\nabla \cdot \boldsymbol{\sigma} + \rho L\mathbf{f} = \mathbf{0}. \tag{6.147}$$

This formally coincides with the equilibrium equations of classical elasticity (cf., (4.9)) but with new external forces. So its solution is especially simple if the external forces are absent. Indeed, if we know the solution $\tilde{\mathbf{u}}$ of $\nabla \cdot \boldsymbol{\sigma} = \mathbf{0}$, then we calculate tensor $\tilde{\boldsymbol{\sigma}}$ and subsequently find the real stresses $\mathbf{T}$ from the equation $L\mathbf{T} = \tilde{\boldsymbol{\sigma}}$.

For example, for (6.142) we have

$$\mathbf{T} - \alpha_1^2 \nabla^2 \mathbf{T} + \alpha_2^4 \nabla^4 \mathbf{T} = \boldsymbol{\sigma} \equiv \lambda \mathbf{E} \operatorname{tr} \boldsymbol{\varepsilon} + 2\mu\boldsymbol{\varepsilon}, \tag{6.148}$$

whereas for (6.142) we get

$$\mathbf{T} - \alpha^2 \nabla^2 \mathbf{T} = \boldsymbol{\sigma}. \tag{6.149}$$

Equation (6.147) transforms into the classical Lamé equation

$$(\lambda + \mu)\nabla\nabla \cdot \mathbf{u} + \mu\nabla \cdot \nabla \mathbf{u} = \mathbf{0}. \tag{6.150}$$

Let $\tilde{\mathbf{u}}$ be its solution. Then we calculate the intrinsic stress tensor $\tilde{\boldsymbol{\sigma}}$ using Hooke's law

$$\tilde{\boldsymbol{\sigma}} = \lambda \mathbf{E} \operatorname{tr} \tilde{\boldsymbol{\varepsilon}} + 2\mu\tilde{\boldsymbol{\varepsilon}}, \qquad \tilde{\boldsymbol{\varepsilon}} = \tfrac{1}{2}(\nabla\tilde{\mathbf{u}} + (\nabla\tilde{\mathbf{u}})^T). \tag{6.151}$$

Finally, for determining the real stresses we use (6.149) or (6.149), where $\boldsymbol{\sigma} = \tilde{\boldsymbol{\sigma}}$. As in linear elasticity many solutions are already derived, the Ru–Aifantis theorem and Eringen's technique enable us to relatively easily obtain many generalizations of classical solutions.

The form of equations (6.148) and (6.149) prompts us to call this model the *stress-gradient elasticity* in the contrast to the earlier considered strain-gradient elasticity. Here the stress tensor can be obtained as a solution of the Helmholtz equation (6.149) or the bi-Helmholtz equation (6.148). So the models based on (6.149) and (6.148) are often called the *nonlocal elasticity of Helmholtz type* and *nonlocal elasticity of bi-Helmholtz type*[43].

Summing up, we have the following constitutive relations for an isotropic solid within the simplest strain-gradient and stress-gradient models, respectively:

$$\mathbf{T} = (1 - \ell^2\nabla^2)(\lambda\mathbf{E}\,\mathrm{tr}\,\boldsymbol{\varepsilon} + 2\mu\boldsymbol{\varepsilon}), \qquad \mathbf{T} - \alpha^2\nabla^2\mathbf{T} = \lambda\mathbf{E}\,\mathrm{tr}\,\boldsymbol{\varepsilon} + 2\mu\boldsymbol{\varepsilon}.$$

Combining these two models, a more general form of the gradient-type elasticity can be derived:

$$\mathbf{T} - \alpha^2\nabla^2\mathbf{T} = (1 - \ell^2\nabla^2)(\lambda\mathbf{E}\,\mathrm{tr}\,\boldsymbol{\varepsilon} + 2\mu\boldsymbol{\varepsilon}). \tag{6.152}$$

Here, in addition to Lamé's moduli, we introduce two length-scale parameters α and ℓ. Using a technique similar to that of Ru–Aifantis and Eringen, we can split the solution of the static problem without external mass forces as follows. First we can find a solution $\tilde{\mathbf{u}}$ to the homogeneous equation (6.150) and the corresponding stress tensor (6.151). Next, using the classical solutions for displacements and stresses, we have to solve two equations for real displacements and real stresses:

$$\mathbf{T} - \alpha^2\nabla^2\mathbf{T} = \tilde{\boldsymbol{\sigma}}, \qquad \left(1 - \ell^2\nabla^2\right)\mathbf{u} = \tilde{\mathbf{u}}. \tag{6.153}$$

Note that both equations in (6.153) are nonhomogeneous *Helmholtz equations*; see Aifantis[44] for details.

6.8 Strain Gradient Fluids

Considering the gradient elasticity it is worth to mention the model of *second gradient fluid.* It presents an example of the strain gradient media

[43] Lazar, M., Maugin, G.A., Aifantis, E.C. (2006). On a theory of nonlocal elasticity of bi-Helmholtz type and some applications. *Int. J. Solids Struct.*, 43(6), 1404–1421.
[44] Aifantis, E.C. (2003). Update on a class of gradient theories. *Mech. Materials*, 35(3), 259–280.

which has a long history. The model was proposed by van der Waals[45] and Korteweg[46] and was later developed by Cahn and Hilliard[47]. So it is also called the Korteweg or Cahn–Hilliard fluid[48]; see also [Maugin (2017)]. Unlike the classical Euler fluid, in this model the strain energy density depends on the mass density and its spatial gradient. Recently the model became part of a so-called molecular theory of capillarity. It is able to predict the existence of an interfacial layer between the fluid and its vapor. Here, using the variational approach, we will present the basic equations of the theory. It is a simple example of the localization phenomena observed in the strain gradient materials.

Strain energy density

As is natural for a fluid, we will use the Eulerian (spatial) description for the kinematics. So all the quantities are defined in the actual configuration. The strain energy of the fluid per unit volume in the actual configuration depends on the mass density ρ and its spatial gradient $\nabla\rho$:

$$W = W(\rho, \nabla\rho). \tag{6.154}$$

Satisfaction of the material frame-indifference principle requires

$$W(\rho, \nabla\rho) = W(\rho, \mathbf{Q}\cdot\nabla\rho)$$

for any orthogonal tensor $\mathbf{Q}$. So W is an isotropic function of $\nabla\rho$ of the form

$$W = W(\rho, \beta), \qquad \beta = \nabla\rho\cdot\nabla\rho. \tag{6.155}$$

As an example, consider the constitutive equation

$$W = W_0(\rho) + \tfrac{1}{2}\lambda\beta^2, \tag{6.156}$$

where λ is the capillary coefficient and $W_0(\rho)$ is a function of ρ. Usually $W_0(\rho)$ is a non-convex function similar to the van der Waals gas constitutive

[45] Van der Waals, J. D. (1893). The thermodynamic theory of capillarity under the hypothesis of a continuous variation of density (English translation by J.S. Rowlinson). *J. Stat. Phys.* **20**, 200–244

[46] Korteweg, D.J. (1901). Sur la forme que prennent les équations des mouvements des fluides si l'on tient compte des forces capillaires par des variations de densité. *Archives Néerlandaises des Sciences exactes et naturelles* **Sér. II**(6), 1–24

[47] Cahn, J.W., Hilliard, J.E. (1958). Free energy of a nonuniform system. I. Interfacial free energy. *J. Chem. Phys.* **28**, 258; (1959). Free energy of a nonuniform system. III. Nucleation in a two-component incompressible fluid. *J. Chem. Phys.* **31**, 688

[48] Seppecher, P. (1996). *Les fluides de Cahn-Hilliard.* Mémoire d'habilitation à diriger des recherches, Université du Sud Toulon

equations. For example, $W_0(\rho) = K\rho_0^{-2}(\rho - \rho_1)^2(\rho - \rho_2)^2 - m\rho - q$, where K is an elastic modulus, m and q are material parameters, ρ_0 is a referential density, and ρ_1 and ρ_2 correspond to two minima of W_0 (Fig. 6.5). Usually these two minima relate to the states of the fluid and its vapor. In what follows, we will retain a general form of W.

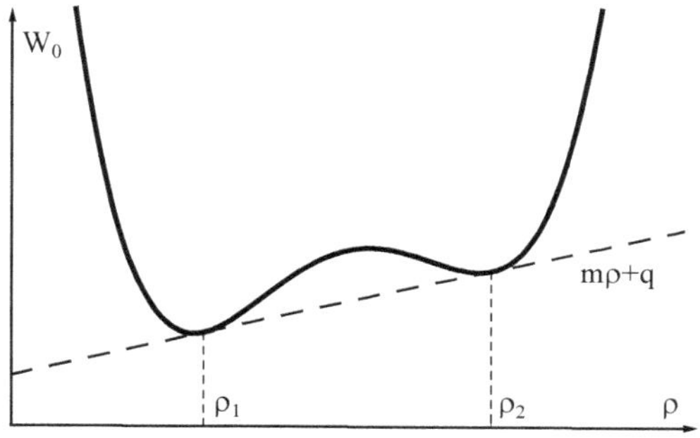

Fig. 6.5 A typical form of $W_0(\rho)$ with two minima at ρ_1 and ρ_2. The dashed line is given by the equation $y = m\rho + q$.

Energy functional and its variation

The energy functional takes the form

$$\mathcal{E} = \int_v W(\rho, \nabla\rho)\, dv, \tag{6.157}$$

where v is the volume occupied by the fluid in the actual configuration at time instant t.

To derive the first variation of $\mathcal{E}$, we transform (6.157) to the Lagrangian form

$$\mathcal{E} = \int_V W(\rho, \beta) J\, dV, \tag{6.158}$$

where V is the volume of the fluid in the reference configuration, $J = \det \mathbf{F}$, $\mathbf{F} = (\nabla_\kappa \mathbf{r})^T$ is the deformation gradient, $\mathbf{r}$ is the position vector of a fluid particle in the actual configuration, and ∇_κ is the nabla-operator in the reference configuration. The first variation of $\mathcal{E}$ is

$$\delta\mathcal{E} = \int_V \delta(JW(\rho,\beta))\, dV = \int_V \left[J\frac{\partial W}{\partial \rho}\delta\rho + J\frac{\partial W}{\partial \beta}\delta\beta + W\delta J \right] dV. \tag{6.159}$$

Introducing the variation of the position vector $\delta\mathbf{r} = \mathbf{w}$ and using

$$\rho J = \rho_0, \qquad \mathbf{F}^T \cdot \nabla = \nabla_\kappa, \qquad \nabla = \mathbf{F}^{-T} \cdot \nabla_\kappa,$$

and

$$\delta\mathbf{F}^{-T} = -\mathbf{F}^{-T} \cdot \delta\mathbf{F}^T \cdot \mathbf{F}^{-T}, \qquad \delta\mathbf{F} = (\nabla_\kappa \mathbf{w})^T,$$

where $\mathbf{F}^{-T} = (\mathbf{F}^{-1})^T$ and ρ_0 is the density in the reference configuration, we obtain the following relations for the variations:

$$\begin{aligned}
\delta J &= J\nabla \cdot \mathbf{w}, \\
\delta\rho &= \delta\frac{\rho_0}{J} = \frac{\rho_0}{J^2}\delta J = \frac{\rho_0}{J}\nabla \cdot \mathbf{w} = -\rho\nabla \cdot \mathbf{w}, \\
\delta(\nabla\rho) &= \delta(\mathbf{F}^{-T} \cdot \nabla_\kappa \rho) \\
&= (\delta\mathbf{F}^{-T}) \cdot \nabla_\kappa \rho + \mathbf{F}^{-T} \cdot \nabla_\kappa \delta\rho \\
&= -\mathbf{F}^{-T} \cdot \delta\mathbf{F}^T \cdot \mathbf{F}^{-T} \cdot \nabla_\kappa \rho - \mathbf{F}^{-T} \cdot \nabla_\kappa(\rho\nabla \cdot \mathbf{w}) \\
&= -\mathbf{F}^{-T} \cdot \delta\mathbf{F}^T \cdot \nabla\rho - \nabla(\rho\nabla \cdot \mathbf{w}) \\
&= -\mathbf{F}^{-T} \cdot \nabla_\kappa \mathbf{w} \cdot \nabla\rho - \nabla(\rho\nabla \cdot \mathbf{w}) \\
&= -\nabla\mathbf{w} \cdot \nabla\rho - \nabla(\rho\nabla \cdot \mathbf{w}), \\
\delta\beta &= 2\nabla\rho \cdot \delta(\nabla\rho) \\
&= 2\nabla\rho \cdot [-\nabla\mathbf{w} \cdot \nabla\rho - \nabla(\rho\nabla \cdot \mathbf{w})].
\end{aligned}$$

Substituting these into (6.159), we deduce

$$\begin{aligned}
\delta\mathcal{E} = &\int_V \left\{ -\frac{\partial W}{\partial\rho}\rho\nabla \cdot \mathbf{w} - 2\frac{\partial W}{\partial\beta}\nabla\rho \cdot [\nabla\mathbf{w} \cdot \nabla\rho + \nabla(\rho\nabla \cdot \mathbf{w}) \right. \\
&\left. + W\nabla \cdot \mathbf{w}] \vphantom{\frac{\partial W}{\partial\rho}} \right\} J\, dV \\
= &\int_v \left\{ \left[W - \rho\frac{\partial W}{\partial\rho} \right] \nabla \cdot \mathbf{w} - 2\frac{\partial W}{\partial\beta}\nabla\rho \cdot [\nabla\mathbf{w} \cdot \nabla\rho + \nabla(\rho\nabla \cdot \mathbf{w})] \right\} dv.
\end{aligned} \tag{6.160}$$

Thus we have expressed $\delta\mathcal{E}$ in terms of $\mathbf{w}$ and its spatial gradient. Applying the integration by parts procedure, based on the spatial divergence theorem, to (6.160), we obtain the following identities:

$$\int_v \left(W - \rho\frac{\partial W}{\partial \rho}\right) \nabla\cdot\mathbf{w}\,dv$$
$$= -\int_v \nabla\left(W - \rho\frac{\partial W}{\partial \rho}\right)\cdot\mathbf{w}\,dv$$
$$+ \int_{\partial v}\left(W - \rho\frac{\partial W}{\partial \rho}\right)\mathbf{n}\cdot\mathbf{w}\,dS, \tag{6.161}$$

$$-\int_v 2\frac{\partial W}{\partial \beta}\nabla\rho\cdot\nabla\mathbf{w}\cdot\nabla\rho\,dv$$
$$= \int_v \nabla\cdot\left(2\frac{\partial W}{\partial \beta}\nabla\rho\nabla\rho\right)\cdot\mathbf{w}\,dv$$
$$- \int_{\partial v} 2\frac{\partial W}{\partial \beta}\mathbf{n}\cdot\nabla\rho\nabla\rho\cdot\mathbf{w}\,dS, \tag{6.162}$$

$$-\int_v 2\frac{\partial W}{\partial \beta}\nabla\rho\cdot\nabla(\rho\nabla\cdot\mathbf{w})\,dv$$
$$= \int_v \nabla\cdot\left(2\frac{\partial W}{\partial \beta}\nabla\rho\right)\rho\nabla\cdot\mathbf{w}\,dv - \int_{\partial v}\mathbf{n}\cdot\left(2\frac{\partial W}{\partial \beta}\nabla\rho\right)\rho\nabla\cdot\mathbf{w}\,dS$$
$$= -\int_v \nabla\left[\rho\nabla\cdot\left(2\frac{\partial W}{\partial \beta}\nabla\rho\right)\right]\cdot\mathbf{w}\,dv$$
$$+ \int_{\partial v}\left[\nabla\cdot\left(2\frac{\partial W}{\partial \beta}\nabla\rho\right)\rho\,\mathbf{n}\cdot\mathbf{w} - \int_{\partial v} 2\frac{\partial W}{\partial \beta}(\mathbf{n}\cdot\nabla\rho)\rho\nabla\cdot\mathbf{w}\right]dS, \tag{6.163}$$

where $\mathbf{n}$ is the outward unit normal to the volume boundary $\partial v = S$ and dS is an elementary area in the actual configuration. Note that (6.163) contains a surface integral with integrand depending on the spatial gradient of $\mathbf{w}$.

Further transformations require the surface divergence theorem, formulated as

$$\int_S (\nabla_s\cdot\mathbf{T} + 2K\mathbf{n}\cdot\mathbf{T})\,dS = \int_{\partial\sigma}\boldsymbol{\nu}\cdot\mathbf{T}\,ds, \tag{6.164}$$

where $\mathbf{T}$ is a tensor field of any order,

$$\nabla_s = \nabla - \mathbf{n}(\mathbf{n}\cdot\nabla)$$

is the surface nabla-operator in the actual configuration, K is the mean curvature of the surface S with the contour $\ell \equiv \partial S$, and $\boldsymbol{\nu}$ is the unit normal vector to ℓ such that $\boldsymbol{\nu}\cdot\mathbf{n} = 0$; see (3.29).

With (6.164), the last surface integral in (6.163) takes the form

$$
\begin{aligned}
-\int_{\partial v} 2\frac{\partial W}{\partial \beta}(\mathbf{n}\cdot\nabla\rho)\rho\nabla\cdot\mathbf{w}\,dS
&= -\int_{\partial v} 2\frac{\partial W}{\partial \beta}(\mathbf{n}\cdot\nabla\rho)\rho\mathbf{n}\cdot(\mathbf{n}\cdot\nabla)\mathbf{w}\;dS\\
&\quad -\int_{\partial v} 2\frac{\partial W}{\partial \beta}(\mathbf{n}\cdot\nabla\rho)\rho\nabla_s\cdot\mathbf{w}\;dS\\
&= -\int_{\partial v} 2\frac{\partial W}{\partial \beta}(\mathbf{n}\cdot\nabla\rho)\rho\mathbf{n}\cdot(\mathbf{n}\cdot\nabla)\mathbf{w}\,dS\\
&\quad +\int_{\partial v} 4\frac{\partial W}{\partial \beta}(\mathbf{n}\cdot\nabla\rho)\rho K\mathbf{n}\cdot\mathbf{w}\,dS\\
&\quad +\int_{\partial v} \nabla_s\left[2\rho\frac{\partial W}{\partial \beta}(\mathbf{n}\cdot\nabla\rho)\right]\cdot\mathbf{w}\,dS\\
&\quad -\int_{\partial S} 2\rho\frac{\partial W}{\partial \beta}(\mathbf{n}\cdot\nabla\rho)\boldsymbol{\nu}\cdot\mathbf{w}\,ds.
\end{aligned}
\tag{6.165}
$$

For a smooth closed surface S the set ∂S is empty, but we retain the last integral in (6.165) to analyze boundary conditions along possible edges.

Summarizing (6.161)–(6.163) and remembering (6.165), we obtain the first variation of $\mathcal{E}$

$$
\begin{aligned}
\delta\mathcal{E} &= \int_v \left\{-\nabla\left[W-\rho\frac{\partial W}{\partial\rho}+\rho\nabla\cdot\left(2\frac{\partial W}{\partial\beta}\nabla\rho\right)\right]\right.\\
&\quad \left. +\nabla\cdot\left(2\frac{\partial W}{\partial\beta}\nabla\rho\nabla\rho\right)\right\}\cdot\mathbf{w}\,dv\\
&\quad +\int_{\partial v}\left\{\left[W-\rho\frac{\partial W}{\partial\rho}+\nabla\cdot\left(2\frac{\partial W}{\partial\beta}\nabla\rho\right)\rho\right]\mathbf{n}-2\frac{\partial W}{\partial\beta}\mathbf{n}\cdot\nabla\rho\nabla\rho\right.\\
&\quad \left. +4\frac{\partial W}{\partial\beta}(\mathbf{n}\cdot\nabla\rho)\rho K\mathbf{n}+\nabla_s\left[2\rho\frac{\partial W}{\partial\beta}(\mathbf{n}\cdot\nabla\rho)\right]\right\}\cdot\mathbf{w}\,dS\\
&\quad -\int_{\partial v} 2\frac{\partial W}{\partial\beta}(\mathbf{n}\cdot\nabla\rho)\rho\mathbf{n}\cdot(\mathbf{n}\cdot\nabla)\mathbf{w}\,dS\\
&\quad -\int_{\partial S} 2\rho\frac{\partial W}{\partial\beta}(\mathbf{n}\cdot\nabla\rho)\boldsymbol{\nu}\cdot\mathbf{w}\,ds.
\end{aligned}
\tag{6.166}
$$

The form of $\delta\mathcal{E}$ dictates a possible form of the virtual work of external loads in case of the second-gradient fluid. Let us take the following form for the virtual work

$$
\delta\mathcal{A} = \int_v \rho\mathbf{f}\cdot\mathbf{w}\,dv + \int_{\partial v}\left[\mathbf{t}\cdot\mathbf{w}+c\,\mathbf{n}\cdot(\mathbf{n}\cdot\nabla)\mathbf{w}\right]dS + \int_{\partial S}\zeta\boldsymbol{\nu}\cdot\mathbf{w}\,ds. \tag{6.167}
$$

In (6.167) $\mathbf{f}$ is the vector of external forces per unit mass, $\mathbf{t}$ is the vector of external surface traction, and c and ζ can be interpreted as the surface double force and the force acting along contour ∂a, respectively.

Now we can formulate the *virtual work principle*:

The fluid is in equilibrium if and only if the relation

$$\delta\mathcal{E} - \delta\mathcal{A} = 0 \tag{6.168}$$

holds for every kinematically admissible virtual displacements $\mathbf{w}$, *i.e., for any sufficiently smooth* $\mathbf{w}$ *satisfying the kinematic boundary conditions.*

Equilibrium equation and natural boundary conditions

The variational equation (6.168) implies the equilibrium equation

$$\nabla\left[W - \rho\frac{\partial W}{\partial\rho} + \rho\nabla\cdot\left(2\frac{\partial W}{\partial\beta}\nabla\rho\right)\right] - \nabla\cdot\left(2\frac{\partial W}{\partial\beta}\nabla\rho\nabla\rho\right) + \rho\mathbf{f} = \mathbf{0} \tag{6.169}$$

and the natural boundary conditions

$$\begin{aligned}
&\left(W - \rho\frac{\partial W}{\partial\rho} + \nabla\cdot\left(2\frac{\partial W}{\partial\beta}\nabla\rho\right)\rho\right)\mathbf{n} - 2\frac{\partial W}{\partial\beta}\mathbf{n}\cdot\nabla\rho\nabla\rho \\
&+ 4\rho K\frac{\partial W}{\partial\beta}(\mathbf{n}\cdot\nabla\rho)\mathbf{n} + \nabla_s\left[2\rho\frac{\partial W}{\partial\beta}(\mathbf{n}\cdot\nabla\rho)\right] = \mathbf{t}, \quad \mathbf{r}\in S\equiv\partial v, \qquad (6.170)\\
&-2\rho\frac{\partial W}{\partial\beta}(\mathbf{n}\cdot\nabla\rho) = c \quad \text{for } \mathbf{r}\in S\equiv\partial v, \qquad (6.171)\\
&-2\rho\frac{\partial W}{\partial\beta}(\mathbf{n}\cdot\nabla\rho) = \zeta \quad \text{for } \mathbf{r}\in \ell\equiv\partial S. \qquad (6.172)
\end{aligned}$$

These equations were established earlier by Eremeyev and Altenbach[49]. Using different techniques, they also were deduced by Auffray *et al*[50].

Introducing the symmetric Cauchy-type stress tensor by the relation

$$\mathbf{T} = -P\mathbf{E} - 2\frac{\partial W}{\partial\beta}\nabla\rho\nabla\rho \tag{6.173}$$

where

$$P = -W + \rho\frac{\partial W}{\partial\rho} - \rho\nabla\cdot\left(2\frac{\partial W}{\partial\beta}\nabla\rho\right),$$

[49]Eremeyev, V.A., and Altenbach, H. (2014). Equilibrium of a second-gradient fluid and an elastic solid with surface stresses. *Meccanica*, 49(11), 2635–2643.

[50]Auffray, N., dell'Isola, F., Eremeyev, V.A., Madeo, A., and Rosi, G. (2015). Analytical continuum mechanics á la Hamilton-Piola least action principle for second gradient continua and capillary fluids. *Math. Mech. Solids*, 20(4), 375–417.

we rewrite (6.169) in the standard compact form

$$\nabla \cdot \mathbf{T} + \rho \mathbf{f} = \mathbf{0}. \tag{6.174}$$

Here P plays the role of a pressure in the second-gradient fluid, whereas $\mathbf{T}$ is often called the *Korteweg stress tensor*. Equations (6.169) or (6.174) constitute the Eulerian equilibrium equations for the second-gradient fluid.

Equation (6.170) becomes

$$\mathbf{n} \cdot \mathbf{T} + 4\rho K \frac{\partial W}{\partial \beta}(\mathbf{n} \cdot \nabla \rho)\mathbf{n} + \nabla_s \left[2\rho \frac{\partial W}{\partial \beta}(\mathbf{n} \cdot \nabla \rho) \right] = \mathbf{t} \quad (\mathbf{r} \in S). \tag{6.175}$$

From (6.175) it follows that in general $\mathbf{n} \cdot \mathbf{T}$ does not coincide with the surface traction $\mathbf{t}$ as in nonlinear elasticity.

Dynamics

The equation of motion for the strain gradient fluid can be easily derived using d'Alembert's principle. Replacing $\mathbf{f}$ by $\mathbf{f} - \dot{\mathbf{v}}$ in (6.174) we obtain

$$\nabla \cdot \mathbf{T} + \rho \mathbf{f} = \rho \dot{\mathbf{v}}. \tag{6.176}$$

This should be supplemented with the classical mass conservation law

$$\frac{\partial \rho}{\partial t} + \nabla \cdot (\rho \mathbf{v}) = 0. \tag{6.177}$$

Equation (6.176) can be simplified and if we specify the constitutive equation for W it can even be integrated. For example, calculating $\mathbf{T}$ with use of the strain energy density (6.156), we find that

$$\mathbf{T} = -p\mathbf{E} + \lambda \nabla \rho \nabla \rho, \quad p = \rho \frac{\partial W_0}{\partial \rho} - W_0 - \frac{\lambda}{2} \nabla \rho \cdot \nabla \rho - \lambda \rho \nabla \cdot \nabla \rho. \tag{6.178}$$

Substituting (6.178) into (6.176) we derive

$$\rho \left(\frac{\partial \mathbf{v}}{\partial t} + \mathbf{v} \cdot \nabla \mathbf{v} \right) + \rho \nabla \left(\frac{\partial W_0}{\partial \rho} \right) - \lambda \rho \nabla \left(\nabla \cdot \nabla \rho \right) = \rho \mathbf{f}. \tag{6.179}$$

This equation can be integrated once with the same technique as in the classical hydrodynamics of an inviscid fluid. Let us assume the external forces to have a potential function Ω, i.e., $\mathbf{f} = \nabla \Omega$. For potential flow we introduce the velocity potential ϕ by the relation $\mathbf{v} = \nabla \phi$. With ϕ, we transform (6.179) into the form

$$\nabla \left[\frac{\partial \phi}{\partial t} + \frac{1}{2} \nabla \phi \cdot \nabla \phi + \frac{\partial W_0}{\partial \rho} - \lambda \nabla \cdot \nabla \rho \right] = \nabla \Omega. \tag{6.180}$$

From (6.180) one may deduce the *unsteady Bernoulli integral*[51]

$$\frac{\partial \phi}{\partial t} + \frac{1}{2}\nabla\phi \cdot \nabla\phi + \frac{\partial W_0}{\partial \rho} - \lambda \nabla \cdot \nabla \rho - \Omega = c(t), \tag{6.181}$$

where $c(t)$ is an unknown time function.

For statics, the Bernoulli integral takes the form of the second-order partial differential equation

$$\frac{\partial W_0}{\partial \rho} - \lambda \nabla \cdot \nabla \rho - \Omega = c, \tag{6.182}$$

where c is now a constant.

Exercise 6.11. Demonstrate that the static Bernoulli integral (6.182) can be derived from the Lagrange variational principle considering only variations of ρ, i.e., from the equation

$$\delta \mathcal{E} = 0, \qquad \mathcal{E} = \int_V [W(\rho, \nabla\rho) - \rho\Omega]\, dV \tag{6.183}$$

with mass conservation as a constraint:

$$\int_V \rho\, dV = \text{constant}.$$

Second- and higher-order gradient fluids

The next extension of the gradient fluid model accounts for higher spatial gradients of ρ. For the second-order gradient fluid, the strain energy density takes the form

$$W = W(\rho, \nabla\rho, \nabla\nabla\rho). \tag{6.184}$$

The principle of material frame-indifference implies the restriction

$$W(\rho, \nabla\rho, \nabla\nabla\rho) = W(\rho, \mathbf{Q} \cdot \nabla\rho, \mathbf{Q} \cdot \nabla\nabla\rho \cdot \mathbf{Q}^T),$$

meaning that W is an isotropic function of the vector $\mathbf{a} \equiv \nabla\rho$ and of the symmetric second-order tensor $\mathbf{S} \equiv \nabla\nabla\rho$. Analyzing the invariants of $\mathbf{a}$ and $\mathbf{S}$ by Table 2.9, we conclude that W depends on

$$\begin{gathered}
\mathbf{a} \cdot \mathbf{a} \equiv \nabla\rho \cdot \nabla\rho, \quad \operatorname{tr}\mathbf{S} \equiv \nabla \cdot \nabla\rho, \quad \operatorname{tr}\mathbf{S}^2 \equiv \nabla\nabla\rho \bullet \nabla\nabla\rho, \\
\operatorname{tr}\mathbf{S}^3 \equiv \operatorname{tr}(\nabla\nabla\rho \cdot \nabla\nabla\rho \cdot \nabla\nabla\rho), \\
\mathbf{a} \cdot \mathbf{S} \cdot \mathbf{a} \equiv \nabla\rho \cdot \nabla\nabla\rho \cdot \nabla\rho, \quad \mathbf{a} \cdot \mathbf{S}^2 \cdot \mathbf{a} \equiv \nabla\rho \cdot \nabla\nabla\rho \cdot \nabla\nabla\rho \cdot \nabla\rho.
\end{gathered}$$

[51] Rosi, G., Giorgio, I., and Eremeyev, V.A. (2013). Propagation of linear compression waves through plane interfacial layers and mass adsorption in second gradient fluids. *ZAMM*, 93(12), 914–927.

So the general form of W is

$$W = W(\rho, \nabla\rho \cdot \nabla\rho, \nabla \cdot \nabla\rho, \nabla\nabla\rho \bullet \nabla\nabla\rho, \operatorname{tr}(\nabla\nabla\rho \cdot \nabla\nabla\rho \cdot \nabla\nabla\rho), \\ \nabla\rho \cdot \nabla\nabla\rho \cdot \nabla\rho, \nabla\rho \cdot \nabla\nabla\rho \cdot \nabla\nabla\rho \cdot \nabla\rho). \qquad (6.185)$$

Considering W to be polynomial up to quadratic terms of $\nabla\rho$ and $\nabla\nabla\rho$ we get its simplified form

$$W = W_0(\rho) + \tfrac{1}{2}\lambda(\rho)\nabla\rho \cdot \nabla\rho + \tfrac{1}{2}K_0(\rho)\nabla \cdot \nabla\rho + \tfrac{1}{2}K_1(\rho)(\nabla \cdot \nabla\rho)^2,$$

where $\lambda(\rho)$, $K_0(\rho)$, and $K_1(\rho)$ are constitutive functions.

The constitutive equations for third- and Nth-order gradient fluids are

$$\begin{aligned} W &= W(\rho, \nabla\rho, \nabla\nabla\rho, \nabla\nabla\nabla\rho), \\ W &= W(\rho, \nabla\rho, \nabla\nabla\rho, \nabla\nabla\nabla\rho, \dots, \underbrace{\nabla \dots \nabla}_{N} \rho). \end{aligned} \qquad (6.186)$$

Here the strain energy densities are also isotropic functions of their arguments including third- and higher-order tensors. However for this theory, the theory of invariants is still under development.

Exercise 6.12. Find the equilibrium equations and natural boundary conditions for a third-order strain gradient fluid with energy density (6.186).

6.9 Liquid Crystals

Earlier we considered various mathematical models for solids and fluids. Now we turn to media whose physical properties stand between those of fluids and solids. In particular, they can flow at least in one direction as a fluid but also demonstrate anisotropy and resistance to shear deformation in some directions. Such materials are now called *liquid crystals*. There are various classes of liquid crystals: nematics, cholesterics, smectics, columnar liquid crystals, and more exotic types [Chandrasekhar (1977); de Gennes and Prost (1993)]. Mechanically, liquid crystals can be represented as a melt or a solution consisting of particles (molecules) having rod- or disk-like shape. In liquid crystals, the particles exhibit preferred directions of orientation. To describe the orientation we introduce a unit vector $\mathbf{d}$ called the *director*, which models the average orientation of the liquid crystal particles. Unlike the micropolar continuum, here we have only one director. In order to derive static equations for liquid crystals we again adopt the variational approach.

Energy functional and its first variation

As for the strain gradient fluids, we again apply the Eulerian description. We introduce the strain energy density per unit volume in the actual configuration as a function of the mass density ρ, the unit director $\mathbf{d}$, and its spatial gradient $\nabla\mathbf{d}$:

$$W = W(\rho, \mathbf{d}, \nabla\mathbf{d}). \tag{6.187}$$

In what follows, we will assume that the requirement of material indifference is fulfilled, i.e., that the equality

$$W(\rho, \mathbf{d}, \nabla\mathbf{d}) = W(\rho, \mathbf{Q}\cdot\mathbf{d}, \mathbf{Q}\cdot\nabla\mathbf{d}\cdot\mathbf{Q}^T) \tag{6.188}$$

holds for all orthogonal tensors $\mathbf{Q}$. We will consider examples of W for nematic and cholesteric liquid crystals later, but here we will consider the general form (6.187) for W.

Thus the energy functional is

$$\mathcal{E} = \int_v W(\rho, \mathbf{d}, \nabla\mathbf{d})\, dv, \tag{6.189}$$

where v is the volume occupied by the liquid crystal in the actual configuration at time t. To derive the first variation of $\mathcal{E}$ we use the same technique as for the strain gradient fluids in the previous section. First we transform (6.189) into Lagrangian form, perform all calculations, then transform the result back to Eulerian form. The Lagrangian form of (6.189) is

$$\mathcal{E} = \int_V W(\rho, \mathbf{d}, \nabla\mathbf{d}) J\, dV, \tag{6.190}$$

where $J = \det\mathbf{F}$, $\mathbf{F} = (\nabla_\kappa\mathbf{r})^T$, and V is the corresponding volume in the reference configuration. As $\mathbf{d}$ is a unit vector, we seek $\delta\mathcal{E}$ considering the constraint $\mathbf{d}\cdot\mathbf{d} = 1$. Following the Lagrange multiplier technique, we now modify the functional as follows:

$$\mathcal{E} = \int_V \left[W(\rho, \mathbf{d}, \nabla\mathbf{d})J - \tfrac{1}{2}\lambda(\mathbf{d}\cdot\mathbf{d} - 1)\right] dV, \tag{6.191}$$

where λ is a Lagrange multiplier. Now considering independent variations $\delta\mathbf{r}$, $\delta\mathbf{d}$, and $\delta\lambda$, we seek the first variation of (6.191).

Using

$$\delta\rho = -\rho\nabla\cdot\mathbf{w}, \qquad \delta J = J\nabla\cdot\mathbf{w}, \qquad \nabla = \mathbf{F}^{-T}\cdot\nabla_\kappa,$$

and

$$\delta(\nabla\mathbf{d}) = \delta(\mathbf{F}^{-T}\cdot\nabla_\kappa\mathbf{d}) = \mathbf{F}^{-T}\cdot\nabla_\kappa\delta\mathbf{d} + \delta(\mathbf{F}^{-T})\cdot\nabla_\kappa\mathbf{d}$$
$$= \nabla\boldsymbol{\omega} - \mathbf{F}^{-T}\cdot\delta(\mathbf{F}^{T})\cdot\mathbf{F}^{-T}\cdot\nabla_\kappa\mathbf{d} = \nabla\boldsymbol{\omega} - \nabla\mathbf{w}\cdot\mathbf{d},$$

where for brevity we denote $\delta\mathbf{r} = \mathbf{w}$ and $\delta\mathbf{d} = \boldsymbol{\omega}$, we get the first variation in the form

$$\delta\mathcal{E} = \int_V \left[-\frac{\partial W}{\partial\rho}\rho\nabla\cdot\mathbf{w} + \frac{\partial W}{\partial\mathbf{d}}\cdot\boldsymbol{\omega} + \frac{\partial W}{\partial\nabla\mathbf{d}}\bullet\delta(\nabla\mathbf{d}) + W\nabla\cdot\mathbf{w} - \lambda\mathbf{d}\cdot\boldsymbol{\omega}\right.$$
$$\left. -\delta\lambda\tfrac{1}{2}(\mathbf{d}\cdot\mathbf{d}-1)]\right] J\,dV$$
$$= \int_v \left\{\left[W - \rho\frac{\partial W}{\partial\rho}\right]\nabla\cdot\mathbf{w} + \left[\frac{\partial W}{\partial\mathbf{d}} - \lambda\mathbf{d}\right]\cdot\boldsymbol{\omega} + \frac{\partial W}{\partial\nabla\mathbf{d}}\bullet[\nabla\boldsymbol{\omega} - \nabla\mathbf{w}\cdot\nabla\mathbf{d}]\right.$$
$$\left. -\delta\lambda\tfrac{1}{2}(\mathbf{d}\cdot\mathbf{d}-1)]\right\} dv.$$

In what follows, we will omit the term $\delta\lambda(\mathbf{d}\cdot\mathbf{d}-1)$ as it brings us the constraint $\mathbf{d}\cdot\mathbf{d} = 1$ from $\delta\mathcal{E} = 0$. Introducing a second-order tensor $\boldsymbol{\mu}$ by

$$\boldsymbol{\mu} = \frac{\partial W}{\partial\nabla\mathbf{d}},$$

which can be called the *Cauchy couple stress tensor*, and using the identity

$$\boldsymbol{\mu}\bullet(\nabla\mathbf{w}\cdot\nabla\mathbf{d}) = \operatorname{tr}\left[\boldsymbol{\mu}\cdot(\nabla\mathbf{w}\cdot\nabla\mathbf{d})^T\right] = \operatorname{tr}\left(\boldsymbol{\mu}\cdot\nabla\mathbf{d}^T\cdot\nabla\mathbf{w}^T\right)$$
$$= \left[\boldsymbol{\mu}\cdot(\nabla\mathbf{d})^T\right]\bullet\nabla\mathbf{w}$$

we get

$$\delta\mathcal{E} = \int_v \left\{\left[W - \rho\frac{\partial W}{\partial\rho}\right]\nabla\cdot\mathbf{w} - \left[\boldsymbol{\mu}\cdot(\nabla\mathbf{d})^T\right]\bullet\nabla\mathbf{w}\right.$$
$$\left. + \left[\frac{\partial W}{\partial\mathbf{d}} - \lambda\mathbf{d}\right]\cdot\boldsymbol{\omega} + \boldsymbol{\mu}\bullet\nabla\boldsymbol{\omega}\right\} dv.$$

Integration by parts gives

$$\delta\mathcal{E} = \int_v \left\{-\nabla\left[W - \rho\frac{\partial W}{\partial\rho}\right]\cdot\mathbf{w} + \nabla\cdot\left[\boldsymbol{\mu}\cdot(\nabla\mathbf{d})^T\right]\cdot\mathbf{w}\right.$$
$$\left. + \left[\frac{\partial W}{\partial\mathbf{d}} - \lambda\mathbf{d}\right]\cdot\boldsymbol{\omega} - (\nabla\cdot\boldsymbol{\mu})\cdot\boldsymbol{\omega}\right\} dv$$
$$+ \int_{\partial v}\left[\left(W - \rho\frac{\partial W}{\partial\rho}\right)\mathbf{n}\cdot\mathbf{w} - \mathbf{n}\cdot\boldsymbol{\mu}\cdot(\nabla\mathbf{d})^T\cdot\mathbf{w} + \mathbf{n}\cdot\boldsymbol{\mu}\cdot\boldsymbol{\omega}\right] dS. \tag{6.192}$$

The form of the first variation (6.192) dictates the following possible form for the virtual work of external loads:

$$\delta\mathcal{A} = \int_v \rho\,(\mathbf{f}\cdot\mathbf{w} + \mathbf{c}\cdot\boldsymbol{\omega})\,dv + \int_{\partial v} (\mathbf{t}\cdot\mathbf{w} + \mathbf{m}\cdot\boldsymbol{\omega})\,dS. \tag{6.193}$$

Here $\mathbf{f}$ is the vector of volumetric forces per unit mass, $\mathbf{t}$ is the vector of surface traction, and $\mathbf{c}$ and $\mathbf{m}$ are volumetric and surface external couples, respectively. As $\mathbf{d}$ is a unit vector, note that $\boldsymbol{\omega}\cdot\mathbf{d} = 0$. Thus $\mathbf{c}$ and $\mathbf{m}$ should be orthogonal to $\mathbf{d}$. In other words, the medium is insensitive with respect to external couples applied along $\mathbf{d}$. From a physical standpoint this is obvious. Indeed, a liquid crystal can be treated as a medium consisting of rigid rod-like particles whose rotations about their axes are free.

Finally, applying the *virtual work principle*

$$\delta\mathcal{E} - \delta\mathcal{A} = 0$$

with (6.192) and (6.193), we derive the equilibrium equations

$$-\nabla\left(W - \rho\frac{\partial W}{\partial\rho}\right) + \nabla\cdot\left[\boldsymbol{\mu}\cdot(\nabla\mathbf{d})^T\right] - \rho\mathbf{f} = \mathbf{0}, \tag{6.194}$$

$$\frac{\partial W}{\partial\mathbf{d}} - \lambda\mathbf{d} - (\nabla\cdot\boldsymbol{\mu}) - \mathbf{c} = \mathbf{0}, \tag{6.195}$$

and the natural boundary conditions

$$\left(W - \rho\frac{\partial W}{\partial\rho}\right)\mathbf{n} - \mathbf{n}\cdot\boldsymbol{\mu}\cdot(\nabla\mathbf{d})^T = \mathbf{t}, \quad \mathbf{n}\cdot\boldsymbol{\mu} = \mathbf{m}, \quad \mathbf{r}\in S = \partial v. \tag{6.196}$$

The term $\lambda\mathbf{d}$ is called the *molecular field* in the hydromechanics of liquid crystals [de Gennes and Prost (1993)].

Introducing the Cauchy-type stress tensor by the formulas

$$\mathbf{T} = -p\mathbf{E} - \boldsymbol{\mu}\cdot(\nabla\mathbf{d})^T, \qquad p = -W + \rho\frac{\partial W}{\partial\rho}, \tag{6.197}$$

we transform (6.194) and $(6.197)_1$ into the more compact classical form

$$\nabla\cdot\mathbf{T} + \rho\mathbf{f} = \mathbf{0} \qquad (\mathbf{r}\in v), \tag{6.198}$$

$$\mathbf{n}\cdot\mathbf{T} = \mathbf{t} \qquad (\mathbf{r}\in S = \partial v). \tag{6.199}$$

The form of $\mathbf{T}$ shows that a liquid crystal possesses shear stresses. As with the Cosserat media, in liquid crystals there are rotational degrees of freedom related to the rotations of $\mathbf{d}$ and couple stresses $\boldsymbol{\mu}$.

We have obtained the natural, static boundary conditions. However in the theory kinematic conditions also arise. The simplest case is when the director is given at the surface, $\mathbf{d} = \mathbf{d}_0$, $\mathbf{d}_0\cdot\mathbf{d}_0 = 0$. More frequently only a preferred direction is given, such as $\mathbf{d} \cdot \mathbf{n} = \cos\psi$. Here the angle ψ may depend on the temperature, chemical compounds of the substrate, how the substrate was polished, or other factors. A more general way of orienting $\mathbf{d}$ at the surface is to consider a surface energy. So instead of (6.189) we consider the functional

$$\mathcal{E} = \int_v W(\rho, \mathbf{d}, \nabla\mathbf{d})\, dv + \int_S U(\mathbf{n} \cdot \mathbf{d})\, dS, \tag{6.200}$$

where U is a surface strain energy given as a function of the contact angle between the director and the surface normal. Note that the kinematic boundary conditions in the theory of liquid crystals are more frequently used than the static conditions, as for almost any surface there is a preferable direction of the director.

Constitutive equations for nematics and cholesterics

As for classical fluids, we assume liquid crystals are incompressible. So the mass density ρ is constant, whereas the pressure p should be determined from the equilibrium equations as in the case of incompressible elastomers. The strain energy density takes the general form

$$W = W(\mathbf{d}, \nabla\mathbf{d})$$

subject to the invariance property (6.188). So W is an isotropic function of a vector and a non-symmetric second-order tensor whose invariants are listed in Table 2.9 and equations (2.118). As $\mathbf{d}$ is a unit vector, here we have a reduction of the number of possible invariants, for example, $\mathbf{d}\cdot\mathbf{d} = 1$, $\nabla\mathbf{d} \cdot \mathbf{d} = \mathbf{0}$, etc.

For nematics we use the following assumptions: both directions $\mathbf{d}$ and $-\mathbf{d}$ are equivalent and non-distinguishable, W is invariant under the space inversion transformations, so

$$W(\mathbf{d}, \nabla\mathbf{d}) = W(-\mathbf{d}, -\nabla\mathbf{d}), \quad W(-\mathbf{d}, \nabla\mathbf{d}) = W(\mathbf{d}, \nabla\mathbf{d}),$$

$W = 0$ if $\mathbf{d} =$ constant, and W takes a minimum at $\nabla\mathbf{d} = \mathbf{0}$. Finally, we assume W is a quadratic function with respect to $\nabla\mathbf{d}$. So we must collect in W all terms of order $\|\nabla\mathbf{d}\|^2$. After some transformations, the following strain energy density can be introduced:

$$W = \tfrac{1}{2}K_1(\nabla \cdot \mathbf{d})^2 + \tfrac{1}{2}K_2\left[\mathbf{d} \cdot (\nabla \times \mathbf{d})\right]^2 + \tfrac{1}{2}K_3\left|\mathbf{d} \times (\nabla \times \mathbf{d})\right|^2, \tag{6.201}$$

where K_1, K_2, and K_3 are positive elastic moduli called *Frank's moduli.* Three terms in (6.201) relate to different deformations which are pure splay, pure twist, and pure bend. Frank's moduli have units of Newtons. For typical liquid crystals, $K_i \sim 10^{-11}$ N.

An acceptable simplification is the *one-constant approximation*

$$W = \tfrac{1}{2}K\left[(\nabla\cdot\mathbf{d})^2 + |\nabla\times\mathbf{d}|^2\right] = \tfrac{1}{2}K\nabla\mathbf{d}\bullet\nabla\mathbf{d},$$

where K is an elastic modulus. This forms follows from (6.201) if we assume $K_1 = K_2 = K_3 = K$. The one-constant approximation density suffices for qualitative analysis of deformations of nematics.

The cholesteric liquid crystals are similar to nematics; directions $\mathbf{d}$ and $-\mathbf{d}$ are also equivalent, but for cholesterics a helical structure is preferable. This means that the strain energy has a minimum of a field

$$\mathbf{d} = \cos(z/\ell)\mathbf{i}_1 + \sin(z/\ell)\mathbf{i}_2,$$

where ℓ is a characteristic length, usually about $10^{-5}cm$, $\mathbf{i}_3$ is the unit vector along z-axis which is the axis of orientation, and $\mathbf{i}_2$ and $\mathbf{i}_3$ are other Cartesian base vectors.

A useful form of the strain energy density for cholesterics is

$$W = \tfrac{1}{2}K_1(\nabla\cdot\mathbf{d})^2 + \tfrac{1}{2}K_2\left[\mathbf{d}\cdot(\nabla\times\mathbf{d}) + \kappa\right]^2 + \tfrac{1}{2}K_3\left|\mathbf{d}\times(\nabla\times\mathbf{d})\right|^2, \quad (6.202)$$

where K_i are elastic moduli and $\kappa = 1/\ell$.

Remembering our previous discussions on pseudoscalars, pseudovectors, and pseudotensors, we must emphasize that κ is an example of a pseudoscalar as is $\mathbf{d}\cdot(\nabla\times\mathbf{d})$. Typically one assumes that $\kappa > 0$, but it should change sign under the inversion transformations; otherwise W becomes non-invariant.

Hydromechanics of liquid crystals, such as the above mentioned classes of nematics and cholesterics, includes other continuum models of smectic and columnar liquid crystals; see, e.g., [de Gennes and Prost (1993); Chandrasekhar (1977)]. These non-classical media demonstrate various properties of fluids and solids. They may contain linear defects — disclinations — like classic solid crystals; in statics, nematics can resist applied torques. On the other side, they may flow as fluids exhibiting interaction between the flow direction and the orientation of the director. Liquid crystals are also sensitive to applied electrical and magnetic fields, which are often used in engineering to control their deformations. All these briefly mentioned phenomena make the continuum theory of liquid crystals more complicated than the theory of classical models or even nonlinear elasticity.

6.10 Other Generalized Models

Here we briefly describe a few other generalized continuum models.

Nematic elastomers

A generalized model of continuum, called the *nematic elastomer* or *optical elastomer*, inherits the properties of nonlinear elasticity and the models of liquid crystals[52]. Kinematics of the model is described by two fields, those of the position vector $\mathbf{r}$ and the unit director $\mathbf{d}$:

$$\mathbf{r} = \mathbf{r}(\mathbf{x}, t), \qquad \mathbf{d} = \mathbf{d}(\mathbf{x}, t), \qquad |\mathbf{d}| = 1.$$

So the strain energy density is a function of $\mathbf{r}$, $\mathbf{d}$, and their gradients

$$W = W(\mathbf{r}, \nabla\mathbf{r}, \mathbf{d}, \nabla\mathbf{d}), \tag{6.203}$$

where ∇ is the gradient operator in the initial configuration. Applying the principle of material-frame-indifference to (6.203), we find that W does not depend on $\mathbf{r}$ and satisfies

$$W(\nabla\mathbf{r}, \mathbf{d}, \nabla\mathbf{d}) = W(\nabla\mathbf{r} \cdot \mathbf{Q}^T, \mathbf{d} \cdot \mathbf{Q}^T, \nabla\mathbf{d} \cdot \mathbf{Q}^T) \tag{6.204}$$

for any orthogonal tensor $\mathbf{Q}$. Using $\mathbf{Q} = \mathbf{R}^T$, where $\mathbf{R}$ is the orthogonal tensor from the polar decomposition $\mathbf{F} = \mathbf{R} \cdot \mathbf{U}$, with positive definite symmetric tensor $\mathbf{U}$, we transform W into

$$W = W(\nabla\mathbf{r} \cdot \mathbf{R}, \mathbf{d} \cdot \mathbf{R}, \nabla\mathbf{d} \cdot \mathbf{R}). \tag{6.205}$$

Substituting $\mathbf{R} = \mathbf{F} \cdot \mathbf{U}^{-1}$ into (6.205), we obtain the constitutive equation consistent with the principle of material frame indifference

$$W = W(\mathbf{C}, \mathbf{d} \cdot \mathbf{F}, \nabla\mathbf{d} \cdot \mathbf{F}), \tag{6.206}$$

where $\mathbf{C} = \mathbf{F}^T \cdot \mathbf{F}$ is the Cauchy–Green strain tensor. Thus W is a function of second-order symmetric tensor $\mathbf{C}$, vector $\mathbf{d} \cdot \mathbf{F}$, and second-order tensor $\nabla\mathbf{d} \cdot \mathbf{F}$. Note that the last tensor is not symmetric, in general. An example of the strain energy density for an incompressible nematic elastomer is presented by Anderson *et al*[52].

The kinetic energy density takes the form

$$K = \tfrac{1}{2}\rho\dot{\mathbf{r}} \cdot \dot{\mathbf{r}} + \tfrac{1}{2}\gamma\dot{\mathbf{d}} \cdot \dot{\mathbf{d}}, \tag{6.207}$$

[52] Anderson, D.R., Carlson, D.E., and Fried, E. (1999). A continuum-mechanical theory for nematic elastomers. *J. Elast.*, 56(1), 33–58.

where ρ and γ are the mass density and the *orientational inertia density*, respectively. Density γ can be represented as $\gamma = \rho\ell^2$ where ℓ is a characteristic length called the *radius of gyration*.

Exercise 6.13. Find the equations of motion and natural boundary conditions for constitutive equations (6.206) and (6.207). Take into account that $\mathbf{d}$ is a unit vector.

Peridynamics

Peridynamics is a continuum model based on the integral representation of interaction forces between parts of solids and oriented toward describing fracture and disintegration of solids and fluids. The equation of motion is

$$\rho\ddot{\mathbf{u}}(\mathbf{x},t) = \int_V \mathbf{f}[\mathbf{u}(\mathbf{x},t) - \mathbf{u}(\mathbf{y},t), \mathbf{x}-\mathbf{y}, \mathbf{x}]\, dV_{\mathbf{y}} + \mathbf{b}(\mathbf{x},t) \quad (\mathbf{x} \in V) \tag{6.208}$$

where ρ is the mass density, $\mathbf{u}$ is the displacement, $\mathbf{f}$ is the interaction force, and $\mathbf{b}$ is the external force density. Also called the *bond* force, $\mathbf{f}$ describes the interaction between material points with positions $\mathbf{x}$ and $\mathbf{y}$. It typically has compact support; i.e., $\mathbf{f} = \mathbf{0}$ if $\|\mathbf{x}-\mathbf{y}\|$ exceeds a certain threshold. Here the strain energy density is similar to the functional considered in Exercise 6.3. Peridynamics clearly belongs to the class of strongly nonlocal continuum models. For details see Silling's paper[53] and Medenci and Oterkus[54], where further references can be found.

Micropolar fluids

The Cosserats' ideas on rotational interactions of particles found also some applications in hydrodynamics. The model of a *viscous micropolar fluid* was proposed by Aero[55] and Eringen[56], see [Eringen (2001)]. As the action of a magnetic field produces a moment that can rotate fluid particles, the model found applications for the description of magnetic fluids. We start by introducing objective *strain rates* in the micropolar continuum:

$$\boldsymbol{\varepsilon} = \nabla\mathbf{v} + \mathbf{I} \times \boldsymbol{\omega}, \qquad \boldsymbol{\kappa} = \nabla\boldsymbol{\omega}, \tag{6.209}$$

[53] Silling, S.A. (2000). Reformulation of elasticity theory for discontinuities and long-range forces. *J. Mech. Phys. Solids*, 48(1), 175–209.
[54] Madenci, E., and Oterkus, E. (2014). *Peridynamic Theory and Its Applications.* (Springer, New York).
[55] Aero, E.L., Bulygin, A.N., Kuvshinskii, E.V. (1965). Asymmetric hydromechanics. *J. Appl. Math. Mech.*, 29(2), 333–346.
[56] Eringen, A.C. (1966). Theory of micropolar fluids. *J. Math. Mech.*, 16(1), 1–18.

where ∇ is the spatial nabla-operator defined in the actual configuration, whereas $\mathbf{v}$ and $\boldsymbol{\omega}$ are the linear and angular velocities, respectively. Note that $\boldsymbol{\varepsilon}$ is an objective tensor while $\boldsymbol{\kappa}$ is an objective pseudotensor.

The Eulerian equations of micropolar dynamics take the form

$$\nabla \cdot \mathbf{T} + \rho \mathbf{f} = \rho \dot{\mathbf{v}}, \tag{6.210}$$

$$\nabla \cdot \mathbf{M} + \mathbf{T}_\times + \rho \mathbf{m} = \rho j \dot{\boldsymbol{\omega}}, \tag{6.211}$$

where $\mathbf{T}$ and $\mathbf{M}$ are the Cauchy stress and couple stress tensors, respectively, ρ is the mass density in an actual configuration, $\mathbf{f}$ and $\mathbf{m}$ are the vectors of external mass forces and couples, and j is a scalar measure of rotatory inertia.

Here we assume $\mathbf{T}$ and $\mathbf{M}$ are the functions of the strain rates:

$$\mathbf{T} = \mathbf{T}(\boldsymbol{\varepsilon}, \boldsymbol{\kappa}), \qquad \mathbf{M} = \mathbf{M}(\boldsymbol{\varepsilon}, \boldsymbol{\kappa}).$$

As $\mathbf{T}$ and $\boldsymbol{\varepsilon}$ are objective tensors while $\mathbf{M}$ and $\boldsymbol{\kappa}$ are objective pseudotensors, the constitutive relations satisfy

$$\mathbf{T}(\mathbf{Q} \cdot \boldsymbol{\varepsilon} \cdot \mathbf{Q}^T, (\det \mathbf{Q})\mathbf{Q} \cdot \boldsymbol{\kappa} \cdot \mathbf{Q}^T) = \mathbf{Q} \cdot \mathbf{T}(\boldsymbol{\varepsilon}, \boldsymbol{\kappa}) \cdot \mathbf{Q}^T, \tag{6.212}$$

$$\mathbf{M}(\mathbf{Q} \cdot \boldsymbol{\varepsilon} \cdot \mathbf{Q}^T, (\det \mathbf{Q})\mathbf{Q} \cdot \boldsymbol{\kappa} \cdot \mathbf{Q}^T) = (\det \mathbf{Q})\mathbf{Q} \cdot \mathbf{M}(\boldsymbol{\varepsilon}, \boldsymbol{\kappa}) \cdot \mathbf{Q}^T, \tag{6.213}$$

for all orthogonal tensors $\mathbf{Q}$. In other words, $\mathbf{T}(\boldsymbol{\varepsilon}, \boldsymbol{\kappa})$ and $\mathbf{M} = \mathbf{M}(\boldsymbol{\varepsilon}, \boldsymbol{\kappa})$ are isotropic tensor-valued functions.

Restricting ourselves to the linear dependencies and applying the representation of an isotropic tensor-valued function discussed in § 2.5, we obtain the constitutive equations of the *linear viscous compressible micropolar fluid*

$$\mathbf{T} = \lambda \mathbf{E} \operatorname{tr} \boldsymbol{\varepsilon} + \mu_1 \boldsymbol{\varepsilon} + \mu_2 \boldsymbol{\varepsilon}^T, \quad \mathbf{M} = \nu_1 (\operatorname{tr} \boldsymbol{\kappa}) \mathbf{E} + \nu_2 \boldsymbol{\kappa} + \nu_3 \boldsymbol{\kappa}^T, \tag{6.214}$$

where λ, μ_1, μ_2, ν_1, ν_2, ν_3 are the viscosities.

For fluids it is convenient to neglect compressibility and use the incompressibility constraint. The latter, in terms of velocities, takes the form

$$\operatorname{tr} \boldsymbol{\varepsilon} \equiv \nabla \mathbf{v} = 0. \tag{6.215}$$

With this constraint $\mathbf{T}$ reduces to

$$\mathbf{T} = -p\mathbf{E} + \mu_1 \boldsymbol{\varepsilon} + \mu_2 \boldsymbol{\varepsilon}^T, \tag{6.216}$$

where p is the pressure, and we get the model of the *linear viscous incompressible micropolar fluid*.

The energy dissipation during fluid flow due to viscosity takes the form $\mathbf{T} \bullet \boldsymbol{\varepsilon} + \mathbf{M} \bullet \boldsymbol{\kappa}$. As the dissipation is nonnegative,

$$\mathbf{T} \bullet \boldsymbol{\varepsilon} + \mathbf{M} \bullet \boldsymbol{\kappa} \geq 0, \quad \forall \ \boldsymbol{\varepsilon}, \boldsymbol{\kappa} \neq \mathbf{0},$$

we get the following constitutive restrictions for the viscosities:

$$\mu_1 + \mu_2 \geq 0, \quad \mu_1 - \mu_2 \geq 0, \quad \nu_1 \geq 0, \quad \nu_2 + \nu_3 \geq 0, \quad \nu_2 - \nu_3 \geq 0.$$

Appendix A

Equation Summary for Tensor Analysis

Kronecker delta symbol

$$\delta_i^j = \delta_{ij} = \begin{cases} 1 \text{ for } j = i \\ 0 \text{ for } j \neq i \end{cases}$$

Cartesian basis

$$\begin{aligned} \mathbf{v} &= \mathbf{i}_1 v_1 + \mathbf{i}_2 v_2 + \mathbf{i}_3 v_3 \\ v_k &= \mathbf{i}_k \cdot \mathbf{v} \end{aligned} \qquad \mathbf{i}_k \cdot \mathbf{i}_n = \delta_{kn}$$

Vectors in reciprocal bases

Definition of reciprocal basis

$$\mathbf{e}_j \cdot \mathbf{e}^i = \delta_j^i \qquad \begin{cases} \mathbf{e}_k = \text{initial basis} \\ \mathbf{e}^k = \text{basis reciprocal to } \mathbf{e}_k \end{cases}$$

Metric coefficients

$$\begin{cases} g^{ji} = \mathbf{e}^j \cdot \mathbf{e}^i \\ g_{ik} = \mathbf{e}_i \cdot \mathbf{e}_k \end{cases} \qquad g_{ij} = g_{ji} \qquad g_{ij} g^{iq} = \delta_j^q$$

$$\begin{cases} \mathbf{e}^j = g^{ji} \mathbf{e}_i \\ \mathbf{e}_j = g_{ji} \mathbf{e}^i \end{cases} \qquad \mathbf{e}_i = \delta_i^k \mathbf{e}_k = g^{kl} g_{li} \mathbf{e}_k$$

Vector representations

$$\mathbf{x} = x^k \mathbf{e}_k = x_k \mathbf{e}^k \qquad \begin{cases} x^k = \mathbf{x} \cdot \mathbf{e}^k = x_j g^{jk} \\ x_k = \mathbf{x} \cdot \mathbf{e}_k = x^i g_{ik} \end{cases}$$

Vector dot product

Commutativity, distributivity

$$\mathbf{b} \cdot \mathbf{a} = \mathbf{a} \cdot \mathbf{b} \qquad (\alpha\mathbf{a} + \mathbf{b}) \cdot \mathbf{c} = \alpha\mathbf{a} \cdot \mathbf{c} + \mathbf{b} \cdot \mathbf{c}$$

Positive definiteness

$$\mathbf{a} \cdot \mathbf{a} \geq 0, \text{ with } \mathbf{a} \cdot \mathbf{a} = 0 \text{ if and only if } \mathbf{a} = \mathbf{0}$$

Expressions

$$\mathbf{x} \cdot \mathbf{y} = x^i y_i = x_i y^i = x^i y^j g_{ij} = x_i y_j g^{ij}$$

Calculation of vector magnitude

$$|\mathbf{x}| = (x_i x^i)^{1/2} = (x^i x^j g_{ij})^{1/2} = (x_i x_j g^{ij})^{1/2}$$

Angle between vectors

$$\cos\phi = \frac{\mathbf{x} \cdot \mathbf{y}}{|\mathbf{x}||\mathbf{y}|} = \frac{x^i y_i}{(x^j x_j y^k y_k)^{1/2}}$$

Vectors under change of basis

Transformation of frame and reciprocal frame

$$\begin{cases} \mathbf{e}_i = A_i^{j'} \mathbf{e}_{j'} \\ \mathbf{e}_{i'} = A_{i'}^{j} \mathbf{e}_j \\ \mathbf{e}^i = A_{j'}^{i} \mathbf{e}^{j'} \\ \mathbf{e}^{i'} = A_j^{i'} \mathbf{e}^j \end{cases} \qquad \begin{cases} A_{i'}^{j} = \mathbf{e}_{i'} \cdot \mathbf{e}^j \\ A_j^{i'} = \mathbf{e}^{i'} \cdot \mathbf{e}_j \end{cases} \qquad \begin{cases} A_{i'}^{j} A_j^{k'} = \delta_{i'}^{k'} \\ A_i^{j'} A_{j'}^{k} = \delta_i^k \end{cases}$$

Transformation of vector components

$$\begin{cases} x^i = A^i_{j'} x^{j'} \\ x^{i'} = A^{i'}_j x^j \end{cases} \qquad \begin{cases} x_i = A^{j'}_i x_{j'} \\ x_{i'} = A^j_{i'} x_j \end{cases}$$

Determinantal identity

$$(\mathbf{a} \cdot (\mathbf{b} \times \mathbf{c}))(\mathbf{u} \cdot (\mathbf{v} \times \mathbf{w})) = \begin{vmatrix} \mathbf{a}\cdot\mathbf{u} & \mathbf{a}\cdot\mathbf{v} & \mathbf{a}\cdot\mathbf{w} \\ \mathbf{b}\cdot\mathbf{u} & \mathbf{b}\cdot\mathbf{v} & \mathbf{b}\cdot\mathbf{w} \\ \mathbf{c}\cdot\mathbf{u} & \mathbf{c}\cdot\mathbf{v} & \mathbf{c}\cdot\mathbf{w} \end{vmatrix}$$

Permutation symbols

Definitions and values

$$\epsilon_{ijk} = \mathbf{e}_i \cdot (\mathbf{e}_j \times \mathbf{e}_k) = \begin{cases} +V \text{ for } (i,j,k) \text{ an even permutation of } (1,2,3) \\ -V \text{ for } (i,j,k) \text{ an odd permutation of } (1,2,3) \\ 0 \text{ for two or more indices equal} \end{cases}$$

$$\epsilon^{ijk} = \mathbf{e}^i \cdot (\mathbf{e}^j \times \mathbf{e}^k) = \begin{cases} +V' \text{ for } (i,j,k) \text{ an even permutation of } (1,2,3) \\ -V' \text{ for } (i,j,k) \text{ an odd permutation of } (1,2,3) \\ 0 \text{ for two or more indices equal} \end{cases}$$

$$|V| \text{ and } |V'| = \text{volumes of frame parallelepipeds}$$

$$V' = 1/V \qquad V^2 = g$$

$$g = \det(g_{ij}) = \begin{vmatrix} g_{11} & g_{12} & g_{13} \\ g_{21} & g_{22} & g_{23} \\ g_{31} & g_{32} & g_{33} \end{vmatrix} \qquad g^{-1} = \det(g^{ij}) = \begin{vmatrix} g^{11} & g^{12} & g^{13} \\ g^{21} & g^{22} & g^{23} \\ g^{31} & g^{32} & g^{33} \end{vmatrix}$$

Relations with Kronecker delta

$$\epsilon_{ijk}\epsilon^{pqr} = \begin{vmatrix} \delta^p_i & \delta^q_i & \delta^r_i \\ \delta^p_j & \delta^q_j & \delta^r_j \\ \delta^p_k & \delta^q_k & \delta^r_k \end{vmatrix} \qquad \begin{cases} \epsilon_{ijk}\epsilon^{pqk} = \delta^p_i\delta^q_j - \delta^q_i\delta^p_j \\ \epsilon_{ijk}\epsilon^{pjk} = 2\delta^p_i \\ \epsilon_{ijk}\epsilon^{ijk} = 6 \end{cases}$$

Vector cross product

Anti-commutativity, distributivity

$$\mathbf{b} \times \mathbf{a} = -\mathbf{a} \times \mathbf{b} \qquad (\alpha\mathbf{a} + \mathbf{b}) \times \mathbf{c} = \alpha\mathbf{a} \times \mathbf{c} + \mathbf{b} \times \mathbf{c}$$

Cartesian basis

$$\mathbf{a} \times \mathbf{b} = \begin{vmatrix} \mathbf{i}_1 & \mathbf{i}_2 & \mathbf{i}_3 \\ a_1 & a_2 & a_3 \\ b_1 & b_2 & b_3 \end{vmatrix} = \mathbf{i}_1(a_2b_3 - a_3b_2) - \mathbf{i}_2(a_1b_3 - a_3b_1) + \mathbf{i}_3(a_1b_2 - a_2b_1)$$

General basis

$$\mathbf{a} \times \mathbf{b} = \mathbf{e}^i \, \epsilon_{ijk} \, a^j \, b^k = \mathbf{e}_i \, \epsilon^{ijk} \, a_j \, b_k$$

Cross products of basis vectors

$$\mathbf{e}^k = \tfrac{1}{2}\epsilon^{kij}(\mathbf{e}_i \times \mathbf{e}_j) \qquad \mathbf{e}_k = \tfrac{1}{2}\epsilon_{kij}(\mathbf{e}^i \times \mathbf{e}^j)$$

$$\mathbf{e}_i \times \mathbf{e}_j = \epsilon_{kij}\mathbf{e}^k \qquad \mathbf{e}^i \times \mathbf{e}^j = \epsilon^{kij}\mathbf{e}_k$$

$$\begin{aligned} \mathbf{e}_i \times \mathbf{e}^j &= \mathbf{e}_i \times g^{jk}\mathbf{e}_k = g^{jk}\epsilon_{sik}\mathbf{e}^s \\ &= g_{ik}\mathbf{e}^k \times \mathbf{e}^j = g_{ik}\epsilon^{skj}\mathbf{e}_s \\ \mathbf{e}^i \times \mathbf{e}_j &= \mathbf{e}^i \times g_{jk}\mathbf{e}^k = g_{jk}\epsilon^{sik}\mathbf{e}_s \\ &= g^{ik}\mathbf{e}_k \times \mathbf{e}_j = g^{ik}\epsilon_{skj}\mathbf{e}^s \end{aligned}$$

$$\mathbf{i}_1 \times \mathbf{i}_2 = \mathbf{i}_3 \qquad \mathbf{i}_2 \times \mathbf{i}_3 = \mathbf{i}_1 \qquad \mathbf{i}_3 \times \mathbf{i}_1 = \mathbf{i}_2$$

Triple products

Index forms

$$\begin{aligned} \mathbf{a} \cdot (\mathbf{b} \times \mathbf{c}) &= \epsilon_{ijk}a^ib^jc^k = \epsilon^{ijk}a_ib_jc_k \\ \mathbf{a} \times (\mathbf{b} \times \mathbf{c}) &= \epsilon^{imn}\epsilon_{jkm}a_ib^jc^k\mathbf{e}_n = \epsilon_{imn}\epsilon^{jkm}a^ib_jc_k\mathbf{e}^n \end{aligned}$$

Triple products in Cartesian system

$$\mathbf{a} \cdot (\mathbf{b} \times \mathbf{c}) = \begin{vmatrix} a_1 & a_2 & a_3 \\ b_1 & b_2 & b_3 \\ c_1 & c_2 & c_3 \end{vmatrix}$$

$$\mathbf{i}_k \times (\mathbf{i}_m \times \mathbf{i}_p) = \delta_{pk}\mathbf{i}_m - \delta_{mk}\mathbf{i}_p$$

$$(\mathbf{i}_k \times \mathbf{i}_m) \times \mathbf{i}_p = \delta_{pk}\mathbf{i}_m - \delta_{mp}\mathbf{i}_k$$

Triple product identities

$$\mathbf{a} \cdot (\mathbf{b} \times \mathbf{c}) = (\mathbf{a} \times \mathbf{b}) \cdot \mathbf{c} \qquad \mathbf{a} \times (\mathbf{b} \times \mathbf{c}) = \mathbf{b}(\mathbf{a} \cdot \mathbf{c}) - \mathbf{c}(\mathbf{a} \cdot \mathbf{b})$$

Tensor product

$$(\lambda\mathbf{a}) \otimes \mathbf{b} = \mathbf{a} \otimes (\lambda\mathbf{b}) = \lambda(\mathbf{a} \otimes \mathbf{b})$$

$$(\mathbf{a} + \mathbf{b}) \otimes \mathbf{c} = \mathbf{a} \otimes \mathbf{c} + \mathbf{b} \otimes \mathbf{c}$$

$$\mathbf{a} \otimes (\mathbf{b} + \mathbf{c}) = \mathbf{a} \otimes \mathbf{b} + \mathbf{a} \otimes \mathbf{c}$$

$$\mathbf{a} \otimes \mathbf{b} = \mathbf{b} \otimes \mathbf{a} \iff \mathbf{a} \text{ is proportional to } \mathbf{b}$$

Dyads

Definition

$$\mathbf{ab} \equiv \mathbf{a} \otimes \mathbf{b}$$

Dot product between dyad and vector

$$\begin{cases} \mathbf{ab} \cdot \mathbf{c} = (\mathbf{b} \cdot \mathbf{c})\mathbf{a} \\ \mathbf{c} \cdot \mathbf{ab} = (\mathbf{c} \cdot \mathbf{a})\mathbf{b} \end{cases} \qquad \begin{cases} \mathbf{wx} \cdot (\lambda\mathbf{a} + \mathbf{b}) = \lambda\mathbf{wx} \cdot \mathbf{a} + \mathbf{wx} \cdot \mathbf{b} \\ (\lambda\mathbf{a} + \mathbf{b}) \cdot \mathbf{wx} = \lambda\mathbf{a} \cdot \mathbf{wx} + \mathbf{b} \cdot \mathbf{wx} \\ \mathbf{a} \cdot (\lambda\mathbf{wx} + \mathbf{yz}) = \lambda\mathbf{a} \cdot \mathbf{wx} + \mathbf{a} \cdot \mathbf{yz} \\ (\lambda\mathbf{wx} + \mathbf{yz}) \cdot \mathbf{a} = \lambda\mathbf{wx} \cdot \mathbf{a} + \mathbf{yz} \cdot \mathbf{a} \end{cases}$$

Projection dyad

$$\mathbf{a} \cdot \mathbf{ee} = \mathbf{ee} \cdot \mathbf{a} = \text{vector component of } \mathbf{a} \text{ along unit vector } \mathbf{e}$$

Cross product between dyad and vector

$$(\mathbf{ab}) \times \mathbf{c} = \mathbf{a}(\mathbf{b} \times \mathbf{c}) \qquad \mathbf{c} \times (\mathbf{ab}) = (\mathbf{c} \times \mathbf{a})\mathbf{b}$$

Dot product between dyads

$$\mathbf{ab} \cdot \mathbf{cd} = (\mathbf{b} \cdot \mathbf{c})\mathbf{ad}$$

Cross product between dyads

$$\mathbf{ab} \times \mathbf{cd} = \mathbf{a}(\mathbf{b} \times \mathbf{c})\mathbf{d}$$

Double dot product between dyads

$$\mathbf{ab} \cdot\cdot\, \mathbf{cd} = (\mathbf{b} \cdot \mathbf{c})(\mathbf{a} \cdot \mathbf{d})$$

Transpose of dyad

$$(\mathbf{ab})^T = \mathbf{ba}$$

Trace of dyad

$$\mathrm{tr}(\mathbf{ab}) = \mathbf{a} \cdot \mathbf{b}$$

Vectorial invariant of dyad

$$(\mathbf{ab})_\times = \mathbf{a} \times \mathbf{b}$$

Scalar product between dyads

$$\mathbf{ab} \bullet \mathbf{cd} = \mathbf{ab} \cdot\cdot\, (\mathbf{cd})^T = \mathbf{ab} \cdot\cdot\, \mathbf{dc} = (\mathbf{a} \cdot \mathbf{c})(\mathbf{b} \cdot \mathbf{d})$$

Full products involving dyads

$$\mathbf{ab} \odot \mathbf{c} = \mathbf{a}(\mathbf{b} \cdot \mathbf{c}) = \mathbf{ab} \cdot \mathbf{c} \qquad \mathbf{ab} \odot \mathbf{cd} = (\mathbf{a} \cdot \mathbf{c})(\mathbf{b} \cdot \mathbf{d}) = \mathbf{ab} \bullet \mathbf{cd}$$

Dot products with basis dyads

$$\mathbf{e}_i\mathbf{e}_j \cdot \mathbf{e}_k = g_{jk}\mathbf{e}_i \qquad \mathbf{e}_i\mathbf{e}_j \cdot \mathbf{e}_k\mathbf{e}_l = g_{jk}\mathbf{e}_i\mathbf{e}_l \qquad \mathbf{e}_i\mathbf{e}_j \cdot\cdot\, \mathbf{e}_k\mathbf{e}_l = g_{jk}g_{il}$$

Identity, unit, or metric tensor

$$\mathbf{E}\cdot\mathbf{x}=\mathbf{x}\cdot\mathbf{E}=\mathbf{x}$$

$$\mathbf{E}\cdot\mathbf{A}=\mathbf{A}\cdot\mathbf{E}=\mathbf{A}$$

$$\mathbf{E}\times\mathbf{x}=\mathbf{x}\times\mathbf{E}$$

Zero tensor

$$\mathbf{0}\cdot\mathbf{a}=\mathbf{0}\qquad \mathbf{0}\cdot\mathbf{A}=\mathbf{A}\cdot\mathbf{0}=\mathbf{0}$$

Second-order tensor components

Types of components

$$\mathbf{A}=a^{ij}\mathbf{e}_i\mathbf{e}_j=a_{kl}\mathbf{e}^k\mathbf{e}^l=a_i{}^j\mathbf{e}^i\mathbf{e}_j=a^k{}_l\,\mathbf{e}_k\mathbf{e}^l$$

$$a^{ij}=\mathbf{e}^i\cdot\mathbf{A}\cdot\mathbf{e}^j\qquad a^i{}_j=\mathbf{e}^i\cdot\mathbf{A}\cdot\mathbf{e}_j$$

$$a_{ij}=\mathbf{e}_i\cdot\mathbf{A}\cdot\mathbf{e}_j\qquad a_i{}^j=\mathbf{e}_i\cdot\mathbf{A}\cdot\mathbf{e}^j$$

Transformation of components

$$\mathbf{A}=a^{i'j'}\mathbf{e}_{i'}\mathbf{e}_{j'}=a_{k'l'}\mathbf{e}^{k'}\mathbf{e}^{l'}=a_{i'}{}^{j'}\mathbf{e}^{i'}\mathbf{e}_{j'}=a^{k'}{}_{l'}\,\mathbf{e}_{k'}\mathbf{e}^{l'}$$

$$a^{i'j'}=A^{i'}_kA^{j'}_l\,a^{kl}\qquad a^{ij}=A^i_{k'}A^j_{l'}a^{k'l'}$$

$$a_{i'j'}=A^k_{i'}A^l_{j'}a_{kl}\qquad a_{ij}=A^{k'}_iA^{l'}_j\,a_{k'l'}$$

$$a_{i'}{}^{j'}=A^k_{i'}A^{j'}_l\,a_k{}^l\qquad a_i{}^j=A^{k'}_iA^j_{l'}a_{k'}{}^{l'}$$

$$a^{i'}{}_{j'}=A^{i'}_kA^l_{j'}a^k{}_l\qquad a^i{}_j=A^i_{k'}A^{l'}_j\,a^{k'}{}_{l'}$$

Transformation to reciprocal frame

$$a^{km}=g^{ki}g^{jm}a_{ij}\qquad a_{km}=a^{ij}g_{ki}g_{jm}$$

Index juggling techniques

$$a_j b^j = a_k \delta^k_j b^j = a_k g^{km} g_{mj} b^j = a^m b_m = a^j b_j$$
$$a_i{}^i = a_k{}^i \delta^k_i = a_k{}^i g^{km} g_{mi} = a^m{}_m = a^i{}_i$$

Identity, unit, or metric tensor

$$\mathbf{E} = \mathbf{e}^i \mathbf{e}_i = \mathbf{e}_j \mathbf{e}^j = g_{ij} \mathbf{e}^i \mathbf{e}^j = g^{ij} \mathbf{e}_i \mathbf{e}_j = \mathbf{i}_n \mathbf{i}_n$$

Some component representations of dot product

$$\begin{aligned}\mathbf{A} \cdot \mathbf{B} &= a^{in} g_{nm} b^{mj} \mathbf{e}_i \mathbf{e}_j = a^i{}_m b^{mj} \mathbf{e}_i \mathbf{e}_j \text{ or } a^{in} b_n{}^j \mathbf{e}_i \mathbf{e}_j \\ &= a_{in} g^{nm} b_{mj} \mathbf{e}^i \mathbf{e}^j = a_i{}^m b_{mj} \mathbf{e}^i \mathbf{e}^j \text{ or } a_{in} b^n{}_j \mathbf{e}^i \mathbf{e}^j \\ &= a^{in} b_{nj} \mathbf{e}_i \mathbf{e}^j \\ &= a_{in} b^{nj} \mathbf{e}^i \mathbf{e}_j\end{aligned}$$

Double dot product of second-order tensors

Component representations

$$\mathbf{A} \cdot\cdot\, \mathbf{B} = a^i{}_j b^j{}_i = a_i{}^j b_j{}^i = a^{ij} b_{ji} = a_{ij} b^{ji}$$

Relation to trace

$$\mathbf{E} \cdot\cdot\, \mathbf{A} = \mathbf{A} \cdot\cdot\, \mathbf{E} = \operatorname{tr} \mathbf{A}$$

Other properties

$$\mathbf{A} \cdot\cdot\, \mathbf{B} = \mathbf{B} \cdot\cdot\, \mathbf{A} = \mathbf{A}^T \cdot\cdot\, \mathbf{B}^T \qquad \mathbf{E} \cdot\cdot\, \mathbf{E} = 3$$
$$\mathbf{B} \cdot\cdot\, \mathbf{A}^T = \mathbf{A} \cdot\cdot\, \mathbf{B}^T \qquad \mathbf{E} \cdot\cdot\, \mathbf{v}\mathbf{v} = \mathbf{v} \cdot \mathbf{v}$$
$$\mathbf{v}\mathbf{v} \cdot\cdot\, \mathbf{A} = \mathbf{v} \cdot \mathbf{A} \cdot \mathbf{v} \qquad \operatorname{tr}(\mathbf{A} \cdot \mathbf{B}) = \mathbf{A} \cdot\cdot\, \mathbf{B} = \mathbf{A}^T \cdot\cdot\, \mathbf{B}^T$$

$$\begin{aligned}(\mathbf{A} \cdot \mathbf{U}) \cdot\cdot\, \mathbf{X} &= (\mathbf{U} \cdot \mathbf{X}) \cdot\cdot\, \mathbf{A} = (\mathbf{X} \cdot \mathbf{A}) \cdot\cdot\, \mathbf{U} \\ &= \mathbf{U} \cdot\cdot\, (\mathbf{X} \cdot \mathbf{A}) = \mathbf{A} \cdot\cdot\, (\mathbf{U} \cdot \mathbf{X}) = \mathbf{X} \cdot\cdot\, (\mathbf{A} \cdot \mathbf{U})\end{aligned}$$

Inverse of nonsingular second-order tensor

$$\mathbf{A}\cdot\mathbf{A}^{-1}=\mathbf{A}^{-1}\cdot\mathbf{A}=\mathbf{E}$$
$$(\mathbf{A}^{-1})^{-1}=\mathbf{A}$$
$$(\mathbf{A}\cdot\mathbf{B})^{-1}=\mathbf{B}^{-1}\cdot\mathbf{A}^{-1}$$
$$\mathbf{A}\text{ nonsingular}\iff(\mathbf{A}\cdot\mathbf{x}=\mathbf{0}\implies\mathbf{x}=\mathbf{0})$$

Determinant of second-order tensor

Definition

$$\det\mathbf{A}=\det(a_i{}^j)=\det(a^i{}_j)$$

Other expressions

$$\begin{aligned}\det\mathbf{A}&=g^{-1}\det(a_{st})=g\det(a^{pq})\qquad g=\det(g_{ij})\\&=\tfrac{1}{6}\epsilon_{ijk}\epsilon^{mnp}a_m{}^i a_n{}^j a_p{}^k\end{aligned}$$

Properties

$$\det(\mathbf{A}\cdot\mathbf{B})=\det\mathbf{A}\det\mathbf{B}\qquad\det\mathbf{ab}=0$$
$$\det\mathbf{A}^{-1}=(\det\mathbf{A})^{-1}\qquad\det\mathbf{E}=1$$

Use in computing inverse

$$\det\mathbf{A}\neq0\iff\mathbf{A}\text{ nonsingular}$$

$$\mathbf{A}^{-1}=\frac{1}{2\det\mathbf{A}}\,\epsilon_{ikl}\epsilon^{jmn}a_m{}^k a_n{}^l\mathbf{e}^i\mathbf{e}_j$$

Transpose of second-order tensor

$$(\mathbf{ab})^T=\mathbf{ba}\qquad\mathbf{A}\cdot\mathbf{x}=\mathbf{x}\cdot\mathbf{A}^T$$

Component forms

$$(a^{ij}\mathbf{e}_i\mathbf{e}_j)^T = a^{ji}\mathbf{e}_i\mathbf{e}_j \qquad (a_i{}^j\mathbf{e}^i\mathbf{e}_j)^T = a^j{}_i\mathbf{e}^i\mathbf{e}_j$$

$$(a_{ij}\mathbf{e}^i\mathbf{e}^j)^T = a_{ji}\mathbf{e}^i\mathbf{e}^j \qquad (a^i{}_j\mathbf{e}_i\mathbf{e}^j)^T = a_j{}^i\mathbf{e}_i\mathbf{e}^j$$

Other properties

$$(\mathbf{A}^T)^T = \mathbf{A} \qquad (\mathbf{A}\cdot\mathbf{B})^T = \mathbf{B}^T\cdot\mathbf{A}^T$$

$$(\mathbf{A}^T)^{-1} = (\mathbf{A}^{-1})^T \qquad \mathbf{a}\cdot\mathbf{A}^T\cdot\mathbf{b} = \mathbf{b}\cdot\mathbf{A}\cdot\mathbf{a}$$

Trace

$$\operatorname{tr}\mathbf{A} = a_i{}^i = a^i{}_i$$

$$\operatorname{tr}\mathbf{A} = \mathbf{E}\cdot\cdot\,\mathbf{A} = \mathbf{A}\cdot\cdot\,\mathbf{E} = \mathbf{A}\bullet\mathbf{E} = \mathbf{A}^T\bullet\mathbf{E}$$

$$\operatorname{tr}(\mathbf{A}\cdot\mathbf{B}) = \mathbf{A}\cdot\cdot\,\mathbf{B} = \mathbf{A}^T\cdot\cdot\,\mathbf{B}^T = \mathbf{A}\bullet\mathbf{B}^T = \mathbf{A}^T\bullet\mathbf{B}$$

$$\operatorname{tr}(\alpha\mathbf{A}+\mathbf{B}) = \alpha\operatorname{tr}\mathbf{A} + \operatorname{tr}\mathbf{B}$$

$$\operatorname{tr}(\mathbf{A}\cdot\mathbf{B}) = \operatorname{tr}(\mathbf{B}\cdot\mathbf{A}) = \operatorname{tr}(\mathbf{A}^T\cdot\mathbf{B}^T) = \operatorname{tr}(\mathbf{B}^T\cdot\mathbf{A}^T)$$

$$\operatorname{tr}(\mathbf{A}\times\mathbf{v}) = \operatorname{tr}(\mathbf{v}\times\mathbf{A})$$

$$\operatorname{tr}\mathbf{A}^T = \operatorname{tr}\mathbf{A}$$

Vectorial invariant

$$(\mathbf{A}+\mathbf{B})_\times = \mathbf{A}_\times + \mathbf{B}_\times \qquad (\alpha\mathbf{A})_\times = \alpha\mathbf{A}_\times$$

Scalar product of second-order tensors

Definition

$$\mathbf{A}\bullet\mathbf{B} = \mathbf{A}\cdot\cdot\,\mathbf{B}^T$$

Properties

$$\mathbf{A} \bullet \mathbf{B} = \mathbf{B} \bullet \mathbf{A} = \mathbf{A}^T \bullet \mathbf{B}^T = \mathbf{A}^T \cdot\cdot\, \mathbf{B}$$
$$\mathbf{A} \bullet (\mathbf{B} + \mathbf{C}) = \mathbf{A} \bullet \mathbf{B} + \mathbf{A} \bullet \mathbf{C}$$
$$\alpha(\mathbf{A} \bullet \mathbf{B}) = (\alpha\mathbf{A}) \bullet \mathbf{B} = \mathbf{A} \bullet (\alpha\mathbf{B})$$
$$\mathbf{A} \bullet \mathbf{A} = 0 \iff \mathbf{A} = \mathbf{0}$$
$$\mathbf{A} \bullet \mathbf{B} = \mathrm{tr}(\mathbf{A} \cdot \mathbf{B}^T)$$

Component forms

$$\mathbf{A} \bullet \mathbf{B} = a_{ij}b^{ij} = a^{ij}b_{ij} = a^i{}_j b_i{}^j = a_i{}^j b^i{}_j$$

Norm induced by scalar product

Definition

$$\|\mathbf{A}\| = (\mathbf{A} \bullet \mathbf{A})^{1/2}$$

Component forms

$$\|\mathbf{A}\| = \sqrt{a_{ij}a^{ij}} = \sqrt{a^i{}_j a_i{}^j}$$

Integer powers of second-order tensors

Definition

$$\mathbf{A}^0 = \mathbf{E} \qquad \mathbf{A}^n = \mathbf{A} \cdot \mathbf{A}^{n-1} \text{ for } n = 1, 2, 3, \ldots$$
$$= \underbrace{\mathbf{A} \cdot \mathbf{A} \cdot \cdots \cdot \mathbf{A}}_{n \text{ times}}$$

Laws of exponents

$$\mathbf{A}^m \cdot \mathbf{A}^n = \mathbf{A}^{m+n} \qquad (\mathbf{A}^m)^n = \mathbf{A}^{mn}$$

Negative powers

$$\begin{aligned}\mathbf{A}^{-n} &= \mathbf{A}^{-n+1}\cdot\mathbf{A}^{-1} \text{ for } n = 2,3,4,\ldots \\ &= \underbrace{\mathbf{A}^{-1}\cdot\mathbf{A}^{-1}\cdot\dots\cdot\mathbf{A}^{-1}}_{n \text{ times}}\end{aligned}$$

Transposes and powers

$$(\mathbf{A}^n)^T = (\mathbf{A}^T)^n \qquad (\mathbf{A}^{-1})^T = (\mathbf{A}^T)^{-1}$$

Symmetric and antisymmetric second-order tensors

Symmetric tensor

$$\mathbf{A} = \mathbf{A}^T \implies \begin{cases}\mathbf{A}\cdot\mathbf{x} = \mathbf{x}\cdot\mathbf{A} \\ (\mathbf{A}\cdot\mathbf{x})\cdot\mathbf{y} = \mathbf{x}\cdot(\mathbf{A}\cdot\mathbf{y}) \\ \mathbf{a}\cdot\mathbf{A}\cdot\mathbf{b} = \mathbf{b}\cdot\mathbf{A}\cdot\mathbf{a} \\ \operatorname{tr}(\mathbf{x}\times\mathbf{A}) = 0 \\ a^{ji} = a^{ij} \\ a_{ji} = a_{ij} \\ a_i^{\ j} = a^j_{\ i}\end{cases}$$

Antisymmetric or skew tensor

$$\mathbf{A} = -\mathbf{A}^T \implies \begin{cases}\mathbf{a}\cdot\mathbf{A}\cdot\mathbf{b} = -\mathbf{b}\cdot\mathbf{A}\cdot\mathbf{a} \\ \operatorname{tr}\mathbf{A} = 0 \\ a^{ji} = -a^{ij} \\ a_{ji} = -a_{ij} \\ a_i^{\ j} = -a^j_{\ i}\end{cases}$$

Symmetric-antisymmetric decomposition

$$\mathbf{A} = \mathbf{A}^S + \mathbf{A}^A \qquad \begin{cases}\mathbf{A}^S = \frac{1}{2}(\mathbf{A}+\mathbf{A}^T) \\ \mathbf{A}^A = \frac{1}{2}(\mathbf{A}-\mathbf{A}^T)\end{cases}$$

$$(\mathbf{A}^S)^T = \mathbf{A}^S \qquad (\mathbf{A}^A)^T = -\mathbf{A}^A$$

$$\mathbf{x}\cdot\mathbf{A}\cdot\mathbf{x}=\mathbf{x}\cdot\mathbf{A}^S\cdot\mathbf{x} \qquad \mathbf{A}^S\cdot\cdot\,\mathbf{B}^A=0$$

$$\mathbf{A}\cdot\cdot\,\mathbf{B}=\mathbf{A}^S\cdot\cdot\,\mathbf{B}^S+\mathbf{A}^A\cdot\cdot\,\mathbf{B}^A \qquad (\mathbf{A}^S)_\times=\mathbf{0}$$

Conjugate vector representation of an antisymmetric tensor

$$\mathbf{C}=\mathbf{E}\times\boldsymbol{\omega}=\boldsymbol{\omega}\times\mathbf{E}$$

$$\boldsymbol{\omega}=-\tfrac{1}{2}\epsilon^{ksp}C_{ks}\mathbf{e}_p=-\tfrac{1}{2}\mathbf{C}_\times=-\tfrac{1}{2}\boldsymbol{\mathcal{E}}\bullet\mathbf{C}$$

$$\mathbf{C}\cdot\mathbf{x}=\boldsymbol{\omega}\times\mathbf{x} \qquad \mathbf{x}\cdot\mathbf{C}=\mathbf{x}\times\boldsymbol{\omega}$$

Eigenpairs of a symmetric second-order tensor
Equation defining eigenpair

$$\mathbf{A}\cdot\mathbf{x}=\lambda\mathbf{x} \quad (\mathbf{x}\neq\mathbf{0})$$

$$(\mathbf{A}-\lambda\mathbf{E})\cdot\mathbf{x}=\mathbf{0}$$

$$({a^i}_k-\lambda\delta^i_k)x^k=0$$

$$[\,\text{or } ({a_i}^k-\lambda\delta^k_i)x_k=0\,]$$

Characteristic equation

$$\begin{vmatrix} {a^1}_1-\lambda & {a^1}_2 & {a^1}_3 \\ {a^2}_1 & {a^2}_2-\lambda & {a^2}_3 \\ {a^3}_1 & {a^3}_2 & {a^3}_3-\lambda \end{vmatrix}=0$$

or

$$-\lambda^3+I_1(\mathbf{A})\lambda^2-I_2(\mathbf{A})\lambda+I_3(\mathbf{A})=0$$

Principal invariants

$$\begin{cases} I_1(\mathbf{A})=\lambda_1+\lambda_2+\lambda_3 \\ I_2(\mathbf{A})=\lambda_1\lambda_2+\lambda_2\lambda_3+\lambda_3\lambda_1 \\ I_3(\mathbf{A})=\lambda_1\lambda_2\lambda_3 \end{cases} \qquad \begin{cases} I_1(\mathbf{A})=\operatorname{tr}\mathbf{A} \\ I_2(\mathbf{A})=\frac{1}{2}(\operatorname{tr}^2\mathbf{A}-\operatorname{tr}\mathbf{A}^2) \\ I_3(\mathbf{A})=\det\mathbf{A} \end{cases}$$

$$I_1(\mathbf{A}^{-1})=\frac{I_2(\mathbf{A})}{I_3(\mathbf{A})} \qquad I_2(\mathbf{A}^{-1})=\frac{I_1(\mathbf{A})}{I_3(\mathbf{A})} \qquad I_3(\mathbf{A}^{-1})=I_3^{-1}(\mathbf{A})$$

Orthogonal (spectral) representation of a tensor

$$\mathbf{A} = \sum_{i=1}^{3} \lambda_i \mathbf{x}_i \mathbf{x}_i \qquad (\lambda_i, \mathbf{x}_i \text{ the eigenpairs})$$

Cayley–Hamilton theorem

$$-\mathbf{A}^3 + I_1(\mathbf{A})\mathbf{A}^2 - I_2(\mathbf{A})\mathbf{A} + I_3(\mathbf{A})\mathbf{E} = \mathbf{0}$$

Consequences

$$\begin{aligned}
\mathbf{A}^3 &= I_1(\mathbf{A})\mathbf{A}^2 - I_2(\mathbf{A})\mathbf{A} + I_3(\mathbf{A})\mathbf{E} \\
\mathbf{A}^4 &= [I_1^2(\mathbf{A}) - I_2(\mathbf{A})]\mathbf{A}^2 + [I_3(\mathbf{A}) - I_1(\mathbf{A})I_2(\mathbf{A})]\mathbf{A} + I_1(\mathbf{A})I_3(\mathbf{A})\mathbf{E} \\
\mathbf{A}^{-1} &= I_3^{-1}(\mathbf{A})[\mathbf{A}^2 - I_1(\mathbf{A})\mathbf{A} + I_2(\mathbf{A})\mathbf{E}] \\
\det \mathbf{A} &= \tfrac{1}{6}(\operatorname{tr}^3 \mathbf{A} - 3 \operatorname{tr} \mathbf{A} \operatorname{tr} \mathbf{A}^2 + 2 \operatorname{tr} \mathbf{A}^3)
\end{aligned}$$

Orthogonal tensors

$$\begin{aligned}
&\mathbf{Q}^T = \mathbf{Q}^{-1} \\
&\det \mathbf{Q} = \pm 1 \quad \text{(proper/improper)} \\
&(\mathbf{Q} \cdot \mathbf{x}) \cdot (\mathbf{Q} \cdot \mathbf{y}) = \mathbf{x} \cdot \mathbf{y} \\
&|\mathbf{Q} \cdot \mathbf{x}| = |\mathbf{x}| \\
&\mathbf{A} \cdot \mathbf{x} = \lambda \mathbf{x} \implies (\mathbf{Q} \cdot \mathbf{A} \cdot \mathbf{Q}^T) \cdot \mathbf{x} = \lambda(\mathbf{Q} \cdot \mathbf{x})
\end{aligned}$$

Representation of proper orthogonal tensor

$$\mathbf{Q} = \mathbf{E} \cos\omega + (1 - \cos\omega)\mathbf{e}\mathbf{e} - \mathbf{e} \times \mathbf{E} \sin\omega$$

Polar decompositions of nonsingular second-order tensor

$$\begin{aligned}
&\mathbf{A} = \mathbf{S} \cdot \mathbf{Q} \qquad && \mathbf{Q} \text{ orthogonal} \\
&\mathbf{A} = \mathbf{Q} \cdot \mathbf{S}' \qquad && \mathbf{S}, \mathbf{S}' \text{ positive definite and symmetric}
\end{aligned}$$

Positive definiteness of symmetric second-order tensor

$$\mathbf{x} \cdot \mathbf{S} \cdot \mathbf{x} > 0 \text{ for all } \mathbf{x} \neq 0 \iff \text{all eigenvalues of } \mathbf{S} \text{ positive}$$

Square root of symmetric, positive definite tensor

$$\mathbf{S} = \sum_{i=1}^{3} \lambda_i \mathbf{x}_i \mathbf{x}_i \ \ (\lambda_i > 0) \implies \mathbf{S}^{1/2} = \sum_{i=1}^{3} \sqrt{\lambda_i}\, \mathbf{x}_i \mathbf{x}_i$$

$$(\mathbf{S}^{1/2})^2 = \mathbf{S}$$

Deviator-ball representation of second-order tensor

$$\mathbf{A} = \tfrac{1}{3} I_1(\mathbf{A})\mathbf{E} + \operatorname{dev}\mathbf{A}$$

Unimodular-ball representation of second-order tensor

$$\mathbf{A} = \mathbf{B} \cdot \mathbf{U} = \mathbf{U} \cdot \mathbf{B} \qquad \begin{cases} \mathbf{B} = |\det \mathbf{A}|\mathbf{E} \\ \mathbf{U} = |\det \mathbf{A}|^{-1}\mathbf{A} \end{cases}$$

Third- and higher-order tensors

Triad

$$\mathbf{abc} = \mathbf{a} \otimes \mathbf{b} \otimes \mathbf{c}$$

Operations with triads

$$\mathbf{abc} \cdot \mathbf{x} = \mathbf{ab}(\mathbf{c} \cdot \mathbf{x})$$

$$\mathbf{abc} \cdot\cdot\, \mathbf{xy} = \mathbf{a}(\mathbf{c} \cdot \mathbf{x})(\mathbf{b} \cdot \mathbf{y})$$

$$\mathbf{abc} \odot \mathbf{xy} = \mathbf{a}(\mathbf{b} \cdot \mathbf{x})(\mathbf{c} \cdot \mathbf{y})$$

$$\mathbf{abc} \cdots \mathbf{xyz} = (\mathbf{c} \cdot \mathbf{x})(\mathbf{b} \cdot \mathbf{y})(\mathbf{a} \cdot \mathbf{z})$$

$$\mathbf{abc} \bullet \mathbf{xyz} = (\mathbf{a} \cdot \mathbf{x})(\mathbf{b} \cdot \mathbf{y})(\mathbf{c} \cdot \mathbf{z})$$

Some component forms for third-order tensors

$$\mathbf{R} = r^{ijk}\mathbf{e}_i\mathbf{e}_j\mathbf{e}_k = r^{i'j'k'}\mathbf{e}_{i'}\mathbf{e}_{j'}\mathbf{e}_{k'}$$

$$\mathbf{R} = r^{ij}{}_k\mathbf{e}_i\mathbf{e}_j\mathbf{e}^k = r^{i'j'}{}_{k'}\mathbf{e}_{i'}\mathbf{e}_{j'}\mathbf{e}^{k'}$$

$$\mathbf{R} = r^i{}_{jk}\mathbf{e}_i\mathbf{e}^j\mathbf{e}^k = r^{i'}{}_{j'k'}\mathbf{e}_{i'}\mathbf{e}^{j'}\mathbf{e}^{k'}$$

$$\mathbf{R} = r_{ijk}\mathbf{e}^i\mathbf{e}^j\mathbf{e}^k = r_{i'j'k'}\mathbf{e}^{i'}\mathbf{e}^{j'}\mathbf{e}^{k'}$$

Some transformation relations for third-order tensors

$$r^{i'j'k'} = r^{mnp} A^{i'}_{m} A^{j'}_{n} A^{k'}_{p} \qquad r^{ijk} = r^{m'n'p'} A^{i}_{m'} A^{j}_{n'} A^{k}_{p'}$$

$$r^{i'j'}{}_{k'} = r^{mn}{}_{p} A^{i'}_{m} A^{j'}_{n} A^{p}_{k'} \qquad r^{ij}{}_{k} = r^{m'n'}{}_{p'} A^{i}_{m'} A^{j}_{n'} A^{p'}_{k}$$

$$r^{i'}{}_{j'k'} = r^{m}{}_{np} A^{i'}_{m} A^{n}_{j'} A^{p}_{k'} \qquad r^{i}{}_{jk} = r^{m'}{}_{n'p'} A^{i}_{m'} A^{n'}_{j} A^{p'}_{k}$$

$$r_{i'j'k'} = r_{mnp} A^{m}_{i'} A^{n}_{j'} A^{p}_{k'} \qquad r_{ijk} = r_{m'n'p'} A^{m'}_{i} A^{n'}_{j} A^{p'}_{k}$$

Levi–Civita tensor

$$\begin{aligned}\boldsymbol{\mathcal{E}} &= -\mathbf{E} \times \mathbf{E} \\ &= \epsilon_{ijk}\, \mathbf{e}^i \mathbf{e}^j \mathbf{e}^k \\ &= \mathbf{i}_1\mathbf{i}_2\mathbf{i}_3 + \mathbf{i}_2\mathbf{i}_3\mathbf{i}_1 + \mathbf{i}_3\mathbf{i}_1\mathbf{i}_2 - \mathbf{i}_1\mathbf{i}_3\mathbf{i}_2 - \mathbf{i}_3\mathbf{i}_2\mathbf{i}_1 - \mathbf{i}_2\mathbf{i}_1\mathbf{i}_3\end{aligned}$$

$$\boldsymbol{\mathcal{E}} \cdots \mathbf{zyx} = \mathbf{x} \cdot (\mathbf{y} \times \mathbf{z})$$

$$\boldsymbol{\mathcal{E}} \cdot\cdot\, \mathbf{xy} = \mathbf{y} \times \mathbf{x}$$

$$\boldsymbol{\mathcal{E}} \cdot \mathbf{x} = -\mathbf{x}\times$$

Tetrad

$$\mathbf{abcd} = \mathbf{a} \otimes \mathbf{b} \otimes \mathbf{c} \otimes \mathbf{d}$$

Fourth-order unit tensor

$$\mathbf{E} = \mathbf{e}_k \mathbf{e}_m \mathbf{e}^k \mathbf{e}^m$$

$$\mathbf{E} \bullet \mathbf{A} = \mathbf{A} \quad (\mathbf{A} \text{ a second-order tensor})$$

Full product

$$\mathbf{X} \odot \mathbf{Y} = x^{i\cdots jmn\cdots s} y_{mn\cdots s}\, \mathbf{e}_i \ldots \mathbf{e}_j \qquad \begin{cases} \mathbf{X} \text{ of order } p \\ \mathbf{Y} \text{ of order } q \le p \\ \mathbf{X} \odot \mathbf{Y} \text{ of order } p - q \end{cases}$$

$$\mathbf{ab} \odot \mathbf{r} = \mathbf{a}(\mathbf{b} \cdot \mathbf{r}) = \mathbf{ab} \cdot \mathbf{r}$$

$$\mathbf{ab} \odot \mathbf{rs} = (\mathbf{a} \cdot \mathbf{r})(\mathbf{b} \cdot \mathbf{s}) = \mathbf{ab} \bullet \mathbf{rs}$$

$$\mathbf{abc} \odot \mathbf{rst} = (\mathbf{a} \cdot \mathbf{r})(\mathbf{b} \cdot \mathbf{s})(\mathbf{c} \cdot \mathbf{t})$$

Quick summary of product operations and some component forms

$\mathbf{ab}\cdot\mathbf{c} = \mathbf{a}(\mathbf{b}\cdot\mathbf{c})$	$\mathbf{ab}\cdot\mathbf{c} = a^k b_m c^m \mathbf{e}_k$
$\mathbf{c}\cdot\mathbf{ab} = (\mathbf{c}\cdot\mathbf{a})\mathbf{b}$	$\mathbf{c}\cdot\mathbf{ab} = c^k a_k b^m \mathbf{e}_m$
$\mathbf{ab}\times\mathbf{c} = \mathbf{a}(\mathbf{b}\times\mathbf{c})$	$\mathbf{ab}\times\mathbf{c} = \epsilon_{sjm} a^k b^j c^m \mathbf{e}_k \mathbf{e}^s$
$\mathbf{c}\times\mathbf{ab} = (\mathbf{c}\times\mathbf{a})\mathbf{b}$	$\mathbf{c}\times\mathbf{ab} = \epsilon_{skj} c^k a^j b^m \mathbf{e}^s \mathbf{e}_m$
$(\mathbf{ab})\mathbf{c} = \mathbf{abc}$	$(\mathbf{ab})\mathbf{c} = a^k b^j c^m \mathbf{e}_k \mathbf{e}_j \mathbf{e}_m$
$\mathbf{c}(\mathbf{ab}) = \mathbf{cab}$	$\mathbf{c}(\mathbf{ab}) = c^k a^j b^m \mathbf{e}_k \mathbf{e}_j \mathbf{e}_m$
$\mathbf{ab}\cdot\mathbf{cd} = \mathbf{a}(\mathbf{b}\cdot\mathbf{c})\mathbf{d}$	$\mathbf{ab}\cdot\mathbf{cd} = a^i b_k c^k d^n \mathbf{e}_i \mathbf{e}_n$
$\mathbf{ab}\times\mathbf{cd} = \mathbf{a}(\mathbf{b}\times\mathbf{c})\mathbf{d}$	$\mathbf{ab}\times\mathbf{cd} = a^k b_j c_n d^m \epsilon^{pjn} \mathbf{e}_k \mathbf{e}_p \mathbf{e}_m$
$(\mathbf{ab})(\mathbf{cd}) = \mathbf{abcd}$	$(\mathbf{ab})(\mathbf{cd}) = a^j b_k c^m d^p \mathbf{e}_j \mathbf{e}_k \mathbf{e}_m \mathbf{e}_p$
$\mathbf{ab}\cdot\cdot\,\mathbf{cd} = (\mathbf{b}\cdot\mathbf{c})(\mathbf{a}\cdot\mathbf{d})$	$\mathbf{ab}\cdot\cdot\,\mathbf{cd} = a^i b^j c_j d_i$
$\mathbf{ab}\bullet\mathbf{cd} = (\mathbf{b}\cdot\mathbf{d})(\mathbf{a}\cdot\mathbf{c})$	$\mathbf{ab}\bullet\mathbf{cd} = a^i b^j c_i d_j$

Tensor-valued functions of tensor arguments

Linearity property

$$f(\lambda\mathbf{a} + \mu\mathbf{b}) = \lambda f(\mathbf{a}) + \mu f(\mathbf{b})$$
$$f(\lambda\mathbf{A} + \mu\mathbf{B}) = \lambda f(\mathbf{A}) + \mu f(\mathbf{B})$$
$$\mathbf{f}(\lambda\mathbf{a} + \mu\mathbf{b}) = \lambda\mathbf{f}(\mathbf{a}) + \mu\mathbf{f}(\mathbf{b})$$

Representation theorems

$$f(\mathbf{x}) = \mathbf{c}\cdot\mathbf{x} \text{ where } \mathbf{c} = f(\mathbf{e}_k)\mathbf{e}^k$$
$$f(\mathbf{A}) = \mathrm{tr}(\mathbf{B}\cdot\mathbf{A}^T) \text{ where } \mathbf{B} = f(\mathbf{e}_i\mathbf{e}_j)\mathbf{e}^i\mathbf{e}^j$$
$$\mathbf{F}(\mathbf{A}) = \mathbf{C}\cdot\cdot\,\mathbf{A}^T \text{ where } \mathbf{C} = \mathbf{F}(\mathbf{e}_m\mathbf{e}_n)\mathbf{e}^m\mathbf{e}^n$$

Derivatives with respect to tensor arguments

$$df = f_{,\mathbf{x}} \cdot d\mathbf{x} = \frac{d}{d\varepsilon} f(\mathbf{x} + \varepsilon\, d\mathbf{x})\bigg|_{\varepsilon=0} \qquad f_{,\mathbf{x}} = \frac{\partial f}{\partial x_k}\mathbf{i}_k$$

$$df = f_{,\mathbf{X}} \odot d\mathbf{X} = \frac{d}{d\varepsilon} f(\mathbf{X} + \varepsilon\, d\mathbf{X})\bigg|_{\varepsilon=0} \qquad f_{,\mathbf{X}} = \frac{\partial f}{\partial x_{mn}}\mathbf{i}_m\mathbf{i}_n$$

$$d\mathbf{f} = \mathbf{f}_{,\mathbf{x}} \odot d\mathbf{x} = \frac{d}{d\varepsilon} \mathbf{f}(\mathbf{x} + \varepsilon\, d\mathbf{x})\bigg|_{\varepsilon=0} \qquad \mathbf{f}_{,\mathbf{x}} = \frac{\partial f_r}{\partial x_m}\mathbf{i}_r\mathbf{i}_m$$

$$d\mathbf{F} = \mathbf{F}_{,\mathbf{X}} \odot d\mathbf{X} = \frac{d}{d\varepsilon} \mathbf{F}(\mathbf{X} + \varepsilon\, d\mathbf{X})\bigg|_{\varepsilon=0} \qquad \mathbf{F}_{,\mathbf{X}} = \frac{\partial f_{rs}}{\partial x_{mn}}\mathbf{i}_r\mathbf{i}_s\mathbf{i}_m\mathbf{i}_n$$

Tensor fields

Curvilinear coordinates and position vector

$$P = (q^1, q^2, q^3) \qquad \frac{\partial q^i}{\partial q^k} = \delta^i_k \qquad \mathbf{r} = \mathbf{r}(q^1, q^2, q^3)$$

Local basis

$$\mathbf{r}_i = \frac{\partial \mathbf{r}}{\partial q^i} = \text{tangent vectors to coordinate lines}$$

Jacobian

$$x^i = x^i(q^1, q^2, q^3) \qquad \sqrt{g} = \mathbf{r}_1 \cdot (\mathbf{r}_2 \times \mathbf{r}_3) = \det\left(\frac{\partial x^i}{\partial q^j}\right)$$

Local reciprocal basis and metric coefficients

$$\mathbf{r}^i \cdot \mathbf{r}_j = \delta^i_j \qquad \begin{cases} g_{ij} = \mathbf{r}_i \cdot \mathbf{r}_j \\ g^{ij} = \mathbf{r}^i \cdot \mathbf{r}^j \\ g^j_i = \mathbf{r}_i \cdot \mathbf{r}^j = \delta^j_i \\ g_{ij}g^{jk} = \delta^k_i \end{cases}$$

Components of a vector field

$$\mathbf{f} = f^i\mathbf{r}_i = f_i\mathbf{r}^i = f^{(i)}\mathbf{r}_{(i)} = f_{(i)}\mathbf{r}^{(i)} \qquad \begin{cases} f^i = \mathbf{f}\cdot\mathbf{r}^i \\ f_i = \mathbf{f}\cdot\mathbf{r}_i \\ f^{(i)} = \sqrt{g_{ii}}\, f^i \text{ (no sum over } i) \\ f_{(i)} = \sqrt{g^{ii}}\, f_i \text{ (no sum over } i) \end{cases}$$

Coordinate transformation

$$q^{j'} = q^{j'}(q^1, q^2, q^3), \quad q^k = q^k(q^{1'}, q^{2'}, q^{3'}) \qquad (j, k = 1, 2, 3)$$

$$\mathbf{r}_i = A_i^{j'}\mathbf{r}_{j'} \qquad A_i^{j'} = \frac{\partial q^{j'}}{\partial q^i}$$

$$\mathbf{r}_{i'} = A_{i'}^{j}\mathbf{r}_j \qquad A_{i'}^{j} = \frac{\partial q^j}{\partial q^{i'}}$$

$$f^{i'} = A_j^{i'} f^j \qquad f_{i'} = A_{i'}^j f_j \qquad f^i = A_{j'}^i f^{j'} \qquad f_i = A_i^{j'} f_{j'}$$

Metric

$$(ds)^2 = d\mathbf{r}\cdot d\mathbf{r} = g_{ij}\, dq^i\, dq^j$$

Covariant differentiation and Christoffel symbols

$$\nabla_k f \equiv \frac{\partial f}{\partial q^k}$$

$$\frac{\partial \mathbf{f}}{\partial q^k} = \mathbf{r}_i \nabla_k f^i \qquad \nabla_k f^i = \frac{\partial f^i}{\partial q^k} + \Gamma_{kt}^i f^t$$

$$\frac{\partial \mathbf{f}}{\partial q^k} = \mathbf{r}^i \nabla_k f_i \qquad \nabla_k f_i = \frac{\partial f_i}{\partial q^k} - \Gamma_{ki}^j f_j$$

$$\frac{\partial \mathbf{r}_i}{\partial q^j} = \Gamma_{ij}^k \mathbf{r}_k \qquad \frac{\partial \mathbf{r}^i}{\partial q^k} = -\Gamma_{jk}^i \mathbf{r}^j \qquad \Gamma_{ijt} = \frac{\partial \mathbf{r}_i}{\partial q^j}\cdot\mathbf{r}_t$$

$$\begin{cases} \Gamma_{ij}^k = \Gamma_{ji}^k \\ \Gamma_{ijk} = \Gamma_{jik} \end{cases} \qquad \begin{cases} \Gamma_{ij}^k g_{kt} = \Gamma_{ijt} \\ \Gamma_{ij}^k = g^{kt}\Gamma_{ijt} \end{cases}$$

$$\Gamma_{ijt} = \frac{1}{2}\left(\frac{\partial g_{it}}{\partial q^j} + \frac{\partial g_{tj}}{\partial q^i} - \frac{\partial g_{ji}}{\partial q^t}\right) \qquad \Gamma^k_{ij} = \frac{g^{kl}}{2}\left(\frac{\partial g_{il}}{\partial x^j} + \frac{\partial g_{jl}}{\partial x^i} - \frac{\partial g_{ij}}{\partial x^l}\right)$$

$$\frac{\partial \mathbf{A}}{\partial q^k} = \mathbf{r}_i\mathbf{r}_j\nabla_k a^{ij} \qquad \nabla_k a^{ij} = \frac{\partial a^{ij}}{\partial q^k} + \Gamma^i_{ks}a^{sj} + a^{is}\Gamma^j_{ks}$$

$$\frac{\partial \mathbf{A}}{\partial q^k} = \mathbf{r}^i\mathbf{r}^j\nabla_k a_{ij} \qquad \nabla_k a_{ij} = \frac{\partial a_{ij}}{\partial q^k} - \Gamma^s_{ki}a_{sj} - a_{is}\Gamma^s_{kj}$$

$$\frac{\partial \mathbf{A}}{\partial q^k} = \mathbf{r}^i\mathbf{r}_j\nabla_k a_i^{\ j} \qquad \nabla_k a_i^{\ j} = \frac{\partial a_i^{\ j}}{\partial q^k} - \Gamma^s_{ki}a_s^{\ j} + a_i^{\ s}\Gamma^j_{ks}$$

$$\frac{\partial \mathbf{A}}{\partial q^k} = \mathbf{r}_i\mathbf{r}^j\nabla_k a^i_{\ j} \qquad \nabla_k a^i_{\ j} = \frac{\partial a^i_{\ j}}{\partial q^k} + \Gamma^i_{ks}a^s_{\ j} - a^i_{\ s}\Gamma^s_{kj}$$

Nabla operator

$$\nabla = \mathbf{r}^i\frac{\partial}{\partial q^i}$$

Gradient of vector field

$$\nabla\mathbf{f} = \mathbf{r}^i\frac{\partial \mathbf{f}}{\partial q^i}$$

$$\nabla\mathbf{f} = \mathbf{i}_k\mathbf{i}_m\frac{\partial f_m}{\partial x_k} \qquad (\nabla\mathbf{f}) = \begin{pmatrix} \frac{\partial f_1}{\partial x_1} & \cdots & \frac{\partial f_n}{\partial x_1} \\ \vdots & \ddots & \vdots \\ \frac{\partial f_1}{\partial x_n} & \cdots & \frac{\partial f_n}{\partial x_n} \end{pmatrix}$$

First differential of vector field

$$d\mathbf{f} = d\mathbf{r}\cdot\nabla\mathbf{f} = (\nabla\mathbf{f})^T\cdot d\mathbf{r}$$

Divergence of vector field

$$\operatorname{div}\mathbf{f} = \nabla\cdot\mathbf{f} = \operatorname{tr}(\nabla\mathbf{f}) = \mathbf{r}^i\cdot\frac{\partial \mathbf{f}}{\partial q^i} = \frac{1}{\sqrt{g}}\frac{\partial}{\partial q^i}(\sqrt{g}f^i)$$

Rotation (curl) of vector field

$$\operatorname{rot}\mathbf{f} = \nabla \times \mathbf{f} = \mathbf{r}^i \times \frac{\partial \mathbf{f}}{\partial q^i} = \mathbf{r}_k \, \epsilon^{ijk} \, \frac{\partial f_j}{\partial q^i}$$

Divergence of second-order tensor field

$$\nabla \cdot \mathbf{A} = \mathbf{r}^i \cdot \frac{\partial \mathbf{A}}{\partial q^i} = \frac{1}{\sqrt{g}} \frac{\partial}{\partial q^i} (\sqrt{g}\, a^{ij} \, \mathbf{r}_j)$$

Rotation of second-order tensor field

$$\nabla \times \mathbf{A} = \mathbf{r}^i \times \frac{\partial \mathbf{A}}{\partial q^i} = \epsilon^{kin} \, \mathbf{r}_n \, \frac{\partial}{\partial q^k} (\mathbf{r}^j \, a_{ij})$$

Identities for divergence and curl of dyad

$$\nabla \cdot (\mathbf{u}\mathbf{v}) = (\nabla \cdot \mathbf{u})\mathbf{v} + \mathbf{u} \cdot (\nabla \mathbf{v})$$

$$\nabla \times (\mathbf{u}\mathbf{v}) = (\nabla \times \mathbf{u})\mathbf{v} - \mathbf{u} \times (\nabla \mathbf{v})$$

Second-order derivatives of vector fields

$$\nabla^j = g^{ij} \, \nabla_i$$

$$\nabla^2 f = \nabla \cdot \nabla f = g^{ij} \left(\frac{\partial^2 f}{\partial q^i \partial q^j} - \Gamma^k_{ij} \, \frac{\partial f}{\partial q^k} \right) = \nabla^j \, \nabla_j \, f$$

$$\nabla^2 \mathbf{f} = \mathbf{r}_j \, \nabla^i \, \nabla_i \, f^j$$

$$\nabla \nabla \cdot \mathbf{f} = \mathbf{r}^i \, \nabla_i \, \nabla_j \, f^j$$

Examples

$$\nabla \mathbf{r} = \mathbf{E} \qquad \nabla \cdot \mathbf{r} = 3 \qquad \nabla \times \mathbf{r} = \mathbf{0}$$

$$\nabla \mathbf{E} = \mathbf{0} \qquad \nabla \boldsymbol{\mathcal{E}} = \mathbf{0}$$

Orthogonal coordinate systems

Lamé coefficients

$$(H_i)^2 = g_{ii}$$
$$\mathbf{r}^i = \mathbf{r}_i/(H_i)^2 \qquad (i = 1, 2, 3)$$
$$\hat{\mathbf{r}}_i = \mathbf{r}_i/H_i$$

Differentiation in the orthogonal basis

$$\nabla = \frac{\hat{\mathbf{r}}_i}{H_i}\frac{\partial}{\partial q^i}$$

$$\nabla \mathbf{f} = \hat{\mathbf{r}}_i \hat{\mathbf{r}}_j \left(\frac{1}{H_i}\frac{\partial f_j}{\partial q^i} - \frac{f_i}{H_i H_j}\frac{\partial H_i}{\partial q^j} + \delta_{ij}\frac{f_k}{H_k}\frac{1}{H_i}\frac{\partial H_i}{\partial q^k} \right)$$

$$\nabla \cdot \mathbf{f} = \frac{1}{H_1 H_2 H_3}\left(\frac{\partial}{\partial q^1}(H_2 H_3 f_1) + \frac{\partial}{\partial q^2}(H_3 H_1 f_2) + \frac{\partial}{\partial q^3}(H_1 H_2 f_3) \right)$$
$$\nabla \times \mathbf{f} = \frac{1}{2}\frac{\hat{\mathbf{r}}_i \times \hat{\mathbf{r}}_j}{H_i H_j}\left(\frac{\partial}{\partial q^i}(H_j f_j) - \frac{\partial}{\partial q^j}(H_i f_i) \right)$$

$$\nabla^2 f = \frac{1}{H_1 H_2 H_3}\left[\frac{\partial}{\partial q^1}\left(\frac{H_2 H_3}{H_1}\frac{\partial f}{\partial q^1} \right) + \frac{\partial}{\partial q^2}\left(\frac{H_3 H_1}{H_2}\frac{\partial f}{\partial q^2} \right) + \frac{\partial}{\partial q^3}\left(\frac{H_1 H_2}{H_3}\frac{\partial f}{\partial q^3} \right) \right]$$

Integration formulas

$$\int_V f(x_1, x_2, x_3)\, dx_1\, dx_2\, dx_3 = \int_V f(q^1, q^2, q^3) J\, dq^1\, dq^2\, dq^3$$

$$J = \sqrt{g} = \left| \frac{\partial x_i}{\partial q^j} \right|$$

$$\int_V \frac{\partial f}{\partial x_k} g\, dV = -\int_V \frac{\partial g}{\partial x_k} f\, dV + \int_S f g\, n_k\, dS$$

$$\int_V \nabla f \, dV = \int_S f\mathbf{n} \, dS \qquad \int_V \nabla \cdot \mathbf{f} \, dV = \int_S \mathbf{n} \cdot \mathbf{f} \, dS$$

$$\int_V \nabla \mathbf{f} \, dV = \int_S \mathbf{n}\mathbf{f} \, dS \qquad \int_V \nabla \times \mathbf{f} \, dV = \int_S \mathbf{n} \times \mathbf{f} \, dS$$

$$\int_V \nabla \mathbf{A} \, dV = \int_S \mathbf{n}\mathbf{A} \, dS$$

$$\int_V \nabla \cdot \mathbf{A} \, dV = \int_S \mathbf{n} \cdot \mathbf{A} \, dS$$

$$\int_V \nabla \times \mathbf{A} \, dV = \int_S \mathbf{n} \times \mathbf{A} \, dS$$

$$\oint_\Gamma \mathbf{f} \cdot d\mathbf{r} = \int_S (\mathbf{n} \times \nabla) \cdot \mathbf{f} \, dS$$

$$\oint_\Gamma d\mathbf{r} \cdot \mathbf{A} = \int_S (\mathbf{n} \times \nabla) \cdot \mathbf{A} \, dS$$

$$\oint_\Gamma \mathbf{A} \cdot d\mathbf{r} = \int_S (\mathbf{n} \times \nabla) \cdot \mathbf{A}^T \, dS$$

Appendix B

Some Formulas for Particular Coordinate Systems

Rectangular Coordinate System

Coordinates, position vector

$$(q^1, q^2, q^3) = (x, y, z)$$

$$\mathbf{r} = \mathbf{i}_1\, x + \mathbf{i}_2\, y + \mathbf{i}_3\, z$$

Element of arc

$$(ds)^2 = (dx)^2 + (dy)^2 + (dz)^2$$

Length of a curve

$$s = \int_a^b \left[\left(\frac{dx}{dt}\right)^2 + \left(\frac{dy}{dt}\right)^2 + \left(\frac{dz}{dt}\right)^2 \right]^{1/2} dt$$

Differential operations

Gradient

$$\nabla f = \mathbf{i}_1\, \frac{\partial f}{\partial x} + \mathbf{i}_2\, \frac{\partial f}{\partial y} + \mathbf{i}_3\, \frac{\partial f}{\partial z}$$

Divergence

$$\nabla \cdot \mathbf{f} = \frac{\partial f_x}{\partial x} + \frac{\partial f_y}{\partial y} + \frac{\partial f_z}{\partial z}$$

Curl

$$\nabla \times \mathbf{f} = \mathbf{i}_1 \left(\frac{\partial f_z}{\partial y} - \frac{\partial f_y}{\partial z} \right) - \mathbf{i}_2 \left(\frac{\partial f_z}{\partial x} - \frac{\partial f_x}{\partial z} \right) + \mathbf{i}_3 \left(\frac{\partial f_y}{\partial x} - \frac{\partial f_x}{\partial y} \right)$$

Laplacian

$$\nabla^2 f = \frac{\partial^2 f}{\partial x^2} + \frac{\partial^2 f}{\partial y^2} + \frac{\partial^2 f}{\partial z^2}$$

Cylindrical Coordinate System

Coordinates, transformation relations, position vector

$$(q^1, q^2, q^3) = (\rho, \phi, z) \qquad \begin{cases} x = \rho \cos \phi \\ y = \rho \sin \phi \\ z = z \end{cases}$$

$$\mathbf{r}(\rho, \phi, z) = \mathbf{i}_1 \, \rho \cos \phi + \mathbf{i}_2 \, \rho \sin \phi + \mathbf{i}_3 \, z$$

Element of arc, Jacobian

$$(ds)^2 = (d\rho)^2 + \rho^2 (d\phi)^2 + (dz)^2 \qquad \left| \frac{\partial(x, y, z)}{\partial(\rho, \phi, z)} \right| = \rho$$

Length of a curve

$$s = \int_a^b \left[\left(\frac{d\rho}{dt} \right)^2 + \rho^2 \left(\frac{d\phi}{dt} \right)^2 + \left(\frac{dz}{dt} \right)^2 \right]^{1/2} dt$$

Curvilinear frame vectors

$$\mathbf{r}_1 = \frac{\partial \mathbf{r}}{\partial \rho} = \mathbf{i}_1 \, \cos \phi + \mathbf{i}_2 \, \sin \phi$$

$$\mathbf{r}_2 = \frac{\partial \mathbf{r}}{\partial \phi} = -\mathbf{i}_1 \, \rho \sin \phi + \mathbf{i}_2 \, \rho \cos \phi$$

$$\mathbf{r}_3 = \frac{\partial \mathbf{r}}{\partial z} = \mathbf{i}_3$$

$$\sqrt{g} = \rho \qquad \begin{cases} |\mathbf{r}_1| = 1 \\ |\mathbf{r}_2| = \rho \\ |\mathbf{r}_3| = 1 \end{cases} \qquad \begin{cases} \mathbf{r}^1 = \mathbf{r}_1 \\ \mathbf{r}^2 = \mathbf{r}_2/\rho^2 \\ \mathbf{r}^3 = \mathbf{r}_3 \end{cases}$$

Metric coefficients

$$(g_{ij}) = \begin{pmatrix} 1 & 0 & 0 \\ 0 & \rho^2 & 0 \\ 0 & 0 & 1 \end{pmatrix} \qquad (g^{ij}) = \begin{pmatrix} 1 & 0 & 0 \\ 0 & 1/\rho^2 & 0 \\ 0 & 0 & 1 \end{pmatrix}$$

Nonzero Christoffel symbols

$$\begin{cases} \Gamma^1_{22} = -\rho \\ \Gamma^2_{12} = \Gamma^2_{21} = 1/\rho \end{cases} \qquad \begin{cases} \Gamma_{221} = -\rho \\ \Gamma_{122} = \Gamma_{212} = \rho \end{cases}$$

Lamé coefficients, normalized base vectors

$$\begin{cases} H_1 = 1 \\ H_2 = \rho \\ H_3 = 1 \end{cases} \qquad \begin{cases} \hat{\mathbf{r}}_1 = \mathbf{i}_1 \cos\phi + \mathbf{i}_2 \sin\phi \\ \hat{\mathbf{r}}_2 = -\mathbf{i}_1 \sin\phi + \mathbf{i}_2 \cos\phi \\ \hat{\mathbf{r}}_3 = \mathbf{i}_3 \end{cases}$$

Differential operations

Gradient

$$\nabla f = \hat{\boldsymbol{\rho}} \frac{\partial f}{\partial \rho} + \hat{\boldsymbol{\phi}} \frac{1}{\rho} \frac{\partial f}{\partial \phi} + \mathbf{i}_3 \frac{\partial f}{\partial z}$$

Divergence

$$\nabla \cdot \mathbf{f} = \frac{1}{\rho} \frac{\partial}{\partial \rho} (\rho f_\rho) + \frac{1}{\rho} \frac{\partial f_\phi}{\partial \phi} + \frac{\partial f_z}{\partial z}$$

Curl

$$\nabla \times \mathbf{f} = \hat{\boldsymbol{\rho}} \left(\frac{1}{\rho} \frac{\partial f_z}{\partial \phi} - \frac{\partial f_\phi}{\partial z} \right) + \hat{\boldsymbol{\phi}} \left(\frac{\partial f_\rho}{\partial z} - \frac{\partial f_z}{\partial \rho} \right) + \mathbf{i}_3 \frac{1}{\rho} \left(\frac{\partial (\rho f_\phi)}{\partial \rho} - \frac{\partial f_\rho}{\partial \phi} \right)$$

Laplacian

$$\nabla^2 f = \frac{1}{\rho}\frac{\partial}{\partial \rho}\left(\rho \frac{\partial f}{\partial \rho}\right) + \frac{1}{\rho^2}\frac{\partial^2 f}{\partial \phi^2} + \frac{\partial^2 f}{\partial z^2}$$

Spherical Coordinate System

Coordinates, transformation relations, position vector

$$(q^1, q^2, q^3) = (r, \theta, \phi) \qquad \begin{cases} x = r\sin\theta\cos\phi \\ y = r\sin\theta\sin\phi \\ z = r\cos\theta \end{cases}$$

$$\mathbf{r}(r, \theta, \phi) = \mathbf{i}_1 r\sin\theta\cos\phi + \mathbf{i}_2 r\sin\theta\sin\phi + \mathbf{i}_3 r\cos\theta$$

Element of arc, Jacobian

$$(ds)^2 = (dr)^2 + r^2(d\theta)^2 + r^2\sin^2\theta(d\phi)^2 \qquad \left|\frac{\partial(x, y, z)}{\partial(r, \theta, \phi)}\right| = r^2\sin\theta$$

Length of a curve

$$s = \int_a^b \left[\left(\frac{dr}{dt}\right)^2 + r^2\left(\frac{d\theta}{dt}\right)^2 + r^2\sin^2\theta\left(\frac{d\phi}{dt}\right)^2\right]^{1/2} dt$$

Curvilinear frame vectors

$$\begin{aligned} \mathbf{r}_1 &= \frac{\partial \mathbf{r}}{\partial r} = \mathbf{i}_1\sin\theta\cos\phi + \mathbf{i}_2\sin\theta\sin\phi + \mathbf{i}_3\cos\theta \\ \mathbf{r}_2 &= \frac{\partial \mathbf{r}}{\partial \theta} = \mathbf{i}_1 r\cos\theta\cos\phi + \mathbf{i}_2 r\cos\theta\sin\phi - \mathbf{i}_3 r\sin\theta \\ \mathbf{r}_3 &= \frac{\partial \mathbf{r}}{\partial \phi} = -\mathbf{i}_1 r\sin\theta\sin\phi + \mathbf{i}_2 r\sin\theta\cos\phi \end{aligned}$$

$$\sqrt{g} = r^2\sin\theta \qquad \begin{cases} |\mathbf{r}_1| = 1 \\ |\mathbf{r}_2| = r \\ |\mathbf{r}_3| = r\sin\theta \end{cases} \qquad \begin{cases} \mathbf{r}^1 = \mathbf{r}_1 \\ \mathbf{r}^2 = \mathbf{r}_2/r^2 \\ \mathbf{r}^3 = \mathbf{r}_3/r^2\sin^2\theta \end{cases}$$

Metric coefficients

$$(g_{ij}) = \begin{pmatrix} 1 & 0 & 0 \\ 0 & r^2 & 0 \\ 0 & 0 & r^2 \sin^2\theta \end{pmatrix} \qquad (g^{ij}) = \begin{pmatrix} 1 & 0 & 0 \\ 0 & 1/r^2 & 0 \\ 0 & 0 & 1/r^2 \sin^2\theta \end{pmatrix}$$

Nonzero Christoffel symbols

$$\begin{cases} \Gamma^1_{22} = -r \\ \Gamma^1_{33} = -r\sin^2\theta \\ \Gamma^3_{31} = \Gamma^3_{13} = 1/r \\ \Gamma^2_{12} = \Gamma^2_{21} = 1/r \\ \Gamma^2_{33} = -\sin\theta\cos\theta \\ \Gamma^3_{23} = \Gamma^3_{32} = \cot\theta \end{cases} \qquad \begin{cases} \Gamma_{221} = -r \\ \Gamma_{331} = -r\sin^2\theta \\ \Gamma_{313} = \Gamma_{133} = r\sin^2\theta \\ \Gamma_{122} = \Gamma_{212} = r \\ \Gamma_{332} = -r^2\sin\theta\cos\theta \\ \Gamma_{233} = \Gamma_{323} = r^2\sin\theta\cos\theta \end{cases}$$

Lamé coefficients, normalized base vectors

$$\begin{cases} H_1 = 1 \\ H_2 = r \\ H_3 = r\sin\theta \end{cases} \qquad \begin{cases} \hat{\mathbf{r}}_1 = \mathbf{i}_1 \sin\theta\cos\phi + \mathbf{i}_2 \sin\theta\sin\phi + \mathbf{i}_3\cos\theta \\ \hat{\mathbf{r}}_2 = \mathbf{i}_1 \cos\theta\cos\phi + \mathbf{i}_2 \cos\theta\sin\phi - \mathbf{i}_3\sin\theta \\ \hat{\mathbf{r}}_3 = -\mathbf{i}_1 \sin\phi + \mathbf{i}_2\cos\phi \end{cases}$$

Differential operations

Gradient

$$\nabla f = \hat{\mathbf{r}}\,\frac{\partial f}{\partial r} + \hat{\boldsymbol{\theta}}\,\frac{1}{r}\frac{\partial f}{\partial \theta} + \hat{\boldsymbol{\phi}}\,\frac{1}{r\sin\theta}\frac{\partial f}{\partial \phi}$$

Divergence

$$\nabla\cdot\mathbf{f} = \frac{1}{r^2}\frac{\partial}{\partial r}\left(r^2 f_r\right) + \frac{1}{r\sin\theta}\frac{\partial}{\partial\theta}\left(\sin\theta f_\theta\right) + \frac{1}{r\sin\theta}\frac{\partial f_\phi}{\partial\phi}$$

Curl

$$\begin{aligned}\nabla \times \mathbf{f} = \hat{\mathbf{r}}\, \frac{1}{r\sin\theta}\left(\frac{\partial}{\partial\theta}(\sin\theta f_\phi) - \frac{\partial f_\theta}{\partial\phi}\right)\\ + \hat{\boldsymbol{\theta}}\, \frac{1}{r\sin\theta}\left(\frac{\partial f_r}{\partial\phi} - \sin\theta\frac{\partial(rf_\phi)}{\partial r}\right)\\ + \hat{\boldsymbol{\phi}}\, \frac{1}{r}\left(\frac{\partial(rf_\theta)}{\partial r} - \frac{\partial f_r}{\partial\theta}\right).\end{aligned}$$

Laplacian

$$\nabla^2 f = \frac{1}{r^2}\frac{\partial}{\partial r}\left(r^2\frac{\partial f}{\partial r}\right) + \frac{1}{r^2\sin\theta}\frac{\partial}{\partial\theta}\left(\sin\theta\frac{\partial f}{\partial\theta}\right) + \frac{1}{r^2\sin^2\theta}\frac{\partial^2 f}{\partial\phi^2}$$

Appendix C

Main Equations of Linear Elasticity

Equations of Motion and Strain

Cartesian system

Equations of motion

$$\frac{\partial \sigma_{ij}}{\partial x_i} + \rho f_j = \rho \frac{\partial^2 u_j}{\partial t^2}$$

$$\frac{\partial \sigma_{11}}{\partial x_1} + \frac{\partial \sigma_{21}}{\partial x_2} + \frac{\partial \sigma_{31}}{\partial x_3} + \rho f_1 = \rho \frac{\partial^2 u_1}{\partial t^2}$$

$$\frac{\partial \sigma_{12}}{\partial x_1} + \frac{\partial \sigma_{22}}{\partial x_2} + \frac{\partial \sigma_{32}}{\partial x_3} + \rho f_2 = \rho \frac{\partial^2 u_2}{\partial t^2}$$

$$\frac{\partial \sigma_{13}}{\partial x_1} + \frac{\partial \sigma_{23}}{\partial x_2} + \frac{\partial \sigma_{33}}{\partial x_3} + \rho f_3 = \rho \frac{\partial^2 u_3}{\partial t^2}$$

Strains

$$\varepsilon_{11} = \frac{\partial u_1}{\partial x_1} \qquad \varepsilon_{12} = \frac{1}{2}\left(\frac{\partial u_1}{\partial x_2} + \frac{\partial u_2}{\partial x_1}\right)$$

$$\varepsilon_{22} = \frac{\partial u_2}{\partial x_2} \qquad \varepsilon_{13} = \frac{1}{2}\left(\frac{\partial u_1}{\partial x_3} + \frac{\partial u_3}{\partial x_1}\right)$$

$$\varepsilon_{33} = \frac{\partial u_3}{\partial x_3} \qquad \varepsilon_{23} = \frac{1}{2}\left(\frac{\partial u_2}{\partial x_3} + \frac{\partial u_3}{\partial x_2}\right)$$

General curvilinear system

Equations of motion

$$\frac{\partial}{\partial q^i}\left(\sqrt{g}\sigma^{ij}\right)+\Gamma^j_{mn}\sigma^{mn}+\rho\sqrt{g}f^j=\rho\sqrt{g}\frac{\partial^2 u^j}{\partial t^2}$$

Strains

$$\boldsymbol{\varepsilon}=\varepsilon_{st}\mathbf{r}^s\mathbf{r}^t \qquad \varepsilon_{st}=\frac{1}{2}\left(\frac{\partial u_s}{\partial q_t}+\frac{\partial u_t}{\partial q_s}\right)-\Gamma^r_{st}u_r$$

Cylindrical coordinates

Equations of motion

$$\frac{\partial\sigma_{rr}}{\partial r}+\frac{\sigma_{rr}-\sigma_{\phi\phi}}{r}+\frac{1}{r}\frac{\partial\sigma_{r\phi}}{\partial\phi}+\frac{\partial\sigma_{zr}}{\partial z}+\rho f_r=\rho\frac{\partial^2 u_r}{\partial t^2}$$

$$\frac{\partial\sigma_{r\phi}}{\partial r}+2\frac{\sigma_{r\phi}}{r}+\frac{1}{r}\frac{\partial\sigma_{\phi\phi}}{\partial\phi}+\frac{\partial\sigma_{z\phi}}{\partial z}+\rho f_\phi=\rho\frac{\partial^2 u_\phi}{\partial t^2}$$

$$\frac{\partial\sigma_{rz}}{\partial r}+\frac{\sigma_{rz}}{r}+\frac{1}{r}\frac{\partial\sigma_{z\phi}}{\partial\phi}+\frac{\partial\sigma_{zz}}{\partial z}+\rho f_z=\rho\frac{\partial^2 u_z}{\partial t^2}$$

Strains

$$\varepsilon_{rr}=\frac{\partial u_r}{\partial r} \qquad \varepsilon_{r\phi}=\frac{1}{2}\left(\frac{\partial u_\phi}{\partial r}+\frac{1}{r}\frac{\partial u_r}{\partial\phi}-\frac{u_\phi}{r}\right)$$

$$\varepsilon_{\phi\phi}=\frac{1}{r}\frac{\partial u_\phi}{\partial\phi}+\frac{u_r}{r} \qquad \varepsilon_{\phi z}=\frac{1}{2}\left(\frac{1}{r}\frac{\partial u_z}{\partial\phi}+\frac{\partial u_\phi}{\partial z}\right)$$

$$\varepsilon_{zz}=\frac{\partial u_z}{\partial z} \qquad \varepsilon_{rz}=\frac{1}{2}\left(\frac{\partial u_r}{\partial z}+\frac{\partial u_z}{\partial r}\right)$$

Equilibrium equations in displacements

$$\mu\left(\Delta u_r - \frac{u_r}{r^2} - \frac{2}{r^2}\frac{\partial u_\phi}{\partial \phi}\right)$$
$$+ (\lambda+\mu)\frac{\partial}{\partial r}\left[\frac{1}{r}\frac{\partial}{\partial r}(r u_r) + \frac{1}{r}\frac{\partial u_\phi}{\partial \phi} + \frac{\partial u_z}{\partial z}\right] + \rho f_r = 0$$

$$\mu\left(\Delta u_\phi - \frac{u_\phi}{r^2} + \frac{2}{r^2}\frac{\partial u_r}{\partial \phi}\right)$$
$$+ (\lambda+\mu)\frac{1}{r}\frac{\partial}{\partial \phi}\left[\frac{1}{r}\frac{\partial}{\partial r}(r u_r) + \frac{1}{r}\frac{\partial u_\phi}{\partial \phi} + \frac{\partial u_z}{\partial z}\right] + \rho f_\phi = 0$$

$$\mu\Delta u_z + (\lambda+\mu)\frac{\partial}{\partial r}\left[\frac{1}{r}\frac{\partial}{\partial r}(r u_r) + \frac{1}{r}\frac{\partial u_\phi}{\partial \phi} + \frac{\partial u_z}{\partial z}\right] + \rho f_z = 0$$

where

$$\Delta = \frac{\partial^2}{\partial r^2} + \frac{1}{r}\frac{\partial}{\partial r} + \frac{1}{r^2}\frac{\partial^2}{\partial \phi^2} + \frac{\partial^2}{\partial z^2}$$

Spherical coordinates

Equations of motion

$$\frac{\partial \sigma_{rr}}{\partial r} + \frac{1}{r}\frac{\partial \sigma_{r\theta}}{\partial \theta} + \frac{1}{r\sin\theta}\frac{\partial \sigma_{r\phi}}{\partial \phi}$$
$$+ \frac{1}{r}\left(2\sigma_{rr} - \sigma_{\theta\theta} - \sigma_{\phi\phi} + \sigma_{r\theta}\cot\theta\right) + \rho f_r = \rho\frac{\partial^2 u_r}{\partial t^2}$$

$$\frac{\partial \sigma_{r\phi}}{\partial r} + \frac{1}{r}\frac{\partial \sigma_{\theta\phi}}{\partial \theta} + \frac{1}{r\sin\theta}\frac{\partial \sigma_{\phi\phi}}{\partial \phi}$$
$$+ \frac{1}{r}\left(3\sigma_{r\phi} + 2\sigma_{\theta\phi}\cot\theta\right) + \rho f_\phi = \rho\frac{\partial^2 u_\phi}{\partial t^2}$$

$$\frac{\partial \sigma_{r\theta}}{\partial r} + \frac{1}{r}\frac{\partial \sigma_{\theta\theta}}{\partial \theta} + \frac{1}{r\sin\theta}\frac{\partial \sigma_{\theta\phi}}{\partial \phi}$$
$$+ \frac{1}{r}\left[(\sigma_{\theta\theta} - \sigma_{\phi\phi})\cot\theta + 3\sigma_{r\theta}\right] + \rho f_\theta = \rho\frac{\partial^2 u_\theta}{\partial t^2}$$

Strains

$$\varepsilon_{rr} = \frac{\partial u_r}{\partial r}$$

$$\varepsilon_{\theta\theta} = \frac{1}{r}\frac{\partial u_\theta}{\partial \theta} + \frac{u_r}{r}$$

$$\varepsilon_{\phi\phi} = \frac{1}{r\sin\theta}\frac{\partial u_\phi}{\partial \phi} + \frac{u_\theta}{r}\cot\theta + \frac{u_r}{r}$$

$$\varepsilon_{\theta\phi} = \frac{1}{2r}\left(\frac{\partial u_\phi}{\partial \theta} + \frac{1}{\sin\theta}\frac{\partial u_\theta}{\partial \phi} - u_\phi\cot\theta\right)$$

$$\varepsilon_{r\phi} = \frac{1}{2}\left(\frac{1}{r\sin\theta}\frac{\partial u_r}{\partial \phi} + \frac{\partial u_\phi}{\partial r} - \frac{u_\phi}{r}\right)$$

$$\varepsilon_{r\theta} = \frac{1}{2}\left(\frac{\partial u_\theta}{\partial r} + \frac{1}{r}\frac{\partial u_r}{\partial \theta} - \frac{u_\theta}{r}\right)$$

Equilibrium equations in displacements

$$\mu\left\{\Delta u_r - \frac{2}{r^2}\left[u_r + \frac{1}{\sin\theta}\frac{\partial}{\partial\theta}(u_\theta\sin\theta) + \frac{1}{\sin\theta}\frac{\partial u_\phi}{\partial\phi}\right]\right\}$$
$$+ (\lambda+\mu)\frac{\partial}{\partial r}\left[\frac{1}{r^2}\frac{\partial}{\partial r}(r^2u_r) + \frac{1}{r\sin\theta}\frac{\partial}{\partial\theta}(u_\theta\sin\theta) + \frac{1}{r\sin\theta}\frac{\partial u_\phi}{\partial\phi}\right] + \rho f_r = 0$$

$$\mu\left\{\Delta u_\theta - \frac{2}{r^2}\left[\frac{\partial u_r}{\partial\theta} - \frac{1}{2\sin^2\theta}u_\theta - \frac{\cos\theta}{\sin^2\theta}\frac{\partial u_\phi}{\partial\phi}\right]\right\}$$
$$+ \frac{\lambda+\mu}{r}\frac{\partial}{\partial\theta}\left[\frac{1}{r^2}\frac{\partial}{\partial r}(r^2u_r) + \frac{1}{r\sin\theta}\frac{\partial}{\partial\theta}(u_\theta\sin\theta) + \frac{1}{r\sin\theta}\frac{\partial u_\phi}{\partial\phi}\right] + \rho f_\theta = 0$$

$$\mu\left\{\Delta u_\phi + \frac{2}{r^2\sin\theta}\left[\frac{\partial u_r}{\partial\phi} + \cot\frac{\partial u_\theta}{\partial\phi} - \frac{u_\phi}{2\sin\theta}\right]\right\}$$
$$+ \frac{\lambda+\mu}{r\sin\theta}\frac{\partial}{\partial\phi}\left[\frac{1}{r^2}\frac{\partial}{\partial r}(r^2u_r) + \frac{1}{r\sin\theta}\frac{\partial}{\partial\theta}(u_\theta\sin\theta) + \frac{1}{r\sin\theta}\frac{\partial u_\phi}{\partial\phi}\right] + \rho f_\phi = 0$$

where

$$\Delta = \frac{1}{r^2}\frac{\partial}{\partial r}\left(r^2\frac{\partial}{\partial r}\right) + \frac{1}{r^2\sin\theta}\frac{\partial}{\partial\theta}\left(\sin\theta\frac{\partial}{\partial\theta}\right) + \frac{1}{r^2\sin\theta}\frac{\partial^2}{\partial\phi^2}$$

Conversions Between Elastic Moduli for Isotropic Body

Moduli	λ, μ	k, μ	μ, ν	E, ν	E, μ
λ	λ	$k - \frac{2}{3}\mu$	$\frac{2\mu\nu}{1-2\nu}$	$\frac{\nu E}{(1+\nu)(1-2\nu)}$	$\frac{(E-2\mu)\mu}{3\mu-E}$
$\mu = G$	μ	μ	μ	$\frac{E}{2(1+\nu)}$	μ
k	$\lambda + \frac{2}{3}\mu$	k	$\frac{2\mu(1+\nu)}{3(1-2\nu)}$	$\frac{E}{3(1-2\nu)}$	$\frac{E\mu}{3(3\mu-E)}$
E	$\frac{\mu(3\lambda+2\mu)}{\lambda+\mu}$	$\frac{9k\mu}{3k+\mu}$	$2\mu(1+\nu)$	E	E
ν	$\frac{\lambda}{2(\lambda+\mu)}$	$\frac{3k-2\mu}{6k+2\mu}$	ν	ν	$\frac{1}{2}\frac{E}{\mu} - 1$

Appendix D

Hints and Answers

Chapter 1 Exercises

Exercise 1.1 (a) Suppose $\mathbf{e}^i$ and $\mathbf{f}^i$ are two vectors such that $x^i = \mathbf{x} \cdot \mathbf{e}^i$ and $x^i = \mathbf{x} \cdot \mathbf{f}^i$ for all $\mathbf{x}$. Then $\mathbf{x} \cdot (\mathbf{e}^i - \mathbf{f}^i) = 0$ for all $\mathbf{x}$, from which it follows that $\mathbf{e}^i - \mathbf{f}^i = \mathbf{0}$. So $\mathbf{f}^i = \mathbf{e}^i$ as required.

Exercise 1.2 Suppose $a^{kj} = a^{jk}$. Then $\epsilon_{ijk} a^{jk} = \epsilon_{ikj} a^{kj} = \epsilon_{ikj} a^{jk} = -\epsilon_{ijk} a^{jk}$ which yields the conclusion.

Chapter 1 Problems

Problem 1.1 The transformation matrix is of the form

$$\begin{pmatrix} \cos\theta & \sin\theta & 0 \\ -\sin\theta & \cos\theta & 0 \\ 0 & 0 & 1 \end{pmatrix},$$

where θ is the angle of rotation.

Problem 1.2 This is a special case of *Lagrange's identity*

$$(\mathbf{a} \times \mathbf{b}) \cdot (\mathbf{c} \times \mathbf{d}) = (\mathbf{a} \cdot \mathbf{c})(\mathbf{b} \cdot \mathbf{d}) - (\mathbf{a} \cdot \mathbf{d})(\mathbf{b} \cdot \mathbf{c})$$

which can be obtained from the scalar and vector triple product identities:

$$(\mathbf{a} \times \mathbf{b}) \cdot (\mathbf{c} \times \mathbf{d}) = \mathbf{a} \cdot [\mathbf{b} \times (\mathbf{c} \times \mathbf{d})] = \mathbf{a} \cdot [(\mathbf{b} \cdot \mathbf{d})\mathbf{c} - (\mathbf{b} \cdot \mathbf{c})\mathbf{d}].$$

Problem 1.3

$$\begin{aligned}\epsilon^{ijk} a_i b_j c_k &= \epsilon^{123} a_1 b_2 c_3 + \epsilon^{231} a_2 b_3 c_1 + \epsilon^{312} a_3 b_1 c_2 \\ &\quad + \epsilon^{321} a_3 b_2 c_1 + \epsilon^{213} a_2 b_1 c_3 + \epsilon^{132} a_1 b_3 c_2 \\ &= V'(a_1 b_2 c_3 + a_2 b_3 c_1 + a_3 b_1 c_2 - a_3 b_2 c_1 - a_2 b_1 c_3 - a_1 b_3 c_2) \\ &= V' \begin{vmatrix} a_1 & a_2 & a_3 \\ b_1 & b_2 & b_3 \\ c_1 & c_2 & c_3 \end{vmatrix}.\end{aligned}$$

Problem 1.4 Use the vector and scalar triple product identities to write

$$\begin{aligned}\mathbf{e}^2 \times \mathbf{e}^3 &= V^{-2}(\mathbf{e}_3 \times \mathbf{e}_1) \times (\mathbf{e}_1 \times \mathbf{e}_2) \\ &= V^{-2}\{[\mathbf{e}_2 \cdot (\mathbf{e}_3 \times \mathbf{e}_1)]\mathbf{e}_1 - [\mathbf{e}_1 \cdot (\mathbf{e}_3 \times \mathbf{e}_1)]\mathbf{e}_2\} \\ &= V^{-2}\,\mathbf{e}_1(\mathbf{e}_2 \cdot \mathbf{e}_3 \times \mathbf{e}_1) = V^{-2}\,\mathbf{e}_1(\mathbf{e}_1 \cdot \mathbf{e}_2 \times \mathbf{e}_3)\end{aligned}$$

hence $V' = \mathbf{e}^1 \cdot \mathbf{e}^2 \times \mathbf{e}^3 = V^{-2}\,\mathbf{e}^1 \cdot \mathbf{e}_1(\mathbf{e}_1 \cdot \mathbf{e}_2 \times \mathbf{e}_3) = V^{-2} V = V^{-1}$.

Problem 1.5 Multiply the identity $\mathbf{e}_i \times \mathbf{e}_j = \epsilon_{sij}\mathbf{e}^s$ by ϵ^{pij} and use (1.26) to get $\epsilon^{pij}(\mathbf{e}_i \times \mathbf{e}_j) = \epsilon^{pij}\epsilon_{sij}\mathbf{e}^s = 2\delta^p_s\mathbf{e}^s = 2\mathbf{e}^p$. Similarly, multiply $\mathbf{e}^i \times \mathbf{e}^j = \epsilon^{sij}\mathbf{e}_s$ by ϵ_{pij} to get $\epsilon_{pij}(\mathbf{e}^i \times \mathbf{e}^j) = \epsilon_{pij}\epsilon^{sij}\mathbf{e}_s = 2\delta^s_p\mathbf{e}_s = 2\mathbf{e}_p$.

Problem 1.6 Use a triple product identity to expand the term $(\mathbf{n}\times\mathbf{x})\times\mathbf{n}$.

Problem 1.7 Sum the three identities

$$\begin{aligned}\mathbf{a} \times (\mathbf{b} \times \mathbf{c}) &= \mathbf{b}(\mathbf{a} \cdot \mathbf{c}) - \mathbf{c}(\mathbf{a} \cdot \mathbf{b}), \\ \mathbf{b} \times (\mathbf{c} \times \mathbf{a}) &= \mathbf{c}(\mathbf{b} \cdot \mathbf{a}) - \mathbf{a}(\mathbf{b} \cdot \mathbf{c}), \\ \mathbf{c} \times (\mathbf{a} \times \mathbf{b}) &= \mathbf{a}(\mathbf{c} \cdot \mathbf{b}) - \mathbf{b}(\mathbf{c} \cdot \mathbf{a}).\end{aligned}$$

Problem 1.8 Denoting $\mathbf{c} \times \mathbf{d} = \mathbf{e}$ we obtain

$$\begin{aligned}(\mathbf{a} \times \mathbf{b}) \cdot (\mathbf{c} \times \mathbf{d}) &= (\mathbf{a} \times \mathbf{b}) \cdot \mathbf{e} \\ &= \mathbf{a} \cdot (\mathbf{b} \times \mathbf{e}) \\ &= \mathbf{a} \cdot [\mathbf{b} \times (\mathbf{c} \times \mathbf{d})] \\ &= \mathbf{a} \cdot [\mathbf{c}(\mathbf{b} \cdot \mathbf{d}) - \mathbf{d}(\mathbf{b} \cdot \mathbf{c})] \\ &= (\mathbf{a} \cdot \mathbf{c})(\mathbf{b} \cdot \mathbf{d}) - (\mathbf{a} \cdot \mathbf{d})(\mathbf{b} \cdot \mathbf{c}),\end{aligned}$$

which is the result.

Problem 1.9

$$\begin{aligned}
\mathbf{e}_k \times \mathbf{e}^k &= \mathbf{e}_k \times \tfrac{1}{2}\epsilon^{kij}(\mathbf{e}_i \times \mathbf{e}_j) \\
&= \tfrac{1}{2}\epsilon^{kij}[\mathbf{e}_i(\mathbf{e}_k \cdot \mathbf{e}_j) - \mathbf{e}_j(\mathbf{e}_k \cdot \mathbf{e}_i)] \\
&= \tfrac{1}{2}\epsilon^{kij}(g_{kj}\mathbf{e}_i - g_{ki}\mathbf{e}_j) = \tfrac{1}{2}\epsilon^{kij}g_{kj}\mathbf{e}_i - \tfrac{1}{2}\epsilon^{kij}g_{ki}\mathbf{e}_j \\
&= \tfrac{1}{2}\epsilon^{kij}g_{kj}\mathbf{e}_i - \tfrac{1}{2}\epsilon^{kji}g_{kj}\mathbf{e}_i = \tfrac{1}{2}(\epsilon^{kij} - \epsilon^{kji})g_{kj}\mathbf{e}_i \\
&= \tfrac{1}{2}(\epsilon^{kij} + \epsilon^{kij})g_{jk}\mathbf{e}_i = \epsilon^{kij}g_{kj}\mathbf{e}_i = \epsilon^{ijk}g_{kj}\mathbf{e}_i \\
&= \mathbf{0}.
\end{aligned}$$

Problem 1.10 Use the result of Problem 1.8.

Chapter 2 Exercises

Exercise 2.1 $\mathbf{ab} \cdot\cdot (\mathbf{cd})^T = \mathbf{ab} \cdot\cdot \mathbf{dc} = (\mathbf{b} \cdot \mathbf{d})(\mathbf{a} \cdot \mathbf{c}) = (\mathbf{a} \cdot \mathbf{c})(\mathbf{b} \cdot \mathbf{d}) = \mathbf{ba} \cdot\cdot \mathbf{cd} = (\mathbf{ab})^T \cdot\cdot \mathbf{cd}$.

Exercise 2.2 $(\mathbf{y} \cdot \mathbf{ab}) \times \mathbf{x} = [(\mathbf{y} \cdot \mathbf{a})\mathbf{b}] \times \mathbf{x} = (\mathbf{y} \cdot \mathbf{a})(\mathbf{b} \times \mathbf{x}) = \mathbf{y} \cdot [\mathbf{a}(\mathbf{b} \times \mathbf{x})] = \mathbf{y} \cdot (\mathbf{ab} \times \mathbf{x})$.

Exercise 2.3 Set $\mathbf{c} \mapsto \mathbf{c} \times \mathbf{d}$ in $(\mathbf{a} \times \mathbf{b}) \times \mathbf{c} = \mathbf{b}(\mathbf{a} \cdot \mathbf{c}) - \mathbf{a}(\mathbf{b} \cdot \mathbf{c})$:

$$\begin{aligned}
(\mathbf{a} \times \mathbf{b}) \times (\mathbf{c} \times \mathbf{d}) &= \mathbf{b}(\mathbf{a} \cdot \mathbf{c} \times \mathbf{d}) - \mathbf{a}(\mathbf{b} \cdot \mathbf{c} \times \mathbf{d}) \\
&= \mathbf{ba} \cdot (\mathbf{c} \times \mathbf{d}) - \mathbf{ab} \cdot (\mathbf{c} \times \mathbf{d}).
\end{aligned}$$

The second part is similar. Subtract the first two results to obtain the third result.

Exercise 2.4 (a) $c_{in} = a_{ij}g^{jk}b_{kn}$, ${c^i}_n = a^{ij}b_{jn}$, and ${c_i}^n = a_{ij}b^{jn}$. (b) $c = {a^i}_m {b^m}_i = {a_i}^m {b_m}^i = a^{in}b_{ni} = a_{in}b^{ni}$. (c) By representing $\mathbf{A}$ in contravariant components as $\mathbf{A} = a^{ij}\mathbf{e}_i\mathbf{e}_j$ we can write $\mathbf{A} \times \mathbf{x} = a^{ij}\mathbf{e}_i(\mathbf{e}_j \times \mathbf{x})$ and $\mathbf{x} \times \mathbf{A} = a^{ij}(\mathbf{x} \times \mathbf{e}_i)\mathbf{e}_j$.

Exercise 2.5 (a) Since a second-order tensor is a linear combination of dyads, it suffices to observe that $\mathbf{ab} \cdot \mathbf{x} = \mathbf{a}(\mathbf{b} \cdot \mathbf{x}) = (\mathbf{x} \cdot \mathbf{b})\mathbf{a} = \mathbf{x} \cdot \mathbf{ba} = \mathbf{x} \cdot (\mathbf{ab})^T$. (b) Show that both sides can be written as $a^{jp}{b_p}^k\mathbf{e}_k\mathbf{e}_j$. (c) We have $\mathbf{b} \cdot \mathbf{C} \cdot \mathbf{a} = \mathbf{b} \cdot (\mathbf{C} \cdot \mathbf{a}) = \mathbf{b} \cdot (\mathbf{a} \cdot \mathbf{C}^T) = (\mathbf{a} \cdot \mathbf{C}^T) \cdot \mathbf{b} = \mathbf{a} \cdot \mathbf{C}^T \cdot \mathbf{b}$. (d) Write $\mathbf{ab} \bullet \mathbf{cd} = (\mathbf{a} \cdot \mathbf{c})(\mathbf{b} \cdot \mathbf{d}) = \mathbf{ba} \cdot\cdot \mathbf{cd} = (\mathbf{ab})^T \cdot\cdot \mathbf{cd}$ and $\mathbf{ab} \bullet \mathbf{cd} =$

$(\mathbf{a}\cdot\mathbf{c})(\mathbf{b}\cdot\mathbf{d}) = (\mathbf{b}\cdot\mathbf{d})(\mathbf{a}\cdot\mathbf{c}) = \mathbf{ab}\cdot\cdot\,\mathbf{dc} = \mathbf{ab}\cdot\cdot\,(\mathbf{cd})^T$. (e) Again we may show this via dyads (with the incidental appearance of a triad) as $(\mathbf{rs}\times\mathbf{uv})\cdot\cdot\,(\mathbf{uv})^T = \mathbf{r}(\mathbf{s}\times\mathbf{u})\mathbf{v}\cdot\cdot\,\mathbf{vu} = (\mathbf{v}\cdot\mathbf{v})\mathbf{r}(\mathbf{s}\times\mathbf{u})\cdot\mathbf{u} = \mathbf{0}$, and then use linearity to infer the result for general second-order $\mathbf{A}$ and $\mathbf{B}$.

Exercise 2.6 $(\mathbf{ab}\cdot\mathbf{uv})\cdot\cdot\,\mathbf{xy} = (\mathbf{b}\cdot\mathbf{u})\mathbf{av}\cdot\cdot\,\mathbf{xy} = (\mathbf{b}\cdot\mathbf{u})(\mathbf{v}\cdot\mathbf{x})(\mathbf{a}\cdot\mathbf{y}) = (\mathbf{v}\cdot\mathbf{x})(\mathbf{y}\cdot\mathbf{a})(\mathbf{u}\cdot\mathbf{b}) = (\mathbf{v}\cdot\mathbf{x})\mathbf{uy}\cdot\cdot\,\mathbf{ab} = (\mathbf{uv}\cdot\mathbf{xy})\cdot\cdot\,\mathbf{ab}$.

Exercise 2.7 (a)

$$T = \begin{pmatrix} -1 & 0 & 0 \\ 0 & 1 & 0 \\ 0 & 0 & 1 \end{pmatrix}.$$

(b) $\mathbf{A}\cdot\cdot\,\mathbf{E} = a^{ij}\mathbf{e}_i\mathbf{e}_j\cdot\cdot\,\mathbf{e}^k\mathbf{e}_k = a^{ij}\delta_j^k g_{ik} = a^{ik}g_{ik} = a_k{}^k = a^i{}_i$. (c) We have $\mathbf{E}\cdot\cdot\,\mathbf{E} = \mathbf{e}^i\mathbf{e}_i\cdot\cdot\,\mathbf{e}^j\mathbf{e}_j = (\mathbf{e}_i\cdot\mathbf{e}^j)(\mathbf{e}^i\cdot\mathbf{e}_j) = \delta_i^j\delta_j^i = \delta_i^i = 3$ and $\mathbf{E}\cdot\cdot\,\mathbf{vv} = \mathbf{e}_k\mathbf{e}^k\cdot\cdot\,\mathbf{vv} = (\mathbf{e}^k\cdot\mathbf{v})(\mathbf{e}_k\cdot\mathbf{v}) = v^k v_k = \mathbf{v}\cdot\mathbf{v}$. (d) One could write $\mathbf{E}\times\mathbf{v} = \mathbf{e}^k\mathbf{e}_k\times v^m\mathbf{e}_m = v^m\mathbf{e}^k(\mathbf{e}_k\times\mathbf{e}_m) = \epsilon_{skm}v^m\mathbf{e}^k\mathbf{e}^s = \epsilon_{kms}v^m\mathbf{e}^k\mathbf{e}^s = \epsilon_{smk}v^m\mathbf{e}^s\mathbf{e}^k = v^m(\mathbf{e}_m\times\mathbf{e}_k)\mathbf{e}^k = v^m\mathbf{e}_m\times\mathbf{e}_k\mathbf{e}^k = \mathbf{v}\times\mathbf{E}$.

Exercise 2.8 Write $\mathbf{A}^T\cdot(\mathbf{A}^{-1})^T = (\mathbf{A}^{-1}\cdot\mathbf{A})^T = \mathbf{E}^T = \mathbf{E}$ and $(\mathbf{A}^{-1})^T\cdot\mathbf{A}^T = (\mathbf{A}\cdot\mathbf{A}^{-1})^T = \mathbf{E}^T = \mathbf{E}$.

Exercise 2.9 (a) $|a_k{}^s| = |g_{kp}a^p{}_t g^{ts}| = |g_{kp}||a^p{}_t||g^{ts}| = g|a^p{}_t|g^{-1} = |a^p{}_t|$ and $|a_i{}^j| = |A_i^{k'}A_{l'}^j a_{k'}{}^{l'}| = |A_i^{k'}||A_{l'}^j||a_{k'}{}^{l'}| = |a_{k'}{}^{l'}|$. (b) From $\mathbf{A}\cdot\mathbf{A}^{-1} = \mathbf{E}$ we get $\det\mathbf{A}\det\mathbf{A}^{-1} = 1$, which yields the result.

Exercise 2.10 (a)

$$\mathbf{E}\cdot\cdot\,\mathbf{A} = \mathbf{e}^i\mathbf{e}_i\cdot\cdot\,a_j{}^k\mathbf{e}^j\mathbf{e}_k = a_j{}^k(\mathbf{e}_i\cdot\mathbf{e}^j)(\mathbf{e}^i\cdot\mathbf{e}_k) = a_j{}^k\delta_i^j\delta_k^i = a_j{}^k\delta_k^j = a_j{}^j,$$

$$\mathbf{E}\bullet\mathbf{A} = \mathbf{e}_i\mathbf{e}^i\bullet a_j{}^k\mathbf{e}^j\mathbf{e}_k = a_j{}^k(\mathbf{e}_i\cdot\mathbf{e}^j)(\mathbf{e}^i\cdot\mathbf{e}_k) = a_j{}^k\delta_i^j\delta_k^i = a_j{}^k\delta_k^j = a_j{}^j.$$

(b) $\operatorname{tr}\mathbf{ab} = a^k b_k = \mathbf{a}\cdot\mathbf{b}$.

Exercise 2.11 The equation $a_{ij} = g_{ki}g_{nj}a^{kn}$ expresses the a_{ij} in terms of the a^{ij}. We could, for instance, set $i = 1$ and $j = 2$ and get $a_{12} = g_{k1}g_{n2}a^{kn} = (g_{11}g_{22} - g_{21}g_{12})a^{12} + (g_{11}g_{32} - g_{31}g_{12})a^{13} + (g_{21}g_{32} - g_{31}g_{22})a^{23}$.

Exercise 2.12 (a) For a symmetric dyad $\mathbf{ab}$ and an antisymmetric dyad $\mathbf{uv}$ we have $\mathbf{ab}\cdot\cdot\,\mathbf{uv} = -\mathbf{ba}\cdot\cdot\,\mathbf{vu} = -(\mathbf{a}\cdot\mathbf{v})(\mathbf{b}\cdot\mathbf{u}) = -(\mathbf{b}\cdot\mathbf{u})(\mathbf{a}\cdot\mathbf{v}) = -\mathbf{ab}\cdot\cdot\,\mathbf{uv}$. Hence $\mathbf{ab}\cdot\cdot\,\mathbf{uv} = 0$. To see that $\mathbf{A}\bullet\mathbf{B} = 0$, use the relation $\mathbf{A}\bullet\mathbf{B} = \mathbf{A}\cdot\cdot\,\mathbf{B}^T$, the fact that $\mathbf{B}^T$ is antisymmetric, and the result of the first part. (b) and

(c) Use the result of Exercise 2.5(c). (d) From $\mathbf{E} \times \mathbf{v} = \epsilon_{skm} v^m \mathbf{e}^k \mathbf{e}^s$ we get $(\mathbf{E} \times \mathbf{v})^T = \epsilon_{skm} v^m \mathbf{e}^s \mathbf{e}^k = \epsilon_{ksm} v^m \mathbf{e}^k \mathbf{e}^s = -\epsilon_{skm} v^m \mathbf{e}^k \mathbf{e}^s = -\mathbf{E} \times \mathbf{v}$.

Exercise 2.13

$$\begin{aligned}\mathbf{x} \cdot \left[\tfrac{1}{2}(\mathbf{A} + \mathbf{A}^T)\right] \cdot \mathbf{x} &= \tfrac{1}{2}\left[\mathbf{x} \cdot \mathbf{A} \cdot \mathbf{x} + \mathbf{x} \cdot \mathbf{A}^T \cdot \mathbf{x}\right] \\ &= \tfrac{1}{2}\left[\mathbf{x} \cdot \mathbf{A} \cdot \mathbf{x} + \mathbf{x} \cdot \mathbf{A} \cdot \mathbf{x}\right] \\ &= \mathbf{x} \cdot \mathbf{A} \cdot \mathbf{x}.\end{aligned}$$

Exercise 2.14 For example,

$$\begin{aligned}\boldsymbol{\omega} \times \mathbf{x} &= -\tfrac{1}{2}\epsilon^{ksp} C_{ks} \mathbf{e}_p \times x^l \mathbf{e}_l = -\tfrac{1}{2}\epsilon^{ksp} C_{ks} x^l \epsilon_{mpl} \mathbf{e}^m \\ &= -\tfrac{1}{2}\epsilon_{lmp}\epsilon^{ksp} C_{ks} x^l \mathbf{e}^m = -\tfrac{1}{2}(\delta_l^k \delta_m^s - \delta_l^s \delta_m^k) C_{ks} x^l \mathbf{e}^m \\ &= -\tfrac{1}{2}(C_{ks} x^k \mathbf{e}^s - C_{ks} x^s \mathbf{e}^k) = -\tfrac{1}{2}(C_{ks} - C_{sk}) x^k \mathbf{e}^s \\ &= C_{sk} \mathbf{e}^s x^k = C_{sk} \mathbf{e}^s \delta_m^k x^m \\ &= C_{sk} \mathbf{e}^s \mathbf{e}^k \cdot x^m \mathbf{e}_m \\ &= \mathbf{C} \cdot \mathbf{x}.\end{aligned}$$

Exercise 2.15 Solve the linear system

$$\begin{pmatrix} \lambda_1^2 & -\lambda_1 & 1 \\ \lambda_2^2 & -\lambda_2 & 1 \\ \lambda_3^2 & -\lambda_3 & 1 \end{pmatrix} \begin{pmatrix} I_1(\mathbf{A}) \\ I_2(\mathbf{A}) \\ I_3(\mathbf{A}) \end{pmatrix} = \begin{pmatrix} \lambda_1^3 \\ \lambda_2^3 \\ \lambda_3^3 \end{pmatrix}$$

by Cramer's rule.

Exercise 2.16 We prove this for two eigenvectors; the reader may readily generalize to any number of eigenvectors. Let $\mathbf{A} \cdot \mathbf{x}_1 = \lambda_1 \mathbf{x}_1$ and $\mathbf{A} \cdot \mathbf{x}_2 = \lambda_2 \mathbf{x}_2$ where $\lambda_2 \neq \lambda_1$. Now suppose

$$\alpha_1 \mathbf{x}_1 + \alpha_2 \mathbf{x}_2 = \mathbf{0}. \tag{D.1}$$

Let us operate on both sides of (D.1) with the tensor $\mathbf{A} - \lambda_2 \mathbf{E}$:

$$\begin{gathered}(\mathbf{A} - \lambda_2 \mathbf{E}) \cdot (\alpha_1 \mathbf{x}_1 + \alpha_2 \mathbf{x}_2) = \mathbf{0}, \\ \alpha_1 \lambda_1 \mathbf{x}_1 + \alpha_2 \lambda_2 \mathbf{x}_2 - \lambda_2 \alpha_1 \mathbf{x}_1 - \lambda_2 \alpha_2 \mathbf{x}_2 = \mathbf{0}, \\ \alpha_1(\lambda_1 - \lambda_2)\mathbf{x}_1 = \mathbf{0}.\end{gathered}$$

But $\mathbf{x}_1 \neq \mathbf{0}$ by definition so $\alpha_1 = 0$. Putting this back into (D.1), we also have $\alpha_2 = 0$.

Exercise 2.17 Assume $\mathbf{x}_1, \mathbf{x}_2, \mathbf{x}_3$ are eigenvectors of $\mathbf{A}$ corresponding to the distinct eigenvalues $\lambda_1, \lambda_2, \lambda_3$, respectively. Let us show that any eigenvector corresponding to λ_1 must be a scalar multiple of $\mathbf{x}_1$. So suppose that

$$\mathbf{A} \cdot \mathbf{x} = \lambda_1 \mathbf{x}. \tag{D.2}$$

By the linear independence of $\mathbf{x}_1, \mathbf{x}_2, \mathbf{x}_3$, these vectors form a basis for three-dimensional space and we can express

$$\mathbf{x} = c_1 \mathbf{x}_1 + c_2 \mathbf{x}_2 + c_3 \mathbf{x}_3.$$

Applying $\mathbf{A}$ to both sides we have, by (D.2),

$$\lambda_1 \mathbf{x} = \lambda_1 c_1 \mathbf{x}_1 + \lambda_1 c_2 \mathbf{x}_2 + \lambda_1 c_3 \mathbf{x}_3 = c_1 \lambda_1 \mathbf{x}_1 + c_2 \lambda_2 \mathbf{x}_2 + c_3 \lambda_3 \mathbf{x}_3.$$

This simplifies to

$$(\lambda_1 - \lambda_2) c_2 \mathbf{x}_2 + (\lambda_1 - \lambda_3) c_3 \mathbf{x}_3 = \mathbf{0}.$$

By linear independence of $\mathbf{x}_2, \mathbf{x}_3$, we get $c_2 = c_3 = 0$. Hence $\mathbf{x} = c_1 \mathbf{x}_1$.

Exercise 2.18 $\mathbf{Q}^n \cdot (\mathbf{Q}^n)^T = \mathbf{Q}^n \cdot (\mathbf{Q}^T)^n = (\mathbf{Q} \cdot \mathbf{Q}^T)^n = \mathbf{E}^n = \mathbf{E}$.

Exercise 2.20 Pre-dot $\mathbf{A} \cdot \mathbf{x} = \lambda \mathbf{x}$ with $\mathbf{x}$ to get

$$\lambda = \frac{\mathbf{x} \cdot (\mathbf{A} \cdot \mathbf{x})}{|\mathbf{x}|^2}.$$

Clearly $\lambda > 0$ under the stated condition.

Exercise 2.21 If $\mathbf{r}$ locates any point on the unit sphere then

$$\mathbf{r} \cdot \mathbf{r} = 1. \tag{D.3}$$

Let $\mathbf{x}$ be the image of $\mathbf{r}$ under $\mathbf{A}$, i.e., $\mathbf{x} = \mathbf{A} \cdot \mathbf{r}$. Then $\mathbf{r} = \mathbf{A}^{-1} \cdot \mathbf{x}$, and substitution into (D.3) along with (2.39) gives

$$\begin{aligned}
1 &= [(\mathbf{S} \cdot \mathbf{Q})^{-1} \cdot \mathbf{x}] \cdot [(\mathbf{S} \cdot \mathbf{Q})^{-1} \cdot \mathbf{x}] = \mathbf{x} \cdot \{[(\mathbf{S} \cdot \mathbf{Q})^{-1}]^T \cdot (\mathbf{S} \cdot \mathbf{Q})^{-1}\} \cdot \mathbf{x} \\
&= \mathbf{x} \cdot \{[\mathbf{Q}^{-1} \cdot \mathbf{S}^{-1}]^T \cdot \mathbf{Q}^{-1} \cdot \mathbf{S}^{-1}\} \cdot \mathbf{x} \\
&= \mathbf{x} \cdot \{(\mathbf{S}^{-1})^T \cdot (\mathbf{Q}^{-1})^T \cdot \mathbf{Q}^{-1} \cdot \mathbf{S}^{-1}\} \cdot \mathbf{x} \\
&= \mathbf{x} \cdot \{(\mathbf{S}^T)^{-1} \cdot \mathbf{Q} \cdot \mathbf{Q}^{-1} \cdot \mathbf{S}^{-1}\} \cdot \mathbf{x} \\
&= \mathbf{x} \cdot \{\mathbf{S}^{-1} \cdot \mathbf{E} \cdot \mathbf{S}^{-1}\} \cdot \mathbf{x} \\
&= \mathbf{x} \cdot \{\mathbf{S}^{-1} \cdot \mathbf{S}^{-1}\} \cdot \mathbf{x}.
\end{aligned} \tag{D.4}$$

Because $\mathbf{S} = \sqrt{\lambda_1}\mathbf{i}_1\mathbf{i}_1 + \sqrt{\lambda_2}\mathbf{i}_2\mathbf{i}_2 + \sqrt{\lambda_3}\mathbf{i}_3\mathbf{i}_3$, expansion in the Cartesian frame $\mathbf{i}_k$ with use of

$$\mathbf{x} = \sum_{i=1}^{3} x_i \mathbf{i}_i, \qquad \mathbf{S}^{-1} = \sum_{j=1}^{3} \frac{1}{\sqrt{\lambda_j}} \mathbf{i}_j \mathbf{i}_j$$

reduces (D.4) to the form

$$\sum_{i=1}^{3} \frac{1}{\lambda_i} x_i^2 = 1.$$

This is the equation of an ellipsoid.

Exercise 2.24 $I_1(\operatorname{dev}\mathbf{A}) = \operatorname{tr}(\mathbf{A} - \frac{1}{3}I_1(\mathbf{A})\mathbf{E}) = \operatorname{tr}\mathbf{A} - \frac{1}{3}I_1(\mathbf{A})\operatorname{tr}\mathbf{E} = 0$. Use this to simplify $\mathbf{A} \cdot\cdot\, \mathbf{B} = [\frac{1}{3}I_1(\mathbf{A})\mathbf{E} + \operatorname{dev}\mathbf{A}] \cdot\cdot\, [\frac{1}{3}I_1(\mathbf{B})\mathbf{E} + \operatorname{dev}\mathbf{B}]$.

Exercise 2.25 The given tensor $\mathbf{U}$ is symmetric; its eigenvalues are λ_1, λ_2, and $1/(\lambda_1\lambda_2)$, the product of which is unity. The last fact follows from the identity $\det(\mathbf{A}\cdot\mathbf{B}) = \det\mathbf{A}\det\mathbf{B}$.

Exercise 2.27 Consider the group Unim. In Exercise 2.25 we noted that a dot product of two unimodular tensors is unimodular, so the closure property is satisfied. The associative property holds for the proposed group operation, which in this case is the dot product between second-order tensors. The unit tensor $\mathbf{E}$ has $\det\mathbf{E} = 1$. Finally, if $\mathbf{A}$ is unimodular then its inverse $\mathbf{A}^{-1}$ exists and is also unimodular because $|\det\mathbf{A}^{-1}| = |\det\mathbf{A}|^{-1} = 1$.

Exercise 2.28 The triclinic group consists of the two elements $\pm\mathbf{E}$ and the operation of tensor dot multiplication. Closure under the operation is seen from the relations $\mathbf{E}\cdot\mathbf{E} = \mathbf{E}$, $\mathbf{E}\cdot(-\mathbf{E}) = -\mathbf{E}$, $-\mathbf{E}\cdot\mathbf{E} = -\mathbf{E}$, and $-\mathbf{E}\cdot(-\mathbf{E}) = \mathbf{E}$. The tensor dot product operation is associative. The identity element for the group is $\mathbf{E}$. Each element in the group is its own inverse.

Exercise 2.29 (a) A square is invariant under rotations with angles $\pm\frac{\pi}{2}$, $\pm\pi$, etc., and under the mirror reflection. Denoting the unit vector normal to the square plane by $\mathbf{i}_3$, and by $\mathbf{i}_1$ and $\mathbf{i}_2$ the unit vectors parallel to the square sides, we have the symmetry group generated by $\{\mathbf{E}, -\mathbf{E}, \mathbf{R}_{\mathbf{i}_1}, \mathbf{R}(\frac{\pi}{2}\mathbf{i}_3)\}$. Note that $\mathbf{R}_{\mathbf{i}_2}$ is not listed here as it is expressed through $\mathbf{R}_{\mathbf{i}_2} = \mathbf{R}(\frac{\pi}{2}\mathbf{i}_3)\cdot\mathbf{R}_{\mathbf{i}_1}\cdot\mathbf{R}(\frac{\pi}{2}\mathbf{i}_3)$. (b) A cone is invariant under rotations about its axis, so the group contains elements $\{\mathbf{R}(\omega\mathbf{e})\}$, where $\mathbf{e}$ is the unit vector of the cone axis. It is also invariant under any reflection about the symmetry plane of the cone. So the group is generated by $\{\mathbf{R}(\omega\mathbf{e}), \mathbf{R}_{\mathbf{i}}\}$, where $\mathbf{i} \cdot \mathbf{e} = 0$. (c) A cylinder is invariant under rotations about its axis $\{\mathbf{R}(\omega\mathbf{e})\}$ and under the reflections $\{\mathbf{R}_\mathbf{e}, \mathbf{R}_\mathbf{i}\}$, where $\mathbf{i} \cdot \mathbf{e} = 0$. So the group is generated by $\{\mathbf{R}(\omega\mathbf{e}), -\mathbf{E}, \mathbf{R}_\mathbf{i}\}$. The main difference between the symmetries of a cone and circular cylinder is that a cylinder is invariant under the mirror reflection while a cone is not.

Exercise 2.30 See the lists on page 333.

Exercise 2.31

(a) $\boldsymbol{\mathcal{E}} = \epsilon_{ijk}\mathbf{e}^i\mathbf{e}^j\mathbf{e}^k = \mathbf{e}^i\epsilon_{ijk}\mathbf{e}^j\mathbf{e}^k = \mathbf{e}^i\epsilon_{jki}\mathbf{e}^j\mathbf{e}^k = \mathbf{e}^i(\mathbf{e}_k \times \mathbf{e}_i)\mathbf{e}^k = -\mathbf{e}^i(\mathbf{e}_i \times \mathbf{e}_k)\mathbf{e}^k = -\mathbf{e}^i\mathbf{e}_i \times \mathbf{e}_k\mathbf{e}^k = -\mathbf{E} \times \mathbf{E}$.

(b) $\boldsymbol{\mathcal{E}} \cdots \mathbf{zyx} = \epsilon_{ijk}\mathbf{e}^i\mathbf{e}^j\mathbf{e}^k \cdots \mathbf{zyx} = \epsilon_{ijk}(\mathbf{e}^k \cdot \mathbf{z})(\mathbf{e}^j \cdot \mathbf{y})(\mathbf{e}^i \cdot \mathbf{x}) = x^i(\epsilon_{ijk}y^jz^k) = \mathbf{x} \cdot (\mathbf{y} \times \mathbf{z})$.

(c) $\boldsymbol{\mathcal{E}} \cdot\cdot\, \mathbf{xy} = \mathbf{e}^i\epsilon_{ijk}(\mathbf{e}^k \cdot \mathbf{x})(\mathbf{e}^j \cdot \mathbf{y}) = \mathbf{e}^i\epsilon_{ijk}y^jx^k = \mathbf{y} \times \mathbf{x}$.

(d) $\boldsymbol{\mathcal{E}} \cdot \mathbf{x} = \epsilon_{ijk}\mathbf{e}^i\mathbf{e}^jx^k$ so that $(\boldsymbol{\mathcal{E}} \cdot \mathbf{x}) \cdot \mathbf{y} = \epsilon_{ijk}\mathbf{e}^i\mathbf{e}^jx^k \cdot \mathbf{y} = \mathbf{e}^i\epsilon_{ijk}y^jx^k = \mathbf{y} \times \mathbf{x} = -\mathbf{x} \times \mathbf{y}$.

Exercise 2.33 First recall the form taken by the mixed components of $\mathbf{E}$:

$$\mathbf{E} = e^i{}_j\mathbf{e}_i\mathbf{e}^j = \mathbf{e}_j\mathbf{e}^j = \delta^i_j\mathbf{e}_i\mathbf{e}^j \implies e^i{}_j = \delta^i_j.$$

Then perform a coordinate transformation:

$$e^{i'}{}_{j'} = A^{i'}_k A^l_{j'} e^k{}_l = A^{i'}_k A^l_{j'} \delta^k_l = A^{i'}_k A^k_{j'} = \delta^{i'}_{j'}.$$

Exercise 2.34 For dyads we have $\mathbf{vw} \cdot\cdot\, (\mathbf{ab})^T = \mathbf{vw} \cdot\cdot\, \mathbf{ba} = (\mathbf{w} \cdot \mathbf{b})(\mathbf{v} \cdot \mathbf{a}) = (\mathbf{b} \cdot \mathbf{w})(\mathbf{a} \cdot \mathbf{v}) = \mathbf{ab} \cdot\cdot\, \mathbf{wv} = \mathbf{ab} \cdot\cdot\, (\mathbf{vw})^T$. A fuller version for second-order tensors might look like $\mathbf{B} \cdot\cdot\, \mathbf{A}^T = b^{ij}\mathbf{e}_i\mathbf{e}_j \cdot\cdot\, a^{mk}\mathbf{e}_k\mathbf{e}_m = b^{ij}a^{mk}g_{jk}g_{im} = a^{mk}\mathbf{e}_m\mathbf{e}_k \cdot\cdot\, b^{ij}\mathbf{e}_j\mathbf{e}_i = \mathbf{A} \cdot\cdot\, \mathbf{B}^T$.

Exercise 2.35 Write $a^{km}\mathbf{e}_k\mathbf{e}_m$ and introduce $\mathbf{B} = f(\mathbf{e}_i\mathbf{e}_j)\mathbf{e}^i\mathbf{e}^j$. Then

$$\begin{aligned}
\operatorname{tr}(\mathbf{B} \cdot \mathbf{A}^T) &= \mathbf{E} \cdot\cdot\, (\mathbf{B} \cdot \mathbf{A}^T) = \mathbf{E} \cdot\cdot\, [f(\mathbf{e}_i\mathbf{e}_j)\mathbf{e}^i\mathbf{e}^j \cdot a^{mk}\mathbf{e}_k\mathbf{e}_m] \\
&= \mathbf{E} \cdot\cdot\, f(\mathbf{e}_i\mathbf{e}_j)a^{mk}\delta^j_k\mathbf{e}^i\mathbf{e}_m = \mathbf{e}^s\mathbf{e}_s \cdot\cdot\, f(\mathbf{e}_i\mathbf{e}_j)a^{mj}\mathbf{e}^i\mathbf{e}_m \\
&= f(\mathbf{e}_i\mathbf{e}_j)a^{mj}\delta^i_s\delta^s_m = f(\mathbf{e}_i\mathbf{e}_i)a^{mj}\delta^i_m = a^{ij}f(\mathbf{e}_i\mathbf{e}_j) \\
&= f(a^{ij}\mathbf{e}_i\mathbf{e}_j) \\
&= f(\mathbf{A}).
\end{aligned}$$

To establish uniqueness, assume $\operatorname{tr}(\mathbf{B}_1 \cdot \mathbf{A}^T) = \operatorname{tr}(\mathbf{B}_2 \cdot \mathbf{A}^T)$ and show that $\mathbf{B}_1 = \mathbf{B}_2$.

Exercise 2.36

$$\begin{aligned}
\mathbf{C} \cdot\cdot\, \mathbf{A}^T &= (\mathbf{B} \cdot \mathbf{e}^n \mathbf{E} \mathbf{e}_n) \cdot\cdot\, \mathbf{A}^T = (b^{ij} \mathbf{e}_i \mathbf{e}_j \cdot \mathbf{e}^n \mathbf{E} \mathbf{e}_n) \cdot\cdot\, \mathbf{A}^T \\
&= b^{ij} \mathbf{e}_i \delta_j^n \mathbf{E} \mathbf{e}_n \cdot\cdot\, \mathbf{A}^T = b^{ij} \mathbf{e}_i \mathbf{E} \mathbf{e}_j \cdot\cdot\, a^{mk} \mathbf{e}_k \mathbf{e}_m \\
&= b^{ij} a^{mk} \mathbf{e}_i \mathbf{E} g_{jk} \cdot \mathbf{e}_m = b^{ij} a^{mk} g_{jk} \mathbf{e}_i \mathbf{e}_l \mathbf{e}^l \cdot \mathbf{e}_m \\
&= b^{ij} a^{mk} g_{jk} \mathbf{e}_i \mathbf{e}_l \delta_m^l = b^{ij} a^{lk} g_{jk} \mathbf{e}_i \mathbf{e}_l = b^{ij} a^{lk} \mathbf{e}_i g_{jk} \mathbf{e}_l \\
&= b^{ij} a^{lk} \mathbf{e}_i (\mathbf{e}_j \cdot \mathbf{e}_k) \mathbf{e}_l = b^{ij} \mathbf{e}_i \mathbf{e}_j \cdot a^{lk} \mathbf{e}_k \mathbf{e}_l \\
&= \mathbf{B} \cdot \mathbf{A}^T.
\end{aligned}$$

Exercise 2.37

$$\mathbf{I} \cdot\cdot\, \mathbf{A}^T = \mathbf{e}_k \mathbf{e}_m \mathbf{e}^k \mathbf{e}^m \cdot\cdot\, a^{ji} \mathbf{e}_i \mathbf{e}_j = a^{ji} \mathbf{e}_k \mathbf{e}_m \delta_i^m \delta_j^k = a^{km} \mathbf{e}_k \mathbf{e}_m = \mathbf{A},$$

$$\mathbf{e}_k \mathbf{E} \mathbf{e}^k \cdot\cdot\, \mathbf{A}^T = \mathbf{e}_k \mathbf{e}_m \mathbf{e}^m \mathbf{e}^k \cdot\cdot\, a^{ji} \mathbf{e}_i \mathbf{e}_j = a^{ji} \mathbf{e}_k \mathbf{e}_m \delta_i^k \delta_j^m = a^{mk} \mathbf{e}_k \mathbf{e}_m = \mathbf{A}^T,$$

$$\mathbf{E}\mathbf{E} \cdot\cdot\, \mathbf{A}^T = \mathbf{E}\, \mathbf{e}_m \mathbf{e}^m \cdot\cdot\, a^{ji} \mathbf{e}_i \mathbf{e}_j = \mathbf{E}\, a^{ji} \delta_i^m g_{jm} = \mathbf{E}\, a_m{}^m = (\operatorname{tr} \mathbf{A}) \mathbf{E}.$$

Exercise 2.38 Use the representation of the invariants through the eigenvalues of $\mathbf{A}$.

Exercise 2.39 By Theorem 2.8 we have $f(\mathbf{A}) = \mathbf{B} \cdot\cdot\, \mathbf{A}^T = \lambda \mathbf{E} \cdot\cdot\, \mathbf{A}^T = \lambda \operatorname{tr}(\mathbf{A})$.

Exercise 2.40 (a) With $\mathbf{A} = \mathbf{ab}$ we have $\mathbf{A} \bullet \mathbf{A} = (\mathbf{a} \cdot \mathbf{a})(\mathbf{b} \cdot \mathbf{b})$, and clearly $\mathbf{A} \bullet \mathbf{A} = 0$ only if $\mathbf{A}$ is the zero dyad. (b) $\|\mathbf{E}\| = (\mathbf{E} \bullet \mathbf{E})^{1/2} = (\mathbf{e}_i \mathbf{e}^i \bullet \mathbf{e}^j \mathbf{e}_j)^{1/2} = ((\mathbf{e}_i \cdot \mathbf{e}^j)(\mathbf{e}^i \cdot \mathbf{e}_j))^{1/2} = (\delta_i^j \delta_j^i)^{1/2} = (\delta_i^i)^{1/2} = 3^{1/2}$.

Exercise 2.41 (a) The proof is by induction. The statement $\|\mathbf{A}^k\| \le \|\mathbf{A}\|^k$ holds trivially for $k = 1$. Suppose it holds for $k = n-1$: $\|\mathbf{A}^{n-1}\| \le \|\mathbf{A}\|^{n-1}$. Then $\|\mathbf{A}^n\| = \|\mathbf{A} \cdot \mathbf{A}^{n-1}\| \le \|\mathbf{A}\| \, \|\mathbf{A}^{n-1}\| \le \|\mathbf{A}\| \, \|\mathbf{A}\|^{n-1} = \|\mathbf{A}\|^n$ and it also holds for $k = n$. (b) Show that

(1) $\|\mathbf{A}^n\| \le q^n \to 0$ as $n \to \infty$;
(2) $(\mathbf{E} - \mathbf{A})(\mathbf{E} + \mathbf{A} + \mathbf{A}^2 + \mathbf{A}^3 + \cdots + \mathbf{A}^n) = \mathbf{E} - \mathbf{A}^{n+1}$;
(3) limit passage may be justified in the last equality.

Exercise 2.42 (a) $\|(\delta_{ij})\|_1 = \|(\delta_{ij})\|_\infty = \|(\delta_{ij})\|_2 = 1$, $\|(\delta_{ij})\|_M = \|(\delta_{ij})\|_{\ell_1} = 3$, $\|(\delta_{ij})\|_E = \sqrt{3}$. (b) $\|(\delta_{ij})\|_1 = \|(\delta_{ij})\|_\infty = \|(\delta_{ij})\|_2 = 1$, $\|(\delta_{ij})\|_M = \|(\delta_{ij})\|_{\ell_1} = n$, $\|(\delta_{ij})\|_E = \sqrt{n}$.

Exercise 2.43

$$\|\mathbf{E}\| = \max_{\|\mathbf{x}\|=1} \|\mathbf{E}\cdot\mathbf{x}\| = \max_{\|\mathbf{x}\|=1} \|\mathbf{x}\| = 1.$$

Exercise 2.44 Proof follows from the identities

$$\left\|(\mathbf{Q}\cdot\mathbf{A}\cdot\mathbf{Q}^T)\cdot\mathbf{x}\right\|^2 = \left[(\mathbf{Q}\cdot\mathbf{A}\cdot\mathbf{Q}^T)\cdot\mathbf{x}\right]\cdot\left[(\mathbf{Q}\cdot\mathbf{A}\cdot\mathbf{Q}^T)\cdot\mathbf{x}\right]$$

$$\begin{aligned}\left\|(\mathbf{Q}\cdot\mathbf{A}\cdot\mathbf{Q}^T)\cdot\mathbf{x}\right\|^2 &= \left[(\mathbf{Q}\cdot\mathbf{A}\cdot\mathbf{Q}^T)\cdot\mathbf{x}\right]\cdot\left[(\mathbf{Q}\cdot\mathbf{A}\cdot\mathbf{Q}^T)\cdot\mathbf{x}\right]\\ &= \mathbf{x}\cdot(\mathbf{Q}\cdot\mathbf{A}\cdot\mathbf{Q}^T)^T\cdot\mathbf{Q}\cdot\mathbf{A}\cdot\mathbf{Q}^T\cdot\mathbf{x}\\ &= \mathbf{y}\cdot\mathbf{A}^T\cdot\mathbf{A}\cdot\mathbf{y},\end{aligned}$$

where $\mathbf{y} = \mathbf{x}\cdot\mathbf{Q}$, and

$$\max_{\|\mathbf{x}\|=1} \|(\mathbf{Q}\cdot\mathbf{A}\cdot\mathbf{Q}^T)\cdot\mathbf{x}\| = \max_{\|\mathbf{y}\|=1} \|\mathbf{A}\cdot\mathbf{y}\| = \max_{\|\mathbf{y}\|=1} \|\mathbf{A}\cdot\mathbf{x}\|.$$

Exercise 2.45

(a) With $f(\mathbf{X}+\varepsilon\mathbf{B}) = \operatorname{tr}\mathbf{X} + \varepsilon\operatorname{tr}\mathbf{B}$, we have

$$\left.\frac{\partial}{\partial\varepsilon} f(\mathbf{X}+\varepsilon\mathbf{B})\right|_{\varepsilon=0} = \operatorname{tr}\mathbf{B} = \mathbf{E}\cdot\cdot\,\mathbf{B}^T$$

so that $f_{,\mathbf{X}} = \mathbf{E}$.

(b) With $f(\mathbf{X}+\varepsilon\mathbf{B}) = \operatorname{tr}\mathbf{X}^2 + \varepsilon\operatorname{tr}(\mathbf{X}\cdot\mathbf{B}) + \varepsilon\operatorname{tr}(\mathbf{B}\cdot\mathbf{X}) + \varepsilon^2\operatorname{tr}\mathbf{B}^2$, we have

$$\begin{aligned}\left.\frac{\partial}{\partial\varepsilon} f(\mathbf{X}+\varepsilon\mathbf{B})\right|_{\varepsilon=0} &= \operatorname{tr}(\mathbf{X}\cdot\mathbf{B}) + \operatorname{tr}(\mathbf{B}\cdot\mathbf{X}) = 2\operatorname{tr}(\mathbf{X}\cdot\mathbf{B})\\ &= 2\mathbf{X}\cdot\cdot\,\mathbf{B} = 2\mathbf{X}^T\cdot\cdot\,\mathbf{B}^T\end{aligned}$$

so that $f_{,\mathbf{X}} = 2\mathbf{X}^T$.

(c) $f_{,\mathbf{X}} = 3(\mathbf{X}^T)^2$.

(d) First differentiate $g(\mathbf{X}) = \operatorname{tr}^2\mathbf{X}$:

$$\begin{aligned}\left.\frac{\partial}{\partial\varepsilon}\left[\operatorname{tr}^2(\mathbf{X}+\varepsilon\mathbf{B})\right]\right|_{\varepsilon=0} &= \left.\frac{\partial}{\partial\varepsilon}\left[(\operatorname{tr}\mathbf{X}+\varepsilon\operatorname{tr}\mathbf{B})^2\right]\right|_{\varepsilon=0}\\ &= \left.\frac{\partial}{\partial\varepsilon}\left(\operatorname{tr}^2\mathbf{X} + 2\varepsilon\operatorname{tr}\mathbf{X}\operatorname{tr}\mathbf{B} + \epsilon^2\operatorname{tr}\mathbf{B}\right)\right|_{\varepsilon=0}\\ &= 2\operatorname{tr}\mathbf{X}\operatorname{tr}\mathbf{B} = 2\operatorname{tr}\mathbf{X}\operatorname{tr}\mathbf{B}^T = 2(\operatorname{tr}\mathbf{X})\mathbf{E}\cdot\cdot\,\mathbf{B}^T\end{aligned}$$

so that $g_{,\mathbf{X}} = 2(\operatorname{tr}\mathbf{X})\mathbf{E}$. Combine this with the result of part (b) to get $f_{,\mathbf{X}} = I_1(\mathbf{X})\mathbf{E} - \mathbf{X}^T$.

(e) Use a consequence of the Cayley–Hamilton theorem:

$$I_3 = \tfrac{1}{3}\operatorname{tr}\left(\mathbf{X}^3 - I_1\mathbf{X}^2 + I_2\mathbf{X}\right).$$

The answer is $f_{,\mathbf{X}} = [\mathbf{X}^2 - I_1(\mathbf{X})\mathbf{X} + I_2(\mathbf{X})\mathbf{E}]^T$.

Exercise 2.47

$$\begin{aligned}
\frac{d}{d\varepsilon}\mathbf{F}(\mathbf{X}+\varepsilon\mathbf{B})\bigg|_{\varepsilon=0} &= \frac{\partial f^{rs}}{\partial x^{mn}}\, b^{mn}\mathbf{e}_r\mathbf{e}_s \\
&= \frac{\partial f^{rs}}{\partial x^{mn}}\, \mathbf{e}_r\mathbf{e}_s(\mathbf{e}^m\cdot\mathbf{e}_u)(\mathbf{e}^n\cdot\mathbf{e}_v)b^{uv} \\
&= \frac{\partial f^{rs}}{\partial x^{mn}}\, \mathbf{e}_r\mathbf{e}_s\mathbf{e}^m\mathbf{e}^n \odot b^{uv}\mathbf{e}_u\mathbf{e}_v.
\end{aligned}$$

Exercise 2.49 Let us derive $f(\mathbf{X}+\varepsilon d\mathbf{X})$. We have

$$\begin{aligned}
f(\mathbf{X}+\varepsilon d\mathbf{X}) = \frac{1}{4}\mathbf{X}\cdot\cdot\,\mathbf{C}\cdot\cdot\,\mathbf{X} + \frac{\varepsilon}{2}d\mathbf{X}\cdot\cdot\,\mathbf{C}\cdot\cdot\,\mathbf{X} \\
+ \frac{\varepsilon}{2}\mathbf{X}\cdot\cdot\,\mathbf{C}\cdot\cdot\,d\mathbf{X} + \frac{\varepsilon^2}{2}d\mathbf{X}\cdot\cdot\,\mathbf{C}\cdot\cdot\,d\mathbf{X}.
\end{aligned}$$

So

$$\begin{aligned}
\frac{\partial}{\partial\varepsilon}f(\mathbf{X}+\varepsilon d\mathbf{X})\bigg|_{\varepsilon=0} &= \tfrac{1}{2}d\mathbf{X}\cdot\cdot\,\mathbf{C}\cdot\cdot\,\mathbf{X} + \tfrac{1}{2}\mathbf{X}\cdot\cdot\,\mathbf{C}\cdot\cdot\,d\mathbf{X} \\
&= \tfrac{1}{2}(dx_{mn}c_{mnpt}x_{pt} + x_{mn}c_{mnpt}dx_{pt}) \\
&= \tfrac{1}{2}(c_{mnpt}x_{pt} + x_{ij}c_{ijmn})dx_{mn}.
\end{aligned}$$

Equating the components of this to the components of $f_{,\mathbf{X}}\cdot\cdot\,d\mathbf{X}$, we get

$$\begin{aligned}
\frac{\partial f}{\partial x_{11}} &= \tfrac{1}{2}(c_{11pt}x_{pt} + x_{mn}c_{mn11}), \\
\frac{\partial f}{\partial x_{12}} &= \tfrac{1}{4}(c_{12pt}x_{pt} + c_{21pt}x_{pt} + x_{mn}c_{mn12} + x_{mn}c_{mn21}), \\
\frac{\partial f}{\partial x_{22}} &= \tfrac{1}{2}(c_{22pt}x_{pt} + x_{mn}c_{mn22}), \\
\frac{\partial f}{\partial x_{13}} &= \tfrac{1}{4}(c_{13pt}x_{pt} + c_{31pt}x_{pt} + x_{mn}c_{mn13} + x_{mn}c_{mn31}), \\
&\ \ \vdots
\end{aligned}$$

In tensor notation these formulas are

$$f_{,\mathbf{X}} = \tfrac{1}{4}(\mathbf{C} \cdot\cdot\, \mathbf{X} + \mathbf{C}'' \cdot\cdot\, \mathbf{X} + \mathbf{X} \cdot\cdot\, \mathbf{C} + \mathbf{X} \cdot\cdot\, \mathbf{C}'),$$

where $\mathbf{C}' = c_{mnpt}\mathbf{i}_m\mathbf{i}_n\mathbf{i}_t\mathbf{i}_p$ and $\mathbf{C}'' = c_{mnpt}\mathbf{i}_n\mathbf{i}_m\mathbf{i}_t\mathbf{i}_p$.

Exercise 2.50 Consider the tensor equality

$$\mathbf{C} \cdot\cdot\, \mathbf{X} = \tfrac{1}{4}(\mathbf{C} \cdot\cdot\, \mathbf{X} + \mathbf{C}'' \cdot\cdot\, \mathbf{X} + \mathbf{X} \cdot\cdot\, \mathbf{C} + \mathbf{X} \cdot\cdot\, \mathbf{C}')$$

in components.

Exercise 2.51 $dV = (\mathbf{r}_1 \cdot \mathbf{r}_2 \times \mathbf{r}_3)\, dq^1\, dq^2\, dq^3 = \sqrt{g}\, dq^1\, dq^2\, dq^3$.

Exercise 2.52

(a) The total differential of φ is

$$d\varphi = \frac{\partial \varphi}{\partial q^i}\, dq^i.$$

If we start at a point on the given surface and move in such a way that we stay on the surface, then $d\varphi = 0$. In the text it is shown that $dq^i = \mathbf{r}^i \cdot d\mathbf{r}$, so if we move along $d\mathbf{r}$ satisfying

$$\frac{\partial \varphi}{\partial q^i}\mathbf{r}^i \cdot d\mathbf{r} = 0$$

then we will stay on the surface. It is clear that the vector

$$\nabla \varphi = \frac{\partial \varphi}{\partial q^i}\mathbf{r}^i$$

is normal to the surface, since it is perpendicular to every tangent direction. We have

$$|\nabla \varphi| = \sqrt{\nabla \varphi \cdot \nabla \varphi} = \sqrt{\frac{\partial \varphi}{\partial q^m}\mathbf{r}^m \cdot \frac{\partial \varphi}{\partial q^n}\mathbf{r}^n} = \sqrt{g^{mn}\frac{\partial \varphi}{\partial q^m}\frac{\partial \varphi}{\partial q^n}},$$

hence

$$\mathbf{n} = \frac{\nabla \varphi}{|\nabla \varphi|} = \frac{\dfrac{\partial \varphi}{\partial q^i}\mathbf{r}^i}{\sqrt{g^{mn}\dfrac{\partial \varphi}{\partial q^m}\dfrac{\partial \varphi}{\partial q^n}}} = \frac{g^{ij}\dfrac{\partial \varphi}{\partial q^i}}{\sqrt{g^{mn}\dfrac{\partial \varphi}{\partial q^m}\dfrac{\partial \varphi}{\partial q^n}}}\mathbf{r}_j.$$

(b) Use the result of part (a).
(c) Use the result of part (b).
(d) Use the result of part (b).

Exercise 2.53 The results are all zero by expansion in Cartesian frame.

Exercise 2.54 By definition of the partial derivative we have $\partial \mathbf{f}/\partial q^i = \mathbf{0}$. But the components of $\mathbf{f}$ are not constant since the $\mathbf{r}_k$ are variable, hence the derivatives of the components of this function are not zero.

We could turn this around and suppose the components f^j of $\mathbf{f}$ are constant with the q^i. Then $\partial f^j/\partial q^i = 0$, but $\mathbf{f}$ can still change since the $\mathbf{r}_k$ can.

Exercise 2.55 In view of (2.78), 18 of the $3^3 = 27$ values are independent.

Exercise 2.56 From $\mathbf{r} = \mathbf{i}_1 c \cosh u \cos v + \mathbf{i}_2 c \sinh u \sin v$ we find

$$\mathbf{r}_u = \frac{\partial \mathbf{r}}{\partial u} = \mathbf{i}_1 c \sinh u \cos v + \mathbf{i}_2 c \cosh u \sin v,$$

$$\mathbf{r}_v = \frac{\partial \mathbf{r}}{\partial v} = -\mathbf{i}_1 c \cosh u \sin v + \mathbf{i}_2 c \sinh u \cos v,$$

and consequently

$$g_{uu} = g_{vv} = c^2(\cosh^2 u - \cos^2 v), \qquad g_{uv} = g_{vu} = 0$$

(the system is orthogonal). Then

$$\mathbf{r}^u = \frac{\mathbf{r}_u}{c^2(\cosh^2 u - \cos^2 v)}, \qquad \mathbf{r}^v = \frac{\mathbf{r}_v}{c^2(\cosh^2 u - \cos^2 v)},$$

from which we get

$$g^{uu} = g^{vv} = \frac{1}{c^2(\cosh^2 u - \cos^2 v)}, \qquad g^{uv} = g^{vu} = 0.$$

An application of (2.83) gives

$$\Gamma_{uuu} = c^2 \cosh u \sinh u, \qquad \Gamma_{vvu} = -c^2 \cosh u \sinh u,$$

$$\Gamma_{uuv} = -c^2 \cos v \sin v, \qquad \Gamma_{vuv} = \Gamma_{uvv} = c^2 \cosh u \sinh u,$$

$$\Gamma_{uvu} = \Gamma_{vuu} = c^2 \cos v \sin v, \qquad \Gamma_{vvv} = c^2 \cos v \sin v.$$

Finally, from (2.85) we obtain

$$\Gamma^u_{uu} = \frac{\cosh u \sinh u}{\cosh^2 u - \cos^2 v}, \qquad \Gamma^u_{vv} = -\frac{\cosh u \sinh u}{\cosh^2 u - \cos^2 v},$$

$$\Gamma^v_{uu} = -\frac{\cos v \sin v}{\cosh^2 u - \cos^2 v}, \qquad \Gamma^v_{vu} = \Gamma^v_{uv} = \frac{\cosh u \sinh u}{\cosh^2 u - \cos^2 v},$$

$$\Gamma^u_{uv} = \Gamma^u_{vu} = \frac{\cos v \sin v}{\cosh^2 u - \cos^2 v}, \qquad \Gamma^v_{vv} = \frac{\cos v \sin v}{\cosh^2 u - \cos^2 v}.$$

Exercise 2.57 We could write

$$\begin{aligned}
\tfrac{1}{2}[\nabla\mathbf{f} + (\nabla\mathbf{f})^T] &= \frac{1}{2}\left[\mathbf{r}^s\frac{\partial\mathbf{f}}{\partial q^s} + \left(\mathbf{r}^s\frac{\partial\mathbf{f}}{\partial q^s}\right)^T\right] \\
&= \frac{1}{2}\left[\mathbf{r}^s\mathbf{r}^k\nabla_s f_k + \left(\mathbf{r}^s\mathbf{r}^k\nabla_s f_k\right)^T\right] \\
&= \frac{1}{2}\mathbf{r}^s\mathbf{r}^k\left(\frac{\partial f_k}{\partial q^s} - \Gamma^r_{sk}f_r\right) + \frac{1}{2}\mathbf{r}^s\mathbf{r}^k\left(\frac{\partial f_s}{\partial q^k} - \Gamma^r_{ks}f_r\right) \\
&= \mathbf{r}^s\mathbf{r}^k\left[\frac{1}{2}\left(\frac{\partial f_k}{\partial q^s} + \frac{\partial f_s}{\partial q^k}\right) - \Gamma^r_{ks}f_r\right]
\end{aligned}$$

to obtain the covariant components of the strain tensor.

Exercise 2.58 We have

$$\begin{aligned}
\frac{\partial}{\partial q^k}(a_i^{\cdot j}\mathbf{r}^i\mathbf{r}_j) &= \frac{\partial a_i^{\cdot j}}{\partial q^k}\mathbf{r}^i\mathbf{r}_j + a_i^{\cdot j}\left(\frac{\partial\mathbf{r}^i}{\partial q^k}\mathbf{r}_j + \mathbf{r}^i\frac{\partial\mathbf{r}_j}{\partial q^k}\right) \\
&= \frac{\partial a_i^{\cdot j}}{\partial q^k}\mathbf{r}^i\mathbf{r}_j + a_i^{\cdot j}\left(-\Gamma^i_{kt}\mathbf{r}^t\mathbf{r}_j + \mathbf{r}^i\Gamma^t_{jk}\mathbf{r}_t\right) \\
&= \frac{\partial a_i^{\cdot j}}{\partial q^k}\mathbf{r}^i\mathbf{r}_j + \left(-\Gamma^i_{kt}a_i^{\cdot j}\mathbf{r}^t\mathbf{r}_j + \Gamma^t_{jk}a_i^{\cdot j}\mathbf{r}^i\mathbf{r}_t\right) \\
&= \frac{\partial a_i^{\cdot j}}{\partial q^k}\mathbf{r}^i\mathbf{r}_j + \left(-\Gamma^s_{ki}a_s^{\cdot j}\mathbf{r}^i\mathbf{r}_j + \Gamma^j_{sk}a_i^{\cdot s}\mathbf{r}^i\mathbf{r}_j\right) \\
&= \left(\frac{\partial a_i^{\cdot j}}{\partial q^k} - \Gamma^s_{ki}a_s^{\cdot j} + \Gamma^j_{sk}a_i^{\cdot s}\right)\mathbf{r}^i\mathbf{r}_j.
\end{aligned}$$

and

$$\begin{aligned}
\frac{\partial}{\partial q^k}(a^i{}_j \mathbf{r}_i \mathbf{r}^j) &= \frac{\partial a^i{}_j}{\partial q^k}\mathbf{r}_i \mathbf{r}^j + a^i{}_j \left(\frac{\partial \mathbf{r}_i}{\partial q^k}\mathbf{r}^j + \mathbf{r}_i \frac{\partial \mathbf{r}^j}{\partial q^k} \right) \\
&= \frac{\partial a^i{}_j}{\partial q^k}\mathbf{r}_i \mathbf{r}^j + a^i{}_j \left(\Gamma^t_{ik}\mathbf{r}_t \mathbf{r}^j - \mathbf{r}_i \Gamma^j_{kt}\mathbf{r}^t \right) \\
&= \frac{\partial a^i{}_j}{\partial q^k}\mathbf{r}_i \mathbf{r}^j + \left(\Gamma^t_{ik} a^i{}_j \mathbf{r}_t \mathbf{r}^j - \Gamma^j_{kt} a^i{}_j \mathbf{r}_i \mathbf{r}^t \right) \\
&= \frac{\partial a^i{}_j}{\partial q^k}\mathbf{r}_i \mathbf{r}^j + (\Gamma^i_{sk} a^s{}_j \mathbf{r}_i \mathbf{r}^j - \Gamma^s_{kj} a^i{}_s \mathbf{r}_i \mathbf{r}^j) \\
&= \left(\frac{\partial a^i{}_j}{\partial q^k} + \Gamma^i_{sk} a^s{}_j - \Gamma^s_{kj} a^i{}_s \right) \mathbf{r}_i \mathbf{r}^j.
\end{aligned}$$

Exercise 2.59 For example,

$$\nabla_k g_{ij} = \frac{\partial g_{ij}}{\partial q^k} - \Gamma^s_{ki} g_{sj} - \Gamma^s_{kj} g_{is} = \frac{\partial g_{ij}}{\partial q^k} - \Gamma_{kij} - \Gamma_{kji}$$

by (2.84). Elimination of Γ_{kij} and Γ_{kji} via (2.83) shows that $\nabla_k g_{ij} = 0$.

Exercise 2.60 Equate components of the first and third members of the trivial identity

$$\nabla(\mathbf{E} \cdot \mathbf{a}) = \nabla \mathbf{a} = \mathbf{E} \cdot \nabla \mathbf{a}.$$

Exercise 2.61

$$\nabla \times \nabla f = \epsilon^{ijk} \mathbf{r}_k \left(\frac{\partial^2 f}{\partial q^i \partial q^j} - \Gamma^n_{ij} \frac{\partial f}{\partial q^n} \right) = \mathbf{0}.$$

(Swapping two adjacent indices on ϵ^{ijk} causes a sign change; hence, whenever the symbol multiplies an expression that is symmetric in two of its subscripts, the result is zero.)

Exercise 2.62

$$\begin{aligned}
\nabla^2 \mathbf{f} &= \mathbf{r}^k \frac{\partial}{\partial q^k} \cdot \mathbf{r}^i \mathbf{r}^j \nabla_i f_j \\
&= \mathbf{r}^k \cdot \frac{\partial}{\partial q^k} \mathbf{r}^i \mathbf{r}^j \left(\frac{\partial f_j}{\partial q^i} - \Gamma^r_{ij} f_r \right) \\
&= \mathbf{r}^k \cdot \left[\frac{\partial}{\partial q^k} \left(\frac{\partial f_j}{\partial q^i} - \Gamma^r_{ij} f_r \right) \right. \\
&\quad \left. - \Gamma^s_{ki} \left(\frac{\partial f_j}{\partial q^s} - \Gamma^r_{sj} f_r \right) - \Gamma^s_{kj} \left(\frac{\partial f_s}{\partial q^i} - \Gamma^r_{is} f_r \right) \right] \mathbf{r}^i \mathbf{r}^j \\
&= g^{ki} \mathbf{r}^j \left[\frac{\partial^2 f_j}{\partial q^k \partial q^i} - \Gamma^r_{ij} \frac{\partial f_r}{\partial q^k} - f_r \frac{\partial \Gamma^r_{ij}}{\partial q^k} \right. \\
&\quad \left. - \Gamma^s_{ki} \frac{\partial f_j}{\partial q^s} + \Gamma^s_{ki} \Gamma^r_{sj} f_r - \Gamma^s_{kj} \frac{\partial f_s}{\partial q^i} + \Gamma^s_{kj} \Gamma^r_{is} f_r \right] \\
&= g^{ki} \mathbf{r}^j \left[\frac{\partial^2 f_j}{\partial q^k \partial q^i} - \Gamma^r_{ij} \frac{\partial f_r}{\partial q^k} - \Gamma^s_{ki} \frac{\partial f_j}{\partial q^s} - \Gamma^s_{kj} \frac{\partial f_s}{\partial q^i} \right. \\
&\quad \left. - f_r \left(\frac{\partial \Gamma^r_{ij}}{\partial q^k} - \Gamma^r_{sj} \Gamma^s_{ki} - \Gamma^r_{is} \Gamma^s_{kj} \right) \right].
\end{aligned}$$

Exercise 2.63 We have

$$\left| \frac{\partial(x^1, x^2, x^3)}{\partial(\tilde{q}^1, \tilde{q}^2, \tilde{q}^3)} \right| = \begin{vmatrix} \frac{\partial x^1}{\partial \tilde{q}^1} & \frac{\partial x^1}{\partial \tilde{q}^2} & \frac{\partial x^1}{\partial \tilde{q}^3} \\ \frac{\partial x^2}{\partial \tilde{q}^1} & \frac{\partial x^2}{\partial \tilde{q}^2} & \frac{\partial x^2}{\partial \tilde{q}^3} \\ \frac{\partial x^3}{\partial \tilde{q}^1} & \frac{\partial x^3}{\partial \tilde{q}^2} & \frac{\partial x^3}{\partial \tilde{q}^3} \end{vmatrix} = \begin{vmatrix} \frac{\partial x^1}{\partial q^i} \frac{\partial q^i}{\partial \tilde{q}^1} & \frac{\partial x^1}{\partial q^i} \frac{\partial q^i}{\partial \tilde{q}^2} & \frac{\partial x^1}{\partial q^i} \frac{\partial q^i}{\partial \tilde{q}^3} \\ \frac{\partial x^2}{\partial q^i} \frac{\partial q^i}{\partial \tilde{q}^1} & \frac{\partial x^2}{\partial q^i} \frac{\partial q^i}{\partial \tilde{q}^2} & \frac{\partial x^2}{\partial q^i} \frac{\partial q^i}{\partial \tilde{q}^3} \\ \frac{\partial x^3}{\partial q^i} \frac{\partial q^i}{\partial \tilde{q}^1} & \frac{\partial x^3}{\partial q^i} \frac{\partial q^i}{\partial \tilde{q}^2} & \frac{\partial x^3}{\partial q^i} \frac{\partial q^i}{\partial \tilde{q}^3} \end{vmatrix}$$

$$= \left| \begin{pmatrix} \frac{\partial x^1}{\partial q^1} & \frac{\partial x^1}{\partial q^2} & \frac{\partial x^1}{\partial q^3} \\ \frac{\partial x^2}{\partial q^1} & \frac{\partial x^2}{\partial q^2} & \frac{\partial x^2}{\partial q^3} \\ \frac{\partial x^3}{\partial q^1} & \frac{\partial x^3}{\partial q^2} & \frac{\partial x^3}{\partial q^3} \end{pmatrix} \begin{pmatrix} \frac{\partial q^1}{\partial \tilde{q}^1} & \frac{\partial q^1}{\partial \tilde{q}^2} & \frac{\partial q^1}{\partial \tilde{q}^3} \\ \frac{\partial q^2}{\partial \tilde{q}^1} & \frac{\partial q^2}{\partial \tilde{q}^2} & \frac{\partial q^2}{\partial \tilde{q}^3} \\ \frac{\partial q^3}{\partial \tilde{q}^1} & \frac{\partial q^3}{\partial \tilde{q}^2} & \frac{\partial q^3}{\partial \tilde{q}^3} \end{pmatrix} \right|.$$

The result follows from the fact that the determinant of a product equals the product of the determinants.

Exercise 2.65 (a) $f(\mathbf{S}) = f_0 + f_1 \operatorname{tr} \mathbf{S} + f_{12} \operatorname{tr}^2 \mathbf{S} + f_{22} \operatorname{tr} \mathbf{S}^2$; (b) $f(\mathbf{\Omega}) = f_0 + f_{12} \operatorname{tr}^2 \mathbf{\Omega} + f_{22} \operatorname{tr} \mathbf{\Omega}^2$. Here f_k and f_{mn} are constants which are assumed to be polar (true) scalars.

Exercise 2.66 (a) $f(\mathbf{S}) = f_0 + f_1 \operatorname{tr} \mathbf{S} + f_{12} \operatorname{tr}^2 \mathbf{S} + f_{22} \operatorname{tr} \mathbf{S}^2 + f_{13} \operatorname{tr}^3 \mathbf{S}^3 + f_{23} \operatorname{tr} \mathbf{S} \operatorname{tr} \mathbf{S}^2 + f_{33} \operatorname{tr} \mathbf{S}^3$; (b) $f(\mathbf{\Omega}) = f_0 + f_{12} \operatorname{tr}^2 \mathbf{\Omega} + f_{22} \operatorname{tr} \mathbf{\Omega}^2$. There is no cubic term.

Chapter 2 Problems

Problem 2.1 $(\mathbf{B}\cdot\mathbf{A}\cdot\mathbf{B})^T = [\mathbf{B}\cdot(\mathbf{A}\cdot\mathbf{B})]^T = (\mathbf{A}\cdot\mathbf{B})^T\cdot\mathbf{B}^T = \mathbf{B}^T\cdot\mathbf{A}^T\cdot\mathbf{B}^T = (-\mathbf{B})\cdot\mathbf{A}\cdot(-\mathbf{B}) = \mathbf{B}\cdot\mathbf{A}\cdot\mathbf{B}$.

Problem 2.2 We have $\mathbf{C}^T = (\mathbf{A}\cdot\mathbf{B} - \mathbf{B}\cdot\mathbf{A})^T = \mathbf{B}^T\cdot\mathbf{A}^T - \mathbf{A}^T\cdot\mathbf{B}^T$. If $\mathbf{A}, \mathbf{B}$ are both symmetric, then $\mathbf{C}^T = \mathbf{B}\cdot\mathbf{A} - \mathbf{A}\cdot\mathbf{B} = -\mathbf{C}^T$. The same is true if they are both antisymmetric.

Problem 2.3 (a)

$$\begin{aligned}
\operatorname{tr}(\mathbf{A}\cdot\mathbf{B}^T) &= \operatorname{tr}[a_i{}^j \mathbf{e}^i\mathbf{e}_j \cdot (b_k{}^m \mathbf{e}^k\mathbf{e}_m)^T] = \operatorname{tr}[a_i{}^j \mathbf{e}^i\mathbf{e}_j \cdot b^m{}_k \mathbf{e}^k\mathbf{e}_m] \\
&= \operatorname{tr}[a_i{}^j b^m{}_k \delta_j^k \mathbf{e}^i\mathbf{e}_m] = \operatorname{tr}[a_i{}^j b^m{}_j \mathbf{e}^i\mathbf{e}_m] = a_i{}^j b^i{}_j \\
&= a^{mj} g_{mi} b^{in} g_{jn} = a^{mj} b^{in} (\mathbf{e}_m\cdot\mathbf{e}_i)(\mathbf{e}_j\cdot\mathbf{e}_n) \\
&= a^{mj}\mathbf{e}_m\mathbf{e}_j \bullet b^{in}\mathbf{e}_i\mathbf{e}_n = \mathbf{A}\bullet\mathbf{B}.
\end{aligned}$$

(b) We have

$$\begin{aligned}
\mathbf{A}\cdot\mathbf{B} &= a_i{}^j \mathbf{e}^i\mathbf{e}_j \cdot b_k{}^l \mathbf{e}^k\mathbf{e}_l = a_i{}^j b_k{}^l \delta_j^k \mathbf{e}^i\mathbf{e}_l = a_i{}^j b_j{}^l \mathbf{e}^i\mathbf{e}_l, \\
\mathbf{B}\cdot\mathbf{A} &= b_k{}^l \mathbf{e}^k\mathbf{e}_l \cdot a_i{}^j \mathbf{e}^i\mathbf{e}_j = b_k{}^l a_i{}^j \delta_l^i \mathbf{e}^k\mathbf{e}_j = b_k{}^l a_l{}^j \mathbf{e}^k\mathbf{e}_j,
\end{aligned}$$

and in each case the trace is $a_i{}^j b_j{}^i$. (c) Compare the calculations

$$\mathbf{A}\cdot\cdot\,\mathbf{B} = a^{ij}\mathbf{e}_i\mathbf{e}_j \cdot\cdot\, b^{kl}\mathbf{e}_k\mathbf{e}_l = a^{ij}b^{kl}g_{jk}g_{il} = a^i{}_k b^k{}_i \text{ or } a_l{}^j b_j{}^l$$

and

$$\begin{aligned}
\operatorname{tr}(\mathbf{A}\cdot\mathbf{B}) &= \mathbf{E}\cdot\cdot\,(\mathbf{A}\cdot\mathbf{B}) = \mathbf{E}\cdot\cdot\,(a^{ij}\mathbf{e}_i\mathbf{e}_j\dot{b}^{kl}\mathbf{e}_k\mathbf{e}_l) = \mathbf{e}^m\mathbf{e}_m \cdot\cdot\,(a^{ij}b^{kl}g_{jk}\mathbf{e}_i\mathbf{e}_l) \\
&= a^{ij}b^{kl}g_{jk}g_{mi}\delta_l^m = a^{ij}b^{kl}g_{jk}g_{li} = a^i{}_k b^k{}_i \text{ or } a_l{}^j b_j{}^l
\end{aligned}$$

and

$$\mathbf{A}^T\cdot\cdot\,\mathbf{B}^T = a^{ji}\mathbf{e}_i\mathbf{e}_j \cdot\cdot\, b^{lk}\mathbf{e}_k\mathbf{e}_l = a^{ji}b^{lk}g_{jk}g_{il} = a_k{}^i b_i{}^k \text{ or } a^j{}_l b^l{}_j.$$

(d)

$$\begin{aligned}\operatorname{tr}(\mathbf{a}\times\mathbf{E}\times\mathbf{b}) &= \operatorname{tr}(a_k\mathbf{i}_k\times\mathbf{i}_j\mathbf{i}_j\times b_n\mathbf{i}_n) = \operatorname{tr}[a_kb_n(\mathbf{i}_k\times\mathbf{i}_j)(\mathbf{i}_j\times\mathbf{i}_n)]\\ &= \operatorname{tr}(a_kb_n\epsilon_{tkj}\mathbf{i}_t\epsilon_{sjn}\mathbf{i}_s) = a_kb_n\epsilon_{tkj}\epsilon_{sjn}\mathbf{i}_t\cdot\mathbf{i}_s\\ &= a_kb_n\epsilon_{tkj}\epsilon_{sjn}\delta_{ts} = a_kb_n\epsilon_{tkj}\epsilon_{tjn} = -a_kb_n\epsilon_{ktj}\epsilon_{ntj}\\ &= -2a_kb_n\delta_{kn} = -2a_kb_k = -2\mathbf{a}\cdot\mathbf{b}.\end{aligned}$$

(e)

$$\begin{aligned}\operatorname{tr}[\mathbf{u}\times(\mathbf{v}\times\mathbf{A})] &= \operatorname{tr}[u_j\mathbf{i}_j\times(v_k\mathbf{i}_k\times a_{mn}\mathbf{i}_m\mathbf{i}_n)]\\ &= u_jv_ka_{mn}\operatorname{tr}[\mathbf{i}_j\times(\mathbf{i}_k\times\mathbf{i}_m\mathbf{i}_m)]\\ &= u_jv_ka_{mn}\operatorname{tr}\{\mathbf{i}_j\times[(\mathbf{i}_k\times\mathbf{i}_m)\mathbf{i}_n]\}\\ &= u_jv_ka_{mn}\operatorname{tr}\{[\mathbf{i}_j\times(\mathbf{i}_k\times\mathbf{i}_m)]\mathbf{i}_n]\}\\ &= u_jv_ka_{mn}\operatorname{tr}\{[\mathbf{i}_k(\mathbf{i}_j\cdot\mathbf{i}_m)-\mathbf{i}_m(\mathbf{i}_j\cdot\mathbf{i}_k)]\mathbf{i}_n\}\\ &= u_jv_ka_{mn}\operatorname{tr}[(\mathbf{i}_k\delta_{jm}-\mathbf{i}_m\delta_{jk})\mathbf{i}_n]\\ &= u_jv_ka_{mn}(\delta_{kn}\delta_{jm}-\delta_{mn}\delta_{jk})\\ &= u_ma_{mn}v_n-u_kv_ka_{nn}\\ &= \mathbf{u}\cdot\mathbf{A}\cdot\mathbf{v}-\mathbf{u}\cdot\mathbf{v}\operatorname{tr}\mathbf{A}.\end{aligned}$$

Problem 2.4

$$\begin{aligned}\operatorname{tr}(\mathbf{A}\times\mathbf{v}) &= \mathbf{E}\cdot\cdot(\mathbf{A}\times\mathbf{v}) = \mathbf{E}\cdot\cdot(a^{ij}\mathbf{e}_i\mathbf{e}_j\times\mathbf{v}) = \mathbf{E}\cdot\cdot[a^{ij}\mathbf{e}_i(\mathbf{e}_j\times v^k\mathbf{e}_k)]\\ &= \mathbf{e}_m\mathbf{e}^m\cdot\cdot(a^{ij}v^k\epsilon_{sjk}\mathbf{e}_i\mathbf{e}^s) = \epsilon_{ijk}a^{ij}v^k = \epsilon_{jki}a^{ij}v^k\\ &= \mathbf{e}^m\mathbf{e}_m\cdot\cdot(a^{ij}v^k\epsilon_{ski}\mathbf{e}^s\mathbf{e}_j) = \mathbf{E}\cdot\cdot[a^{ij}(v^k\mathbf{e}_k\times\mathbf{e}_i)\mathbf{e}_j]\\ &= \mathbf{E}\cdot\cdot(a^{ij}(\mathbf{v}\times\mathbf{e}_i)\mathbf{e}_j) = \mathbf{E}\cdot\cdot(\mathbf{v}\times\mathbf{A}) = \operatorname{tr}(\mathbf{v}\times\mathbf{A}).\end{aligned}$$

For $\mathbf{A}$ symmetric, $\epsilon_{ijk}a^{ij}v^k$ vanishes because ϵ_{ijk} multiplies an expression a^{ij} symmetric in its indices.

Problem 2.5 First calculate the trace of the given tensor and thereby find its ball portion. Then subtract the ball portion from the tensor to find the deviator portion. Answers: (a) $\mathbf{0}$ and $\mathbf{i}_1\mathbf{i}_2$; (b) $\frac{1}{3}(\mathbf{a}\cdot\mathbf{a})\mathbf{E}$ and $\mathbf{aa}-\frac{1}{3}(\mathbf{a}\cdot\mathbf{a})\mathbf{E}$.

Problem 2.6 The first and third relations are analogous to the corresponding relations for matrices. Before proving the second relation, we note that if λ is an eigenvalue of $\mathbf{A}$, then λ^2 is an eigenvalue of $\mathbf{A}^2$:

$$\mathbf{A} \cdot \mathbf{x} = \lambda \mathbf{x} \implies \mathbf{A}^2 \cdot \mathbf{x} = \mathbf{A} \cdot \mathbf{A} \cdot \mathbf{x} = \mathbf{A} \cdot \lambda \mathbf{x} = \lambda^2 \mathbf{x}.$$

Now

$$\begin{aligned} I_2(\mathbf{A}) &= \lambda_1\lambda_2 + \lambda_2\lambda_3 + \lambda_3\lambda_1 \\ &= \tfrac{1}{2}[\lambda_1^2 + \lambda_2^2 + \lambda_3^2 + 2(\lambda_1\lambda_2 + \lambda_2\lambda_3 + \lambda_3\lambda_1) - (\lambda_1^2 + \lambda_2^2 + \lambda_3^2)] \\ &= \tfrac{1}{2}[(\lambda_1 + \lambda_2 + \lambda_3)^2 - (\lambda_1^2 + \lambda_2^2 + \lambda_3^2)] = \tfrac{1}{2}[\operatorname{tr}^2 \mathbf{A} - \operatorname{tr} \mathbf{A}^2]. \end{aligned}$$

Problem 2.7 Dot multiplying the relation

$$\mathbf{A}^{-1} = I_3^{-1}(\mathbf{A})[\mathbf{A}^2 - I_1(\mathbf{A})\mathbf{A} + I_2(\mathbf{A})\mathbf{E}]$$

with $\mathbf{A}^{-2}$ to get

$$\mathbf{A}^{-3} = I_3^{-1}(\mathbf{A})[\mathbf{E} - I_1(\mathbf{A})\mathbf{A}^{-1} + I_2(\mathbf{A})\mathbf{A}^{-2}]$$

which can be rearranged as

$$-(\mathbf{A}^{-1})^3 + \frac{I_2(\mathbf{A})}{I_3(\mathbf{A})}(\mathbf{A}^{-1})^2 - \frac{I_1(\mathbf{A})}{I_3(\mathbf{A})}\mathbf{A}^{-1} + \frac{1}{I_3(\mathbf{A})}\mathbf{E} = \mathbf{0}.$$

Now appeal to the Cayley–Hamilton theorem for the tensor $\mathbf{A}^{-1}$.

Problem 2.8 (a)

$$\begin{aligned} I_1(\mathbf{aa}) &= \mathbf{E} \cdot\cdot\, \mathbf{aa} = \mathbf{a} \cdot \mathbf{a}, \\ I_2(\mathbf{aa}) &= \tfrac{1}{2}[\operatorname{tr}^2(\mathbf{aa}) - \operatorname{tr}(\mathbf{aa} \cdot \mathbf{aa})] = \tfrac{1}{2}[\operatorname{tr}^2(\mathbf{aa}) - (\mathbf{a} \cdot \mathbf{a}) \operatorname{tr}(\mathbf{aa})] = 0, \\ I_3(\mathbf{aa}) &= \tfrac{1}{6}\epsilon_{ijk}\epsilon^{mnp}(a_m a^i)(a_n a^j)(a_p a^k) = 0. \end{aligned}$$

(b)

$$\begin{aligned} I_1(\lambda\mathbf{E}) &= \mathbf{E} \cdot\cdot\, \lambda\mathbf{E} = 3\lambda, \\ I_2(\lambda\mathbf{E}) &= \tfrac{1}{2}[\operatorname{tr}^2(\lambda\mathbf{E}) - \operatorname{tr}(\lambda\mathbf{E} \cdot \lambda\mathbf{E})] = \tfrac{1}{2}[(3\lambda)^2 - \operatorname{tr}(\lambda^2\mathbf{E})] = 3\lambda^2, \\ I_3(\lambda\mathbf{E}) &= \tfrac{1}{6}\epsilon_{ijk}\epsilon^{mnp}\lambda\delta_m^i\lambda\delta_n^j\lambda\delta_p^k = \tfrac{1}{6}\lambda^3\epsilon_{ijk}\epsilon^{ijk} = \lambda^3. \end{aligned}$$

Problem 2.9

(a) $\mathbf{S} = \mathbf{S}' = |\lambda|\mathbf{E}$ and $\mathbf{Q} = \operatorname{sgn}(\lambda)\mathbf{E}$.

(b) $\mathbf{S} = \mathbf{S}' = \mathbf{aa} + \mathbf{bb} + \mathbf{cc}$ and $\mathbf{Q} = \frac{1}{\mathbf{a}\cdot\mathbf{a}}\mathbf{aa} + \frac{1}{\mathbf{b}\cdot\mathbf{b}}\mathbf{bb} + \frac{1}{\mathbf{c}\cdot\mathbf{c}}\mathbf{cc} \equiv \mathbf{E}$;

Problem 2.10

$$\begin{aligned} 2I_2(\operatorname{dev}\mathbf{A}) &= \operatorname{tr}^2 \operatorname{dev}\mathbf{A} - \operatorname{tr}\left[\operatorname{dev}\mathbf{A}\cdot\operatorname{dev}\mathbf{A}\right] = -\operatorname{tr}\left[\operatorname{dev}\mathbf{A}\cdot(\operatorname{dev}\mathbf{A})^T\right] \\ &= -\|\operatorname{dev}\mathbf{A}\|^2 \le 0. \end{aligned}$$

Problem 2.11 (a) We have

$$(\mathbf{ab})_\times = (a_m b_n \mathbf{e}^m \mathbf{e}^n)_\times = a_m b_n \mathbf{e}^m \times \mathbf{e}^n = a_m \mathbf{e}^m \times b_n \mathbf{e}^n = \mathbf{a}\times\mathbf{b}.$$

(b) $(\mathbf{aa})_\times = \mathbf{0}$ by the result of part (a). (c) One could write

$$\begin{aligned} (\mathbf{E}\times\mathbf{v})_\times &= [(g^{ij}\mathbf{e}_i\mathbf{e}_j)\times\mathbf{v}]_\times = [g^{ij}\mathbf{e}_i(\mathbf{e}_j\times\mathbf{v})]_\times = g^{ij}\mathbf{e}_i\times(\mathbf{e}_j\times\mathbf{v}) \\ &= \mathbf{e}^j\times(\mathbf{e}_j\times\mathbf{v}) = \mathbf{e}_j(\mathbf{e}^j\cdot\mathbf{v}) - \mathbf{v}(\mathbf{e}^j\cdot\mathbf{e}_j) = \omega^j\mathbf{e}_j - 3\mathbf{v} = -2\mathbf{v}. \end{aligned}$$

(d) Write

$$\begin{aligned} (\mathbf{ab}-\mathbf{ba})_\times &= [(a_m b_n - b_m a_n)\mathbf{e}^m\mathbf{e}^n]_\times = (a_m b_n - b_m a_n)\mathbf{e}^m\times\mathbf{e}^n \\ &= a_m\mathbf{e}^m\times b_n\mathbf{e}^n - b_m\mathbf{e}^m\times a_n\mathbf{e}^n = \mathbf{a}\times\mathbf{b} - \mathbf{b}\times\mathbf{a} = 2\mathbf{a}\times\mathbf{b}. \end{aligned}$$

(e) We have

$$\mathbf{A}_\times = (a_{mn}\mathbf{e}^m\mathbf{e}^n)_\times = a_{mn}\mathbf{e}^m\times\mathbf{e}^n = a_{mn}\epsilon^{smn}\mathbf{e}_s = \mathbf{0}$$

since ϵ^{smn} multiplies an expression symmetric in its indices m and n. (f) $\mathbf{E}_\times = \mathbf{0}$ since $\mathbf{E}$ is symmetric.

Problem 2.12 (a) Decompose $\mathbf{A}$ into symmetric and antisymmetric terms. As shown in Problem 2.11, the symmetric term will have $\frac{1}{2}(\mathbf{A}+\mathbf{A}^T)_\times = \mathbf{0}$. (b) Write

$$\begin{aligned} (\mathbf{v}\times\mathbf{A})_\times &= (\mathbf{v}\times a^{ij}\mathbf{e}_i\mathbf{e}_j)_\times = [a^{ij}(\mathbf{v}\times\mathbf{e}_i)\mathbf{e}_j]_\times = [a^{ij}(v^k\mathbf{e}_k\times\mathbf{e}_i)\mathbf{e}_j]_\times \\ &= (a^{ij}v^k\epsilon_{ski}\mathbf{e}^s\mathbf{e}_j]_\times = a^{ij}v^k\epsilon_{ski}\mathbf{e}^s\times\mathbf{e}_j = a^{ij}v^k\epsilon_{ski}\mathbf{e}^s\times g_{pj}\mathbf{e}^p \\ &= a^{ij}v^k g_{pj}\epsilon_{ski}\epsilon^{tsp}\mathbf{e}_t = a^{ij}v^k g_{pj}\epsilon_{kis}\epsilon^{pts}\mathbf{e}_t \\ &= a^{ij}v^k g_{pj}(\delta_k^p\delta_i^t - \delta_k^t\delta_i^p)\mathbf{e}_t = a^{ij}\mathbf{e}_i v_j - v^k\mathbf{e}_k a^{ij}g_{ij} \\ &= a^{ij}\mathbf{e}_i\mathbf{e}_j\cdot v_k\mathbf{e}^k - v^k\mathbf{e}_k a_j{}^j \\ &= \mathbf{A}\cdot\mathbf{v} - \mathbf{v}\operatorname{tr}\mathbf{A}. \end{aligned}$$

(c) One could write

$$
\begin{aligned}
(\mathbf{v} \times \mathbf{E})_\times &= [\mathbf{v} \times \mathbf{e}^k \mathbf{e}_k]_\times = [(\mathbf{v} \times \mathbf{e}^k)\mathbf{e}_k]_\times = (\mathbf{v} \times \mathbf{e}^k) \times \mathbf{e}_k \\
&= -\mathbf{e}_k \times (\mathbf{v} \times \mathbf{e}^k) = -\mathbf{v}(\mathbf{e}_k \cdot \mathbf{e}^k) + \mathbf{e}^k(\mathbf{e}_k \cdot \mathbf{v}) \\
&= -\mathbf{v}\delta_k^k + \mathbf{e}^k v_k = -3\mathbf{v} + \mathbf{v} \\
&= -2\mathbf{v}
\end{aligned}
$$

or simply set $\mathbf{A} = \mathbf{E}$ in the result of part (b), noting that $\operatorname{tr} \mathbf{E} = 3$.

Problem 2.13 We have

$$
\begin{aligned}
(\mathbf{A} \cdot \mathbf{B})_\times + (\mathbf{B} \cdot \mathbf{A})_\times &= (a_{ij}\mathbf{e}^i\mathbf{e}^j \cdot b_{km}\mathbf{e}^k\mathbf{e}^m)_\times + (b_{ij}\mathbf{e}^i\mathbf{e}^j \cdot a_{km}\mathbf{e}^k\mathbf{e}^m)_\times \\
&= (a_{ij}b_{km}g^{jk}\mathbf{e}^i\mathbf{e}^m)_\times + (b_{ij}a_{km}g^{jk}\mathbf{e}^i\mathbf{e}^m)_\times \\
&= g^{jk}(a_{ij}b_{km} + b_{ij}a_{km})\mathbf{e}^i \times \mathbf{e}^m \\
&= g^{jk}(a_{ij}b_{km} + a_{km}b_{ij})\epsilon^{sim}\mathbf{e}_s \\
&= g^{jk}(a_{ij}b_{km} + a_{mk}b_{ji})\epsilon^{sim}\mathbf{e}_s \\
&= (a_{ij}b^j{}_m + a_{mk}b^k{}_i)\epsilon^{sim}\mathbf{e}_s \\
&= (a_{ij}b^j{}_m + a_{mj}b^j{}_i)\epsilon^{sim}\mathbf{e}_s.
\end{aligned}
$$

The parenthetical expression is symmetric in i and m.

Problem 2.14

$$
\begin{aligned}
(\mathbf{E} \times \mathbf{v}) \cdot \mathbf{A} &= (g^{ij}\mathbf{e}_i\mathbf{e}_j \times \mathbf{v}) \cdot \mathbf{A} = g^{ij}\mathbf{e}_i(\mathbf{e}_j \times \mathbf{v}) \cdot \mathbf{A} \\
&= g^{ij}\mathbf{e}_i(\mathbf{e}_j \times v^k\mathbf{e}_k) \cdot \mathbf{A} = g^{ij}v^k\mathbf{e}_i\epsilon_{sjk}\mathbf{e}^s \cdot \mathbf{A} \\
&= g^{ij}v^k\epsilon_{sjk}\mathbf{e}_i\mathbf{e}^s \cdot a^{ml}\mathbf{e}_m\mathbf{e}_l = g^{ij}v^k\epsilon_{sjk}a^{ml}\delta_m^s\mathbf{e}_i\mathbf{e}_l \\
&= g^{ij}v^k\epsilon_{sjk}a^{sl}\mathbf{e}_i\mathbf{e}_l = v^k a^{sl}\epsilon_{jks}\mathbf{e}^j\mathbf{e}_l = v^k a^{sl}(\mathbf{e}_k \times \mathbf{e}_s)\mathbf{e}_l \\
&= a^{sl}(v^k\mathbf{e}_k \times \mathbf{e}_s)\mathbf{e}_l = a^{sl}(\mathbf{v} \times \mathbf{e}_s)\mathbf{e}_l = \mathbf{v} \times a^{sl}\mathbf{e}_s\mathbf{e}_l \\
&= \mathbf{v} \times \mathbf{A}.
\end{aligned}
$$

Problem 2.15

$$\begin{aligned}
-(\mathbf{v}\times\mathbf{A}^T)^T &= -[a^{ji}(\mathbf{v}\times\mathbf{e}_i)\mathbf{e}_j]^T = -[a^{ji}(v^m\mathbf{e}_m\times\mathbf{e}_i)\mathbf{e}_j]^T\\
&= -[a^{ji}v^m\epsilon_{smi}\mathbf{e}^s\mathbf{e}_j]^T = -a^{ji}v^m\epsilon_{smi}\mathbf{e}_j\mathbf{e}^s\\
&= -a^{ij}v^m\epsilon_{smj}\mathbf{e}_i\mathbf{e}^s = a^{ij}v^m\epsilon_{sjm}\mathbf{e}_i\mathbf{e}^s = a^{ij}v^m\mathbf{e}_i\epsilon_{sjm}\mathbf{e}^s\\
&= a^{ij}v^m\mathbf{e}_i(\mathbf{e}_j\times\mathbf{e}_m) = a^{ij}\mathbf{e}_i(\mathbf{e}_j\times v^m\mathbf{e}_m) = a^{ij}\mathbf{e}_i(\mathbf{e}_j\times\mathbf{v})\\
&= \mathbf{A}\times\mathbf{v}.
\end{aligned}$$

$$\begin{aligned}
-(\mathbf{A}^T\times\mathbf{v})^T &= -[a^{ji}\mathbf{e}_i(\mathbf{e}_j\times\mathbf{v})]^T = -[a^{ji}\mathbf{e}_i(\mathbf{e}_j\times v^m\mathbf{e}_m)]^T\\
&= -[a^{ji}v^m\mathbf{e}_i\epsilon_{sjm}\mathbf{e}^s]^T = -a^{ji}v^m\epsilon_{sjm}\mathbf{e}^s\mathbf{e}_i\\
&= -a^{ij}v^m\epsilon_{sim}\mathbf{e}^s\mathbf{e}_j = a^{ij}v^m\epsilon_{smi}\mathbf{e}^s\mathbf{e}_j\\
&= a^{ij}v^m(\mathbf{e}_m\times\mathbf{e}_i)\mathbf{e}_j = a^{ij}(v^m\mathbf{e}_m\times\mathbf{e}_i)\mathbf{e}_j\\
&= a^{ij}(\mathbf{v}\times\mathbf{e}_i)\mathbf{e}_j\\
&= \mathbf{v}\times\mathbf{A}.
\end{aligned}$$

Problem 2.16 $\mathbf{Q}$ is symmetric, so

$$\begin{aligned}
\mathbf{Q}\cdot\mathbf{Q}^T &= \mathbf{Q}^T\cdot\mathbf{Q} = \mathbf{Q}\cdot\mathbf{Q} = (\mathbf{E}-2\mathbf{ee})\cdot(\mathbf{E}-2\mathbf{ee})\\
&= \mathbf{E}\cdot\mathbf{E}-2\mathbf{ee}\cdot\mathbf{E}-2\mathbf{E}\cdot\mathbf{ee}+4\mathbf{ee}\cdot\mathbf{ee} = \mathbf{E}.
\end{aligned}$$

Problem 2.17 More generally we have

$$\begin{aligned}
\mathbf{u}\times\mathbf{E}\times\mathbf{v} &= \mathbf{u}\times\mathbf{e}_k\mathbf{e}^k\times\mathbf{v} = (u^i\mathbf{e}_i\times\mathbf{e}_k)(\mathbf{e}^k\times v_n\mathbf{e}^n) = u^i\mathbf{e}^t\epsilon_{tik}v_n\mathbf{e}_s\epsilon^{skn}\\
&= u^iv_n\epsilon_{tik}\epsilon^{nsk}\mathbf{e}^t\mathbf{e}_s = u^iv_n(\delta_t^n\delta_i^s-\delta_i^n\delta_t^s)\mathbf{e}^t\mathbf{e}_s\\
&= u^iv_n\delta_t^n\delta_i^s\mathbf{e}^t\mathbf{e}_s - u^iv_n\delta_i^n\delta_t^s\mathbf{e}^t\mathbf{e}_s = v_n\mathbf{e}^nu^i\mathbf{e}_i - u^iv_i\mathbf{e}^s\mathbf{e}_s\\
&= \mathbf{vu}-(\mathbf{u}\cdot\mathbf{v})\mathbf{E}.
\end{aligned}$$

Problem 2.18 First note that the tensors $\mathbf{E}+\mathbf{\Omega}$ and $\mathbf{E}-\mathbf{\Omega}$ commute:

$$(\mathbf{E}+\mathbf{\Omega})\cdot(\mathbf{E}-\mathbf{\Omega}) = \mathbf{E}-\mathbf{\Omega}\cdot\mathbf{\Omega} = (\mathbf{E}-\mathbf{\Omega})\cdot(\mathbf{E}+\mathbf{\Omega}).$$

We have

$$\begin{aligned}\mathbf{Q}^T &= [(\mathbf{E}+\boldsymbol{\Omega})\cdot(\mathbf{E}-\boldsymbol{\Omega})^{-1}]^T = [(\mathbf{E}-\boldsymbol{\Omega})^{-1}]^T\cdot(\mathbf{E}+\boldsymbol{\Omega})^T\\ &= [(\mathbf{E}-\boldsymbol{\Omega})^T]^{-1}\cdot(\mathbf{E}+\boldsymbol{\Omega})^T = (\mathbf{E}-\boldsymbol{\Omega}^T)^{-1}\cdot(\mathbf{E}+\boldsymbol{\Omega}^T)\\ &= (\mathbf{E}+\boldsymbol{\Omega})^{-1}\cdot(\mathbf{E}-\boldsymbol{\Omega}),\end{aligned}$$

so

$$\begin{aligned}\mathbf{Q}^T\cdot\mathbf{Q} &= (\mathbf{E}+\boldsymbol{\Omega})^{-1}\cdot(\mathbf{E}-\boldsymbol{\Omega})\cdot(\mathbf{E}+\boldsymbol{\Omega})\cdot(\mathbf{E}-\boldsymbol{\Omega})^{-1}\\ &= (\mathbf{E}+\boldsymbol{\Omega})^{-1}\cdot(\mathbf{E}+\boldsymbol{\Omega})\cdot(\mathbf{E}-\boldsymbol{\Omega})\cdot(\mathbf{E}-\boldsymbol{\Omega})^{-1} = \mathbf{E}\end{aligned}$$

and

$$\begin{aligned}\mathbf{Q}\cdot\mathbf{Q}^T &= (\mathbf{E}+\boldsymbol{\Omega})\cdot(\mathbf{E}-\boldsymbol{\Omega})^{-1}\cdot(\mathbf{E}+\boldsymbol{\Omega})^{-1}\cdot(\mathbf{E}-\boldsymbol{\Omega})\\ &= (\mathbf{E}+\boldsymbol{\Omega})\cdot[(\mathbf{E}+\boldsymbol{\Omega})\cdot(\mathbf{E}-\boldsymbol{\Omega})]^{-1}\cdot(\mathbf{E}-\boldsymbol{\Omega})\\ &= (\mathbf{E}+\boldsymbol{\Omega})\cdot[(\mathbf{E}-\boldsymbol{\Omega})\cdot(\mathbf{E}+\boldsymbol{\Omega})]^{-1}\cdot(\mathbf{E}-\boldsymbol{\Omega})\\ &= (\mathbf{E}+\boldsymbol{\Omega})\cdot(\mathbf{E}+\boldsymbol{\Omega})^{-1}\cdot(\mathbf{E}-\boldsymbol{\Omega})^{-1}\cdot(\mathbf{E}-\boldsymbol{\Omega}) = \mathbf{E}.\end{aligned}$$

Problem 2.19

$$\begin{aligned}(\mathbf{Q}-\mathbf{E})\cdot\cdot(\mathbf{Q}^T-\mathbf{E}) &= \mathbf{Q}\cdot\cdot\mathbf{Q}^T - \mathbf{E}\cdot\cdot\mathbf{Q}^T - \mathbf{Q}\cdot\cdot\mathbf{E} + \mathbf{E}\cdot\cdot\mathbf{E}\\ &= q^{mn}\mathbf{e}_m\mathbf{e}_n\cdot\cdot q_{st}\mathbf{e}^t\mathbf{e}^s - \mathbf{e}^k\mathbf{e}_k\cdot\cdot q_{st}\mathbf{e}^t\mathbf{e}^s\\ &\quad - q_{st}\mathbf{e}^s\mathbf{e}^t\cdot\cdot\mathbf{e}_k\mathbf{e}^k + \mathbf{e}_k\mathbf{e}^k\cdot\cdot\mathbf{e}_p\mathbf{e}^p\\ &= q^{mn}q_{st}\delta_n^t\delta_m^s - q_{st}\delta_k^t g^{ks} - q_{st}\delta_k^t g^{sk} + \delta_p^k\delta_k^p\\ &= q^{mn}q_{mn} - q_{sk}g^{ks} - q_{sk}g^{sk} + \delta_k^k\\ &= 3 - q_s{}^s - q_s{}^s + 3 = 6 - 2\operatorname{tr}\mathbf{Q}.\end{aligned}$$

Problem 2.20 Using (2.32) and the results of Problem 2.11, we get

$$\begin{aligned}\mathbf{Q}_\times &= [\mathbf{E}\cos\omega + (1-\cos\omega)\mathbf{ee} - \mathbf{e}\times\mathbf{E}\sin\omega]_\times\\ &= \mathbf{E}_\times\cos\omega + (1-\cos\omega)(\mathbf{ee})_\times - (\mathbf{e}\times\mathbf{E})_\times\sin\omega = -(\mathbf{e}\times\mathbf{E})_\times\sin\omega.\end{aligned}$$

But

$$\begin{aligned}(\mathbf{e}\times\mathbf{E})_\times &= [\mathbf{e}\times(g^{ij}\mathbf{e}_i\mathbf{e}_j)]_\times = [g^{ij}(\mathbf{e}\times\mathbf{e}_i)\mathbf{e}_j]_\times = g^{ij}(\mathbf{e}\times\mathbf{e}_i)\times\mathbf{e}_j\\ &= -g^{ij}\mathbf{e}_j\times(\mathbf{e}\times\mathbf{e}_i) = -\mathbf{e}^i\times(\mathbf{e}\times\mathbf{e}_i) = -\mathbf{e}(\mathbf{e}^i\cdot\mathbf{e}_i)+\mathbf{e}_i(\mathbf{e}^i\cdot\mathbf{e})\\ &= -\mathbf{e}\delta_i^i+\mathbf{e}_ie^i = -3\mathbf{e}+\mathbf{e} = -2\mathbf{e}\end{aligned}$$

and we have the result.

Problem 2.21 The Cayley–Hamilton theorem states that

$$\mathbf{A}^3 = I_1(\mathbf{A})\mathbf{A}^2 - I_2(\mathbf{A})\mathbf{A} + I_3(\mathbf{A})\mathbf{E}.$$

Take the trace of both sides (i.e., apply $\mathbf{E}\cdot\cdot$ to both sides) to get

$$\begin{aligned}\operatorname{tr}\mathbf{A}^3 &= I_1(\mathbf{A})\operatorname{tr}\mathbf{A}^2 - I_2(\mathbf{A})\operatorname{tr}\mathbf{A} + 3I_3(\mathbf{A})\\ &= \operatorname{tr}\mathbf{A}\operatorname{tr}\mathbf{A}^2 - \tfrac{1}{2}(\operatorname{tr}^2\mathbf{A} - \operatorname{tr}\mathbf{A}^2)\operatorname{tr}\mathbf{A} + 3I_3(\mathbf{A})\\ &= \tfrac{3}{2}\operatorname{tr}\mathbf{A}\operatorname{tr}\mathbf{A}^2 - \tfrac{1}{2}\operatorname{tr}^3\mathbf{A} + 3I_3(\mathbf{A}),\end{aligned}$$

then solve for $I_3(\mathbf{A})$.

Problem 2.25 Using the identity $\mathbf{a}\cdot\mathbf{B}\cdot\mathbf{b} = \mathbf{ab}\cdot\cdot\,\mathbf{B}^T$ (verify!) we get

$$\frac{\partial}{\partial\varepsilon}f(\mathbf{X}+\varepsilon\mathbf{B}) = \frac{\partial}{\partial\varepsilon}\left[\mathbf{a}\cdot(\mathbf{X}+\varepsilon\mathbf{B})\cdot\mathbf{b}\right]\Big|_{\varepsilon=0} = \mathbf{a}\cdot\mathbf{B}\cdot\mathbf{b} = \mathbf{ab}\cdot\cdot\,\mathbf{B}^T.$$

Hence $f_{,\mathbf{X}} = \mathbf{ab}$.

Problem 2.28 (a) $\mathbf{EA}$; (b) $\mathbf{EA}^T$; (c) $\mathbf{cdab}$.

Problem 2.31 If V is a linear space, then a subset U of V is a subspace of V if U is also a linear space (under the same operations of addition of elements and multiplication of elements by scalars). In linear algebra books it is shown that U is a subspace of V if and only if the following three conditions hold: (1) $0\in U$, where 0 is the zero element of V; (2) $x+y\in U$ whenever $x,y\in U$; (3) $\alpha x\in U$ whenever $x\in U$ and α is a scalar. Note that the second-order zero tensor $\mathbf{0}$ is both symmetric and antisymmetric.

Problem 2.37 Each identity can be proved by treating it in a Cartesian system and then returning to the general form. The technique is illustrated in the hints to Problem 2.38, below.

Problem 2.38 An example is

$$\begin{aligned}\nabla\cdot(\mathbf{A}\cdot\mathbf{f}) &= \mathbf{i}_k\cdot\frac{\partial}{\partial x_k}(a_{mn}\mathbf{i}_m\mathbf{i}_n\cdot f_s\mathbf{i}_s)\\ &= \mathbf{i}_k\cdot(\mathbf{i}_m\mathbf{i}_n\cdot\mathbf{i}_s)\left(\frac{\partial a_{mn}}{\partial x_k}f_s + a_{mn}\frac{\partial f_s}{\partial x_k}\right)\\ &= \mathbf{i}_k\cdot(\mathbf{i}_m\mathbf{i}_n\cdot\mathbf{i}_s)\frac{\partial a_{mn}}{\partial x_k}f_s + (\mathbf{i}_k\cdot\mathbf{i}_m)(\mathbf{i}_n\cdot\mathbf{i}_s)a_{mn}\frac{\partial f_s}{\partial x_k}\\ &= \mathbf{i}_k\cdot\frac{\partial(a_{mn}\mathbf{i}_m\mathbf{i}_n)}{\partial x_k}\cdot(f_s\mathbf{i}_s) + (a_{mn}\mathbf{i}_n\mathbf{i}_m)\cdot\cdot\,\mathbf{i}_k\frac{\partial(f_s\mathbf{i}_s)}{\partial x_k}\\ &= (\nabla\cdot\mathbf{A})\cdot\mathbf{f} + \mathbf{A}^T\cdot\cdot\,\nabla\mathbf{f}.\end{aligned}$$

Problem 2.39 Answers: (a) $\mathbf{E}$; (b) $-6\mathbf{A}$; (c) $\operatorname{tr}\mathbf{A}$; (d) ∇f; (e) $3\mathbf{E}$; (f) $4\mathbf{r}$. Sample calculations done in a Cartesian system:

$$\begin{aligned}\nabla\cdot(\mathbf{E}\mathbf{r}) &= \mathbf{i}_k\cdot\frac{\partial}{\partial x_k}(\mathbf{i}_j\mathbf{i}_j x_m\mathbf{i}_m) = (\mathbf{i}_k\cdot\mathbf{i}_j)\mathbf{i}_j\mathbf{i}_m\frac{\partial x_m}{\partial x_k}\\ &= \delta_{kj}\mathbf{i}_j\mathbf{i}_m\delta_{mk} = \mathbf{i}_k\mathbf{i}_k = \mathbf{E}.\end{aligned}$$

$$\begin{aligned}\nabla\cdot(\mathbf{A}\cdot\mathbf{r}) &= \mathbf{i}_k\cdot\frac{\partial}{\partial x_k}(a_{mn}\mathbf{i}_m\mathbf{i}_n\cdot x_p\mathbf{i}_p)\\ &= (\mathbf{i}_k\cdot\mathbf{i}_m)(\mathbf{i}_n\cdot\mathbf{i}_p)a_{mn}\frac{\partial x_p}{\partial x_k}\quad(\mathbf{A}\text{ constant})\\ &= \delta_{km}\delta_{np}a_{mn}\delta_{pk} = a_{kk} = \operatorname{tr}\mathbf{A}.\end{aligned}$$

$$\begin{aligned}\nabla\cdot(\mathbf{r}\mathbf{E}) &= \mathbf{i}_k\cdot\frac{\partial}{\partial x_k}(x_j\mathbf{i}_j\mathbf{i}_m\mathbf{i}_m) = (\mathbf{i}_k\cdot\mathbf{i}_j)\frac{\partial x_j}{\partial x_k}\mathbf{i}_m\mathbf{i}_m\\ &= \delta_{kj}\delta_{jk}\mathbf{i}_m\mathbf{i}_m = \delta_{kk}\mathbf{i}_m\mathbf{i}_m = 3\mathbf{E}.\end{aligned}$$

$$\begin{aligned}\nabla\cdot(\mathbf{r}\mathbf{r}) &= \mathbf{i}_k\cdot\frac{\partial}{\partial x_k}(x_j\mathbf{i}_j x_m\mathbf{i}_m) = (\mathbf{i}_k\cdot\mathbf{i}_j)\frac{\partial(x_j x_m)}{\partial x_k}\mathbf{i}_m\\ &= \delta_{kj}\left(x_j\frac{\partial x_m}{\partial x_k} + x_m\frac{\partial x_j}{\partial x_k}\right)\mathbf{i}_m = \delta_{kj}(x_j\delta_{mk} + x_m\delta_{jk})\mathbf{i}_m\\ &= x_j\delta_{kj}\delta_{mk}\mathbf{i}_m + x_m\delta_{kj}\delta_{jk}\mathbf{i}_m = x_k\mathbf{i}_k + \delta_{kk}x_m\mathbf{i}_m\\ &= x_k\mathbf{i}_k + 3x_k\mathbf{i}_k = 4x_k\mathbf{i}_k = 4\mathbf{r}.\end{aligned}$$

Problem 2.40

(a) $f'\mathbf{e}_r + \frac{f-g}{r}\mathbf{e}_r + h'\mathbf{e}_z$;
(b) $f'\mathbf{e}_r + 2\frac{f-g}{r}\mathbf{e}_r$;
(c) $f'\mathbf{e}_r + \frac{f-g}{r}\mathbf{e}_r$;
(d) $f'\mathbf{i}_1 + g'\mathbf{i}_2 + h'\mathbf{i}_3$;
(e) $\mathbf{0}$;
(f) $f'\mathbf{e}_\phi + \frac{f+g}{r}\mathbf{e}_\phi + h'\mathbf{e}_z$;
(g) $\frac{f}{r}\mathbf{e}_z + \frac{g}{r}\mathbf{e}_\phi + h'\mathbf{e}_r$.

Problem 2.41 $\mathbf{f} \times \mathbf{r}$.

Problem 2.42 Find the coordinate vectors for the new coordinates and demonstrate their orthogonality. We have

(a)

$$H_\sigma = a^2\frac{\sigma^2 - \tau^2}{\sigma^2 - 1}, \quad H_\tau = a^2\frac{\sigma^2 - \tau^2}{1 - \tau^2}, \quad H_z = 1;$$

(b)

$$H_\sigma = H_\tau = \sigma^2 + \tau^2, \quad H_\phi = \sigma^2\tau^2;$$

(c)

$$H_\sigma = H_\tau = \frac{a^2}{(\cosh\tau - \cos\sigma)^2}, \quad H_z = 1;$$

(d)

$$H_\sigma = H_\tau = \frac{a^2}{(\cosh\tau - \cos\sigma)^2}, \quad H_\phi = \frac{a^2\sin^2\sigma}{(\cosh\tau - \cos\sigma)^2};$$

(e)

$$H_\sigma = H_\tau = \frac{a^2}{(\cosh\tau - \cos\sigma)^2}, \quad H_\phi = \frac{a^2\sinh^2\tau}{(\cosh\tau - \cos\sigma)^2}.$$

Problem 2.43

$$\begin{aligned}\int_V \nabla\times\mathbf{f}\,dV &= \int_V \mathbf{i}_k \times \frac{\partial(f_t\mathbf{i}_t)}{\partial x_k}\,dV = \mathbf{i}_k\times\mathbf{i}_t\int_V \frac{\partial f_t}{\partial x_k}\,dV \\ &= \mathbf{i}_k\times\mathbf{i}_t\int_S f_t n_k\,dS = \int_S n_k\mathbf{i}_k\times f_t\mathbf{i}_t\,dS \\ &= \int_S \mathbf{n}\times\mathbf{f}\,dS.\end{aligned}$$

$$\int_V (\nabla \mathbf{f})^T \, dV = \int_V \left[\mathbf{i}_k \frac{\partial (f_t \mathbf{i}_t)}{\partial x_k} \right]^T dV = \mathbf{i}_t \mathbf{i}_k \int_V \frac{\partial f_t}{\partial x_k} \, dV$$

$$= \mathbf{i}_t \mathbf{i}_k \int_S f_t n_k \, dS = \int_S f_t \mathbf{i}_t \, n_k \mathbf{i}_k \, dS$$

$$= \int_S \mathbf{f}\mathbf{n} \, dS.$$

$$\int_V \nabla \mathbf{A} \, dV = \int_V \mathbf{i}_k \, \frac{\partial (a_{mn} \mathbf{i}_m \mathbf{i}_n)}{\partial x_k} \, dV = \mathbf{i}_k \mathbf{i}_m \mathbf{i}_n \int_V \frac{\partial a_{mn}}{\partial x_k} \, dV$$

$$= \mathbf{i}_k \mathbf{i}_m \mathbf{i}_n \int_S a_{mn} n_k \, dS = \int_S n_k \mathbf{i}_k \, a_{mn} \mathbf{i}_m \mathbf{i}_n \, dS$$

$$= \int_S \mathbf{n}\mathbf{A} \, dS.$$

$$\int_V \nabla \cdot \mathbf{A} \, dV = \int_V \mathbf{i}_k \cdot \frac{\partial (a_{mn} \mathbf{i}_m \mathbf{i}_n)}{\partial x_k} \, dV = \mathbf{i}_k \cdot \mathbf{i}_m \mathbf{i}_n \int_V \frac{\partial a_{mn}}{\partial x_k} \, dV$$

$$= \mathbf{i}_k \cdot \mathbf{i}_m \mathbf{i}_n \int_S a_{mn} n_k \, dS = \int_S n_k \mathbf{i}_k \cdot a_{mn} \mathbf{i}_m \mathbf{i}_n \, dS$$

$$= \int_S \mathbf{n} \cdot \mathbf{A} \, dS.$$

$$\int_V \nabla \times \mathbf{A} \, dV = \int_V \mathbf{i}_k \times \frac{\partial (a_{mn} \mathbf{i}_m \mathbf{i}_n)}{\partial x_k} \, dV = \mathbf{i}_k \times \mathbf{i}_m \mathbf{i}_n \int_V \frac{\partial a_{mn}}{\partial x_k} \, dV$$

$$= \mathbf{i}_k \times \mathbf{i}_m \mathbf{i}_n \int_S a_{mn} n_k \, dS = \int_S n_k \mathbf{i}_k \times a_{mn} \mathbf{i}_m \mathbf{i}_n \, dS$$

$$= \int_S \mathbf{n} \times \mathbf{A} \, dS.$$

Problem 2.44

$$\int_S \mathbf{n}\mathbf{r} \, dS = \int_S n_k \mathbf{i}_k \, x_t \mathbf{i}_t \, dS = \mathbf{i}_k \mathbf{i}_t \int_S x_t n_k \, dS = \mathbf{i}_k \mathbf{i}_t \int_V \frac{\partial x_t}{\partial x_k} \, dV$$

$$= \mathbf{i}_k \mathbf{i}_t \delta_{tk} \int_V dV = \mathbf{i}_k \mathbf{i}_k V = V\mathbf{E}.$$

Problem 2.46

$$\int_V (\nabla f)\cdot \mathbf{a}\, dV = \int_V [f\nabla\cdot\mathbf{a} + (\nabla f)\cdot\mathbf{a}]\, dV = \int_V \nabla\cdot(f\mathbf{a})\, dV$$
$$= \int_V \mathbf{i}_k \cdot \frac{\partial(fa_t\mathbf{i}_t)}{\partial x_k}\, dV = \mathbf{i}_k\cdot\mathbf{i}_t \int_V \frac{\partial(fa_t)}{\partial x_k}\, dV$$
$$= \mathbf{i}_k\cdot\mathbf{i}_t \int_S fa_t n_k\, dS = \int_S f\, n_k\mathbf{i}_k\cdot a_t\mathbf{i}_t\, dS = \int_S f\mathbf{n}\cdot\mathbf{a}\, dS.$$

Problem 2.47

$$\int_S \mathbf{n}\, dS = \int_S n_k\mathbf{i}_k\, dS = \mathbf{i}_k \int_S n_k\, dS = \mathbf{i}_k \int_V \frac{\partial(1)}{\partial x_k}\, dV = \mathbf{0}.$$

Problem 2.50 Consider the integral

$$\int_V \nabla\cdot(\mathbf{Ar})\, dV.$$

Using Gauss–Ostrogradsky theorem we get

$$\int_V \nabla\cdot(\mathbf{Ar})\, dV = \int_S \mathbf{n}\cdot\mathbf{Ar}\, dS = \int_S \mathbf{gr}\, dS.$$

On the other hand, with regard for the identity

$$\nabla\cdot(\mathbf{Ar}) = (\nabla\cdot\mathbf{A})\mathbf{r} + \mathbf{A}^T\cdot\nabla\mathbf{r} = (\nabla\cdot\mathbf{A})\mathbf{r} + \mathbf{A}^T$$

we have

$$\int_V \nabla\cdot(\mathbf{Ar})\, dV = \int_V (\mathbf{fr} + \mathbf{A}^T)\, dV.$$

Comparing these two expressions we find

$$\int_V \mathbf{A}^T\, dV = \int_S \mathbf{gr}\, dS - \int_V \mathbf{fr}\, dV,$$

from which the answer follows. The answer is

$$\int_S \mathbf{rg}\, dS - \int_V \mathbf{rf}\, dV.$$

Chapter 3 Exercises

Exercise 3.1 This time $s(t) = t\sqrt{\alpha^2 + \beta^2}$, hence

$$\mathbf{r}(s) = \mathbf{i}_1 \alpha \cos\left(\frac{s}{\sqrt{\alpha^2+\beta^2}}\right) + \mathbf{i}_2 \alpha \sin\left(\frac{s}{\sqrt{\alpha^2+\beta^2}}\right) + \mathbf{i}_3 \beta \frac{s}{\sqrt{\alpha^2+\beta^2}}.$$

This gives

$$\mathbf{r}'(s) = \frac{1}{\sqrt{\alpha^2+\beta^2}}\left[- \mathbf{i}_1 \alpha \sin\left(\frac{s}{\sqrt{\alpha^2+\beta^2}}\right) + \mathbf{i}_2 \alpha \cos\left(\frac{s}{\sqrt{\alpha^2+\beta^2}}\right) + \mathbf{i}_3 \beta \right],$$

$$\mathbf{r}''(s) = \frac{\alpha}{\alpha^2+\beta^2}\left[-\mathbf{i}_1 \cos\left(\frac{s}{\sqrt{\alpha^2+\beta^2}}\right) - \mathbf{i}_2 \sin\left(\frac{s}{\sqrt{\alpha^2+\beta^2}}\right) \right],$$

and hence

$$k = |\mathbf{r}''(s)| = \frac{\alpha}{\alpha^2+\beta^2}.$$

Exercise 3.2 The desired plane at point P has equation

$$(\mathbf{r} - \mathbf{r}_P) \cdot \boldsymbol{\beta} = 0$$

where $\mathbf{r} = \mathbf{i}_1 x + \mathbf{i}_2 y + \mathbf{i}_3 z$ and $\mathbf{r}_P = \mathbf{r}(1) = \mathbf{i}_1 + \mathbf{i}_2 2 + \mathbf{i}_3$. Note that, because of the zero on the right-hand side, we can use any vector parallel to $\boldsymbol{\beta}$. We have

$$\mathbf{r}'(1) = \mathbf{i}_1 + \mathbf{i}_2 4 + \mathbf{i}_3 3, \qquad \mathbf{r}''(1) = \mathbf{i}_2 4 + \mathbf{i}_3 6,$$

and thus

$$\boldsymbol{\beta} \sim \mathbf{r}'(1) \times \mathbf{r}''(1) = \mathbf{i}_1 12 - \mathbf{i}_2 6 + \mathbf{i}_3 4.$$

The desired plane is

$$[\mathbf{i}_1(x-1) + \mathbf{i}_2(y-2) + \mathbf{i}_3(z-1)] \cdot (\mathbf{i}_1 12 - \mathbf{i}_2 6 + \mathbf{i}_3 4) = 0$$

which reduces to $6x - 3y + 2z = 2$.

Exercise 3.3 The principal normal is

$$\boldsymbol{\nu} = -\mathbf{i}_1 \cos\left(\frac{s}{\sqrt{\alpha^2+\beta^2}}\right) - \mathbf{i}_2 \sin\left(\frac{s}{\sqrt{\alpha^2+\beta^2}}\right).$$

The binormal is obtained as

$$\boldsymbol{\beta} = \begin{vmatrix} \mathbf{i}_1 & \mathbf{i}_2 & \mathbf{i}_3 \\ -\frac{\alpha}{\sqrt{\alpha^2+\beta^2}} \sin\left(\frac{s}{\sqrt{\alpha^2+\beta^2}}\right) & \frac{\alpha}{\sqrt{\alpha^2+\beta^2}} \cos\left(\frac{s}{\sqrt{\alpha^2+\beta^2}}\right) & \frac{\beta}{\sqrt{\alpha^2+\beta^2}} \\ -\cos\left(\frac{s}{\sqrt{\alpha^2+\beta^2}}\right) & -\sin\left(\frac{s}{\sqrt{\alpha^2+\beta^2}}\right) & 0 \end{vmatrix}$$

and is

$$\boldsymbol{\beta} = \frac{1}{\sqrt{\alpha^2+\beta^2}} \left[\mathbf{i}_1 \beta \sin\left(\frac{s}{\sqrt{\alpha^2+\beta^2}}\right) - \mathbf{i}_2 \beta \cos\left(\frac{s}{\sqrt{\alpha^2+\beta^2}}\right) + \mathbf{i}_3 \alpha \right].$$

We can determine a plane in space by specifying the normal to the plane and a single point through which the plane passes. If the plane passes through the point located by the position vector $\mathbf{r}_0$, and $\mathbf{N}$ is any normal vector, then the plane is described by the vector equation

$$\mathbf{N} \cdot (\mathbf{r} - \mathbf{r}_0) = 0$$

(that is, all points whose position vectors $\mathbf{r}$ satisfy the above equation will lie in the plane). Hence the equations for the three fundamental planes are

$$\begin{aligned} &\boldsymbol{\beta} \cdot (\mathbf{r} - \mathbf{r}_0) = 0 && \text{osculating plane,} \\ &\boldsymbol{\tau} \cdot (\mathbf{r} - \mathbf{r}_0) = 0 && \text{normal plane,} \\ &\boldsymbol{\nu} \cdot (\mathbf{r} - \mathbf{r}_0) = 0 && \text{rectifying plane.} \end{aligned}$$

The rectifying plane for the helix is found to be $x \cos t_0 + y \sin t_0 = \alpha$.

Exercise 3.4 First show that $\boldsymbol{\beta}$ is a constant $\boldsymbol{\beta}_0$. Then use the definition of a plane as a set of points described by a radius vector $\mathbf{r}$ such that $\mathbf{r} \cdot \mathbf{n} = c$, where c is a constant.

Exercise 3.5 Straightforward using equations (3.3) and (3.5).

Exercise 3.6 The vectors $\mathbf{r}'(t)$ and $\mathbf{r}'''(t)$ reverse their directions, while $\mathbf{r}''(t)$ does not. So $\boldsymbol{\tau}$ reverses direction, $\boldsymbol{\nu}$ maintains its direction, and therefore $\boldsymbol{\beta}$ reverses direction. The moving trihedron changes its handedness. The formulas for k_1 and k_2 show that these quantities do not change sign.

Exercise 3.7 We have

$$k_2 = -\frac{\beta}{\alpha^2+\beta^2}.$$

Note that the ratio of the curvature to the torsion is constant for the helix.

Exercise 3.8

$$\frac{d^3\mathbf{r}}{ds^3} = -k_1^2\boldsymbol{\tau} + \frac{dk_1}{ds}\boldsymbol{\nu} - k_1k_2\boldsymbol{\beta}.$$

Exercise 3.9

$$\boldsymbol{\tau}' = \boldsymbol{\delta} \times \boldsymbol{\tau} = (k_1\boldsymbol{\beta} - k_2\boldsymbol{\tau}) \times \boldsymbol{\tau} = k_1\boldsymbol{\beta} \times \boldsymbol{\tau} = k_1\boldsymbol{\nu},$$
$$\boldsymbol{\nu}' = \boldsymbol{\delta} \times \boldsymbol{\nu} = (k_1\boldsymbol{\beta} - k_2\boldsymbol{\tau}) \times \boldsymbol{\nu} = k_1\boldsymbol{\beta} \times \boldsymbol{\nu} - k_2\vec{\boldsymbol{\tau}} \times \boldsymbol{\nu} = -k_1\boldsymbol{\tau} - k_2\boldsymbol{\beta},$$
$$\boldsymbol{\beta}' = \boldsymbol{\delta} \times \boldsymbol{\beta} = (k_1\boldsymbol{\beta} - k_2\boldsymbol{\tau}) \times \boldsymbol{\beta} = -k_2(-\boldsymbol{\nu}) = k_2\boldsymbol{\nu}.$$

Exercise 3.10 We have

$$\frac{\partial\mathbf{r}}{\partial\theta} = \mathbf{i}_1 a\cos\theta\cos\phi + \mathbf{i}_2 a\cos\theta\sin\phi - \mathbf{i}_3 a\sin\theta,$$
$$\frac{\partial\mathbf{r}}{\partial\phi} = -\mathbf{i}_1 a\sin\theta\sin\phi + \mathbf{i}_2 a\sin\theta\cos\phi,$$

hence

$$E = a^2, \qquad F = 0, \qquad G = a^2\sin^2\theta,$$

so that $(ds)^2 = a^2(d\theta)^2 + a^2\sin^2\theta(d\phi)^2$.

Exercise 3.11

(a) The differentials of coordinates for elementary curves on the coordinate lines are $(du^1, du^2 = 0)$ and $(du^1 = 0, du^2)$. The answer is given by $\cos\theta = F/\sqrt{EG}$.
(b) The desired angle is $\psi = \pi/4$.

Exercise 3.12 We find

$$\mathbf{r}_1 = \mathbf{i}_1\cos u^2 + \mathbf{i}_2\sin u^2 + \mathbf{i}_3,$$
$$\mathbf{r}_2 = -\mathbf{i}_1 u^1\sin u^2 + \mathbf{i}_2 u^1\cos u^2,$$

hence $E = 2$, $F = 0$, $G = (u^1)^2$. Then

$$S = \int_0^{2\pi}\int_0^a \sqrt{2}u^1\, du^1\, du^2 = \pi\sqrt{2}a^2.$$

Exercise 3.15 We have

$$\frac{\partial \mathbf{r}}{\partial u} = \mathbf{i}_1 \cos\phi + \mathbf{i}_2 \sin\phi \qquad \frac{\partial \mathbf{r}}{\partial \phi} = -\mathbf{i}_1 u \sin\phi + \mathbf{i}_2 u \cos\phi + \mathbf{i}_3 c$$

First fundamental coefficients:

$$E = 1 \qquad F = 0 \qquad G = u^2 + c^2$$

Unit normal vector:

$$\mathbf{N} = \frac{1}{\sqrt{EG - F^2}} \begin{vmatrix} \mathbf{i}_1 & \mathbf{i}_2 & \mathbf{i}_3 \\ \cos\phi & \sin\phi & 0 \\ -u\sin\phi & u\cos\phi & c \end{vmatrix} = \frac{1}{\sqrt{u^2+c^2}}[\mathbf{i}_1 c \sin\phi - \mathbf{i}_2 c \cos\phi + \mathbf{i}_3 u]$$

We also have

$$\frac{\partial^2 \mathbf{r}}{\partial u^2} = 0 \qquad \frac{\partial^2 \mathbf{r}}{\partial u \partial \phi} = -\mathbf{i}_1 \sin\phi + \mathbf{i}_2 \cos\phi \qquad \frac{\partial^2 \mathbf{r}}{\partial \phi^2} = -\mathbf{i}_1 u \cos\phi - \mathbf{i}_2 u \sin\phi$$

Second fundamental coefficients:

$$L = 0 \qquad M = -\frac{c}{\sqrt{u^2+c^2}} \qquad N = 0$$

Set

$$\begin{vmatrix} \kappa & \dfrac{c}{\sqrt{u^2+c^2}} \\ \dfrac{c}{\sqrt{u^2+c^2}} & (u^2+c^2)\kappa \end{vmatrix} = 0$$

to get the principal curvatures

$$\kappa_{1,2} = \pm \frac{c}{u^2+c^2}$$

Mean curvature and Gaussian curvatures:

$$H = \frac{1}{2}\left(\frac{c}{u^2+c^2} - \frac{c}{u^2+c^2}\right) = 0 \qquad K = -\frac{c^2}{(u^2+c^2)^2}$$

Chapter 3 Problems

Problem 3.1 The curve is an astroid. Its parametric representation is $x = a\cos^3 t$, $y = a\sin^3 t$. The singular points are $(0, a)$, $(0, -a)$, $(a, 0)$, and $(-a, 0)$.

Problem 3.2 The curve is a cycloid. The singular points are $(0, 2\pi k)$ for $k = 0, \pm 1, \pm 2, \ldots$.

Problem 3.3

$$\frac{x}{0} = \frac{y}{1} = \frac{z-1}{0}.$$

Problem 3.4 $\mathbf{r}'(t) = \mathbf{i}_1 2\sinh 2t + \mathbf{i}_2 2\cosh 2t + \mathbf{i}_3 2$ so

$$\begin{aligned} s &= \int_0^{\pi} [(2\sinh 2t)^2 + (2\cosh 2t)^2 + (2)^2]^{1/2}\,dt \\ &= 2\int_0^{\pi} [(\sinh 2t)^2 + (\cosh 2t)^2 + 1]^{1/2}\,dt \\ &= 2\int_0^{\pi} [2\cosh^2 2t]^{1/2}\,dt = 2\sqrt{2}\int_0^{\pi} \cosh 2t\,dt = \sqrt{2}\sinh 2\pi. \end{aligned}$$

Problem 3.5 (a) $6a$, (b) $8a$.

Problem 3.6

$$k_1 = \frac{1}{4}\sqrt{1+\sin^2\frac{t}{2}}.$$

Problem 3.7

$$k_1 = \frac{1}{2a\cosh^2 t}, \qquad k_2 = \frac{1}{2a\cosh^2 t}.$$

Problem 3.8 $k_2 = a\cosh t$.

Problem 3.10 Differentiating the relation $\mathbf{Q}\cdot\mathbf{Q}^T = \mathbf{E}$ with respect to s, we obtain $\mathbf{Q}'\cdot\mathbf{Q}^T + \mathbf{Q}\cdot\mathbf{Q}'^T = \mathbf{0}$. Hence $\mathbf{Q}'\cdot\mathbf{Q}^T$ is an antisymmetric tensor, so it takes the form $\mathbf{Q}' = \mathbf{d}\times\mathbf{Q}$ where $\mathbf{d}$ is the corresponding conjugate vector. Thus $\mathbf{Q}'\cdot\mathbf{Q}^T = \mathbf{d}\times\mathbf{E}$. Now refer to the result of Problem 2.11(c).

Problem 3.11

$$\frac{-2du\,dv}{\sqrt{u^2+1}}.$$

Problem 3.15 $2H = \operatorname{tr}\mathbf{B} = -\operatorname{tr}\widetilde{\nabla}\mathbf{n} = -\widetilde{\nabla}\cdot\mathbf{n}$.

Problem 3.16

$$\mathbf{0} = \widetilde{\nabla}\cdot\mathbf{E} = \widetilde{\nabla}\cdot(\mathbf{A}+\mathbf{nn}) = \widetilde{\nabla}\cdot\mathbf{A} + (\widetilde{\nabla}\cdot\mathbf{n})\mathbf{n} = \widetilde{\nabla}\cdot\mathbf{A} - 2H\mathbf{n}.$$

Problem 3.18

$$K = \frac{f_{xx} g_{yy}}{\left(1 + f_x^2 + g_y^2\right)^2}.$$

Problem 3.19

$$K = -\frac{1}{AB}\left[\left(\frac{A_v}{B}\right)_v + \left(\frac{B_u}{A}\right)_u\right],$$

where indices u and v denote partial derivatives with respect to u and v respectively.

Problem 3.20 Let us consider Stokes' formula for a tensor field $\mathbf{X}$ and a vector field $\mathbf{x}$:

$$\oint_\Gamma \boldsymbol{\tau} \cdot \mathbf{X}\, ds = \int_S \mathbf{n} \cdot (\nabla \times \mathbf{X})\, dS, \quad \oint_\Gamma \boldsymbol{\tau} \cdot \mathbf{x}\, ds = \int_S \mathbf{n} \cdot (\nabla \times \mathbf{x})\, dS, \quad \text{(D.5)}$$

where we have used the relations $d\mathbf{r} = \boldsymbol{\tau}\, ds$ and $(\mathbf{n} \times \nabla) \cdot \mathbf{X} = \mathbf{n} \cdot (\nabla \times \mathbf{X})$. Using the identities

$$\mathbf{A} = -\mathbf{n} \times (\mathbf{n} \times \mathbf{A})$$

and

$$\mathbf{A} = \mathbf{E} \cdot \mathbf{X} = (\mathbf{A} + \mathbf{nn}) \cdot \mathbf{X} = -\mathbf{n} \times (\mathbf{n} \times \mathbf{A}) \cdot \mathbf{X} + \mathbf{nn} \cdot \mathbf{X},$$

we represent the tensor $\mathbf{X}$ as

$$\mathbf{X} = \mathbf{n} \times \mathbf{X}_1 + \mathbf{n}\mathbf{x}_2 \quad \text{(D.6)}$$

where $\mathbf{X}_1 = -\mathbf{n} \times \mathbf{X}$ and $\mathbf{x}_2 = \mathbf{n} \cdot \mathbf{X}$. Note that the surface and spatial gradient operators $\widetilde{\nabla}$ and ∇ are related by

$$\nabla = \widetilde{\nabla} + \mathbf{n}\frac{\partial}{\partial z},$$

where z is the distance coordinate along the normal to S. Then

$$\mathbf{n} \cdot (\nabla \times \mathbf{X}_1) = \mathbf{n} \cdot (\widetilde{\nabla} \times \mathbf{X}_1).$$

Formula (D.5) for $\mathbf{X}_1$ yields

$$\int_S \mathbf{n} \cdot (\widetilde{\nabla} \times \mathbf{X}_1)\, dS = \oint_\Gamma \boldsymbol{\tau} \cdot \mathbf{X}_1\, ds. \quad \text{(D.7)}$$

In view of the identities $\boldsymbol{\tau} = \boldsymbol{\nu} \times \mathbf{n}$ and $(\boldsymbol{\nu} \times \mathbf{n}) \cdot \mathbf{X}_1 = \boldsymbol{\nu} \cdot (\mathbf{n} \times \mathbf{X}_1)$, equation (D.7) reduces to

$$\int_S \widetilde{\nabla} \cdot (\mathbf{n} \times \mathbf{X}_1)\, dS = \oint_\Gamma \boldsymbol{\nu} \cdot (\mathbf{n} \times \mathbf{X}_1)\, ds. \tag{D.8}$$

Using the relation

$$\widetilde{\nabla} \cdot (\mathbf{n}\mathbf{x}_2) = (\widetilde{\nabla} \cdot \mathbf{n})\mathbf{x}_2 = -2H\mathbf{x}_2,$$

from (D.6) and (D.8) we derive the final formula

$$\int_S \left(\widetilde{\nabla} \cdot \mathbf{X} + 2H\mathbf{n} \cdot \mathbf{X}\right) dS = \oint_\Gamma \boldsymbol{\nu} \cdot \mathbf{X}\, ds.$$

Problem 3.21 Similar to Problem 3.20.

Problem 3.22 Applying (3.29) to the tensor $\mathbf{AX}$, we obtain the first formula

$$\int_S \left(\widetilde{\nabla}\mathbf{X} + 2H\mathbf{n}\mathbf{X}\right) dS = \oint_\Gamma \boldsymbol{\nu}\mathbf{X}\, ds.$$

The second one,

$$\int_S \left(\widetilde{\nabla} \times \mathbf{X} + 2H\mathbf{n} \times \mathbf{X}\right) dS = \oint_\Gamma \boldsymbol{\nu} \times \mathbf{X}\, ds,$$

follows from this. Substituting the tensor $\mathbf{nX}$ into the last formula, we obtain the third formula

$$\int_S \widetilde{\nabla} \times (\mathbf{nX})\, dS = \oint_\Gamma \boldsymbol{\tau}\mathbf{X}\, ds.$$

Chapter 4 Exercises

Exercise 4.1 Use integration by parts and treat n_k as 0, 1, or -1.

Exercise 4.2 In the identity $d\mathbf{f} = d\mathbf{x} \cdot \nabla\mathbf{f}$, set $\mathbf{x} = \mathbf{r}_0$ and $\mathbf{f} = \mathbf{r}$ to get $d\mathbf{r} = d\mathbf{r}_0 \cdot \nabla\mathbf{r} = d\mathbf{r}_0 \cdot \nabla(\mathbf{r}_0 + \mathbf{u}) = d\mathbf{r}_0 \cdot (\nabla\mathbf{r}_0 + \nabla\mathbf{u}) = d\mathbf{r}_0 \cdot (\mathbf{E} + \nabla\mathbf{u})$. Here we had $\nabla\mathbf{r}_0 = \mathbf{E}$ because ∇ is written in the undeformed state.

Exercise 4.5 Start with Hooke's law: $\boldsymbol{\sigma} = \lambda\mathbf{E}\,\mathrm{tr}\,\boldsymbol{\varepsilon} + 2\mu\boldsymbol{\varepsilon}$. Apply the trace operation to it and find $\mathrm{tr}\,\boldsymbol{\varepsilon}$:

$$\mathrm{tr}\,\boldsymbol{\varepsilon} = \frac{1}{3\lambda + 2\mu}\,\mathrm{tr}\,\boldsymbol{\sigma}.$$

Substituting this into Hooke's law, obtain $\boldsymbol{\varepsilon}$:

$$\boldsymbol{\varepsilon} = \frac{1}{2\mu}\left[\boldsymbol{\sigma} - \frac{\lambda}{3\lambda + 2\mu}\mathbf{E}\operatorname{tr}\boldsymbol{\sigma}\right] = \frac{1}{2\mu}\left[\boldsymbol{\sigma} - \frac{\nu}{1+\nu}\mathbf{E}\operatorname{tr}\boldsymbol{\sigma}\right].$$

Now

$$W = \frac{1}{2}\boldsymbol{\sigma}\cdot\cdot\,\boldsymbol{\varepsilon} = \frac{1}{4\mu}\left[\boldsymbol{\sigma}\cdot\cdot\,\boldsymbol{\sigma} - \frac{\nu}{1+\nu}(\operatorname{tr}\boldsymbol{\sigma})^2\right].$$

Exercise 4.8 Use d'Alembert's principle, which states that in dynamics the system of all forces — including the inertia forces — are in equilibrium. So in the equilibrium equations, change

$$f_i \to f_i - \rho\,\frac{d^2u_i}{dt^2}.$$

Exercise 4.12 Consider $\mathbf{A}(\mathbf{k})$ and $\mathbf{C}$ in a Cartesian frame. By definition,

$$\begin{aligned}\mathbf{A}(\mathbf{k}) &= \mathbf{k}\cdot\mathbf{C}\cdot\mathbf{k} = k_m\mathbf{i}_m\cdot c_{pqrt}\mathbf{i}_p\mathbf{i}_q\mathbf{i}_r\mathbf{i}_t\cdot k_s\mathbf{i}_s = k_mk_sc_{mqrs}\mathbf{i}_q\mathbf{i}_r \\ &= k_mk_sc_{mrqs}\mathbf{i}_q\mathbf{i}_r = k_mk_sc_{mqrs}\mathbf{i}_r\mathbf{i}_q = \mathbf{A}(\mathbf{k})^T.\end{aligned}$$

Exercise 4.13 It suffices to show that $\mathbf{a}\cdot\mathbf{A}(\mathbf{k})\cdot\mathbf{a} > 0$ for $\mathbf{a} \neq \mathbf{0}$. The tensor $\mathbf{C}$ is such that $\boldsymbol{\varepsilon}\cdot\cdot\,\mathbf{C}\cdot\cdot\,\boldsymbol{\varepsilon} > 0$ whenever $\boldsymbol{\varepsilon} \neq \mathbf{0}$. Substitute $\boldsymbol{\varepsilon} = \mathbf{ka}$. Because $\mathbf{C}$ is symmetric, $(\mathbf{ka})\cdot\cdot\,\mathbf{C} = (\mathbf{ak})\cdot\cdot\,\mathbf{C}$. Thus $\boldsymbol{\varepsilon}\cdot\cdot\,\mathbf{C}\cdot\cdot\,\boldsymbol{\varepsilon} = (\mathbf{ka})\cdot\cdot\,\mathbf{C}\cdot\cdot\,(\mathbf{ka}) = \mathbf{a}\cdot(\mathbf{k}\cdot\mathbf{C}\cdot\mathbf{k})\mathbf{a} = \mathbf{a}\cdot\mathbf{A}(\mathbf{k})\cdot\mathbf{a} > 0$.

Exercise 4.14 From Hooke's law it follows that

$$\operatorname{tr}\boldsymbol{\varepsilon} = \varepsilon_{11} + \varepsilon_{22} = \frac{\sigma_{11}+\sigma_{22}}{\lambda + 2\mu}.$$

So we have

$$\begin{aligned}\varepsilon_{11} &= \frac{1}{2\mu}\sigma_{11} - \frac{\lambda}{2\mu(\lambda+2\mu)}(\sigma_{11}+\sigma_{22}),\\ \varepsilon_{22} &= \frac{1}{2\mu}\sigma_{22} - \frac{\lambda}{2\mu(\lambda+2\mu)}(\sigma_{11}+\sigma_{22}),\\ \varepsilon_{12} &= \frac{\sigma_{12}}{\mu}.\end{aligned}$$

Exercise 4.15 By Hooke's law we get

$$\begin{aligned}&\sigma_{11} = \lambda\operatorname{tr}\boldsymbol{\varepsilon} + 2\mu\varepsilon_{11}, &&\sigma_{22} = \lambda\operatorname{tr}\boldsymbol{\varepsilon} + 2\mu\varepsilon_{22}, &&0 = \lambda\operatorname{tr}\boldsymbol{\varepsilon} + 2\mu\varepsilon_{33},\\ &\sigma_{12} = \mu\varepsilon_{12}, &&0 = \mu\varepsilon_{13}, &&0 = \mu\varepsilon_{23}.\end{aligned}$$

It follows that

$$\varepsilon_{13} = 0, \quad \varepsilon_{23} = 0, \quad \varepsilon_{12} = \frac{\sigma_{12}}{\mu}, \quad \varepsilon_{33} = -\frac{\lambda}{\lambda + 2\mu}(\varepsilon_{11} + \varepsilon_{22}).$$

Substituting ε_{33} into the first two equations, we obtain

$$\sigma_{11} = \frac{2\lambda\mu}{\lambda + 2\mu}(\varepsilon_{11} + \varepsilon_{22}) + 2\mu\varepsilon_{11}, \quad \sigma_{22} = \frac{2\lambda\mu}{\lambda + 2\mu}(\varepsilon_{11} + \varepsilon_{22}) + 2\mu\varepsilon_{22}.$$

These equations coincide with the equations of the plane deformation problem when we change λ to $\lambda^\star = \frac{2\lambda\mu}{\lambda+2\mu}$. Then we have

$$\varepsilon_{11} + \varepsilon_{22} = \frac{\sigma_{11} + \sigma_{22}}{\lambda^\star + 2\mu} = \frac{\lambda + 2\mu}{4\mu(\lambda + \mu)}(\sigma_{11} + \sigma_{22}),$$

$$\varepsilon_{11} = \frac{1}{2\mu}\sigma_{11} - \frac{\lambda^\star}{2\mu(\lambda^\star + 2\mu)}(\sigma_{11} + \sigma_{22}),$$
$$\varepsilon_{22} = \frac{1}{2\mu}\sigma_{22} - \frac{\lambda^\star}{2\mu(\lambda^\star + 2\mu)}(\sigma_{11} + \sigma_{22}).$$

Chapter 4 Problems

Problem 4.1

(a) On sides AB and CD: $\sigma_{11} = \sigma$, $\sigma_{12} = 0$; on sides AD and BC: $\sigma_{22} = 2\sigma$, $\sigma_{12} = 0$.
(b) On sides AB and CD: $\sigma_{11} = 0$, $\sigma_{12} = -\tau$; on sides AD and BC: $\sigma_{22} = 0$, $\sigma_{12} = -\tau$.
(c) On sides AB and CD: $\sigma_{11} = -p$, $\sigma_{12} = 0$; on sides AD and BC: $\sigma_{22} = \sigma$, $\sigma_{12} = 0$.
(d) On sides AB and CD: $\sigma_{11} = 0$, $\sigma_{12} = \tau$; on sides AD and BC: $\sigma_{22} = -p$, $\sigma_{12} = \tau$.

Problem 4.3 On AB we have $\mathbf{n} = -\mathbf{i}_2$. The boundary condition is $\mathbf{n}\cdot\boldsymbol{\sigma} = \mathbf{0}$. Using the representation

$$\boldsymbol{\sigma} = \sigma_{11}\mathbf{i}_1\mathbf{i}_1 + \boldsymbol{\sigma}_{12}(\mathbf{i}_1\mathbf{i}_2 + \mathbf{i}_2\mathbf{i}_1) + \sigma_{22}\mathbf{i}_2\mathbf{i}_2,$$

we get

$$\mathbf{n}\cdot\boldsymbol{\sigma} = -\sigma_{22}\mathbf{i}_2 - \sigma_{12}\mathbf{i}_1 = \mathbf{0}.$$

So the conditions on side AB are: $\sigma_{22} = 0$, $\sigma_{12} = 0$.

On BC we have $\mathbf{n} = \mathbf{i}_1$. The conditions on side BC are

$$\mathbf{n} \cdot \boldsymbol{\sigma} = \sigma_{11}\mathbf{i}_1 + \sigma_{12}\mathbf{i}_2 = \mathbf{0}.$$

On AC we have

$$\mathbf{n} = -\sin\alpha\mathbf{i}_1 + \cos\alpha\mathbf{i}_2.$$

The boundary condition is $\mathbf{n} \cdot \boldsymbol{\sigma} = \mathbf{0}$. Using the representation

$$\boldsymbol{\sigma} = \sigma_{11}\mathbf{i}_1\mathbf{i}_1 + \boldsymbol{\sigma}_{12}(\mathbf{i}_1\mathbf{i}_2 + \mathbf{i}_2\mathbf{i}_1) + \sigma_{22}\mathbf{i}_2\mathbf{i}_2,$$

we get

$$\begin{aligned}\mathbf{n} \cdot \boldsymbol{\sigma} &= (-\sin\alpha\mathbf{i}_1 + \cos\alpha\mathbf{i}_2) \cdot (\sigma_{11}\mathbf{i}_1\mathbf{i}_1 + \boldsymbol{\sigma}_{12}(\mathbf{i}_1\mathbf{i}_2 + \mathbf{i}_2\mathbf{i}_1) + \sigma_{22}\mathbf{i}_2\mathbf{i}_2) \\ &= \mathbf{i}_1(-\sigma_{11}\sin\alpha + \sigma_{12}\cos\alpha) + \mathbf{i}_2(-\sigma_{12}\sin\alpha + \sigma_{22}\cos\alpha) = \mathbf{0},\end{aligned}$$

which yields the needed result. The final answers are as follows. Side AB: $\sigma_{22} = 0$, $\sigma_{12} = 0$. Side BC: $\sigma_{11} = -\gamma y$, $\sigma_{12} = 0$. Side AC: $-\sigma_{11}\sin\alpha + \sigma_{12}\cos\alpha = 0$, $-\sigma_{12}\sin\alpha + \sigma_{22}\cos\alpha = 0$.

Problem 4.3 On side BC, the normal is

$$\mathbf{n} = \sin\beta\mathbf{i}_1 + \cos\beta\mathbf{i}_2.$$

Then

$$\begin{aligned}\mathbf{n} \cdot \boldsymbol{\sigma} &= (\sin\beta\mathbf{i}_1 + \cos\beta\mathbf{i}_2) \cdot (\sigma_{11}\mathbf{i}_1\mathbf{i}_1 + \boldsymbol{\sigma}_{12}(\mathbf{i}_1\mathbf{i}_2 + \mathbf{i}_2\mathbf{i}_1) + \sigma_{22}\mathbf{i}_2\mathbf{i}_2) \\ &= \mathbf{i}_1(\sigma_{11}\sin\beta + \sigma_{12}\cos\beta) + \mathbf{i}_2(\sigma_{12}\sin\beta + \sigma_{22}\cos\beta).\end{aligned}$$

The force on BC is

$$-p_3\mathbf{n} = -p_3(\sin\beta\mathbf{i}_1 + \cos\beta\mathbf{i}_2).$$

The conditions on side BC follow from the relation $\mathbf{n} \cdot \boldsymbol{\sigma} = -p_3\mathbf{n}$.

The normal to AC is

$$\mathbf{n} = -\sin\alpha\mathbf{i}_1 + \cos\alpha\mathbf{i}_2.$$

The rest is similar to the above solution for BC.

The final answers are as follows. Side AB:

$$\sigma_{22} = -p_1, \qquad \sigma_{12} = 0.$$

Side BC:

$$\begin{aligned}\sigma_{11}\sin\beta + \sigma_{12}\cos\beta &= -p_3\sin\beta, \\ \sigma_{12}\sin\beta + \sigma_{22}\cos\beta &= -p_3\cos\beta.\end{aligned}$$

Side AC:

$$-\sigma_{11}\sin\alpha + \sigma_{12}\cos\alpha = p_2\sin\alpha,$$
$$-\sigma_{12}\sin\alpha + \sigma_{22}\cos\alpha = -p_2\cos\alpha.$$

Problem 4.4

(a) On side AB: $\sigma_{\phi\phi} = \sigma$, $\sigma_{\phi r} = 0$.
On a part of circle BC: $\sigma_{rr} = -p_2$, $\sigma_{r\phi} = 0$.
On a part of circle AC: $\sigma_{rr} = -p_1$, $\sigma_{r\phi} = 0$.
On side CD: $\sigma_{\phi\phi} = \sigma$, $\sigma_{\phi r} = 0$.

(b) On side AB: $\sigma_{\phi\phi} = -p$, $\sigma_{\phi r} = 0$.
On a part of circle BC: $\sigma_{rr} = 0$, $\sigma_{r\phi} = -\tau_2$.
On a part of circle AC: $\sigma_{rr} = 0$, $\sigma_{r\phi} = -\tau_1$.
On side CD: $\sigma_{\phi\phi} = -p$, $\sigma_{\phi r} = 0$.

Problem 4.5

(a) On side ABCD: $\sigma_{11} = \sigma$, $\sigma_{12} = 0$, $\sigma_{13} = \tau$.
On a part of circle BEFC: $\sigma_{33} = 2\sigma$, $\sigma_{32} = 0$, $\sigma_{31} = \tau$.
On a part of circle DCFG: $\sigma_{22} = \sigma$, $\sigma_{21} = 0$, $\sigma_{23} = 0$.

(b) On side ABCD: $\sigma_{11} = -p_1$, $\sigma_{12} = \tau$, $\sigma_{13} = 0$.
On a part of circle BEFC: $\sigma_{33} = -p_3$, $\sigma_{32} = 0$, $\sigma_{31} = 0$.
On a part of circle DCFG: $\sigma_{22} = -p_2$, $\sigma_{21} = \tau$, $\sigma_{23} = 0$.

Problem 4.6 Denote the principal axes by $\mathbf{i}_1, \mathbf{i}_2, \mathbf{i}_3$. In this Cartesian system, the normal to triangle ABC is

$$\mathbf{n} = \frac{1}{\sqrt{3}}(\mathbf{i}_1 + \mathbf{i}_2 + \mathbf{i}_3).$$

The stress tensor is

$$\boldsymbol{\sigma} = \sigma_1\mathbf{i}_1\mathbf{i}_1 + \sigma_2\mathbf{i}_2\mathbf{i}_2 + \sigma_3\mathbf{i}_3\mathbf{i}_3.$$

The required stress vector is

$$\mathbf{t} = \mathbf{n}\cdot\boldsymbol{\sigma} = \frac{1}{\sqrt{3}}(\sigma_1\mathbf{i}_1 + \sigma_2\mathbf{i}_2 + \sigma_3\mathbf{i}_3) = \frac{1}{\sqrt{3}}(50\mathbf{i}_1 - 50\mathbf{i}_2 + 75\mathbf{i}_3).$$

Problem 4.7

(a) $\boldsymbol{\varepsilon} = \frac{1}{2}\gamma(\mathbf{i}_1\mathbf{i}_2 + \mathbf{i}_2\mathbf{i}_1)$;

(b) $\boldsymbol{\varepsilon} = \lambda_1 \mathbf{i}_1 \mathbf{i}_1$;
(c) $\boldsymbol{\varepsilon} = \lambda \mathbf{E}$;
(d) $\boldsymbol{\varepsilon} = u'(r)\mathbf{e}_r\mathbf{e}_r + u(r)/r\,\mathbf{e}_\phi\mathbf{e}_\phi + k\mathbf{e}_z\mathbf{e}_z$;
(e) $\boldsymbol{\varepsilon} = (u'(r) - u(r)/r)(\mathbf{e}_r\mathbf{e}_\phi + \mathbf{e}_\phi\mathbf{e}_r) + k\mathbf{e}_z\mathbf{e}_z$;
(f) $\boldsymbol{\varepsilon} = u'(r)(\mathbf{e}_r\mathbf{e}_z + \mathbf{e}_z\mathbf{e}_r) + k\mathbf{e}_z\mathbf{e}_z$;
(g) $\boldsymbol{\varepsilon} = u'(r)\mathbf{e}_r\mathbf{e}_r + u(r)/r\,\mathbf{e}_\phi\mathbf{e}_\phi + u(r)/r\,\mathbf{e}_\theta\mathbf{e}_\theta$.

Chapter 5 Exercises

Exercise 5.6 Hooke's law is given by the formula

$$\boldsymbol{\sigma} = \lambda \mathbf{E} \operatorname{tr} \boldsymbol{\varepsilon} + 2\mu\boldsymbol{\varepsilon}.$$

Dot-multiply this by $\mathbf{n}$ from the left and the right. From $\sigma_{33} = 0$ it follows that

$$\lambda \operatorname{tr} \boldsymbol{\varepsilon} + 2\mu\varepsilon_{33} = 0.$$

Because $\operatorname{tr} \boldsymbol{\varepsilon} = \operatorname{tr} \tilde{\boldsymbol{\varepsilon}} + \varepsilon_{33}$ where $\operatorname{tr} \tilde{\boldsymbol{\varepsilon}} = \varepsilon_1^1 + \varepsilon_2^2$, we get

$$\varepsilon_{33} = -\frac{\lambda}{\lambda + 2\mu} \operatorname{tr} \tilde{\boldsymbol{\varepsilon}}.$$

So

$$\operatorname{tr} \boldsymbol{\varepsilon} = \frac{2\mu}{\lambda + 2\mu} \operatorname{tr} \tilde{\boldsymbol{\varepsilon}}.$$

It follows that

$$\boldsymbol{\sigma} = 2\mu \left[\boldsymbol{\varepsilon} + \frac{\lambda}{\lambda + 2\mu} \mathbf{E} \operatorname{tr} \tilde{\boldsymbol{\varepsilon}} \right] \equiv \frac{E}{1+\nu} \left[\boldsymbol{\varepsilon} + \frac{\nu}{1-\nu} \mathbf{E} \operatorname{tr} \tilde{\boldsymbol{\varepsilon}} \right].$$

Exercise 5.7 In Cartesian coordinates we get

$$\boldsymbol{\epsilon} \cdot\cdot \boldsymbol{\epsilon} = \epsilon_{11}^2 + \epsilon_{22}^2 + 2\epsilon_{12}^2, \qquad \operatorname{tr}^2 \boldsymbol{\epsilon} = (\epsilon_{11} + \epsilon_{22})^2 = \epsilon_{11}^2 + \epsilon_{22}^2 + 2\epsilon_{11}\epsilon_{22}.$$

So,

$$\begin{aligned} \boldsymbol{\epsilon} \cdot\cdot \boldsymbol{\epsilon} + \frac{\nu}{1-\nu} \operatorname{tr}^2 \boldsymbol{\epsilon} &= \left(1 + \frac{\nu}{1-\nu}\right) \epsilon_{11}^2 + \left(1 + \frac{\nu}{1-\nu}\right) \epsilon_{22}^2 \\ &\quad + \frac{2\nu}{1-\nu} \epsilon_{11}\epsilon_{22} + 2\epsilon_{12}^2 \\ &= \frac{1}{1-\nu} \left[\epsilon_{11}^2 + \epsilon_{22}^2 + 2\nu\epsilon_{11}\epsilon_{22}\right] + 2\epsilon_{12}^2. \end{aligned}$$

If $|\nu| < 1$ and $\epsilon_{11} \neq 0$, $\epsilon_{22} \neq 0$ then $\epsilon_{11}^2 + \epsilon_{22}^2 + 2\nu\epsilon_{11}\epsilon_{22} > 0$. This proves the required positivity.

Exercise 5.8 Doubly dot-multiplying $\mathbf{T}$ by $\boldsymbol{\epsilon}$, we get

$$\mathbf{T} \cdot\cdot\, \boldsymbol{\epsilon} = (\mathbf{T} \cdot \mathbf{A} + (\mathbf{T} \cdot \mathbf{n})\mathbf{n}) \cdot\cdot\, \boldsymbol{\epsilon} = (\mathbf{T} \cdot \mathbf{A}) \cdot\cdot\, \boldsymbol{\epsilon} = T_{\alpha\beta}\epsilon^{\alpha\beta}.$$

Substituting (5.27) into this, we get

$$\mathbf{T} \cdot\cdot\, \boldsymbol{\epsilon} = \frac{Eh}{1+\nu}\left[\boldsymbol{\epsilon} + \frac{\nu}{1-\nu}\mathbf{A}\operatorname{tr}\boldsymbol{\epsilon}\right] \cdot\cdot\, \boldsymbol{\epsilon} = \frac{Eh}{1+\nu}\left[\boldsymbol{\epsilon} \cdot\cdot\, \boldsymbol{\epsilon} + \frac{\nu}{1-\nu}\operatorname{tr}^2 \boldsymbol{\epsilon}\right].$$

Using the equality $\mathbf{M} = -M_{\alpha\beta}\boldsymbol{\rho}^\alpha\boldsymbol{\rho}^\beta \times \mathbf{n}$ we find that

$$M_{\alpha\beta} = -\boldsymbol{\rho}_\alpha \cdot \mathbf{M} \cdot (\boldsymbol{\rho}_\beta \times \mathbf{n}).$$

Then

$$\begin{aligned} M_{\alpha\beta}æ^{\alpha\beta} &= -\boldsymbol{\rho}_\alpha \cdot \mathbf{M} \cdot (\boldsymbol{\rho}_\beta \times \mathbf{n})æ^{\alpha\beta} = -\boldsymbol{\rho}_\alpha \cdot \mathbf{M} \cdot (\boldsymbol{\rho}_\beta \times \mathbf{n})\boldsymbol{\rho}^\alpha \cdot æ \cdot \boldsymbol{\rho}^\beta \\ &= -\mathbf{M} \cdot\cdot \left[(\boldsymbol{\rho}_\beta \times \mathbf{n})\boldsymbol{\rho}_\alpha\boldsymbol{\rho}^\alpha \cdot æ \cdot \boldsymbol{\rho}^\beta\right] = \mathbf{M} \cdot\cdot \left[(\mathbf{n} \times \boldsymbol{\rho}_\beta)æ \cdot \boldsymbol{\rho}^\beta\right] \\ &= \mathbf{M} \cdot\cdot \left[(\mathbf{n} \times \boldsymbol{\rho}_\beta)\boldsymbol{\rho}^\beta \cdot æ\right] = \mathbf{M} \cdot\cdot (\mathbf{n} \times æ) = (\mathbf{M} \times \mathbf{n}) \cdot\cdot\, æ. \end{aligned}$$

In addition, we obtain

$$\mathbf{M} \cdot\cdot (\mathbf{n} \times æ) = \frac{Eh^3}{12(1+\nu)}\left[æ \cdot\cdot\, æ + \frac{\nu}{1-\nu}\operatorname{tr}^2 æ\right].$$

Exercise 5.10 $\operatorname{tr}\tilde{\nabla}\tilde{\mathbf{v}} = \operatorname{tr}(\tilde{\nabla}\tilde{\mathbf{v}})^T = \tilde{\nabla} \cdot \tilde{\mathbf{v}}$ and $\operatorname{tr}\tilde{\nabla}\tilde{\nabla}w = \tilde{\nabla} \cdot \tilde{\nabla}w$. It follows that $\operatorname{tr}\boldsymbol{\varepsilon} = \tilde{\nabla} \cdot \tilde{\mathbf{v}} - z\tilde{\nabla} \cdot \tilde{\nabla}w$.

Exercise 5.11 Let the equation of the bent plate surface be $z = w(x_1, x_2)$. In vector form, the surface is given by

$$\mathbf{r} = x_1\mathbf{i}_1 + x_2\mathbf{i}_2 + \mathbf{n}w(x_1, y_1).$$

With slightly different notation, this was analyzed in Exercise 3.16; it was shown that

$$L = \frac{w_{xx}}{\sqrt{1 + w_x^2 + w_y^2}}, \qquad M = \frac{w_{xy}}{\sqrt{1 + w_x^2 + w_y^2}}, \qquad N = \frac{w_{yy}}{\sqrt{1 + w_x^2 + w_y^2}}.$$

We recall that $L = b_{11}$, $M = b_{12}$, $N = b_{22}$ are the components of the curvature tensor $\mathbf{B}$. If w and its derivatives are small, we get

$$\mathbf{B} = w_{xx}\mathbf{i}_1\mathbf{i}_1 + w_{xy}(\mathbf{i}_1\mathbf{i}_2 + \mathbf{i}_2\mathbf{i}_1) + w_{yy}\mathbf{i}_2\mathbf{i}_2 = -æ.$$

In other words, the second fundamental tensor of the bent plate surface coincides with æ up to a difference in algebraic sign.

Exercise 5.13 Modify the proof in three-dimensional elasticity, changing $\boldsymbol{\sigma}$ to $\mathbf{T} \cdot \mathbf{A}$, and $\mathbf{u}$ to $\tilde{\mathbf{v}}$.

Chapter 6 Exercises

Exercise 6.1 To prove Nanson's formula, first recall that the oriented elementary area $\mathbf{n}\, dS$ can be treated as the area of the parallelogram spanned by two elementary vectors $d\mathbf{x}_1$ and $d\mathbf{x}_2$ (Fig. D.1):

$$\mathbf{n}\, dS = d\mathbf{x}_1 \times d\mathbf{x}_2.$$

Here $d\mathbf{x}_1 = \mathbf{e}_1\, dx_1$ and $d\mathbf{x}_1 = \mathbf{e}_1\, dx_1$ where $\mathbf{e}_1$ and $\mathbf{e}_2$ are unit vectors tangent to S such that $\mathbf{e}_1 \times \mathbf{e}_2 = \mathbf{n}$.

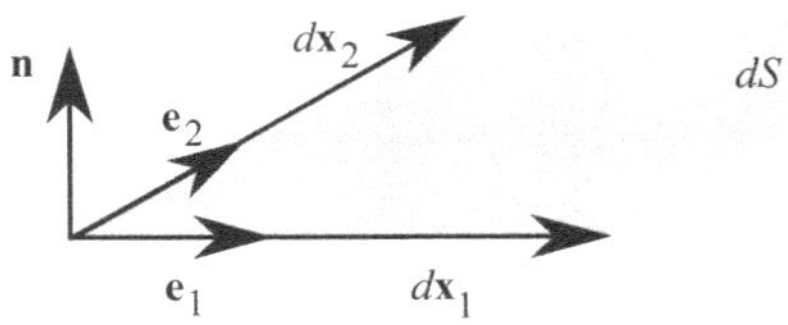

Fig. D.1 An oriented area.

Similarly $\widehat{\mathbf{n}}\, \widehat{dS} = d\mathbf{r}_1 \times \mathbf{r}_2$. We have $d\mathbf{r}_1 = \mathbf{F} \cdot d\mathbf{x}_1$ and $d\mathbf{r}_2 = \mathbf{F} \cdot d\mathbf{x}_2$ where $\mathbf{F} = (\nabla \mathbf{r})^T$ is the deformation gradient. Thus

$$\widehat{\mathbf{n}}\, \widehat{dS} = d\mathbf{r}_1 \times \mathbf{r}_2 = (\mathbf{F} \cdot d\mathbf{x}_1) \times (\mathbf{F} \cdot d\mathbf{x}_2).$$

For any nonsingular tensor $\mathbf{F}$ and two vectors $\mathbf{a}$ and $\mathbf{b}$ the formula

$$(\mathbf{F} \cdot \mathbf{a}) \times (\mathbf{F} \cdot \mathbf{b}) = (\det \mathbf{F})\mathbf{F}^{-T} \cdot (\mathbf{a} \times \mathbf{b})$$

holds. We thereby get

$$\widehat{\mathbf{n}}\, \widehat{dS} = (\mathbf{F} \cdot d\mathbf{x}_1) \times (\mathbf{F} \cdot d\mathbf{x}_2) = (\det \mathbf{F})\mathbf{F}^{-T} \cdot (d\mathbf{x}_1 \times d\mathbf{x}_2) = (\det \mathbf{F})\mathbf{F}^{-T} \cdot \mathbf{n}\, dS.$$

Exercise 6.2 Take any volume V and integrate $\nabla \cdot [J(\nabla \mathbf{r})^{-T}]$ over V. By the divergence theorem

$$\int_V \nabla \cdot [J(\nabla \mathbf{r})^{-T}]\, dV = \int_{\partial V} \mathbf{n} \cdot J(\nabla \mathbf{r})^{-T}\, dS.$$

Applying Nanson's formula, we transform the result into a surface integral over the boundary ∂v of the volume $v = \mathbf{r}(V)$:

$$\int_{\partial V} \mathbf{n} \cdot J(\nabla \mathbf{r})^{-T} \, dS = \int_{\partial v} \hat{\mathbf{n}} \, \widehat{dS}.$$

Clearly, for every closed smooth surface we have the identity

$$\int_{\partial v} \hat{\mathbf{n}} \, \widehat{dS} = \mathbf{0}.$$

Thus

$$\int_V \nabla \cdot [J(\nabla \mathbf{r})^{-T}] \, dV = \mathbf{0}$$

for any V, and $(6.13)_1$ holds. The second Piola identity can be proved in a similar way.

Exercise 6.3 Since W depends on $\mathbf{r}$ at the current time instant, it is obvious that the functional satisfies the requirements of the principle of determinism. From the identity

$$|\mathbf{r}^*(\mathbf{x}') - \mathbf{r}^*(\mathbf{x})| = |\mathbf{r}(\mathbf{x}') \cdot \mathbf{Q} - \mathbf{r}(\mathbf{x}) \cdot \mathbf{Q}| = |[\mathbf{r}(\mathbf{x}') - \mathbf{r}(\mathbf{x})] \cdot \mathbf{Q}| = |\mathbf{r}(\mathbf{x}') - \mathbf{r}(\mathbf{x})|$$

for all orthogonal tensors, it also follows that the principle of material frame-indifference is satisfied. Let $f(s)$ be an arbitrary scalar-valued function. Prove that the functional

$$W(\mathbf{r}(\mathbf{x})) = \int_V f(|\mathbf{r}(\mathbf{x}') - \mathbf{r}(\mathbf{x})|) \, dV',$$

where the integration is with respect to $\mathbf{x}'$, satisfies the requirements of the principles of determinism and material frame-indifference.

Exercise 6.4 To find $\delta \det(\nabla \mathbf{r})$ recall the formula

$$\delta f(\mathbf{X}) = \frac{df}{d\mathbf{X}} \bullet \delta \mathbf{X}.$$

Using the formula $(\det \mathbf{X})_{,\mathbf{X}} = (\det \mathbf{X}) \mathbf{X}^{-T}$ (Problem 2.29) we get

$$\delta \det(\nabla \mathbf{r}) = \det(\nabla \mathbf{r})(\nabla \mathbf{r})^{-T} \bullet \delta \nabla \mathbf{r}$$

So

$$\delta \int_V p \det(\nabla \mathbf{r}) \, dV = p \int_V \delta \det(\nabla \mathbf{r}) \, dV = p \int_V \det(\nabla \mathbf{r})(\nabla \mathbf{r})^{-T} \bullet \delta \nabla \mathbf{r} \, dV.$$

Integration by parts gives

$$\delta \int_V p \det(\nabla \mathbf{r})\, dV = -p \int_V \nabla \cdot [(\det \nabla \mathbf{r})(\nabla \mathbf{r})^{-T}] \cdot \delta \mathbf{r}\, dV$$
$$+ p \int_{\partial V} \mathbf{n} \cdot [(\det \nabla \mathbf{r})(\nabla \mathbf{r})^{-T}] \cdot \delta \mathbf{r}\, dS$$
$$= \int_{\partial V} p\mathbf{n} \cdot [(\det \nabla \mathbf{r})(\nabla \mathbf{r})^{-T}] \cdot \delta \mathbf{r}\, dS.$$

Note that here we used Piola's identity $\nabla \cdot [(\det \nabla \mathbf{r})(\nabla \mathbf{r})^{-T}] = 0$. So the natural boundary conditions follow from the relations

$$\delta \int_V [W + p \det(\nabla \mathbf{r})]\, dV = - \int_V (\nabla \cdot \mathbf{S}) \cdot \delta \mathbf{r}\, dV$$
$$+ \int_{\partial V} \{\mathbf{n} \cdot \mathbf{S} + p\mathbf{n} \cdot [(\det \nabla \mathbf{r})(\nabla \mathbf{r})^{-T}]\} \cdot \delta \mathbf{r}\, dS = 0.$$

Exercise 6.5 Let κ_1 and κ_2 be reference configurations with strain energy densities W_1 and W_2 and symmetry groups $\mathcal{G}_1$ and $\mathcal{G}_2$, respectively. $\mathbf{H} = \nabla_1 \mathbf{x}_2$ is the gradient of the mapping $\kappa_1 \to \kappa_2$. There are the following relations between W_1 and W_2, and $\mathbf{C}_1$ and $\mathbf{C}_2$:

$$W = W_1(\mathbf{C}_1) = W_2(\mathbf{C}_2), \quad \mathbf{C}_1 = \mathbf{H} \cdot \mathbf{C}_2 \cdot \mathbf{H}^T, \quad \mathbf{C}_2 = \mathbf{H}^{-1} \cdot \mathbf{C}_1 \cdot \mathbf{H}^{-T}.$$

So

$$W_1(\mathbf{C}_1) = W_2(\mathbf{H}^{-1} \cdot \mathbf{C}_1 \cdot \mathbf{H}^{-T}).$$

Let $\mathbf{H}_1 \in \mathcal{G}_1$. So the relation

$$W_1(\mathbf{C}_1) = W_1(\mathbf{H}_1 \cdot \mathbf{C}_1 \cdot \mathbf{H}_1^T)$$

holds for all arguments $\mathbf{C}_1$. Substituting $\mathbf{C}_1 = \mathbf{H} \cdot \mathbf{C}_2 \cdot \mathbf{H}^T$ we find

$$W_2(\mathbf{C}_2) = W_1(\mathbf{C}_1) = W_1(\mathbf{H}_1 \cdot \mathbf{H} \cdot \mathbf{C}_2 \cdot \mathbf{H}^T \cdot \mathbf{H}_1^T)$$
$$= W_2(\mathbf{H}^{-1} \cdot \mathbf{H}_1 \cdot \mathbf{H} \cdot \mathbf{C}_2 \cdot \mathbf{H}^T \cdot \mathbf{H}_1^T \cdot \mathbf{H}^{-T}).$$

This means that

$$W_2(\mathbf{C}_2) = W_2(\mathbf{H}_2 \cdot \mathbf{C}_2 \cdot \mathbf{H}_2^T),$$

where

$$\mathbf{H}_2 = \mathbf{H}^{-1} \cdot \mathbf{H}_1 \cdot \mathbf{H}. \tag{D.9}$$

In other words $\mathbf{H}_2$ belongs to $\mathcal{G}_2$ for any $\mathbf{H}_1 \in \mathcal{G}_1$. Relation (D.9) constitutes Noll's rule and can be symbolically written as $\mathcal{G}_2 = \mathbf{H}^{-1} \cdot \mathcal{G}_1 \cdot \mathbf{H}$.

Exercise 6.6 The derivation mimics the standard procedure described in the section on nonlinear elasticity.

Exercise 6.7 To prove (6.101) we represent all tensors in a Cartesian frame as

$$\mathbf{A} = A_{mnk}\mathbf{i}_m\mathbf{i}_n\mathbf{i}_k, \quad \mathbf{B} = B_{pqr}\mathbf{i}_p\mathbf{i}_q\mathbf{i}_r, \quad \mathbf{X} = X_{ts}\mathbf{i}_t\mathbf{i}_s, \quad \mathbf{Y} = Y_{ij}\mathbf{i}_i\mathbf{i}_j.$$

Then

$$\begin{aligned}
\mathbf{A} \bullet (\mathbf{B} \cdot \mathbf{X} \cdot \mathbf{Y}) &= A_{mnk}B_{mnr}X_{rs}Y_{sk} \\
&= (B_{mnr}\mathbf{i}_r A_{mnk}Y_{sk}\mathbf{i}_s) \bullet (X_{pq}\mathbf{i}_p\mathbf{i}_q) \\
&= (B_{mnr}\mathbf{i}_r A_{mnk}\mathbf{i}_k \cdot Y_{st}\mathbf{i}_t\mathbf{i}_s) \bullet \mathbf{X} \\
&= (B_{mnr}\mathbf{i}_r A_{mnk}\mathbf{i}_k \cdot \mathbf{Y}^T) \bullet \mathbf{X} \\
&= (\mathbf{B} : \mathbf{A} \cdot \mathbf{Y}^T) \bullet \mathbf{X}.
\end{aligned}$$

Exercise 6.8 For deriving $\mathbf{T}$ and $\mathbf{M}$ from (6.90) we consider dW. We already have (6.99), which says that

$$dW = \mathbf{T} \bullet d\mathbf{F} + \mathbf{M} \bullet d(\nabla\mathbf{P}) + \frac{\partial W}{\partial \mathbf{P}} \bullet d\mathbf{P}.$$

Calculating dW using (6.90), i.e., using $W = W(\mathbf{C}, \mathbf{P} \cdot \mathbf{F}^{-1}, \nabla\mathbf{P} \cdot \mathbf{F}^{-1})$, we get

$$dW = \frac{\partial W}{\partial \mathbf{C}} \bullet d\mathbf{C} + \frac{\partial W}{\partial(\mathbf{P} \cdot \mathbf{F}^{-1})} \bullet d(\mathbf{P} \cdot \mathbf{F}^{-1}) + \frac{\partial W}{\partial(\nabla\mathbf{P} \cdot \mathbf{F}^{-1})} \bullet d(\nabla\mathbf{P} \cdot \mathbf{F}^{-1}).$$

Substituting into this equation the formulas for differentials

$$\begin{aligned}
&d\mathbf{C} = d\mathbf{F} \cdot \mathbf{F}^T + \mathbf{F} \cdot d\mathbf{F}^T, \\
&d(\mathbf{P} \cdot \mathbf{F}^{-1}) = d\mathbf{P} \cdot \mathbf{F}^{-1} - \mathbf{P} \cdot \mathbf{F}^{-1} \cdot d\mathbf{F} \cdot \mathbf{F}^{-1}, \\
&d(\nabla\mathbf{P} \cdot \mathbf{F}^{-1}) = d(\nabla\mathbf{P}) \cdot \mathbf{F}^{-1} - \nabla\mathbf{P} \cdot \mathbf{F}^{-1} \cdot d\mathbf{F} \cdot \mathbf{F}^{-1},
\end{aligned}$$

we arrive at

$$dW = \left[2\frac{\partial W}{\partial \mathbf{C}} \cdot \mathbf{F} - \mathbf{F}^{-T} \cdot \mathbf{P}^T \cdot \frac{\partial W}{\partial(\mathbf{P} \cdot \mathbf{F}^{-1})} \cdot \mathbf{F}^{-T} \right.$$
$$\left. - (\nabla \mathbf{P} \cdot \mathbf{F}^{-1}) : \frac{\partial W}{\partial(\nabla \mathbf{P} \cdot \mathbf{F}^{-1})} \cdot \mathbf{F}^{-T}\right] \bullet d\mathbf{F}$$
$$+ \left[\frac{\partial W}{\partial(\nabla \mathbf{P} \cdot \mathbf{F}^{-1})} \cdot \mathbf{F}^{-T}\right] \bullet d(\nabla \mathbf{P}) + \left[\frac{\partial W}{\partial(\mathbf{P} \cdot \mathbf{F}^{-1})} \cdot \mathbf{F}^{-T}\right] \bullet d\mathbf{P}.$$

As a result, we obtain

$$\mathbf{T} = 2\frac{\partial W}{\partial \mathbf{C}} \cdot \mathbf{F} - \mathbf{F}^{-T} \cdot \mathbf{P}^T \cdot \frac{\partial W}{\partial(\mathbf{P} \cdot \mathbf{F}^{-1})} \cdot \mathbf{F}^{-T}$$
$$- (\nabla \mathbf{P} \cdot \mathbf{F}^{-1}) : \frac{\partial W}{\partial(\nabla \mathbf{P} \cdot \mathbf{F}^{-1})} \cdot \mathbf{F}^{-T},$$
$$\mathbf{M} = \frac{\partial W}{\partial(\nabla \mathbf{P} \cdot \mathbf{F}^{-1})} \cdot \mathbf{F}^{-T},$$
$$\frac{\partial W}{\partial \mathbf{P}} = \frac{\partial W}{\partial(\mathbf{P} \cdot \mathbf{F}^{-1})} \cdot \mathbf{F}^{-T}.$$

Note that the expression for $\mathbf{T}$ is more complex than $(6.102)_1$.

Exercise 6.9 First, we have the identity

$$\nabla_S \cdot (\mathbf{T} \cdot \mathbf{w}) = (\nabla_S \cdot \mathbf{T}) \cdot \mathbf{w} + \mathbf{T} \bullet \nabla_S \mathbf{w}.$$

Then, applying (6.122) to $\nabla_S \cdot (\mathbf{T} \cdot \mathbf{w})$ we get

$$\int_\Sigma (\nabla_S \cdot (\mathbf{T} \cdot \mathbf{w}) + 2H\mathbf{n} \cdot \mathbf{T} \cdot \mathbf{w})\, dS = \oint_{\partial\Sigma} \boldsymbol{\nu} \cdot \mathbf{T} \cdot \mathbf{w}\, ds.$$

The requested formula follows.

Exercise 6.10 For brevity we introduce the following notation:

$$\mathbf{F}^{(k)} = \nabla^k \mathbf{r}, \quad \mathbf{F}^{(1)} = \mathbf{F} = \nabla \mathbf{r}, \quad \mathbf{F}^{(0)} = \mathbf{E}, \quad \mathbf{T}^{(k)} = \frac{\partial W}{\partial \mathbf{F}^{(k)}}, \quad \mathbf{T} = \frac{\partial W}{\partial \mathbf{F}}.$$

Here $\mathbf{T}$ is the stress tensor and $\mathbf{T}^{(k)}$ $(k = 2, \ldots, N)$ are the hyper-stress tensors of $k+1$-order. The first variation takes the form

$$\delta \mathcal{E} = \int_V \sum_{k=1}^N \frac{\partial W}{\partial \mathbf{F}^{(k)}} \bullet \nabla^k \mathbf{w}\, dV$$
$$= \int_V \left[\mathbf{T} \bullet \nabla \mathbf{w} + \mathbf{T}^{(2)} \bullet \nabla^2 \mathbf{w} + \mathbf{T}^{(3)} \bullet \nabla^3 \mathbf{w} + \ldots + \mathbf{T}^{(N)} \bullet \nabla^N \mathbf{w}\right] dV,$$

where $\mathbf{w} = \delta\mathbf{r}$. Since we are interested in equilibrium equations we assume that $\mathbf{w} = \mathbf{0}$ at the boundary with all necessary derivatives. In other words, unlike the previous exercise we can forget boundary conditions.

Integrating by parts once, we get

$$\delta\mathcal{E} = \int_V \Big[-(\nabla\cdot\mathbf{T})\cdot\mathbf{w} - (\nabla\cdot\mathbf{T}^{(2)})\bullet\nabla\mathbf{w} - (\nabla\cdot\mathbf{T}^{(3)})\bullet\nabla^2\mathbf{w} - \cdots \\ -(\nabla\cdot\mathbf{T}^{(N)})\bullet\nabla^{N-1}\mathbf{w}\Big]\,dV.$$

Repeating the integration by parts, we have

$$\delta\mathcal{E} = \int_V \Big\{-(\nabla\cdot\mathbf{T})\cdot\mathbf{w} + \Big[\nabla\cdot(\nabla\cdot\mathbf{T}^{(2)})\Big]\cdot\mathbf{w} + \Big[\nabla\cdot(\nabla\cdot\mathbf{T}^{(3)})\Big]\bullet\nabla\mathbf{w} + \cdots \\ + \Big[\nabla\cdot(\nabla\cdot\mathbf{T}^{(N)})\Big]\bullet\nabla^{N-2}\mathbf{w}\Big\}\,dV$$

$$= \int_V \Big\{\Big[-\nabla\cdot\mathbf{T} + \nabla\cdot(\nabla\cdot\mathbf{T}^{(2)}) - \nabla\cdot\Big(\nabla\cdot(\nabla\cdot\mathbf{T}^{(3)})\Big)\Big]\cdot\mathbf{w} - \cdots \\ -\nabla\cdot\Big(\nabla\cdot(\nabla\cdot\mathbf{T}^{(N)})\Big)\bullet\nabla^{N-3}\mathbf{w}\Big\}\,dV$$

$$= \int_V \left[\sum_{k=1}^{N}(-1)^k \underbrace{\nabla\cdot\ldots(\nabla}_{k\ \text{times}}\cdot\mathbf{T}^{(k)})\ldots)\right]\cdot\mathbf{w}\,dV.$$

Thus, from $\delta\mathcal{E} = 0$ it follows that

$$\sum_{k=1}^{N}(-1)^{k-1} \underbrace{\nabla\cdot\ldots(\nabla}_{k\ \text{times}}\cdot\mathbf{T}^{(k)})\ldots) = \mathbf{0},$$

or in a recurrent form

$$\nabla\cdot(\mathbf{T} - \nabla\cdot(\mathbf{T}^{(2)} - \nabla\cdot(\mathbf{T}^{(3)} - \nabla\cdot(\mathbf{T}^{(4)} - \cdots - \nabla\cdot\mathbf{T}^{(N)})\ldots) = \mathbf{0}.$$

The second-order tensor $\mathbf{T} - \nabla\cdot(\mathbf{T}^{(2)} - \nabla\cdot(\mathbf{T}^{(3)} - \nabla\cdot(\mathbf{T}^{(4)} - \cdots - \nabla\cdot\mathbf{T}^{(N)})\ldots$ can be called the effective first Piola–Kirchhoff stress tensor for the Nth gradient continuum.

Exercise 6.11 Let us find the first variation of

$$\mathcal{E} = \int_v [W(\rho,\nabla\rho) - \rho\Omega]\,dv, \quad W = W_0(\rho) + \tfrac{1}{2}\nabla\rho\cdot\nabla\rho$$

with the integral constraint

$$\int_v \rho\,dV = \text{constant}.$$

Here the constant is the mass of the fluid in volume v. Note also that Ω is given and does not depend on ρ. For this constraint the Lagrange multiplier is denoted by c. So we have to calculate the Euler–Lagrange equation for the functional

$$\mathcal{E} - c\left(\int_v \rho\, dv - \text{constant}\right).$$

We get

$$\begin{aligned}
&\delta\left[\mathcal{E} - c\left(\int_v \rho\, dv - \text{constant}\right)\right] \\
&= \int_v \delta\left[W(\rho, \nabla\rho) - \rho\Omega\right] dv - c\int_v \delta\rho\, dv \\
&= \int_v \left(\frac{\partial W_0}{\partial \rho}\delta\rho + \lambda\nabla\rho\cdot\nabla\delta\rho - \delta\rho\Omega - c\delta\rho\right) dv = 0.
\end{aligned}$$

Integrating by parts and keeping only the volume integral we have

$$\int_v \left(\frac{\partial W_0}{\partial \rho} - \lambda\nabla\cdot\nabla\rho - \Omega - c\right)\delta\rho\, dv = 0.$$

This yields the static Bernoulli integral (6.182).

Exercise 6.12 The equilibrium conditions follow from the variational equation

$$\delta\int_v W\, dv = \int_v \left[\delta W + W\nabla\cdot\mathbf{w}\right] dv = 0,$$

where $\mathbf{w} = \delta\mathbf{r}$ and

$$\delta W = \frac{\partial W}{\partial \rho}\delta\rho + \frac{\partial W}{\partial \nabla\rho}\cdot\delta\nabla\rho + \frac{\partial W}{\partial \nabla\nabla\rho}\bullet\delta\nabla\nabla\rho + \frac{\partial W}{\partial \nabla\nabla\nabla\rho}\bullet\delta\nabla\nabla\nabla\rho.$$

For the variations we have the formulae

$$\begin{aligned}
\delta\nabla &= -\nabla\mathbf{w}\cdot\nabla, \\
\delta\rho &= -\rho\nabla\cdot\mathbf{w}, \\
\delta\nabla\rho &= -\nabla\mathbf{w}\cdot\nabla\rho - \nabla(\rho\nabla\cdot\mathbf{w}), \\
\delta\nabla\nabla\rho &= -\nabla\mathbf{w}\cdot\nabla\nabla\rho - \nabla(\nabla\mathbf{w}\cdot\nabla)\nabla\rho - \nabla\nabla(\rho\nabla\cdot\mathbf{w}), \\
\delta\nabla\nabla\nabla\rho &= -\nabla\mathbf{w}\cdot\nabla\nabla\nabla\rho - \nabla(\nabla\mathbf{w}\cdot\nabla)\nabla\nabla\rho \\
&\quad - \nabla\nabla(\nabla\mathbf{w}\cdot\nabla)\nabla\rho - \nabla\nabla\nabla(\rho\nabla\cdot\mathbf{w}).
\end{aligned}$$

Then, applying integration by parts up to four times we can derive the equilibrium conditions.

Exercise 6.13 First, let us consider the Hamilton variational principle

$$\delta\mathcal{H} = \delta\int_{t_1}^{t_2}\int_V \left[K - W + \tfrac{1}{2}\lambda(\mathbf{d}\cdot\mathbf{d} - 1)\right] dV\,dt = 0.$$

Using integration by parts we get

$$\begin{aligned}
\delta\mathcal{H} &= \int_{t_1}^{t_2}\int_V \Big[\rho\dot{\mathbf{r}}\cdot\delta\dot{\mathbf{r}} + \gamma\dot{\mathbf{d}}\cdot\delta\dot{\mathbf{d}} - \frac{\partial W}{\partial\nabla\mathbf{r}}\bullet\nabla\delta\mathbf{r} \\
&\quad - \frac{\partial W}{\partial\nabla\mathbf{d}}\bullet\nabla\delta\mathbf{d} - \frac{\partial W}{\partial\mathbf{d}}\cdot\delta\mathbf{d} + \lambda\mathbf{d}\cdot\delta\mathbf{d}\Big]\, dV\,dt \\
&= \int_{t_1}^{t_2}\int_V \Big[-\rho\ddot{\mathbf{r}}\cdot\delta\mathbf{r} - \gamma\ddot{\mathbf{d}}\cdot\delta\mathbf{d} + \left(\nabla\cdot\frac{\partial W}{\partial\nabla\mathbf{r}}\right)\cdot\delta\mathbf{r} \\
&\quad + \left(\nabla\cdot\frac{\partial W}{\partial\nabla\mathbf{d}}\right)\cdot\delta\mathbf{d} - \frac{\partial W}{\partial\mathbf{d}}\cdot\delta\mathbf{d} + \lambda\mathbf{d}\cdot\delta\mathbf{d}\Big]\, dV\,dt \\
&\quad - \int_{t_1}^{t_2}\int_V \left[\left(\mathbf{n}\cdot\frac{\partial W}{\partial\nabla\mathbf{r}}\right)\cdot\delta\mathbf{r} + \left(\mathbf{n}\cdot\frac{\partial W}{\partial\nabla\mathbf{d}}\right)\cdot\delta\mathbf{d}\right] dS\,dt.
\end{aligned}$$

Introducing the nominal stress tensor $\mathbf{S}$ and the nominal couple stress tensor $\mathbf{M}$ by the formulas

$$\mathbf{S} = \frac{\partial W}{\partial\nabla\mathbf{r}}, \qquad \mathbf{M} = \frac{\partial W}{\partial\nabla\mathbf{d}},$$

from $\delta\mathcal{H} = 0$ we have the equations of motion and the homogeneous natural boundary conditions

$$\begin{aligned}
&\nabla\cdot\mathbf{S} = \rho\ddot{\mathbf{r}}, \quad \nabla\cdot\mathbf{M} - \frac{\partial W}{\partial\mathbf{d}} - \lambda\mathbf{d} = \gamma\ddot{\mathbf{d}} \qquad (\mathbf{x}\in V), \\
&\mathbf{n}\cdot\mathbf{S} = \mathbf{0}, \quad \mathbf{n}\cdot\mathbf{M} = \mathbf{0} \qquad (\mathbf{x}\in\partial V).
\end{aligned}$$

Here λ is a Lagrange multiplier corresponding to the constraint $\mathbf{d}\cdot\mathbf{d} = 1$.

Now let us find $\mathbf{S}$ and $\mathbf{M}$ using (6.206). First we have

$$dW = \mathbf{S}\bullet d\nabla\mathbf{r} + \mathbf{M}\bullet d\nabla\mathbf{d} + \frac{\partial W}{\partial\mathbf{d}}\cdot d\mathbf{d}.$$

On the other hand, from objective representation (6.206) we get

$$dW = \frac{\partial W}{\partial \mathbf{C}} \bullet d\mathbf{C} + \frac{\partial W}{\partial(\mathbf{d}\cdot\mathbf{F})} \bullet d(\mathbf{d}\cdot\mathbf{F}) + \frac{\partial W}{\partial(\nabla\mathbf{d}\cdot\mathbf{F})} \bullet d(\nabla\mathbf{d}\cdot\mathbf{F})$$

$$= \left[2\frac{\partial W}{\partial \mathbf{C}}\cdot\nabla\mathbf{r} + \left(\frac{\partial W}{\partial(\nabla\mathbf{d}\cdot\mathbf{F})}\right)^T\cdot\nabla\mathbf{d} + \frac{\partial W}{\partial(\mathbf{d}\cdot\mathbf{F})}\mathbf{d}\right] \bullet d\nabla\mathbf{r}$$

$$+ \left[\frac{\partial W}{\partial(\nabla\mathbf{d}\cdot\mathbf{F})}\cdot\nabla\mathbf{r}\right] \bullet d\nabla\mathbf{d} + \left[\frac{\partial W}{\partial(\mathbf{d}\cdot\mathbf{F})}\cdot\nabla\mathbf{r}\right]\cdot d\mathbf{d}.$$

Comparing these expressions we obtain

$$\mathbf{S} = 2\frac{\partial W}{\partial \mathbf{C}}\cdot\mathbf{F}^T + \left(\frac{\partial W}{\partial(\nabla\mathbf{d}\cdot\mathbf{F})}\right)^T\cdot\nabla\mathbf{d} + \frac{\partial W}{\partial(\mathbf{d}\cdot\mathbf{F})}\mathbf{d},$$

$$\mathbf{M} = \frac{\partial W}{\partial(\nabla\mathbf{d}\cdot\mathbf{F})}\cdot\mathbf{F}^T.$$

Bibliography

Antman, S. S. (2005). *Nonlinear Problems of Elasticity, 2nd ed.* (Springer Science Media, New York).

Arnold, V. I. (1992). *Ordinary Differential Equations* (Springer, Berlin).

Astarita, G. and Marrucci, G. (1974). *Principles of non-Newtonian Fluid Mechanics* (McGraw-Hill, Maidenhead).

Bauchau, O. A. (2010) *Flexible Multibody Dynamics*, volume 176 of *Solid Mechanics and its Applications* (Springer, Dordrecht).

Bertram, A. (2008) *Elasticity and Plasticity of Large Deformations: An Introduction*, 2nd ed. (Springer, Berlin).

Boehler, J. P. (Ed.) (1987). *Applications of Tensor Functions in Solid Mechanics* (Springer, Wien).

Chandrasekhar, S. (1977). *Liquid Crystals* (Cambridge University Press, Cambridge, UK).

Chróścielewski, J., Makowski, J. and Pietraszkiewicz, W. (2004). *Statyka i dynamika powłok wielopłatowych. Nieliniowa teoria i metoda elementòw skończonych* (Wydawnictwo IPPT PAN, Warszawa).

Ciarlet, Ph. G. (1988). *Mathematical Elasticity. Vol. I. Three-Dimensional Elasticity* (North-Holland, Amsterdam).

Ciarlet, Ph. G. (1997). *Mathematical Elasticity. Vol. II. Theory of Plates* (North-Holland, Amsterdam).

Ciarlet, Ph. G. (2000). *Mathematical Elasticity. Vol. III. Theory of Shells* (North-Holland, Amsterdam).

Cosserat, E. and Cosserat, F. (1909). *Théorie des corps déformables* (Herman et Fils, Paris). English translation: NASA TT F-11, 561 (NASA, Washington, DC).

Courant R., Hilbert D. (1989). *Methods of Mathematical Physics, Vol 2: Partial Differential Equations* (Wiley, New York).

de Gennes P. G. and Prost J. (1993). *The Physics of Liquid Crystals. 2nd ed.* (Clarendon Press, Oxford).

Donnell, L. H. (1976). *Beams, Plates and Shells* (McGraw-Hill, New York).

Eremeyev, V. A., Lebedev, L. P., and Altenbach, H. (2013). *Foundations of Micropolar Mechanics* (Springer, Heidelberg).

Eremeyev, V. A. and Zubov L. M. (2008). *Mechanics of Elastic Shells (in Russian)* (Nauka, Moscow).

Eremeyev, V. A. Lebedev, L. P., and Cloud, M. J. The Rayleigh and Courant variational principles in the six-parameter shell theory. *Math. Mech. of Solids*, **20**(7) 806-822, 2015.

Eringen, A. C. (1999) *Microcontinuum Field Theory. I. Foundations and Solids* (Springer-Verlag, New York).

Eringen, A. C. (2001) *Microcontinuum Field Theory. II. Fluent Media* (Springer-Verlag, New York).

Eringen, A. C. (2002) *Nonlocal Continuum Field Theories* (Springer-Verlag, New York).

Eringen, A. C., Maugin, G. A. (1990) *Electrodynamics of Continua* (Springer-Verlag, New York).

Feynman, R. P., Leighton, R. B., Sands, M. (1977). *The Feynman Lectures on Physics. Vol. 1, 6th ed.* (Addison-Wesley, Reading, Massachusetts).

Goldenveizer, A. L. (1976). *Theory of Thin Elastic Shells (in Russian)* (Moscow, Nauka). English translation by G. Herrmann of 1953 ed. Gol'denveizer, A. L. (1961). *Theory of Thin Elastic Shells* (Pergamon Press, Oxford).

Green, A. E. and Adkins, J. E. (1960). *Large Elastic Deformations and Non-Linear Continuum Mechanics* (Clarendon Press, Oxford).

Green, A. E. and Zerna, W. (1954). *Theoretical Elasticity* (Clarendon Press, Oxford).

Gurtin, M. E. (1983). *Topics in Finite Elasticity* (SIAM, Philadelphia).

Horn, R.A., and Johnson, C.R. (1990). *Matrix Analysis.* (Cambridge University Press, Cambridge, UK).

Jeffreys, H. (1931). *Cartesian Tensors* (Cambridge University Press, Cambridge, UK).

Kafadar, C. B., Eringen, A. C. (1971). *Micropolar media – I. The classical theory.* Int. J. Engng Sci. **9**(3), 271–305.

Koiter, W. T. On the foundations of the linear theory of thin elastic shells. I. *Proc. Kon. Ned. Ak. Wet.* B73, 169–195.

Korn, G. A. and Korn, T. M. (1968). *Mathematical Handbook for Scientists and Engineers: Definitions, Theorems and Formulas for Reference and Review*, 2nd ed. (McGraw-Hill, New York).

Lebedev, L. P. and Cloud, M. J. (2003). *The Calculus of Variations and Functional Analysis with Optimal Control and Applications in Mechanics* (World Scientific, Singapore).

Libai, A. and Simmonds, J. G. (1998). *The Nonlinear Theory of Elastic Shells*, 2nd ed (Cambridge University Press, Cambridge, UK).

Lurie, A. I. (1990). *Non-Linear Theory of Elasticity*, Series in Applied Mathematics and Mechanics, 36 (North-Holland, Amsterdam).

Lurie, A. I. (2005). *Theory of Elasticity* (Springer, Berlin).

Maugin, G. A. (1988). *Continuum Mechanics of Electromagnetic Solids* (Elsevier, Oxford).

Maugin, G. A. (2017). *Non-Classical Continuum Mechanics. A Dictionary* (Springer Nature, Singapore).

Muskhelishvili N. I. (1966). *Some Basic Problems of the Mathematical Theory of Elasticity; Fundamental Equations, Plane Theory of Elasticity, Torsion and Bending, 5th ed. (in Russian)*, Nauka, Moscow. Transl. P. Noordhoff, Groningen, 1953 (Translation of the 3rd ed., Izd. Akad. Nauk SSSR, Moscow-Leningrad, 1949).

Naumenko, K. and Altenbach, H. (2016). *Elements of Tensor Analysis.* In: Modeling High Temperature Materials Behavior for Structural Analysis, Advanced Structured Materials, Vol. 28, Springer Switzerland, pp. 343–363.

Novozhilov, V. V., Chernykh, K. Ph. and Mikhailovskiy E. M. (1991). *Linear Theory of Thin Shells (in Russian)* (Politechnika, Leningrad).

Nye, J. F. (1957). *Physical Properties of Crystals: Their Representation by Tensors and Matrices* (Oxford, UK).

Ogden, R. W. (1997). *Non-Linear Elastic Deformations* (Dover, New York).

Ogden, R. W. (2001). *Elements of the theory of finite elasticity.* In: Fu, Y. B., Ogden, R. W. (Eds), Nonlinear Elasticity: Theory and Applications (Cambridge University Press, Cambridge, pp. 1–58).

Ogden, R. W. and Holzapfel, G. A. (Eds) (2006). *Mechanics of Biological Tissue* (Springer, Berlin).

Pietraszkiewicz, W., Eremeyev, V. A. (2009). On vectorially parameterized natural strain measures of the non-linear Cosserat continuum. *Int. J. Solids Struct.*, 46(11), 2477–2480.

Pogorelov, A. V. (1957). *Differential Geometry.* Translated from the first Russian ed. by L. F. Boron (P. Noordhoff, Groningen).

Spencer, A. J. M. (1971). *Theory of invariants.* In: Eringen, A. C. (Ed.), Continuum Physics, Vol. 1. (Academic Press, New York, pp. 239–353).

Stephenson, R.A. (1980). On the uniqueness of the square root of a symmetric, positive definite tensor. *J. Elast.*, 10(2), 213–214.

Timoshenko, S. P., Woinowsky-Krieger, S. (1985). *Theory of Plates and Shells* (McGraw Hill, New York).

Truesdell, C. and Noll, W. (2004). *The Nonlinear Field Theories of Mechanics.* 3rd ed (Springer, Berlin).

Vlasov, V. Z. (1949). *General Theory of Shells and its Applications in Technics (in Russian)* (Moscow, Saint-Petersburg, Gostekhizdat). German translation: Wlassow, W. S. (1958). *Allgemeine Schalentheorie und ihre Anwendung in der Technik.* (Berlin, Akademie-Verlag).

Vorovich, I. I. (1999). *Nonlinear Theory of Shallow Shells* (Springer, New York).

Wang, C. M., Reddy, J. N. and Lee, K. H. (2000). *Shear Deformable Beams and Shells* (Elsevier, Amsterdam).

Wiśniewski, K. (2010). *Finite Rotation Shells: Basic Equations and Finite Elements for Reissner Kinematics* (Springer, Berlin).

Young, E. (1993). *Vector and Tensor Analysis* (Marcel Dekker, New York).

Yosida, K. (1980). *Functional Analysis* (Springer, Berlin).

Zheng, Q. S. (1994). Theory of representations for tensor functions — a unified invariant approach to constitutive equations. *Appl. Mech. Rev.* 47(11), 545–587.

Zhilin, P. A. (2006). *Applied Mechanics. Foundations of the Theory of Shells (in*

Russian) (St. Petersburg State Polytechnical University, St. Petersburg).
Zubov, L. M. and Karyakin M. I. (2006). *Tensor Calculus (in Russian)* (Vuzovskaya Kniga, Moscow).
Zubov, L. M. (1982). *Methods of Nonlinear Elasticity in the Theory of Shells (in Russian)* (Rostov State University, Rostov on Don).
Zubov, L. M., and Rudev, A. N. (1996). An explicit expression for the elements of the polar representation of a second-rank tensor. *Doklady Physics*, 41(11), 544–547.
Zubov, L. M., and Rudev, A. N. (1998). On canonical resolution for the deviator of a symmetric tensor. *Doklady Akademii Nauk*, 358(1), 44–47.

Index

www.ingramcontent.com/pod-product-compliance
Ingram Content Group UK Ltd.
Pitfield, Milton Keynes, MK11 3LW, UK
UKHW020402150926
8624IPUK00004B/295

9 789813 238961